List of abbreviations

Abbreviations, commonly used in Astronomy, and not always explained explicitly in this book.

AU	Astronomical Unit (= Distance Earth–Sun)
B.C.	Bolometric Correction
BD	Bonner Durchmusterung
CLV	Center-limb variation
CMD	Colour-magnitude-diagram
CNO	Carbon, Nitrogen, and Oxygen (*not* as molecule) e.g. CNO cycle, CNO anomalies
ESA	European Space Agency
ESO	European Southern Observatory
ET	or E.T. Ephemeris Time
EUV	Extreme ultraviolet
FWHW	Full Width of Half Maximum
HD	Henry Draper Catalogue
HR	Harvard Revised Catalogue
HRD	Hertzsprung-Russell Diagram
IAU	International Astronomical Union
IR	Infrared
ISM	Interstellar Matter
JD	Julian Date
LB, NS or LB	Landolt-Börnstein, Numerical Data and Functional Relationships in Science and Technology, New Series or: Landolt-Börnstein, NS
LC	Luminosity Class
LF	Luminosity Function
LMC	Large Magellanic Cloud
LTE	Local thermodynamic equilibrium
M	Messier Catalogue
MHD	Magneto-hydrodynamics
MMT	Multi-Mirror-Telescope
MPI	Max-Planck-Institut
NASA	National Aeronautics and Space Administration
NEP	Noise Equivalent Power
NGC	New General Catalogue
NLTE	Non-local thermodynamic equilibrium
NRAO	National Radio Astronomy Observatory, Green Banks, W. Va., USA
POSS	Palomar Observatory Sky Survey
RV	Radial velocity
SMC	Small Magellanic Cloud
Sp	Spectral type
URSI	International Union of Radio Science
UT	Universal time
UV	Ultraviolet
VLBI	Very Long Baseline Interferometry
XUV	X-ray and ultraviolet region
ZAMS	Zero Age Main Sequence

Abbreviations of further Star Catalogues: see 8.1.1
For abbreviations of special star types (e.g. WR stars), see "Spectralclassification" (4.1.1), "Variable stars" (5.1), "Peculiar stars" (5.2) and subject index.

Some important Astronomical Artificial Satellites, mentioned in this book

ANS	Astronomical Netherlands Satellite (The Netherlands NASA)
ATS	Applications Technology Satellite
COS	Cosmic Ray Satellite (ESA)
GIRL	German Infrared Laboratory
HEAO	High Energy Astrophysical Observatory (NASA)
HEOS	High Eccentricity Earth-Orbiting Satellite (ESA)
IMP	Interplanetary Monitoring Platform
IRAS	Infrared Astronomical Satellite
IUE	International Ultraviolet Explorer (NASA–UK–ESA)
OAO	Orbiting Astronomical Observatory (NASA)
OGO	Orbiting Geophysical Observatory
OSO	Orbiting Solar Observatory
MTS	Meteoroid Technology Satellite (NASA)
RAE	Radio Astronomy Explorer
SAS	Small Astronomy Satellite (NASA)

LANDOLT-BÖRNSTEIN

Numerical Data and Functional Relationships in Science and Technology

New Series

Editor in Chief: K.-H. Hellwege

Group VI: Astronomy · Astrophysics and Space Research

Volume 2

Astronomy and Astrophysics

Extension and Supplement to Volume 1

Subvolume a

Methods · Constants · Solar System

W. I. Axford · A. Behr · A. Bruzek · C. J. Durrant · H. Enslin · H. Fechtig · W. Fricke
F. Gondolatsch · H. Grün · O. Hachenberg · W.-H. Ip · E. K. Jessberger · T. Kirsten · Ch. Leinert
D. Lemke · H. Palme · W. Pilipp · J. Rahe · G. Schmahl · M. Scholer · J. Schubart
J. Solf · R. Staubert · H. E. Suess · J. Trümper · G. Weigelt · R. M. West · R. Wolf · H. D. Zeh

Editors: K. Schaifers and H. H. Voigt

Springer-Verlag Berlin · Heidelberg · New York 1981

LANDOLT-BÖRNSTEIN

Zahlenwerte und Funktionen aus Naturwissenschaften und Technik

Neue Serie

Gesamtherausgabe: K.-H. Hellwege

Gruppe VI: Astronomie · Astrophysik und Weltraumforschung

Band 2

Astronomie und Astrophysik

Weiterführung und Ergänzung von Band 1

Teilband a

Methoden · Konstanten · Sonnensystem

W. I. Axford · A. Behr · A. Bruzek · C. J. Durrant · H. Enslin · H. Fechtig · W. Fricke
F. Gondolatsch · H. Grün · O. Hachenberg · W.-H. Ip · E. K. Jessberger · T. Kirsten · Ch. Leinert
D. Lemke · H. Palme · W. Pilipp · J. Rahe · G. Schmahl · M. Scholer · J. Schubart
J. Solf · R. Staubert · H. E. Suess · J. Trümper · G. Weigelt · R. M. West · R. Wolf · H. D. Zeh

Herausgeber: K. Schaifers und H. H. Voigt

Springer-Verlag Berlin · Heidelberg · New York 1981

CIP-Kurztitelaufnahme der Deutschen Bibliothek *Zahlenwerte und Funktionen aus Naturwissenschaften und Technik* / Landolt-Börnstein. – Berlin; Heidelberg; New York: Springer. Parallelt.: Numerical data and functional relationships in science and technology.
NE: Landolt-Börnstein, ...; PT. N.S./Gesamthrsg.: K.-H. Hellwege. N.S., Gruppe 6, Astronomie, Astrophysik und Weltraumforschung.
N.S., Gruppe 6, Bd. 2. Astronomie und Astrophysik: Erg. u. Erw. zu Bd. 1. N.S., Gruppe 6, Bd. 2, Teilbd. a. Methoden, Konstanten, Sonnensystem / W. I. Axford ... Hrsg.: K. Schaifers u. H. H. Voigt. – 1981. – ISBN 3-540-10054-7 (Berlin, Heidelberg, New York).
ISBN 0-387-10054-7 (New York, Heidelberg, Berlin) [Erscheint: November 1981].
NE: Axford, William I. [Mitverf.]; Schaifers, Karl [Hrsg.]; Hellwege, Karl-Heinz [Hrsg.].

Printed in Germany

Typesetting, printing and bookbinding: Brühlsche Universitätsdruckerei, 6300 Giessen
2163/3020—543210

Additions

In the meantime published:

p. 167 ref. 44a: **86** (1981) 3097.

Corrections

List of contributors, line 2 from the bottom: *instead of* W. Philipp *read* W. Pilipp

p. 3 line 5 from the bottom: *delete the whole line and replace it by*
...necessary for tracking (compensation of the earth's rotation). But the variable direction of gravity results in posi-...

p. 18 ref. 33: *instead of* 237 *read* 240

p. 41 ref. 37: *instead of* Symp. No. **51** *read* Symp. No. **41**

p. 63 Table 1, line 1 and 2: *interchange ref.* 18a *and* 18b

p. 70 2nd line: *instead of* 14.2-months. Chandler Excitations...
read 14.2-months Chandler.Excitations...

p. 73 line 5 from the bottom in 2.2.3.1: *instead of* [18, 40] *read* [18a, 40]

p. 78 ref. 34: *delete* (1981)

p. 87 ref. 4: *delete* **61** (1979) 9,
ref. 6: *insert* **61** (1979) 9.

p. 95 ref. 3: *instead of* **275** *read* **225**

p. 101 ref. 41a: *instead of* Ichiro, K. *read* Kawaguchi, I.
ref. 47: *instead of* (1973) *read* (1974)
ref. 63: *instead of* Mon. Not. R. Astron. Soc. **68** (1951) 89.
read Mem. R. Astron. Soc. **68** (1961) 89.
ref. 83: *instead of* (1977) *read* (1976)

p. 104 ref. 51: *instead of* 83 *read* 81

p. 203 last line before Table 1: *supply* p. 375 in Subvol. c

p. 219 in the head of Table 18: *instead of* [1] *read* [l]

p. 223 line 7 in 3.3.3.6: *instead of* Mechanical concepts do not apply well to dust tails, nor to...
read Mechanical concepts apply well to dust tails, not to...

p. 230 line 4 from the bottom: *instead of* $r^{-}\gamma$ *read* $r^{-\gamma}$

Editors

K. Schaifers, Landessternwarte, Königstuhl, 6900 Heidelberg, FRG

H. H. Voigt, Universitätssternwarte, Geismarlandstraße 11, 3400 Göttingen, FRG

Contributors

W. I. Axford, Max-Planck-Institut für Aeronomie, 3411 Lindau/Harz, FRG

A. Behr, Eschenweg 3, 3406 Bovenden, FRG

A. Bruzek, Kiepenheuer-Institut für Sonnenphysik, Schöneckstraße 6, 7800 Freiburg, FRG

C. J. Durrant, Kiepenheuer-Institut für Sonnenphysik, Schöneckstraße 6, 7800 Freiburg, FRG

H. Enslin, Deutsches Hydrographisches Institut, Bernhard-Nocht-Straße 78, 2000 Hamburg 4, FRG

H. Fechtig, Max-Planck-Institut für Kernphysik, Saupfercheckweg, 6900 Heidelberg, FRG

W. Fricke, Astronomisches Rechen-Institut, Mönchhofstraße 12–14, 6900 Heidelberg, FRG

F. Gondolatsch, Astronomisches Rechen-Institut, Mönchhofstraße 12–14, 6900 Heidelberg, FRG

H. Grün, Max-Planck-Institut für Kernphysik, Saupfercheckweg, 6900 Heidelberg, FRG

O. Hachenberg, Radioastronomisches Institut der Universität, Auf dem Hügel 71, 5300 Bonn 1, FRG

W.-H. Ip, Max-Planck-Institut für Aeronomie, 3411 Lindau/Harz, FRG

E. K. Jessberger, Max-Planck-Institut für Kernphysik, Saupfercheckweg, 6900 Heidelberg, FRG

T. Kirsten, Max-Planck-Institut für Kernphysik, Saupfercheckweg, 6900 Heidelberg, FRG

Ch. Leinert, Max-Planck-Institut für Astronomie, Königstuhl, 6900 Heidelberg, FRG

D. Lemke, Max-Planck-Institut für Astronomie, Königstuhl, 6900 Heidelberg, FRG

H. Palme, Max-Planck-Institut für Chemie, Saarstraße 23, 6500 Mainz, FRG

W. Pilipp, Max-Planck-Institut für Physik und Astrophysik, Institut für Extraterrestrische Physik, 8046 Garching b. München, FRG

J. Rahe, Dr. Remeis-Sternwarte, Sternwartstraße 7, 8600 Bamberg, FRG

G. Schmahl, Universitätssternwarte, Geismarlandstraße 11, 3400 Göttingen, FRG

M. Scholer, Max-Planck-Institut für Physik und Astrophysik, Institut für Extraterrestrische Physik, 8046 Garching b. München, FRG

J. Schubart, Astronomisches Rechen-Institut, Mönchhofstraße 12–14, 6900 Heidelberg, FRG

J. Solf, Max-Planck-Institut für Astronomie, Königstuhl, 6900 Heidelberg, FRG

R, Staubert, Astronomisches Institut der Universität, Waldhäuserstraße 64, 7400 Tübingen, FRG

H. E. Suess, Univ. of California, Chemistry Department, La Jolla/Calif. 92093, USA

J. Trümper, Max-Planck-Institut für Physik und Astrophysik, Institut für Extraterrestrische Physik, 8046 Garching b. München, FRG

G. Weigelt, Physikalisches Institut der Universität, Erwin-Rommel-Straße 1, 8520 Erlangen, FRG

R. M. West, European Southern Observatory, Karl-Schwarzschild-Straße 2, 8046 Garching b. München, FRG

R. Wolf, Max-Planck-Institut für Astronomie, Königstuhl, 6900 Heidelberg, FRG

H. D. Zeh, Institut für Theoretische Physik der Universität, Philosophenweg 19, 6900 Heidelberg, FRG

Vorwort

In allen Fachwissenschaften führt das ständige Ansteigen einer Flut von immer spezialisierter und unübersichtlicher werdenden Publikationen zu dem Bedürfnis, in gewissen zeitlichen Abständen die für die Weiterarbeit gebrauchten neuen Fakten, Zahlenwerte und Funktionen möglichst vollständig, kritisch und übersichtlich zusammenzustellen. Das gilt nicht nur für die klassischen Laborwissenschaften, sondern auch für die Astronomie.

Im „Landolt-Börnstein" wurde die Astronomie erstmals in dem 1952 von J. Bartels und P. ten Bruggencate herausgegebenen III. Band „Astronomie und Geophysik" der 6. Auflage dargestellt. In der „Neuen Serie" erschien dann 1965 in der Gruppe VI der von H. H. Voigt edierte Band VI/1 „Astronomie und Astrophysik", dem nun nach 16 Jahren dieser Band VI/2 (in drei Teilbänden 2a, 2b, 2c) folgt. Der Aufbau dieses neuen Bandes entspricht in großen Zügen dem vorhergehenden von 1965.

Konnte im Jahre 1952 die Astronomie auf 255 Seiten von 25 Autoren und 1965 auf etwa 700 Seiten von 39 Autoren abgehandelt werden, so sind jetzt mehr als 60 Kollegen an der Erstellung der drei Teilbände beteiligt.

Dieses starke Wachsen des Umfangs in 30 Jahren hat zahlreiche Gründe, die im einzelnen hier nicht angesprochen werden sollen, die sich aber kaum schöner aufzeigen lassen, als am Inhalt der drei, über 30 Jahre zeitlich auseinanderliegenden Astronomie-Bände des Landolt-Börnsteins. Gerade das Verschwinden von Teilabschnitten und das Auftauchen ganz neuer Kapitel, auch der Wandel in der Betrachtungsweise – von der Statistik hin zum Einzelobjekt – (dadurch das Anschwellen einzelner Kapitel, z. B. Peculiar stars) zeigen die rasche Entwicklung unserer Wissenschaft in den letzten Jahrzehnten. Aber nicht nur für den Wissenschaftsgeschichtler behält der vorhergehende Band seine Bedeutung, sondern er ist gleichzeitig eine wichtige Quelle für die vor 1965 erbrachten Zahlenwerte und Funktionen, da in dem jetzt vorgelegten Band VI/2 bei weiter zurückliegenden Fakten lediglich Hinweise auf die entsprechende Darstellung in Band VI/1 gegeben werden. Trotzdem hat auch in diesem Band die Bibliographie einen nicht unerheblichen Umfang genommen, obwohl in erster Linie Review-Artikel und Monographien vor Primärliteratur zitiert werden.

Wegen des Umfangs dieses neuen Bandes "Astronomy and Astrophysics" mußte dieser – wie schon gesagt – in drei Teilbände aufgeteilt werden:

a) Methoden. Konstanten. Sonnensystem.
b) Sterne und Sternhaufen.
c) Interstellare Materie. Die Galaxis. Universum.

Ein Gesamtregister für alle drei Teilbände erscheint am Ende des letzten Teilbandes VI/2c.

Vor etwa drei Jahrzehnten schrieben die Herausgeber des Bandes „Astronomie und Geophysik" der 6. Auflage des Landolt-Börnstein: Verfasser, Verlag und Herausgeber glauben ihr Ziel erreicht zu haben, wenn die Beurteilung des Bandes lautet „Mit dem Abschnitt über mein Spezialgebiet bin ich nicht zufrieden, aber die übrigen Teile des Buches sind recht nützlich". Diesem Motto schließen wir uns auch heute noch an.

Dabei gilt unser Dank in erster Linie den Verfassern der einzelnen Abschnitte. In ihren Händen lag die eigentliche wissenschaftliche Arbeit und Verantwortung. Wir danken ihnen, daß sie sich bei der Auswahl des Stoffes und der Art der Darstellung weitgehend unseren Vorstellungen und Richtlinien angepaßt haben.

Ferner haben wir zu danken der Gesamtredaktion dieses Werks in Darmstadt, besonders Frau G. Burfeindt, in deren Händen die eigentliche redaktionelle Arbeit lag, und Herrn Dr. C. J. Durrant in Freiburg für die Durchsicht des englischen Textes. Wir danken ferner dem Verlag, der unseren Vorstellungen im Rahmen des Möglichen stets gefolgt ist, für die gewohnt hervorragende Buch-Ausstattung gesorgt hat und der diesen Teilband, wie alle anderen Bände des Landolt-Börnstein, ohne finanzielle Hilfe von anderer Seite veröffentlicht.

Heidelberg, Göttingen, August 1981 **Die Herausgeber**

Preface

In all fields of science the steady increase in the number of ever more specialized and intricate publications calls from time to time for a complete, critical and well-arranged compilation of facts, numerical values and functions. This not only applies to the classical laboratory sciences, but also to astronomy and astrophysics.

In "Landolt-Börnstein" astronomy was first treated as part of the third volume of the sixth edition (1952): "Astronomie und Geophysik" edited by J. Bartels and P. ten Bruggencate. In Group VI of the New Series this field was treated anew in 1965 by Volume VI/1 "Astronomy and Astrophysics", edited by H. H. Voigt, and now sixteen years later extended and supplemented by the present VI/2 (in three subvolumes 2a, 2b, 2c), the structure of which largely follows that of the 1965 volume.

Where in 1952 astronomy could be treated by 25 authors in 255 pages and in 1965 by 39 authors in 700 pages, now there are more than 60 experts at work on the three subvolumes.

This increase in size within 30 years has numerous causes that need not be discussed here in detail, but that are obvious when comparing the contents of these three volumes on astronomy that have appeared over a period of 30 years. The disappearance of some sections and emergency of whole new topics, as well as a change in approach – from statistics to the individual object – (and consequently the enlarging of some chapters, for instance "Peculiar stars") show the development of our science in the last decades. However, the previous volume retains its importance not only for the historian of science, it is also the main source for the numerical values and functions published before 1965, since the present Volume VI/2 refers in cases of older data back to the discussions in this previous volume. In spite of this, the present bibliography's bulk has grown considerably, although citation of review articles and monographs is generally preferred to that of primary literature.

The size of this new volume "Astronomy and Astrophysics", required a division into three subvolumes:

a) Methods. Constants. Solar System.
b) Stars and Star Clusters.
c) Interstellar Matter. Galaxy. Universe.

A comprehensive index for all three subvolumes is included at the end of Subvolume VI/2c.

About three decades ago the editors of the volume "Astronomie und Geophysik" in the 6th edition wrote: "Authors, publishers and editors believe they have succeeded, if each reader responds with: I'm not satisfied with the chapter on my speciality, but the other sections are quite useful." We, the present editors, can only adhere to this motto.

Our thanks are due first of all to the authors of the individual chapters. They had to do the scientific work and bear the final responsibility, and they usually followed our ideas and suggestions with regard to the selection and presentation of the material.

We also want to thank the Landolt-Börnstein editorial staff in Darmstadt, especially Mrs. G. Burfeindt, who was responsible for the actual editing, and Dr. Durrant in Freiburg for checking the English text. Thanks are also due to the publishers – always following our wishes if at all possible – for the high quality presentation of this volume which, as with all Landolt-Börnstein volumes, is published without financial support from outside sources.

Heidelberg, Göttingen, August 1981 **The Editors**

Survey

Subvolume a: Methods. Constants. Solar system
1 Astronomical instruments
2 Positions and time determination, astronomical constants
3 The solar system

Subvolume b: Stars and Star Clusters
4 The stars
5 Special types of stars
6 Double stars and star clusters

Subvolume c: Interstellar Matter. Galaxy. Universe
7 Interstellar matter
8 Our Galaxy
9 Galaxies and universe
Comprehensive index

Übersicht

Teilband a: Methoden. Konstanten. Sonnensystem
1 Astronomische Instrumente
2 Orts- und Zeitbestimmung, astronomische Konstanten
3 Das Sonnensystem

Teilband b: Sterne und Sternhaufen
4 Die Sterne
5 Spezielle Sterntypen
6 Doppelsterne und Sternhaufen

Teilband c: Interstellare Materie. Die Galaxis. Universum
7 Interstellare Materie
8 Unsere Galaxis
9 Galaxien und Universum
Gesamtregister

Contents

1 Astronomical instruments

2 Positions and time determination, astronomical constants

3 The solar system

List of abbreviations

Abbreviations, commonly used in Astronomy, and not always explained explicitly in this book.

AU	Astronomical Unit (= Distance Earth–Sun)
B.C.	Bolometric Correction
BD	Bonner Durchmusterung
CLV	Center-limb variation
CMD	Colour-magnitude-diagram
CNO	Carbon, Nitrogen, and Oxygen (*not* as molecule) e.g. CNO cycle, CNO anomalies
ESA	European Space Agency
ESO	European Southern Observatory
ET	or E.T. Ephemeris Time
EUV	Extreme ultraviolet
FWHW	Full Width of Half Maximum
HD	Henry Draper Catalogue
HR	Harvard Revised Catalogue
HRD	Hertzsprung-Russell Diagram
IAU	International Astronomical Union
IR	Infrared
ISM	Interstellar Matter
JD	Julian Date
LB, NS or LB	Landolt-Börnstein, Numerical Data and Functional Relationships in Science and Technology, New Series or: Landolt-Börnstein, NS
LC	Luminosity Class
LF	Luminosity Function
LMC	Large Magellanic Cloud
LTE	Local thermodynamic equilibrium
M	Messier Catalogue
MHD	Magneto-hydrodynamics
MMT	Multi-Mirror-Telescope
MPI	Max-Planck-Institut
NASA	National Aeronautics and Space Administration
NEP	Noise Equivalent Power
NGC	New General Catalogue
NLTE	Non-local thermodynamic equilibrium
NRAO	National Radio Astronomy Observatory, Green Banks, W. Va., USA
POSS	Palomar Observatory Sky Survey
RV	Radial velocity
SMC	Small Magellanic Cloud
Sp	Spectral type
URSI	International Union of Radio Science
UT	Universal time
UV	Ultraviolet
VLBI	Very Long Baseline Interferometry
XUV	X-ray and ultraviolet region
ZAMS	Zero Age Main Sequence

Abbreviations of further Star Catalogues: see 8.1.1
For abbreviations of special star types (e.g. WR stars), see "Spectralclassification" (4.1.1), "Variable stars" (5.1), "Peculiar stars" (5.2) and subject index.

Some important Astronomical Artificial Satellites, mentioned in this book

ANS	Astronomical Netherlands Satellite (The Netherlands NASA)
ATS	Applications Technology Satellite
COS	Cosmic Ray Satellite (ESA)
GIRL	German Infrared Laboratory
HEAO	High Energy Astrophysical Observatory (NASA)
HEOS	High Eccentricity Earth-Orbiting Satellite (ESA)
IMP	Interplanetary Monitoring Platform
IRAS	Infrared Astronomical Satellite
IUE	International Ultraviolet Explorer (NASA–UK–ESA)
OAO	Orbiting Astronomical Observatory (NASA)
OGO	Orbiting Geophysical Observatory
OSO	Orbiting Solar Observatory
MTS	Meteoroid Technology Satellite (NASA)
RAE	Radio Astronomy Explorer
SAS	Small Astronomy Satellite (NASA)

1 Astronomical instruments

1.1 Optical telescopes

1.1.1 Introduction

Ground-based optical astronomy of the last decade can be characterized by the development, more or less simultaneously, of more new telescopes than in any decade before. Above all, the number of large telescopes with apertures over 3 m has increased considerably. This is demonstrated in Fig. 1.

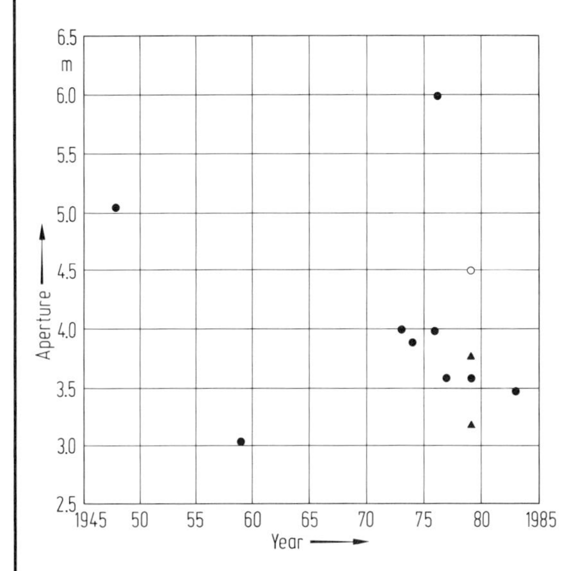

Fig. 1. Aperture versus begin of operation of the largest reflectors.
● general purpose telescopes
▲ infrared telescopes
○ Multiple Mirror Telescope

Most of the new telescopes are constructed as general purpose instruments with reflective optics. The larger ones have several focus stations with different optical systems to satisfy the needs of astronomers. Also there are highly specialized instruments for infrared observations. See Ring et al., and Becklin in [e], see also 1.8.

For wide-field photographic purposes some new Schmidt cameras or related types have been installed.

New technologies in telescope design have been applied with the Multiple Mirror Telescope, a possible prototype of future telescopes. See Strittmatter and Hoffman in [e].

In Table 3 telescopes erected after 1960 are listed. It includes reflectors with a minimum aperture of 1.2 m and wide-field cameras with a minimum aperture of 0.6 m. Older telescopes, see [a].

1.1.2 Optics

All major new telescopes have reflective optics, in some cases with additional refractive elements to reduce the aberrations.

Wide-field cameras: Schmidt telescopes have a single concave spherical mirror and a diaphragm at the center of curvature of the mirror.

Such telescopes are free of coma and astigmatism. The remaining spherical aberration is compensated by a thin aspherical correction plate at the center of curvature. Because of the focus position within the tube, it is suitable for photographic work only. Typical field diameters are 5°.

Various modifications have been made to shorten the tube length and to produce a better focus access [a, g].

Maksutov (Bouwers) telescopes use a large meniscus lens to compensate the aberration of the spherical mirror. With a properly chosen distance between the mirror and the meniscus, the coma can be corrected. The distance is of the order of 1.3···1.4 times the focal length. Typical field diameters are 4°.

Modified systems with better focus access are possible [g].

Reflectors have one or two mirrors which form the image. Plane mirrors may be added for better focus access or for a stationary focus position [a, g]. Aberrations, see [b, g].

The preferred optical system is the aplanatic two-mirror system of the Ritchey-Chrétien type which is free of coma and spherical aberration. The usable field of typically 0°.5 is limited by astigmatism. Both mirrors have a hyperboloidal shape. The deformation of the secondary is stronger than in the true Cassegrain system. Prime focus work requires a corrector also on the optical axis (spherical aberration).

Focus stations: Usually the different optical systems are named after the focus station.

Prime focus: use of the main mirror only; *f*-ratio 1:2.5···1:5. Larger fields require correctors, in the case of hyperboloids even for on-axis work.

Newtonian focus: as prime focus but with an additional flat mirror which reflects the beam at 90° toward the side of the tube giving a better focus access in the case of smaller telescopes.

Cassegrain focus: two-mirror system with increased focal length. A convex hyperboloidal secondary mirror reflects the beam back into the tube and forms an image behind the primary mirror which has a central hole.

The true Cassegrain system has a paraboloidal primary; *f*-ratio: 1:10···1:20. The field is small, limited by coma.

Ritchey-Chrétien focus: same as Cassegrain focus but for telescopes with Ritchey-Chrétien optics; *f*-ratio: 1:7··· 1:10. The field of typically 0°.5 is curved. It is limited by astigmatism.

Gregory focus: an increased focal length can be achieved likewise with a concave ellipsoidal secondary mirror. This system is occasionally used for solar telescopes.

Nasmyth focus: Cassegrain or Ritchey-Chrétien system with an additional flat mirror to bring the light beam toward the side of the tube.

Coudé focus: in principle, a two-mirror Cassegrain system of extended back focal length with 1···3 additional flat mirrors, depending on the mounting type, which reflect the light beam in a fixed direction to produce a stationary focus position; *f*-ratio: 1:30···1:45. Image rotation must be compensated when observing extended objects.

Richardson type coudé focus: to reduce the light losses by the numerous reflections of a conventional coudé system, all mirrors except the primary are provided with highly reflecting dielectric multilayers. At least 3 band-pass mirrors respectively are necessary to cover the optical spectral range. *f*-ratio: 1:100···1:150. See Richardson (1971) in [c].

Mirrors: The mirror material used is mainly low expansion glass (trade names: Pyrex, Duran), fused silica, or, recently, glass ceramics (trade names: Cer-Vit, Zerodur) which have an extremely low thermal expansion coefficient, Table 1.

A few telescopes have metal mirrors consisting of an aluminum alloy with a hard nickel phosphide surface layer which makes polishing easier.

Table 1. Thermal properties of glass ceramics and fused silica. (All data converted to SI-units.)

	Cer-Vit [1]	Zerodur [2]	Fused silica [1]
Coefficient of thermal expansion α $[10^{-7}\,°C^{-1}]$	0 ± 1.5 (0···300 °C)	-0.5 ± 1.5 (−195···20 °C)	5.5
Specific heat $[J\,kg^{-1}\,°C^{-1}]$	51.83	46.81	42.99
Thermal conductivity $[J\,m^{-1}\,s^{-1}\,°C^{-1}]$	0.0955	0.0931 (80···100 °C)	0.0788
Thermal diffusivity $[10^{6}\,m^{2}\,s^{-1}]$	0.80	0.79	0.82

[1]) Monnier, R.C., Appl. Opt. **6** (1967) 1437.
[2]) Supplied by Jenaer Glaswerke Schott & Gen. Mainz, West Germany.

The actual reflecting surface is a thin vacuum-deposited aluminum layer with a thickness of 100···200 nm. In special infrared arrangements also gold or silver surface layers are used.

A high reflectivity up to 99% is achieved with dielectric multilayers. The bandwidth of such high reflectance coatings is of the order of 100···200 nm. Only relatively small mirrors can be coated with layers of good quality; the maximum mirror diameter seems to be 0.5 m.

Protective layers are not usual yet for aluminum films. For silver or dielectric coatings overcoatings of SiO_2 (reactive evaporation of SiO with O_2), MgF_2, or polymerized siloxane are used.

Auxiliary optics: For larger fields it is necessary to correct the aberrations off the optical axis with additional optical elements, namely lens correctors. The prime focus image of a Ritchey-Chrétien system must be corrected also on the optical axis.

Depending on the desired field angle one or more optical elements are required:

Singlets are used for correction of field curvature or for small prime focus applications.

The spherical aberration and coma of a hyperboloidal primary mirror can be removed with a single aspherical plate, the Gascoigne plate. The field diameter of about 10 arc min is limited by astigmatism.

Doublets give field diameters of 0°.5 in the prime focus and of 1° in the Ritchey-Chrétien focus.

Triplets give 1° field diameter in the prime focus.

Detailed discussion with extensive reference list: see Wilson (1971) in [c]; [g].

1.1.3 Mounting

The mounting is a rotatable structure to support the optical elements and the detector in a fixed relationship so that the optical axis can be pointed in all directions. There are three main parts to telescope mounts:

Mirror support system: Gravity compensated systems to keep the optical figure of the mirror surface and the mirror position within the specified tolerances.

For large mirrors closed-loop servo controlled systems (pneumatic, hydrostatic systems) are used, see Pearson (1971) in [c] and references in [a].

Tube mount: Structural frame balanced about the rotation axis. At the lower end is the cell with the mirror and its support system; at the upper end is the focusable prime focus station (cage) or, alternately, secondary mirrors for Cassegrain or coudé systems.

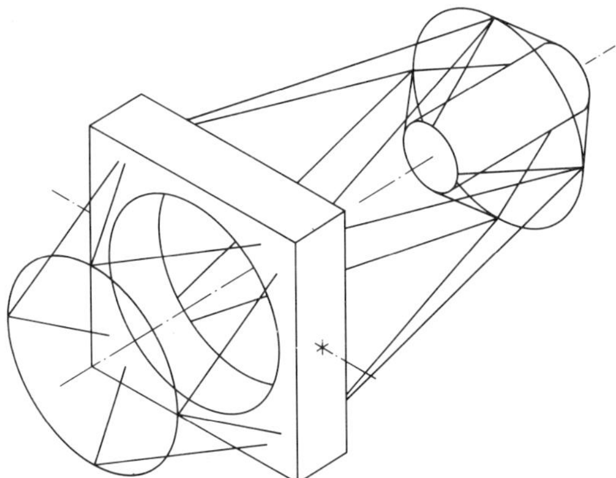

Fig. 2. Principle of the Serrurier tube frame.

In general the frame is a four-sided parallelogram type truss (M. Serrurier truss) as shown in Fig. 2, which allows reasonably large deflections of the tube ends without tilting the mirrors, see Pope (1971) in [c].

The diverse secondary mirrors and the prime focus equipment are either completely exchangeable (mirrors with cell structure or the complete front end) in a horizontal or vertical telescope position (for vertical change-over a special dome crane must be part of the system), or the mirrors are mounted in a common flip-over ring. In the latter case the obscuration is given by the largest element.

For Schmidt telescopes with the plateholder in the prime focus, closed tubes with temperature compensation systems (invar or quartz rods) between the mirror and the focal surface are used.

Axis mount: There are two basic types of mount in existing telescopes. The one most commonly used is the equatorial mounting which has the polar axis parallel to the earth's rotation axis and the declination axis perpendicular to it. The other type is the altazimuth mounting having a vertical azimuth axis and a horizontal elevation axis.

The advantage of the latter mounting type is the constant effect of the direction of gravity upon the bearings and the structural elements of the mount. Disadvantages are the non-uniform motions in both axes for tracking (which again is no problem with modern drives), the field rotation, and the singularity in the zenith which makes observing in a certain area around the zenith impossible.

The 6 m telescope at Zelenchuk and the Multiple Telescope are so mounted.

In the case of equatorially mounted telescopes, in principle only a uniform motion around the polar axis is necessary for tracking (compensation of the earth's rotation). But the variable direction of gravity results in position dependent deflections of the mounting structure of a large telescope. Therefore small position and velocity corrections in both axes are required. Likewise, misalignment of the polar axis demands instrumental corrections.

The design of equatorial mountings is either symmetrical, such as the fork, yoke, or horseshoe types, or asymmetrical, such as the cross axis, polar axis, german mounting types, with a counterweight for tube balancing. [a, c, g].

1.1.4 Drive and control

The total speed range required of a large telescope drive is about 1:3000 for both axes as shown in Table 2, to perform all driving functions from guiding and tracking to slewing. In older telescope drives this could be realized only with several motors and complicated gear trains. The drive can be simplified and improved with modern DC torque motors operating in a servo system with high resolution digital tachometers. Because of the wide dynamic range of the torque motor, a single motor and gearing combination for all speeds is possible. A second torque motor preloads the main gear to suppress the backlash.

Table 2. Summary of speeds of telescope drives.

Speed	Hour angle	Declination
Slewing	±1°/s	±1°/s
Setting	±120″/s	±120″/s
Tracking	15″/s	–
Guiding	±(0···5)″/s	±(0···5)″/s
Additional angular rates	±(0···2)″/s	±(0···2)″/s

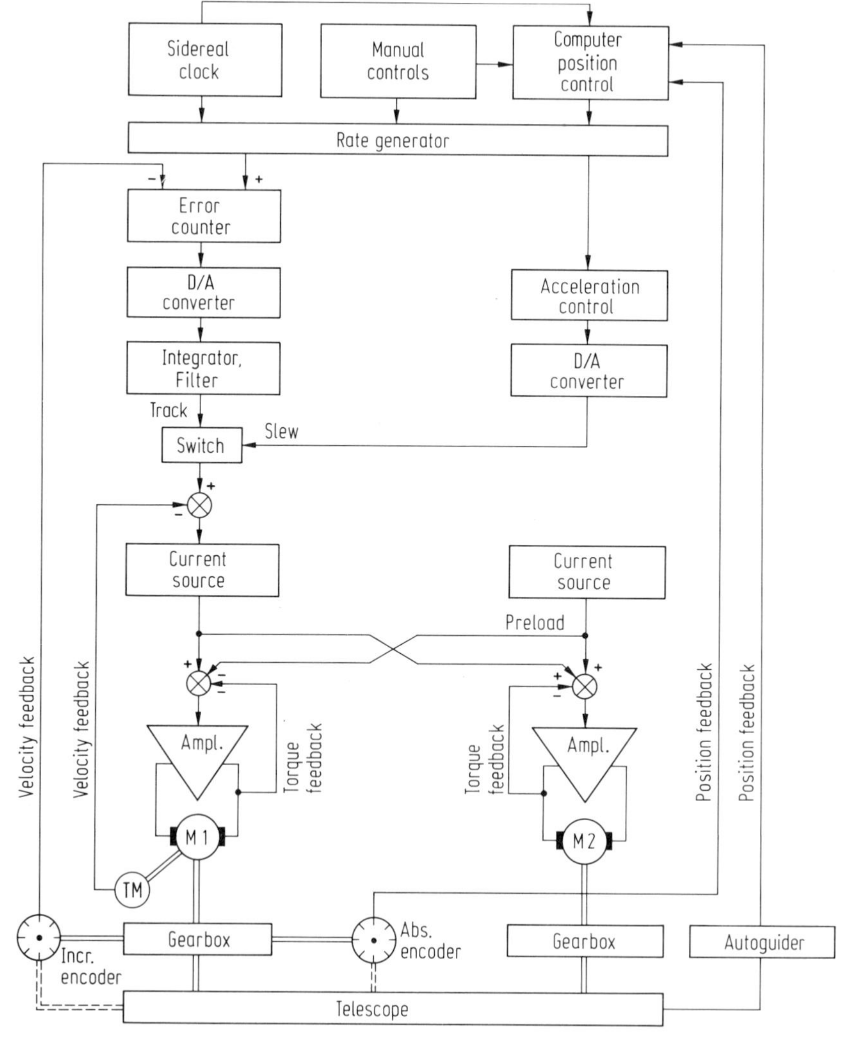

Fig. 3. Block diagram of a servo-controlled telescope drive (dashed lines indicate direct coupling of the encoders to the axis).

The servo control system provides for several different feedbacks. A simplified block diagram is shown in Fig. 3. The first feedback is the current feedback to the power amplifier of the torque motor to get a current-controlled or torque-controlled amplification. The next is a velocity feedback. This is different for high and for low speeds. For high speeds a voltage from the DC tachometer, proportional to its rotation rate, is compared with the voltage demanded for the preselected rate. For the low speeds this procedure is not sensitive enough. In that case the velocity feedback is derived from an incremental encoder (digital tachometer). Its pulses decrement an up/down error counter which is incremented by the pulses of an accurately timed rate generator which is set with the desired speed. The output from the counter represents the instantaneous position error. It is converted into an analogue voltage which is integrated and applied to the motor amplifier as a velocity demand.

The next order feedback is that from the absolute encoder of the position readout system. Any difference between that and the preselected coordinate gives via the servo loop an instantaneous velocity correction. With the absolute encoders coupled directly to the axes, the telescope mounting is part of the servo system. This is not the case if the encoders are coupled to the worms or pinions.

With a photoelectric autoguiding system in the focal plane, error signals derived from image motions can be fed back into the control system. So the tube also is included in the whole servo loop.

1.1.5 Building and dome

Building and dome are integral parts of any major telescope. The building houses the telescope and its control and service equipment. The dome shelters the telescope during day-time and protects it against wind during the observation. It must keep the night temperature during the day which requires a good isolation of the dome surface and of the observing floor.

The telescope stands on a pier separated from the remaining building. Thus vibrations from the rotating dome are not transmitted to the telescope.

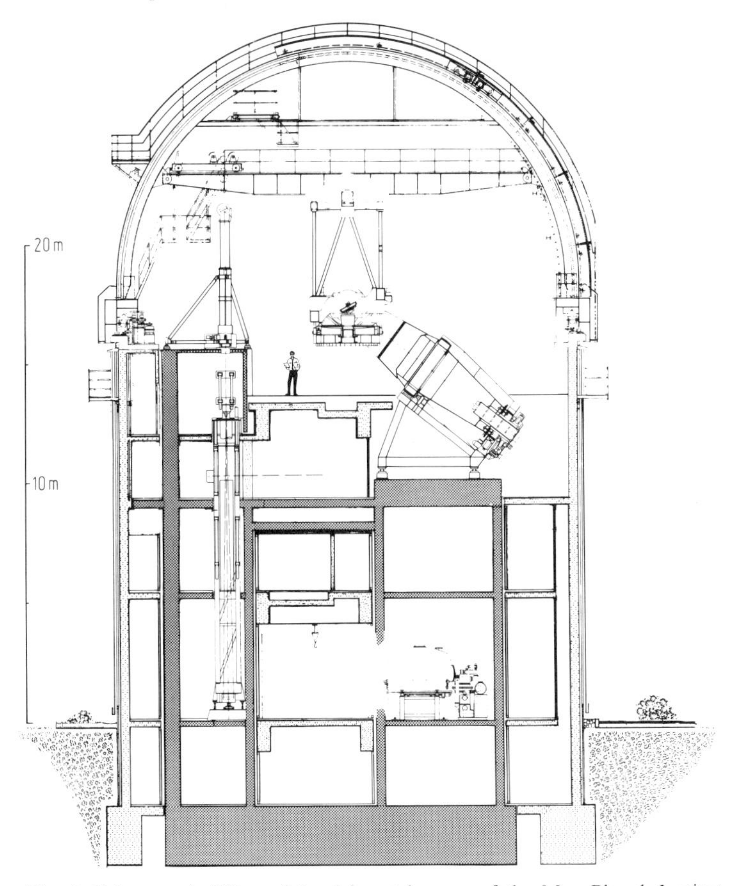

Fig. 4. Telescope building of the 2.2 m telescope of the Max-Planck-Institut für Astronomie with vertical coudé spectrograph.

The interior of the building is primarily determined by the shape of the telescope pier which again depends on the mounting type, by the scientific requirements (laboratories, darkrooms, observing stations), by the necessary space for large permanently installed instrumentation (coudé spectrographs), and by the handling of telescope parts (front end change-over, transportation of the primary mirror for re-aluminization). In Fig. 4, a building for a medium sized telescope used mainly for spectroscopy is shown. For references, see Baustian (1971) in [c].

1.1.6 Future developments

For future developments of substantially larger telescopes radically different possibilities are under discussion. These include mosaic telescopes with a segmented primary mirror and telescope arrays. The latter may consist of several complete telescopes with a combined focus position or of several mittor systems in a common mounting with a combined Cassegrain focus.

The Multiple Mirror Telescope at Mt. Hopkins is an experimental telescope of this new technology. It is an array of six identical Cassegrain telescopes, each with a diameter of 1.82 m. The light collecting area is that of a 4.5 m telescope. The optical alignment is maintained by a laser servo control system [e, f, h].

1.1.7 List of large optical telescopes erected after 1960

Table 3. List of large optical telescopes erected after 1960. It includes reflectors with a minimum aperture of 1.2 m and wide-field cameras with a minimum aperture of 0.6 m. For telescopes erected before 1960, see the previous volume, [a]; for 1960···1965 there is some overlap [indicated by *)].

Abbreviations:

alt-az	altazimuth mounting	Nas	Nasmyth focus
Cas	Cassegrain focus	New	Newtonian focus
Cou	coudé focus	Pr	prime focus
doubl.	doublet corrector	RC	Ritchey Chrétien focus
Gasc.	aspheric Gascoigne plate	Re	reflector
IR	infrared telescope facility	Sch	Schmidt camera
MMT	Multiple Mirror Telescope	tripl.	triplet corrector
mod.	modified		

Location (Observatory)	Type	Mount	Aperture [m]	Optical system	Focal length [m]	Year	Ref.
Asiago (Padua Univ.)	Sch*)	yoke	0.65/0.92 [1])	Pr	2.15	1964	a
	Re	fork	1.82	Cas	16.4	1973	1, 2
Baja California (Univ. of Mexico)	Re		1.5			1971	
Brazopolis (Brazilian Nat. Obs.)	Re	off-axis	1.6	RC Cou [2])	16.0 240.0/49.9	[13])	
Budapest (Budapest)	Sch*)	fork	0.6/0.9 [1])	Pr	1.8	1963	a

For footnotes, see p. 11.

continued

Table 3 (continued)

Location (Observatory)	Type	Mount	Aperture [m]	Optical system	Focal length [m]	Year	Ref.
Byurakan (Byurakan)	Sch*)	fork	1.0/1.5 [1])	Pr	2.13	1961	a
	Re	fork	2.6	Pr	9.4	1975	
				Nas	41.6		
				Cou	104.0		
Calar Alto (Max-Planck-Inst. for Astronomy, German-Spanish Astronomical Center)	Re	pole universal	1.23	mod. RC [3])	9.86	1975	3
				mod. RC/doubl.	9.81		
	Re	fork	2.2	RC	17.6	1979	4
				RC/doubl.	17.0		
				Cou [4])	88.0		
	Sch[5,*])	fork	0.8/1.2 [1])	Pr	2.4	1980	a
	Re	horse-shoe	3.5	Pr/doubl.	12.2	1983	5
				Pr/tripl.	13.8		
				RC	35.0		
				Cou	122.5		
Calar Alto (Spanish Nat. Obs. Madrid)	Re	cross axis	1.5	RC [4])	12.0	1978	
				Cou	45.0		
Castel Gandolfo (Specola Vaticana)	Sch*)	fork	0.64/0.98 [1])	Pr	2.4	1961	a
Cerro Las Campanas (Carnegie Southern Obs.)	Re	fork	2.54	RC/Gasc.	19.05	1976	6, 7
				Cou [4])	76.2		
Cerro La Silla (European Southern Obs.)	Re	cross axis	1.5	Cas	22.4	1968	8
				Cou [4])	46.9		
	Sch	fork	1.0/1.6 [1])	Pr	3.06	1969	8
	Re	horseshoe and fork	3.6	Pr/Gasc.	10.9	1976	8
				Pr/tripl.	11.3		
				RC	28.6		
				Cou	114.6		
Cerro La Silla (Univ. of Copen-hagen/ESO)	Re	off-axis	1.5	RC	13.1	1979	8
Cerro Tololo (Inter-American Obs.)	Sch[6,*])	cross-axis	0.61/0.91 [1])	Pr	2.13	1967	a
	Re	off-axis	1.52	RC/Gasc.	11.4		9
				Cas	20.5		
				IR	45.6		
				Cou [4])	47.4		
	Re	horsehoe	4.0	Pr	10.6		9, 18
				RC	31.2		

continued

Table 3 (continued)

Location (Observatory)	Type	Mount	Aperture [m]	Optical system	Focal length [m]	Year	Ref.
Coonabarabran (Siding Spring Obs.) [7])	Sch	fork	1.2/1.8 [1])	Pr	3.06	1973	10
Coonabarabran (Siding Spring Obs./Anglo Australian Obs.)	Re	horseshoe	3.9	Pr/Gasc.	12.7	1975	11
				Pr/doubl.	12.7		
				Pr/tripl.	13.5		
				RC	30.8		
				Cas	57.9		
				Cou	140.2		
Crimea (Crimean Astrophys. Obs.)	Re*)	fork	2.64	Pr	10.0	1961	a
				Cas	43.0		
				Nas	41.0		
				Cou	105.0		
Crimea (Sternberg South Station)	Re*)		1.25	Pr	5.0	1960	a
				New	5.0		
				Cas	21.0		
Flagstaff (Perkins Obs.)	Re[8,*])	cross axis	1.83	Cas	31.0	1961	a, 12
Flagstaff (U.S. Naval Obs.)	Re[9,*])	fork	1.55	Pr	15.0	1964	a
Fort Davis (McDonald Obs.) Univ. of Texas)	Re	cross axis	2.7	RC	24.0	1969	13
				Cas	48.6		
				Cou [4])	89.1		
Hamburg-Bergedorf (Hamburg Obs.)	Re	fork	1.25	RC	15.6	1976	
Helwan (Helwan Obs.)	Re*)	cross axis	1.88	New	9.14	1963	a
				Cas	34.0		
				Cou	56.0		
Herstmonceux (Royal Greenwich Obs.)	Re [10])	fork with polar disk	2.49	Pr	7.5	1967	14
				Pr/doubl.	8.2		
				Cas	36.8		
				Cou	82.0		
Hsing-lung (Peking Obs.)	Sch } *)	fork	0.6/0.9 [1])	Pr	1.8	1963	a
	Re }		0.9	Cas	13.5		
Hyderabad (Nizamiah Obs.)	Re*)	cross axis	1.22	Pr	4.9	1963	a
				Cas	18.0		
				Cou	37.0		
Jelm Mt. (Univ. of Wyoming)	IR	yoke	2.3	Pr	4.8	1977	15
				Cas	62.1		

continued

Table 3 (continued)

Location (Observatory)	Type	Mount	Aperture [m]	Optical system	Focal length [m]	Year	Ref.
Jena (Jena Obs.)	Sch } *)	fork	0.6/0.9 [1])	Pr	1.8	1963	a
	Re }		0.9	Cas	13.5		
Kiaton (Univ. of Athens)	Re	off-axis	1.2	Cas	15.6	1975	16
Kiso Mts. (Kiso Obs.)	Sch	fork	1.05/1.5 [1])	Pr	3.25	1974	
				Cas	34.5		
Kitt Peak (Kitt Peak Nat. Obs.)	Re*)	fork	2.13	RC	16.2	1963	a, 17
				IR	57.3		
				Cou [4])	66.5		
	Re	horseshoe	4.0	Pr/tripl.	11.1	1973	17, 18
				RC	30.8		
				Cou	652.0		
	Re		1.27	IR	18.8		17
Kitt Peak (McGraw-Hill Obs. Univ. of Michigan)	Re [6])	cross axis	1.32	Cas	10.0	1975	19
				Cas	17.8		
				Cou	44.2		
Kitt Peak (Steward Obs., Univ. of Arizona)	Re	fork	2.28	RC	20.5	1969	20
				Cou	70.7		
Kvistaberg Station (Uppsala Univ.)	Sch*)	fork	1.0/1.35 [1])	Pr	3.0	1964	a
Llano del Hato (Univ. of the Andes, Merida)	Sch*)	bent yoke	1.0/1.52 [1])	Pr	3.0	1978	a
London, Ontario (Univ. of Western Ontario)	Re		1.22			1968	
Mauna Kea (Canada, France, Hawaii)	Re	horseshoe	3.6	Pr/tripl.	13.7	1979	21
				Cou [11])	72.0		
Mauna Kea (NASA)	IR	yoke	3.0 [12])	Cas	105.0	1979	22
				Cou	360.0		
Mauna Kea (United Kingdom)	IR	yoke	3.8	Cas	34.2	1978	23, 24
				Cas	133.0		
				Cou	76.0		
Mauna Kea (Univ. of Hawaii)	Re	fork	2.2	RC	22.0	1970	25
				Cou	72.6		

continued

Table 3 (continued)

Location (Observatory)	Type	Mount	Aperture [m]	Optical system	Focal length [m]	Year	Ref.
Mendoza (La Plata Obs.)	Re	fork	2.13	RC Cou	16.2 66.5	[13]	
Merate (Milan-Merate)	Re	fork	1.37	Cas [14]	20.1	1972	26
Mt. Chikurin (Okayama Obs.)	Re*)	cross axis	1.88	New Cas Cou	9.2 33.9 54.3	1960	a
Mt. Hopkins (Smithonian Astrophys. Obs.)	Re		1.52	Cas Cou	15.2 36.6	1970	27
Mt. Hopkins (Smithonian Astrophys. Obs. and Univ. of Arizona)	MMT	alt-az	4.46 (6 × 1.82)	Nas/Cas	57.7 [15] 49.9 [16]	1979	28, 29
Mt. Lemmon (NASA)	IR		1.52			1974	
Mt. Lemmon (Univ. of Minnesota and Univ. of California)	IR [17]	yoke	1.52	RC	12.2	1972	
Mt. Megantic (Univ. of Montreal)	Re	off-axis	1.6	Cas Cas	12.8 24.0	1978	
Ondrejov	Re	off-axis	2.0	Pr Cas Cou	9.0 29.6 72.0	1967	30, 31
Palomar Mt. (Hale Obs.)	Re	fork	1.52	RC Cou	13.3 45.6	1970	32
Pic du Midi (Pic du Midi)	Re [18]	horseshoe	2.0	Pr Cas	9.98 50.0	1979	33
Rattlesnake Mt. (Penn State Univ.)	Re	yoke	1.52	Cas		1974	34
Saltsjöbaden (Stockholm Univ.)	Sch*)		0.65/1.0 [1]	Pr	3.0	1964	a
Shemakha (Shemakha Astrophys. Obs.)	Re	off-axis	2.0	Pr Cas Cou	9.0 29.6 72.0	1967	30
Sutherland (South African Astronomical Obs.)	Re [19,*]	cross axis	1.88	New Cas Cou	9.15 34.0 53.0	1974	a

continued

Table 3 (continued)

Location (Observatory)	Type	Mount	Aperture [m]	Optical system	Focal length [m]	Year	Ref.
Tautenburg (Karl-Schwarz-schild-Obs.)	Sch } *)	fork	1.34/2.0 [1])	Pr	4.0	1960	a
	Re }		2.0	Cas	21.0		
Torun	Sch } *)	fork	0.6/0.9 [1])	Pr	1.8	1962	a
	Re }		0.9	Cas	13.5		
Victoria, British Columbia (Dominion Astro-phys. Obs.)	Re*)	off-axis	1.22	Pr	4.88	1962	a
				Cas	22.0		
				Cou	36.5		
	[20])			Cou	177.0/36.5		35
Vienna (Leopold-Figl-Astrophys. Obs.)	Re	fork	1.52	RC/doubl.	12.5	1969	36
				Cas	22.5		
				Cou	45.0		
Zelenchuk (Special Astro-phys. Obs.)	Re	alt-az	6.0	Pr	24.0	1976	37
				Nas	180.0		

[1]) Clear aperture/aperture of mirror.
[2]) Richardson coudé with conversion lens.
[3]) And Nasmyth focus.
[4]) Spectrograph.
[5]) From Hamburg with new mounting.
[6]) Moved from Portage Lake Obs.
[7]) Operated by Royal Obs. Edinburgh.
[8]) New mirror.
[9]) Flat secondary.
[10]) Will be moved to La Palma with new primary mirror (2.52 m).
[11]) Richardson coudé with conversion lens.
[12]) Mirror diameter 3.2 m.
[13]) Not installed.
[14]) And Nasmyth focus.
[15]) Focus mode 1.
[16]) Focus mode 2.
[17]) Moved from Catalina, new optics.
[18]) Flat secondary.
[19]) Moved from Radcliffe Obs.
[20]) Richardson coudé with conversion lens.

1.1.8 References for 1.1.7

1 Barbieri, C., Galazzi, A.: Descrizione del Telescopio Copernico in: Atti delle Celebrazione del V⁰ Centenario della Nascita di Nicolo' Copernico e Inaugurazione e Convegno Scientifico all' Osservatorio di Cima Ekar, (Rosino, L., Barbieri, C., eds.), Padua-Asiago (1973) p. 45.
2 Barbieri, C., Rosino, L., Stagni, R.: Sky Telesc. **47** (1974) 298.
3 Schlegelmilch, R.: Mitt. Astron. Ges. **30** (1971) 84.
4 Bahner, K.: Sterne Weltraum **12** (1973) 103.
5 Bahner, K.: Mitt. Astron. Ges. **36** (1975) 57.
6 Bowen, I.S., Vaughan, A.H.: Appl. Opt. **12** (1973) 1430.
7 Carnegie Inst. Washington, Year Book **74** (1974/75) 366.
8 User's Manual, European Southern Observatory, Garching (1980).
9 The Facilities Book of Cerro Tololo Inter-American Observatory.
10 Reddish, V.C. in: Conf. on the Role of Schmidt Telescopes in Astronomy, Proc. ESO/SRC/Hamburger Sternwarte (Haug, U., ed.), (1972) p. 135.

11 Morton, P.C. (ed.): Anglo Australian Telescopes Observer's Guide (1976).
12 Hall, J.S., Slettebak, A.: Sky Telesc. **37** (1969) 222.
13 Smith, H.J.: Sky Telesc. **36** (1968) 360.
14 Brown, P.L.: Sky Telesc. **34** (1967) 356.
15 Gehrz, R.D., Hackwell, J.A.: Sky Telesc. **55** (1978) 467.
16 Kotsakis, D. in: In Honorem S. Plakidis, (Kotsakis, D., ed.), Athens (1974) p. 161.
17 The Facilities Book of Kitt Peak National Observatory, Tucson (1977).
18 Crawford, D.L.: J. Opt. Soc. Am. **61** (1971) 682.
19 Wehinger, P.A., Mohler, O.C.: Sky Telesc. **41** (1971) 72.
20 Sky Telesc. **38** (1969) 164.
21 Odgers, G.J., Richardson, E.H., Grundman, W.A. in: [e] of 1.1.9. p. 79.
22 The Infrared Telescope Facility Observer's Manual, Hawaii (1980).
23 Carpenter, G.C., Ring, J., Long, J.F. in: [e] of 1.1.9. p. 47.
24 Brown, D.S., Humphries, C.M. in: [e] of 1.1.9. p. 55.
25 Sky Telesc. **40** (1970) 276.
26 de Moltoni, G.: Sky Telesc. **43** (1972) 296.
27 Schild, R.E.: Smithonian Astrophysical Observatory Special Report No. 355, 5+23+A5 (1973).
28 Strittmatter, P.A. in: [e] of 1.1.9. p. 165.
29 Hoffman, T.E. in: [e] of 1.1.9. p. 185.
30 Jena Review (Jenaer Rundschau) **13** (1968).
31 Grygar, J., Koubsky, P.: Bull. d'Information, Ass. Développement International Obs. Nice No. **8** (1971) 5.
32 Bowen, I.S., Rule, B.H.: Sky Telesc. **32** (1966) 185.
33 Rösch, J., Dragesco, J.: Sky Telesc. **59** (1980) 6.
34 Zabriskie, F.R.: Sky Telesc. **49** (1975) 219.
35 Richardson, E.H. in: [c] of 1.1.9. p. 179.
36 Meurers, J.: Sterne Weltraum **8** (1969) 195.
37 Ioannisiani, B.K.: Sky Telesc. **54** (1977) 356.

1.1.9 General references for 1.1

a Bahner, K. in: Landolt-Börnstein, NS, Vol. VI/1 (1965) p. 1.
b Siedentopf, H. in: Landolt-Börnstein, NS, Vol. VI/1 (1965) p. 35.
c Proc. ESO/CERN Conf. on Large Telescope Design (West, R.M., ed.), Geneva (1971).
d Proc. ESO/SRC/CERN Conf. on Research Programmes for the New Large Telescopes (Reiz, A., ed.), Geneva (1974).
e Proc. ESO Conf. Optical Telescopes of the Future (Pacini, F., Richter, W., Wilson, R.N., eds.), Geneva (1977).
f Next Generation Telescope Reports, Kitt Peak National Observatory, Tucson (1977/1978).
g Bahner, K.: Teleskope, in Handb. Physik (Flügge, S., ed.), Springer, Berlin-Heidelberg-New York **29** (1967).
h Proc. Optical and Infrared Telescopes of the 1990s (Hewitt, A., ed.), Kitt Peak National Observatory, Tucson (1980).

1.2 Solar telescopes

Principles of telescope and spectrograph design: see LB, NS VI/1 (1965) [4]. Since 1965, most development has been in the fields of narrow-band filters and of areal detectors as alternatives to photographic plates.

Table 1. Solar telescopes with free aperture ∅≧25 cm.
Explanation of columns

1. Location of institution (name of institute), [location of field station]
2. Type of telescope
   - T tower telescope
   - H horizontal system
   - E equatorial mounting
   - A altazimuth mounting
3. Coel: Coelostat
   Diameter of coelostat and auxiliary separated by/
   Other systems denoted by H (heliostat), S (siderostat)
4. Opt: Optics
   - L lens optics (refractor)
   - M mirror optics (reflector)
5. ∅: aperture of telescope objective
6. f_T: Effective focal length(s) of telescope
7. Arr: arrangement
   - Cor coronograph
   - C Cassegrain
   - G Gregorian
   - N Newtonian
   - Cde coudé
8. Equipment
   - Sp spectrograph
   - G grating
   - CG concave grating
   - EG Echelle grating

   (Sp, G, CG, EG: focal length or radius of curvature of camera [cm])
   - ORSp optical resonance spectrograph
   - IRSp infrared spectrograph
   - FTSp Fourier transform spectrograph
   - Mag magnetograph
   - VMag vector magnetograph/Stokesmeter
   - Shg spectroheliograph
   - F( ) filter (type: Hα, K, UBF = universal birefringent filter)
   - CPol coronal polarimeter
   - CPh coronal photometer
9. Ref.: reference

Location	Type	Coel cm	Opt	∅ cm	f_T cm	Arr	Equipment	Ref.
Abastumani (Abastumani Astrophys. Obs.)	H E	44/44	M L	44 52	1750/6000	N, C Cor	GSp	21
Alma-Ata (Astrophys. Inst. of the Academy of Sciences, Kazakh)	H E	44/44	M L	44 53	1750/6000 1250	N, C Cor, Cde	GSp(700); Mag GSp(800)	35
Arcetri (Arcetri Astrophys. Obs.)	T E	42/42	L L	30 36	1800 540		GSp(400, 800); Shg	37
Athens (National Obs. of Athens) [Pentele]	E		L	25	225		F(Hα)	25
Baku (Shemakha Astrophys. Obs.)	H	44/44	M	44	1750/6000	N, C	GSp	16

continued

Table 1, continued

Location	Type	Coel cm	Opt	∅ cm	f_T cm	Arr	Equipment	Ref.
Bangalore (Indian Inst. of Astrophys.)	T	61/61	L	38	3600		GSp(1830); Mag; Shg; F(Hα)	5
			M	30	1830	G, N	F(Hα)	
[Kodaikanal]	H	30(S)	L	30	640		Shg	
[Kavalur]	T		M	45	3300	N, Cde	EGSp(640); F(Hα)	
Boulder (High Altitude Obs.) [Sacramento Peak]	E		L	40	630		CPol	3
Catania (Catania Astrophys. Obs.)	E		L	33	347		Shg	13
Crimea (Crimean Astrophys. Obs.)	T	120/110	M	100	5000		GSp(2000, 1000); VMag; Shg	39
	T	60/50	M	45	2000/3500/1200		GSp(800, 1600); EGSp(800); Mag	
	A		L	50	800	Cor	GSp(800); F(Hα)	
Debrecen (Debrecen Heliophysical Obs.)	E		L	25	450		F(Hα)	11
Freiburg (Kiepenheuer Inst.)								
[Schauinsland]	T	55/55	L	45	800		GSp(720)	22
[Anacapri]	E		L	35	1600	Cde	GSp(1720); VMag	
[Izaña]	E		M	40	3500	N	F(Hα, K)	
Göttingen (Göttingen Obs.)								
[Hainberg]	T	65/65	M	45	1650/2400	C, Cde	GSp(800), CGSp(665)	6
[Locarno][6])	E		M	45	2400	G, Cde	EGSp(1000), CGSp(665); VMag	
[Muchachos][1])	E		M	45	2400	G	F(UBF)	38a
Herstmonceaux (Royal Greenwich Obs.)	H	41/41	L	27	775		F(Hα)	
Honolulu (Univ. of Hawaii) [Haleakala]	E		L	25	750	Cor	GSp, EGSp(300); F(Hα)	19
Irkutsk (SibIZMIR) [Lake Baikal]	T	120(H)	L	100			Mag; F	
	E		L	52		Cor, Cde	GSp	

For footnotes, see p. 17. continued

Table 1, continued

Location	Type	Coel cm	Opt	∅ cm	f_T cm	Arr	Equipment	Ref.
Kiev (Main Astron. Obs. of the Ukrainian Acad. of Sci.)	H	44/44	M	44	1750/6000	N, C	GSp(700)	15
Kunming (Yunnan Obs.)	H			30			GSp; Shg	14
Kyoto								
(Kwasan Obs.)	H	70/70		50	2000		GSp(1500); Shg	
(Hida Obs.)	A		M	60	3200	G	GSp; Shg	
Liège	H	30/30	M	50	1220		GSp(730)	
(Univ. of Liège) [Jungfraujoch]	E [2])		M	76		Cde	IRSp(250)	
Moscow (Izmiran)	T	55/55	M	65	1700/2700	N, C	GSp(1000); Shg; VMag; F(Hα, K)	32
(Sternberg Astron. Inst.)	T	44/44	M	30	1510/1905		GSp(997); CGSp(200); Shg	
[Kutschino]	H	30/26	M	30	1500		GSp(499)	
Nanking (Purple Mountain Obs.)	H	40		30			GSp; Shg	14
Nice (Nice Obs.)	E		L	40	1000	Cde		1
Northridge (San Fernando Obs.)	E		M	61/28	1200/550	C	Shg; F(Hα, K)	26
Ondrejov	H	54/54	M	44	3500		GSp(960, 900); Mag	7
Oslo (Inst. of Astrophys.) [Harestna]	T	46/46	M	30	2920/963/2293		GSp(1050, 2150)	20
Ottawa (Ottawa River Obs.)	E		L	25	1300		F(Hα)	12
Oxford	T	41/41	M	40	1200	C	GSp(1000)	36, 40
(Univ. of Oxford)	E		M	51	3500	C, Cde	GSp(1500)	
Paris-Meudon	H	50/40	L	25	401		Shg	31, 27
(Obs. of Paris-Meudon)	H	75(S)	M	40	2300	N, G	GSp(700); VMag	
	T	80/70	M	60	4500		GSp(1400)	
	E		L	30	900		F(UBF)	
[Pic-du-Midi]	H	60/50	M	40	2237/1085	N	GSp(900, 400); F(Hα)	30, 31

continued

Table 1, continued

Location	Type	Coel cm	Opt	Ø cm	f_T cm	Arr	Equipment	Ref.
Pasadena (Hale Obs.) [Mt. Wilson]	T	91/76	M	66	1900/3800/6000			17, 18
	H	76/61	M	61	1825		GSp(460)	
	T	43/32	L }	30	1830		Shg	
			M }	25	900			
	T	51/41	L	30	4575		GSp(2250); Mag	
[Big Bear]	E		L }	25	3600/10800		Mag; F(Hα, K)	
			M }	65	3250	G	GSp(400); F(UBF)	
Peking [Sha-ho]			M	60	3750	C, Cde	GSp	14
Pic-du-Midi	E		L	50	645		Shg; F(Hα); GSp(800) [3]	
(Obs. of Pic- du-Midi)	E		L	26	400	Cor	CPol	
Potsdam (Heinrich-Hertz Inst.) [Telegrafenberg]	T	60/60	L	60	1400		GSp(1200); VMag	10
Pulkovo	H	67/50	M	50	1750/6200		GSp; Mag	23
	H		M	44			GSp	
[Kislovodsk]	E		L	53	800	Cor, Cde	GSp(800)	
Rome	T	70/65	L	45	2800		GSp(1800); Mag	8, 9
Sunspot (Sacramento Peak Obs.)	H	41/41	L	30 }	762/1130/2750	Cor	{ GSp(1300, 152); Shg;	3
	E		{ L	40 }			{ F(Hα, K); CPh; VMag [4]	
			{ L	40	615		GSp(156); Mag	
	T	112/112	M	76	5500		EGSp(1200); F(UBF)	
Sydney (CSIRO) [Culgoora]	E		L	30	305		F(Hα)	24
Tashkent	H	44/44	M	44	1750/6000		GSp(700); F(Hα)	38
Tucson (Kitt Peak National Obs.)	T	203(H)	M	152	8246		GSp(1370, 200); VMag;	2
		91(H)	M	81	4037		IRSp(2000); Shg;	
		91(H)	M	81	3580		FTSp	
	T	104/91	M	70	3639		GSp(1039); Mag	
Tokyo (Univ. of Tokyo) [Norikura]	E		L	25	880	Cor	GSp(700); EGSp(350)	33, 34
[Okayama]	E		M	65	600/3700	Cde	GSp(1000); EGSp(400)	
Uttar Pradesh	H	46	M	27	1750		GSp	41
Wroclaw	T	40	M	30				28

continued

Table 1, continued

Location	Type	Coel cm	Opt	∅ cm	f_T cm	Arr	Equipment	Ref.
Zürich (Swiss Federal Obs.)	T	30	L	25	1100		F(Hα, K)	42
[Arosa]	H	30	L	25	2950		GSp(1250)	
SOON (USAF Weather Service) [5]								
Holloman AFB, US	E		L	25	533		GSp(277); Mag; F(Hα)	
Palehua, Hawaii; Ramey AFB, Puerto Rico; Learmonth, Australia	each with the same instrument							

[1]) Temporary location.
[2]) Primarily non-solar instrument.
[3]) Instrument of Obs. of Paris.
[4]) Instrument of High Altitude Obs.
[5]) A fifth site planned.
[6]) Will be moved to Izaña.

References for 1.2

1 Aime, C., Demarcq, J., Fossat, E., Ricort, G.: Nouv. Rev. Opt. **5** (1974) 257.
2 Anon.: Kitt Peak National Observatory Facilities Book, Kitt Peak National Observatory, Tucson (1977).
3 Anon.: Sacramento Peak Observatory Users Manual, Sacramento Peak Observatory, Sunspot (1978).
4 Bahner, K.: Landolt-Börnstein, NS, Vol. VI/1 (1965) p. 14.
5 Bappu, M.K.V.: Sol. Phys. **1** (1967) 151.
6 Brückner, G.E., Schröter, E.H., Voigt, H.H.: Sol. Phys. **1** (1967) 487.
7 Bumba, V., Klvana, M., Macak, P.: Bull. Astron. Inst. Czech. **27** (1976) 257.
8 Cimino, M.: Sol. Phys. **2** (1967) 375.
9 Cimino, M., Cacciani, A., Fofi, M.: Sol. Phys. **11** (1970) 319.
10 Daene, H., Jäger, F.W.: Sol. Phys. **4** (1968) 489.
11 Dezsö, L.: Sol. Phys. **2** (1967) 129.
12 Gaizauskas, V.: J. R. Astron. Soc. Can. **70** (1976) 1.
13 Godoli, G.: Sol. Phys. **9** (1969) 246.
14 Goldberg, L.: Sky Telesc. **56** (1978) 383.
15 Gurtovenko, E.A.: Sol. Phys. **4** (1968) 108.
16 Guseinov, R.E.: Sol. Phys. **16** (1971) 490.
17 Howard, R.: Sol. Phys. **7** (1969) 153.
18 Howard, R.: Sol. Phys. **38** (1974) 283.
19 Jefferies, J.T.: Sol. Phys. **2** (1967) 369.
20 Jensen, E.: Sol. Phys. **4** (1968) 114.
21 Khetsuriani, Ts.S.: Sol. Phys. **2** (1967) 237.
22 Kiepenheuer, K.O.: Sol. Phys. **1** (1967) 162.
23 Krat, V.A.: Sol. Phys. **4** (1968) 118.
24 Loughhead, R.E., Bray, R.J., Tappere, E.J., Winter, J.G.: Sol. Phys. **4** (1968) 185.
25 Macris, C.J.: Sol. Phys. **2** (1967) 125.
26 Mayfield, E.B., Vrabec, D., Rogers, E., Janssens, T., Becker, R.A.: Sky Telesc. **37** (1969) 208.
27 Mein, P.: Sol. Phys. **54** (1977) 45.
28 Mergentaler, J.: Sol. Phys. **10** (1969) 229.
29 Michard, R.: Astrophys. J. **127** (1958) 504.
30 Michard, R.: Ann. Astrophys. **22** (1959) 185.

31 Michard, R.: Sol. Phys. **1** (1967) 498.
32 Mogilevsky, E.I.: Sol. Phys. **10** (1969) 231.
33 Nagasawa, S.: Sol. Phys. **2** (1967) 240
34 Nishi, K., Makita, M.: Publ. Astron. Soc. Jpn. **25** (1973) 51.
35 Obashev, S.: Sol. Phys. **16** (1971) 493.
36 Plaskett, H.H.: Mon. Not. R. Astron. Soc. **115** (1955) 542.
37 Righini, G.: Sol. Phys. **1** (1967) 494.
38 Scheglov, A., Slonim, Yu.: Sol. Phys. **11** (1970) 157.
38a Schröter, E.H., Wöhl, H.: Joint Organization for Solar Observations (JOSO), Annu. Rep. (1977) p. 62.
39 Severny, A., Stepanyan, N.: Sol. Phys. **1** (1967) 484.
40 Shallis, M.J.: Mon. Not. R. Astron. Soc. **183** (1978) 1.
41 Sinvhal, S.D.: Int. Astron. Union Trans. **XIV A** (1970) 65.
42 Waldmeier, M.: Sol. Phys. **5** (1968) 423.

1.3 Photoelectric photometry

1.3.0 Symbols and definitions

T [K]	temperature of the star
D [cm]	aperture of the telescope
λ[Å]	wavelength
$\Delta\lambda$ [Å]	wavelength bandwidth
$\Delta f = 1/RC$ [Hz]	frequency bandwidth
RC [s]	time constant
I [A]	photocurrent
$\overline{i^2}$	r.m.s. fluctuation of photocurrent
W[W]	radiant flux
S [A/W]	spectral response
D [W^{-1}]	detectivity [28]
D^* [cm Hz$^{1/2}$/W]	D for $\Delta f = 1$ Hz and receiver surface area = 1 cm^2
$n_{h\nu}$	number of incident light quanta (photons)
n_e	number of photoelectrons emitted from the cathode
q	relative quantum efficiency $= n_e/n_{h\nu}$
e	electron charge

$S = q\frac{e}{h\nu} = q \cdot 8.066 \cdot \lambda \cdot 10^{-5}$ [A/W]

$D = I/\sqrt{\overline{i^2}}$ [W^{-1}]

1.3.1 Acronyms and nomenclature

PMT	photomultiplier tube
MOS	metal oxide semiconductor
MCP	microchannel plate
SEC	secondary electron conduction (target)
ST	silicon target
SIT	silicon intensifier target } same device
EBS	electron bombarded silicon target } same device
IDS	image dissector scanner
IDA	self scanned or integrated diode array
CCD	charge couplet device
CID	charge injection device
RBV	return beam vidicon
EMI	Industrial Electronics Ltd., Ruislip, Middlesex, England
WE	Westinghouse Electric (Electron Tube Division), Elmira, New York
RCA	Radio Corporation of America (Electronic Components), Harrison, New Jersey
ITT	International Telephone and Telegraph (Electron Tube Division), Fort Wayne, Indiana

1.3.2 Photoelectric radiation detectors

A. Vacuum photocells, Gas-filled photocells, Photomultiplier tubes } sensitive from the near ultraviolet to the near infrared [21, 25, 30]

B. Photoconductive cells (MOS), Photodiodes, Photovoltaic cells, Thermal detectors } sensitive primarily to infrared radiation [3, 11, 23, 34]

For technical details, see information sheets regularly published by RCA, EMI, ITT, WE, and other manufacturers.

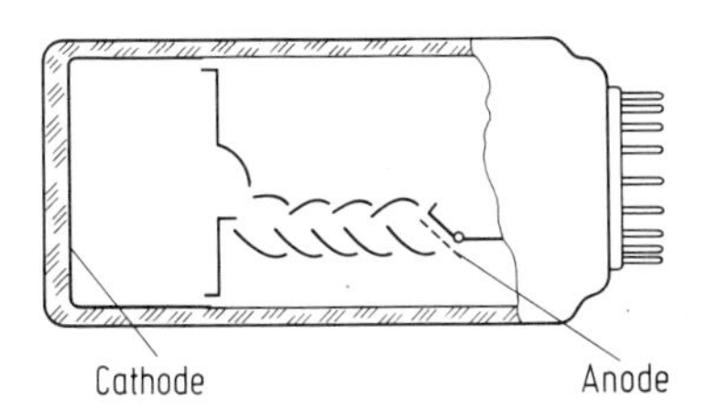

Fig. 1. Linear focused dynode configuration with fast response for photon counting in EMI-photomultipliers (e.g. D 341, D 347) [other dynode arrangements see, LB, NS, Vol. VI/1 (1965) p. 44].

Table 1. Typical examples of photomultipliers used in photoelectric photometry.

Multiplier type	S-type	Photocathode	Window	Anode pulse rise time [ns]
RCA 7102	S-1	Ag–O–Cs semi-transparent	glass, head on	≦ 2.5
RCA 1P21	S-4	Ni–Cs–Sb opaque	glass, side on	≦ 2.0
RCA 1P28	S-5	Ni–Cs–Sb opaque	UV-transmitting glass, side on	≦ 2.0
RCA 6199	S-11	Cs–Sb semi-transparent	glass, head on	≦ 2.5
EMI 6094	S-11	Cs–Sb semi-transparent	glass, head on	10
EMI 6256 SA	S-11	Cs–Sb semi-transparent	glass, head on	10
EMI 9558	S-20 (trialkali)	Na–Ka–Cs–Sb semi-transparent	glass, head on	10
RCA 7326	S-20	Na–Ka–Cs–Sb semi-transparent	glass, head on	≦ 2.5
RCA 7265	S-20	Na–Ka–Cs–Sb semi-transparent	glass, head on	≦ 3.0
EMI 9558 QB	S-20 (UV ext.)	Na–Ka–Cs–Sb semi-transparent	quartz, head on	15
EMI 9659	S-20 (red ext.)	Na–Ca–Cs–Sb semi-transparent	glass, head on	10
EMI 9827	S-11	Cs–Sb semi-transparent	glass, head on	1.8
EMI 9881	bialkali	K–Cs–Sb semi-transparent	glass, head on	1.8
EMI D341	S-20	Na–Ka–Cs–Sb semi-transparent	glass, head on	< 1.3
EMI D347	bialkali	K–Cs–Sb semi-transparent	glass, head on	< 1.3
RCA 8575	bialkali	K–Cs–Sb semi-transparent	glass, head on	≦ 2.5
RCA C32025C	Quantacon	Ga–As opaque	UV-transmitting glass, side on	≦ 1.5
RCA C31034A	Quantacon	Ga–As opaque	UV-transmitting glass, head on	≦ 2.5

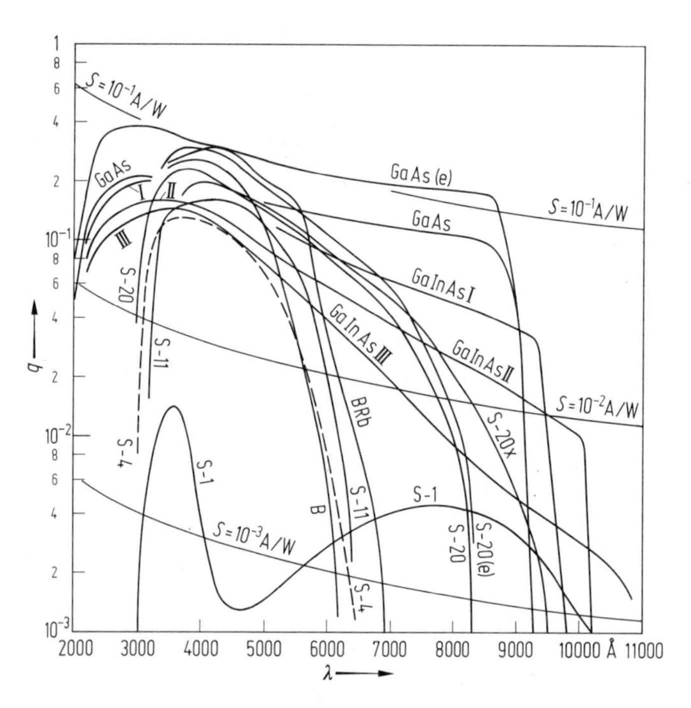

Fig. 2. Relative quantum efficiency, q, and absolute spectral response, S, of various photomultiplier cathodes.
B = bialkali, BRb = enhanced bialkali (Rb), S – 20 x = S – 20 red extended, (e) = selected for extremely high sensitivity.

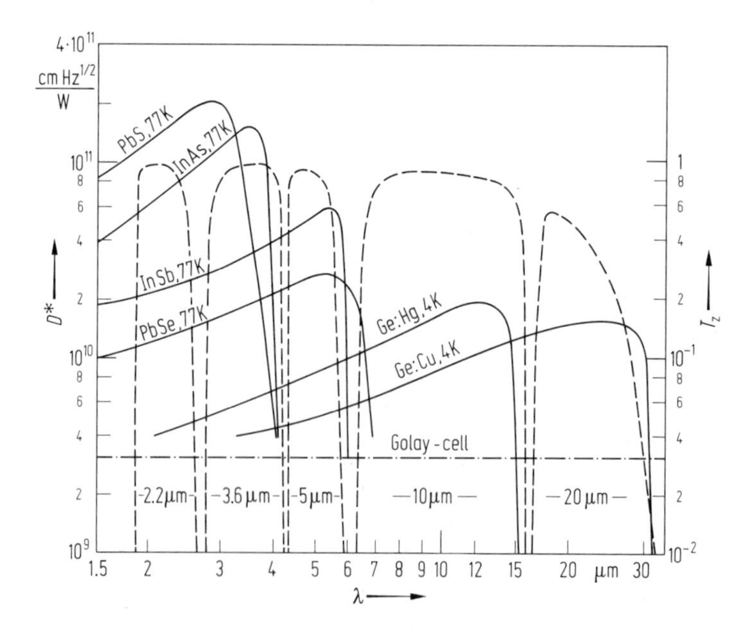

Fig. 3. Detectivity, D^*, of various infrared detectors at temperatures as indicated, measured with a bandwidth $\Delta f = 1$ Hz and relative to a receiver surface area of 1 cm^2 (solid line), and transmission T_z in zenith for 1 mm precipitable water at 500 mbar (dashed line) [23].

1.3.3 Measuring techniques

a) DC-measurements: standard procedure with a minimum of problems. Especially useful for PMT with large anode pulse rise time, e.g. EMI-types with Venetian blind configuration of dynodes [25, 30].

b) AC-measurements: procedure for special applications.

c) Photon-counting techniques: especially used for objects at low light levels. In this case superior to all other techniques if correction for non-linear response due to the limited resolving power (dead time correction) is applied. Need PMT with short anode pulse rise time, e.g. EMI D 341 or RCA Quantacon [5, 36].

Dead time correction

commonly used: $N = n/(1 - \varrho n)$ — N = true rate, n = apparent rate, ϱ = resolving time

better: $\varrho = an^b$; $N = n/(1 - an^{(1+b)})$ allows extension to brighter stars. — a and b depend on the individual photometric arrangement [27].

1.3.4 High speed photometry

The photometry of astronomical objects with a time resolution between 10 μs and 10 s is possible with computer-directed observing programs with photon counting techniques. Different procedures are possible.

a) Sampling: taking measurements during a certain time interval (standard photometry).

b) Integral: continuous sampling divided into consecutive intervals (registration of light variations).

c) Cyclic: cycling around within a limited storage region, discarding the oldest reading and replacing it by the newest one until the significant event is found (occultations).

d) Synchronous: the newest reading is always added to the oldest reading in any particular cycle; the total cycle is synchronized with the modulation of the object under study [33].

1.3.5 Detectors for two-dimensional (imaging) photometry

See [9]
Electronographic cameras [4, 31]
Image intensifiers [6, 10]
Microchannel plates [10]
Silicon diode arrays (Reticon), image dissector scanner [35, 37, 38]
Integrating television tubes [7, 8, 10, 24]
Image photon counting systems [10, 12, 29, 32, 39]
Hybrid systems [10, 12, 39]

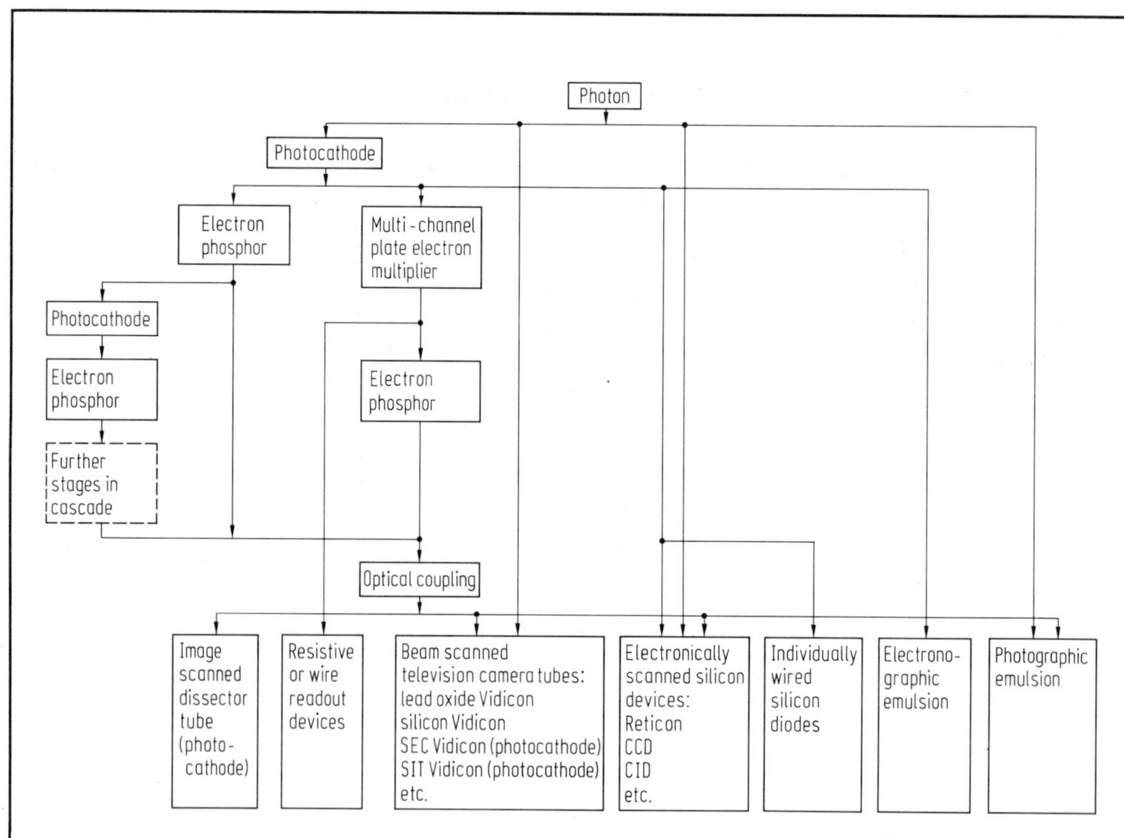

Fig. 4. Diagram showing how detector components are used in photometric imaging techniques [10].

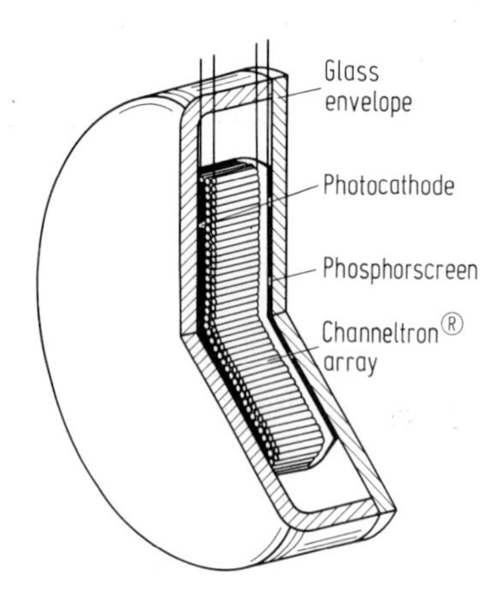

Fig. 5. A wafer intensifier incorporating a microchannel plate imaging electron multiplier [10].

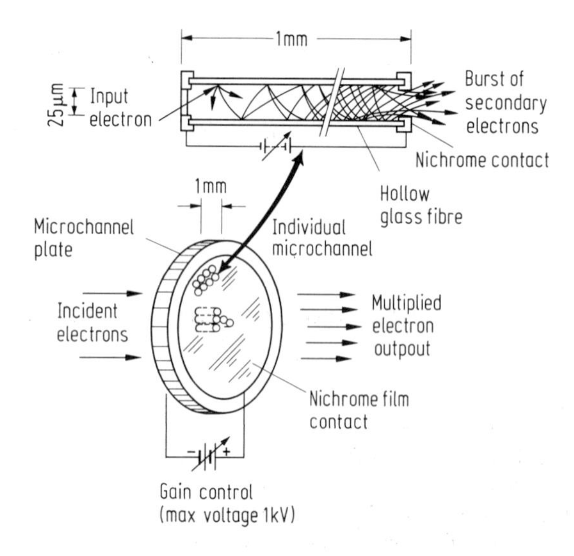

Fig. 6. The operation of a microchannel plate imaging electron multiplier [10].

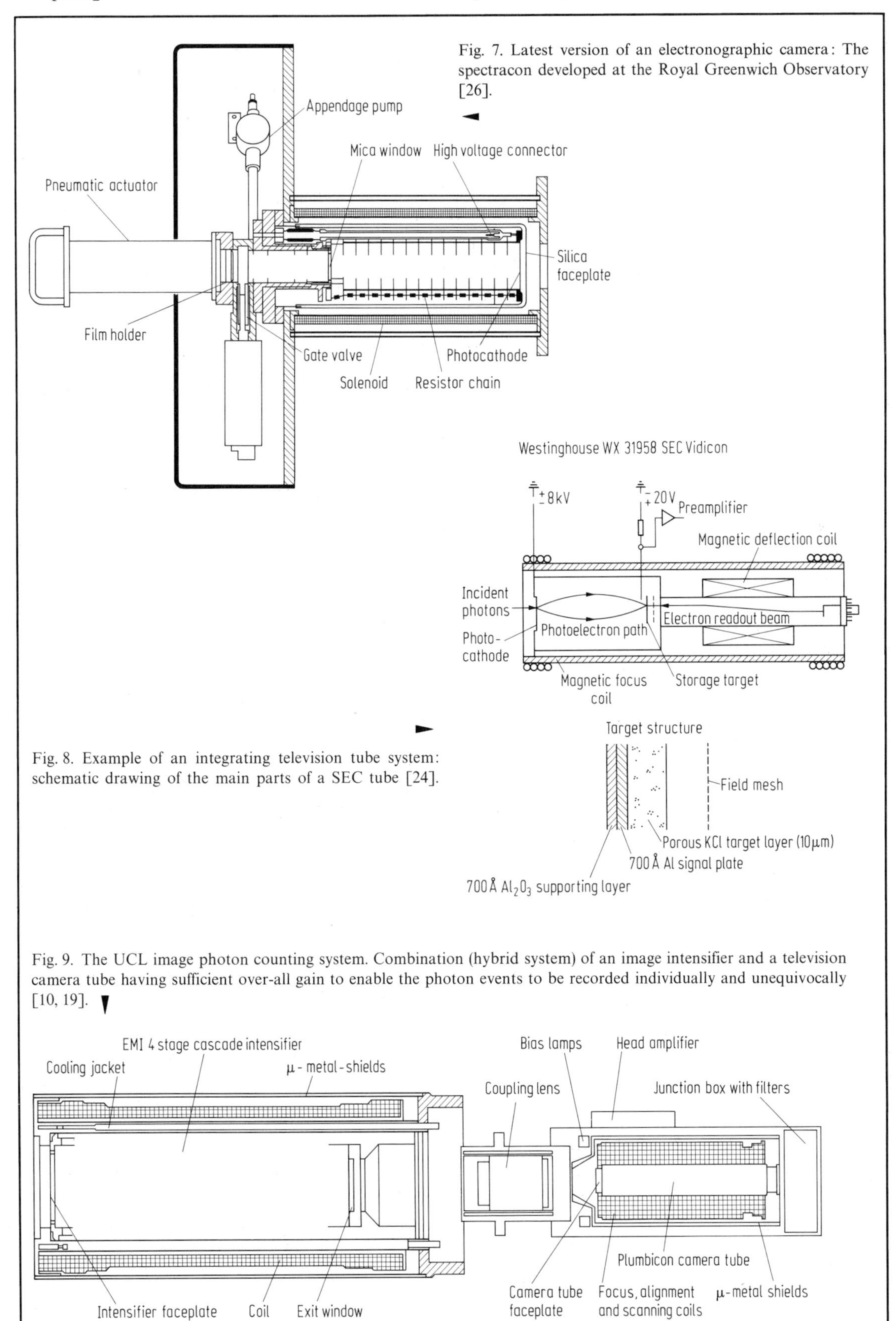

Fig. 7. Latest version of an electronographic camera: The spectracon developed at the Royal Greenwich Observatory [26].

Fig. 8. Example of an integrating television tube system: schematic drawing of the main parts of a SEC tube [24].

Fig. 9. The UCL image photon counting system. Combination (hybrid system) of an image intensifier and a television camera tube having sufficient over-all gain to enable the photon events to be recorded individually and unequivocally [10, 19].

1.3.6 The measured radiant flux

The measured radiant flux, W, from a star is given by

$$\log W(\lambda, T) = 2 \log D - 0.4\, m_v + \log \Delta\lambda - 15.51 - 5 \log \frac{\lambda}{5500} - \frac{6240}{T}\left(\frac{1}{\lambda} - 0.000182\right) \cdot 10^4 ,$$

with W in [W], D in [cm], λ in [Å] and T in [K].

This value must be multiplied by the weight function, $g(\lambda)$, to correct for the extinction in the earth's atmosphere, and for the absorption in the optical system and in the colour filters used.

The corresponding photocurrent measured at the cathode of a photomultiplier is shown in Table 2.

Table 2. Radiant flux W, the corresponding photocurrent I, and count rate n_e as function of wavelength λ for various stellar temperatures T (black body radiation) and different photocathodes (S 11: Cs–Sb; S 20: trialkali (multialkali) Na–K–Cs–Sb; Qu: RCA-Quantacon Ga–As, selected for extremely high sensitivity). For $S, q, g(\lambda)$, see 1.3.0 and text. The values have been calculated for a telescope of $D = 100$ cm with a central obturation of 20 %, a star of $m_v = 10^m.0$, and $\Delta\lambda = 500$ Å.

					T=3000 K			T=10000 K			T=25000 K		
	λ Å	S 10^{-2} A/W	q $n_e/n_{h\nu}$	$g(\lambda)$	W 10^{-14} W	I 10^{-16} A	n_e 10^3 s^{-1}	W 10^{-14} W	I 10^{-16} A	n_e 10^3 s^{-1}	W 10^{-14} W	I 10^{-16} A	n_e 10^3 s^{-1}
S 11	3500	3.73	0.132	0.15	0.82	0.46	0.29	26.69	14.92	9.31	65.26	36.48	22.77
	4500	5.52	0.152	0.20	4.91	5.42	3.38	18.92	20.87	13.03	26.76	29.45	18.38
	5500	3.19	0.072	0.25	12.47	9.96	6.22	12.39	9.90	6.18	12.37	9.88	6.17
S 20	3500	5.76	0.204	0.15	0.82	0.71	0.44	26.69	23.06	14.39	65.26	56.38	35.19
	4500	6.46	0.178	0.20	4.91	6.35	3.96	18.92	24.44	15.26	26.76	34.57	21.58
	5500	4.53	0.102	0.25	12.47	14.11	8.80	12.39	14.02	8.75	12.37	14.00	8.74
	6500	3.15	0.060	0.30	20.65	19.49	12.16	8.04	7.58	4.73	6.30	5.95	3.71
Qu	3500	9.88	0.350	0.15	0.82	1.22	0.76	26.69	39.56	24.69	65.26	96.73	60.37
	4500	10.03	0.283	0.20	4.91	9.85	6.15	18.92	37.93	23.68	26.76	53.66	33.49
	5500	10.20	0.230	0.25	12.47	31.81	19.85	12.39	31.61	19.73	12.37	31.56	19.70
	6500	10.49	0.200	0.30	20.65	64.96	40.54	8.04	25.28	15.78	6.30	19.83	12.38

1.3.7 Accuracy of measurements

The accuracy of a photoelectric measurement is limited by the statistical emission of photoelectrons from the cathode, and the resulting fluctuation of the photocurrent $\overline{i^2}$

$$\overline{i^2} = 2eI\Delta f \quad \text{(Schottky: shot noise)}$$

The photocurrent I has in general three components:

$$I = I_{St} + I_H + I_{th}$$

I_{St} photocurrent of the star
I_H photocurrent of sky background
I_{th} thermionic emission (dark current).

Each component has a statistical fluctuation

$$\overline{i^2_{St}},\ \overline{i^2_H},\ \text{and}\ \overline{i^2_{th}}.$$

A further source of noise, independent of the magnitude of the photocurrent, is due to astronomical "seeing", with $\overline{i^2_{Sz}}$, (scintillation).

The mean error of a photoelectric measurement with a multiplier phototube can be expressed by:

$$\text{m.e.}\,[\%] = a\,\frac{100\sqrt{\Delta f}}{I_{St}}\sqrt{2e(I_{St}+I_H+I_{th})+\overline{i^2_{Sz}}}$$

a = noise amplification by secondary electron emission (≈ 1.3)

I_{St}, I_H, and I_{th}, measured at the photocathode.

The thermionic emission is

$$I_{th} = A \cdot F \cdot T^2 e^{\omega/kT} \quad \text{(Richardson-Law)}$$

A constant
F [cm^2] area of the cathode
ω [eV] work function.

I_{th} may be reduced to a negligible value by cooling with dry ice or liquid nitrogen.

Some PMT do not allow cooling below a specified value for mechanical reasons; bialkali cathodes show a significant drop of sensitivity when cooled below 273 K.

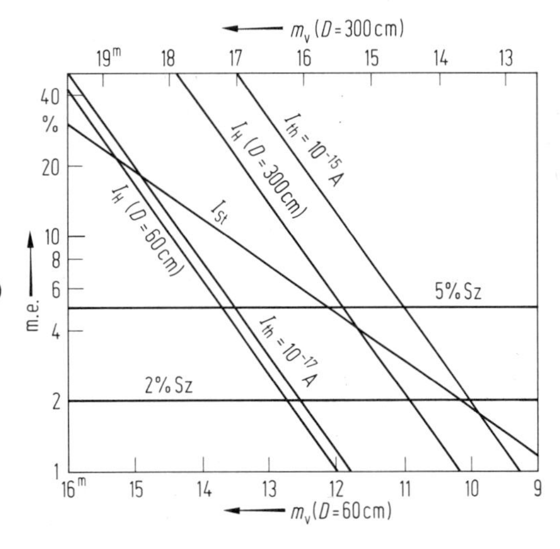

Fig. 10. Contribution of the different sources of noise to the mean error (m.e.) as a function of the visual magnitude, m_v.

Lower scale: telescope of 60 cm aperture	
upper scale: telescope of 300 cm aperture	
diameter of measuring diaphragm 20″	
photocurrent of the star, I_{St}	
sky background, I_H (1 star of $m_v = 21^m\!.5$)/□″	
photocathode	S-11 (see Fig. 2)
thermionic emission	
without cooling	$I_{th} = 10^{-15}$ A
cooled with dry ice	$I_{th} = 10^{-17}$ A
seeing (scintillation)	Sz
weight function	$g(\lambda) = 0.20$
(extinction, absorption ...) see Table 2	
adopted temperature of the star	$T = 10000$ K
effective wavelength	$\lambda_{eff} = 4500$ Å
band width	$\Delta\lambda = 500$ Å
time constant	$RC = 1$ s

To obtain the actual error, the ordinate values should be divided by the square root of integrated time (duration of the measurements in [s]).

1.3.8 References for 1.3

Review articles and books

1 Zworykin, V.K., Ramberg, E.G.: Photoelectricity and its Application, J. Wiley & Sons, New York (1949).
2 Simon, H., Suhrman, R.: Der lichtelektrische Effekt und seine Anwendungen, 2. Aufl. Berlin, Göttingen, Heidelberg: Springer (1958).
3 Smith, R.A., Jones, F.E., Cashmar, R.P.: The Detection and Measurement of Infrared Radiation, Clarendon Press, Oxford (1975); 2nd ed. (1968).
4 Whitford, A.E.: Photoelectric Techniques, Hdb. Phys. **54** (1962) 240.
5 Hiltner, W.A.: Astronomical Techniques, in: Kuiper, G.P.: Stars and Stellar Systems. Univ. of Chicago Press, Chicago, Vol. II (1962).
6 Ford, W.K.,jr.: Electronic Image Intensification, Annu. Rev. Astron. Astrophys. **6** (1968) 1.
7 Livingston, W.C.: Image Tube Systems, Annu. Rev. Astron. Astrophys. **11** (1973) 95.
8 McGee, J.D.: Image Tubes in Astronomy, Vistas Astron. **15** (1973) 61.
9 Duchesne, M., Lelievre, G. (ed.): Astronomical Application of Image Detectors with Linear Response, Proc. Int. Astron. Union Coll. No. 40., Paris-Meudon (1976).
10 Boksenberg, A.: Review of Trends in Detector Development, Proc. ESO Conference on Optical Telescopes of the Future (Pacini, F., Richter, W., Wilson, R.N., eds.), Geneva (1978) 497.
11 Soifer, B.T., Pipher, J.L.: Instrumentation for Infrared Astronomy, Annu. Rev. Astron. Astrophys. **16** (1978) 335.
12 Ford, W.K.,jr.: Digital Imaging Techniques, Annu. Rev. Astron. Astrophys. **17** (1979) 189.

Special papers

21 Baum, W.A.: Annu. Rev. Astron. Astrophys. **2** (1964) 165.
22 Behr, A.: Landolt-Börnstein, NS, Vol. VI/1 (1965) 44.
23 Borgman, J., Andriesse, C.D., Van Duinen, R.J.: Infrared Astronomy and ESO, Groningen (1972).
24 Crane, P., Nees, W.: ESO-Messenger **17** (1979) 34.
25 Engstrom, R.W.: J. Opt. Soc. Am. **37** (1947) 420.
26 Gyldenkerne, K., Florentin Nielson, F., McMullan, D.: ESO-Messenger **17** (1979) 36.
27 Fernie, J.D.: Publ. Astron. Soc. Pacific **88** (1976) 696.
28 Jones, R.C.: Proc. Inst. Rad. Eng. **47** (1959) 1495.
29 Kinman, T.D., Green, M.: Publ. Astron. Soc. Pacific **86** (1974) 334.
30 Kron, G.E.: Astrophys. J. **103** (1946) 326.
31 Lallemand, A., Duchesne, M., Walker, M.F.: Publ. Astron. Soc. Pacific **72** (1960) 268.
32 Mende, S.B., Chaffee, F.H.: Appl. Opt. **16** (1977) 2698.
33 Nather, R.E.: Vistas Astron. **15** (1973) 91.
34 Nayar, P.S., Hamilton, W.O.: Appl. Opt. **16** (1977) 2942.
35 Robinson, L.B., Wampler, E.J.: Publ. Astron. Soc. Pacific **84** (1972) 161.
36 Rodman, J.P., Smith, H.J.: Appl. Opt. **2** (1963) 181.
37 Rudolf, R., Schlosser, W., Schmidt-Kaler, Th., Tüg, H.: Astron. Astrophys. **65** (1978) L5.
38 Voigt, S.S., Tull, R.G., Kelton, Ph.: Appl. Opt. **17** (1978) 574.
39 Ulrich, M.-H.: ESO-Messenger **20** (1980) 8.

1.4 Photographic emulsions

1.4.1 Introduction

The photographic emulsion has been in use as a photon detector in astronomy for more than one hundred years; a historical review is given by Miller [10]. It has not been made obsolete by advances in electronic detectors because of its unsurpassed information storage capacity ($>10^{10}$ bits on a $30\,\mathrm{cm}\times 30\,\mathrm{cm}$ plate) together with a good spatial resolution ($\approx 10\,\mu\mathrm{m}$), a relatively low cost and, not the least, simple handling. The main drawbacks are low quantum efficiency (max. 4%) and non-linear response.

All recently constructed telescopes in the 3···4 m class have been equipped with prime focus correctors, giving well defined images over a field of diameter $\gtrsim 1°$, specifically for the use of photographic plates. A 4 m telescope ($f/3$) will reach $24^{m}\!.5$ on IIIa–J plates in 60···90 minutes.

The use of photographic materials involves exposure at the telescope and calibration, processing (development, fixing, rinsing, drying) and measurement. Recent improvements in astronomical photography are discussed in the Am. Astron. Soc. Photo-Bulletin, in the Proceedings of an ESO Workshop [21], by Miller [11], and Smith and Hoag [17]. A bibliography has been prepared by Sim [15]. Technical data are to be found in [7].

1.4.2 Definitions

The following definitions and units are used in astronomical photography:

Intensity I: photons $\cdot 1000\,\mu m^{-2} \cdot s^{-1}$, at specified wavelengths (standard: 460, 600, 800 nm)

Exposure $E = I \cdot t$: photons $\cdot 1000\,\mu m^{-2}$

Transmission T: intensity of light transmitted through an exposed emulsion relative to the intensity of incident light (note that $T = 1.0$ for "no plate").

Density $D = -\log_{10} T$: measured in "ANSI diffuse densities", i.e. 180° cone angle of light beam ($D = 0.0$ for "no plate").

Characteristic curve: plot of $\log E$ versus D, cf. Fig. 1.

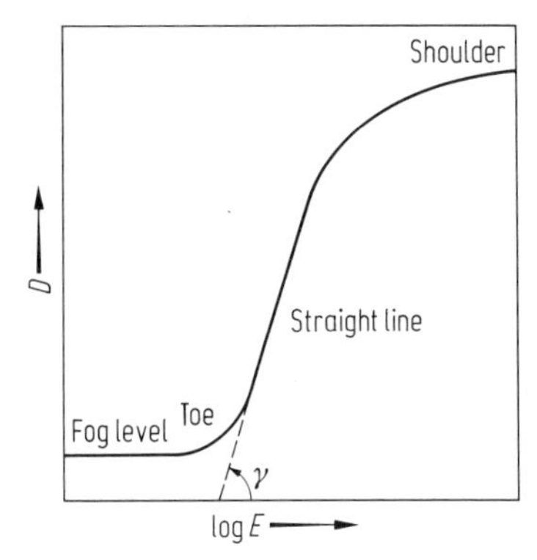

Fig. 1. The characteristic curve of a photographic emulsion.

Gradient $\gamma = \dfrac{dD}{d(\log E)}$: inclination of tangent to characteristic curve.

Contrast: γ of the straight part of the characteristic curve.

LIRF: low-intensity reciprocity failure; the density depends on both I and t, not just on E. When I is very small, the t necessary to give a certain D increases more rapidly than I^{-1}.

Resolving power: number of line pairs/mm that can be resolved by the emulsion. Usually given for the standard values of the contrast 1:1000 and 1:1.6.

Graininess: subjective impression of grain (from "very coarse" to "very fine").

Granularity σ_D: rms of density, as measured through a circular aperture of area $1000\,\mu m^2$.

$(S/N)_{in}$: signal-to-noise of incoming signal ($= Q^{0.5}$ for Q photons).

$(S/N)_{out} = \dfrac{0.4343 \cdot \gamma}{\sigma_D}$: signal-to-noise of resulting image. For a particular emulsion, σ_D may be found from D by means of Table 1 in [3].

$DQE = \dfrac{(S/N)_{out}}{(S/N)_{in}}$: detective quantum efficiency.

1.4.3 Emulsions in use in astronomy – hypersensitization

Until 1973, most astronomical emulsions were low-contrast and fast, but rather grainy (e.g. 103a–O). Since then, great advances have occurred in terms of limiting magnitude and (S/N)-ratio due to the introduction of high-contrast, slow fine-grain emulsions (e.g. IIIa–J). However, to achieve a reasonable speed (and to overcome LIRF), emulsions must be hypersensitized before use; summaries are given in [12, 13, 16]. The most common emulsions, their contrast and resolving power, and the corresponding hypersensitization methods are compiled in Table 1.

Table 1. Emulsions in use in astronomy.

()	Preferred by at least one user	me	moderately effective
VE	very effective	se	small effect (not very useful or better alternatives exist)
E	effective	A	to be avoided
		blank	no result known

	Emulsion characteristics			Hypersensitization											
	Emulsion type	Contrast class	Resolving power (1:1000) linepairs/mm	Gas treatments					Liquid treatments				Others		
				Baking (usually 55 to 75 °C)			Soaking (20 °C)		N_2 bake + H_2 soak	Bathing (usually 5 to 20 °C)			Evacuate	Cool	H_2O bathe + H_2 soak
				Air	N_2	Forming gas	N_2	H_2		H_2O	NH_4OH	$AgNO_3$			
Kodak	103a–O	medium	80	se	E	E	E	(E)	(VE)	se			E	se	
Kodak	103a–D	medium	80		E										
Kodak	103a–E	medium	80		E		se	E							
Kodak	098	high	100		E		E	VE	(VE)	E	se		E		
Kodak	IIa–O	medium	100	se	(E)	(VE)	E	(E)		se			me		
Kodak	IIa–D	medium	100		E		E	VE							
Kodak	IIa–F	medium	100		E			VE					me	E	
Kodak	IIIa–J	very high	200	se	(VE)	(VE)	E	(VE)	VE	A			me	E	
Kodak	IIIa–F	very high	200	se	E	(VE)	E	(VE)	(VE)	A	me		me		
Kodak	IV–N	very high	200	se	se	me	se	E		me	(E)	(VE)			E
Kodak	I–N	high	100			(VE)				E	E	(VE)	me	E	E
Kodak	I–Z	high	125							E	E	(VE)			
Orwo	ZU–2	medium			E										
Orwo	ZU–1	medium	80		E										
Orwo	ZP–3	medium			E										

1.4.4 Exposure and calibration

Photographical emulsions should preferably be exposed to the optimum background (sky) density, i.e. the maximum value of $(S/N)_{out}$, cf. [17]. For most emulsions, this corresponds to $D_{sky} \approx 1.0 \cdots 1.5$.

Direct exposures are normally made through glass or interference filters in order to approximate the passbands of standard photometric systems (Table 2).

Table 2. Common emulsion-filter combinations.
Colours and magnitudes measured with a particular (telescope-emulsion-filter) combination must be transformed to the standard system (see 4.2) by means of measurements of standard objects (cf. 1.3).

System	Kodak type	Schott filter	Approximate λ-interval
U	O, J	2 mm UG1	(3000)···3900
B	O	2 mm GG385	3800···5000
V	D	2 mm GG495	4900···6400
R	F	2 mm RG630	6250···7000
I	N	2 mm RG715	7100···9000
R	E	2 mm RG610	6050···6700
G	O	2 mm GG455	4500···5000
U	O	2 mm UG1	(3000)···3900

Calibration marks are projected on the emulsion outside the field (or the spectrum), e.g. with a spot sensitometer, as described by Schoening [14] and Hoag [4]. This enables the measurement of the gradient at the sky level. Careful calibration has made it possible to obtain photometric accuracies of 1% or better.

1.4.5 Processing

The most common developers for astronomical emulsions are Kodak D 19 and D 76 [6] and MWP 2 [2]. Large photographic plates are most uniformly developed in a tray-rocker [9]. In order to obtain archival quality, it is necessary to use two fixing baths and at least two rinsing baths. The plates are finally dipped in a detergent (Photoflo) and dried in a dust-free cabinet.

1.4.6 Storage

Astronomical photographs must be protected from dust and gases which influence the image silver. This is best done by placing them in Tyvek envelopes [18, 19] and in steel cabinets at a controlled temperature (15···20 °C) and humidity (40···60%).

1.4.7 Measurements

Astronomical photographs may be blinked (to detect variable objects), measured with an Iris photometer (stellar magnitude) or with a modern, automatic, scanning microphotometer (eg. PDS, COSMOS, etc.). Many measuring centers have been set up during the past few years. A summary of image processing in astronomy is to be found in the Proceedings of the Utrecht meeting [1]. For the best accuracy, it is important to have standard objects on the same plate as the objects to be measured.

Direct photographs

On these plates extended and "stellar" images may be measured. The aim is to determine the positions, the surface intensity distribution (density profiles), and integrated magnitudes. With the use of advanced algorithms, accuracies of ± 0.3 μm can be obtained for the positions of stellar images and $\pm 0^{m}.07$ or better for integrated magnitudes.

Spectra

Astronomical spectra may be obtained through a slit (one object at a time) or through an objective prism (many objects on one plate). The measurement aims at the identification of spectral lines by wavelength, determination of line profiles and equivalent widths and radial velocities.

1.4.8 Copying

It is often desirable to copy astronomical photographs. Special methods have been developed to enhance details [5, 8]. Large-scale copying of large plates for photographic atlases has been discussed by West and Dumoulin [20].

1.4.9 Outlook

Current research mainly concerns the improvement of the speed of photographical plates by means of alternative hypersensitization methods. It is unlikely that radically new emulsions will become available within the next few years, with the exception of infrared emulsions.

1.4.10 References for 1.4

1 De Jager, C., Nieuwenhuijzen, H.: Image Processing Techniques in Astronomy, Reidel, Dordrecht (1975).
2 Difley, J.A.: Astron. J. **73** (1968) 762.
3 Furenlid, I. in: [21] p. 153.
4 Hoag, A.A.: Am. Astron. Soc. Photo-Bull. **13** (1976) 14.
5 Högner, W. in: [21] p. 175.
6 Kodak: Processing Chemicals and Formulae, J–1, Rochester (1963).
7 Kodak: Kodak Plates and Films for Scientific Photography, P–315, Rochester (1973).
8 Malin, D.F.: Am. Astron. Soc. Photo-Bull. **16** (1977) 10.
9 Miller, Wm.C.: Am. Astron. Soc. Photo-Bull. **4** (1971) 762.
10 Miller, Wm.C. in: [21] p. 1.
11 Miller, Wm.C.: A Darkroom Manual, Hale Observatories, Pasadena (1978).
12 Millikan, A.G.: Am. Astron. Soc. Photo-Bull. **18** (1978) 10.
13 Millikan, A.G., Sim, M.E. in: [21] p. 294.
14 Schoening, Wm.: Am. Astron. Soc. Photo-Bull. **11** (1976) 8.
15 Sim, M.E.: Astronomical Photography: A Bibliography, IAU Working Group on Photographic Problems, Edinburgh (1977).
16 Sim, M.E. in: [21] p. 23.
17 Smith, A.G., Hoag, A.A.: Advances in Astronomical Photography at Low Light Levels, Annu. Rev. Astron. Astrophys. **17** (1979) 43.
18 Van Altena, W.F.: Am. Astron. Soc. Photo-Bull. **6** (1972) 15.
19 Van Altena, W.F.: Am. Astron. Soc. Photo-Bull. **8** (1975) 18.
20 West, R.M., Dumoulin, B.: Photographic Reproduction of Large Astronomical Plates, ESO, Geneva (1974).
21 West, R.M., Heudier, J.-L., (eds.): Modern Techniques in Astronomical Photography, ESO, Geneva (MTAP) (1978).

1.5 Spectrometers and spectrographs

1.5.1 Definitions

Spectrometers are employed to measure the brightness of one or more elements of the spectrum from one or more elements of a source observed with a telescope. Their basic components consist of a spectrometric element (e.g. prism, grating, étalon), imaging elements and a photon detector. The spectrograph is a spectrometer equipped with an image detector capable of measuring more than one spectral element simultaneously. The monochromator measures only one spectral element at one time, so that a spectrum must be scanned.

The following parameters are useful to define a spectrometer [1]:

The spectral resolution $R=\lambda/\delta\lambda$ is determined by the width of the spectral element $\delta\lambda$ ("instrumental profile"). In most spectrometers the resultant R is smaller than the maximum resolution obtainable by the spectrometric element, alone.

The luminosity [2] of a spectrometer $L=S\Omega\tau$ is determined by the area S of the spectrometric element, the acceptance cone Ω for the radiation analysed, and the brightness transmission factor τ of the optical components.

The luminosity resolution product $L\cdot R$ can be considered as a figure of merit when comparing different spectrometers. For many particular spectrometers this product is a fixed constant; therefore the spectral resolution can be increased only at the expense of luminosity and vice versa.

The spectral simultaneity gain and the spatial simultaneity gain [1] denote the number of spectral elements or the number of different spatial elements of the source, respectively, analysed simultaneously by the spectrometer either by a multiplicity of detector elements or by multiplexing.

The detective quantum efficiency of the detector determines the fraction of incident photons resulting in a measurable event and, therefore, directly contributes to the performance of the spectrometer.

In the design and use of spectrometers, these various parameters have been optimized with particular weighting, depending on the specific application.

1.5.2 Grating spectrometers

In the last decades, the prism has been widely replaced by the blazed diffraction grating which can achieve a $L \cdot R$ product 50 times higher than the prism. The grooves of these gratings are cut with the same inclination angle, resulting in a "blaze angle" (normal to the grooves) at which the grating efficiency may reach 80% for the corresponding wavelengths. Grating spectrometers [3] are among the most frequently used auxiliary instruments in optical astronomy. A large variety of gratings (accurate replicas from master gratings with sizes up to 40 cm × 60 cm) are now offered commercially. Ruled gratings can be produced mechanically or holographically [9].

The spectral resolution of a grating spectrometer, fed by a telescope of aperture D, is

$$R = \frac{2W \sin \Theta}{\alpha D}.$$

(W = ruled width of grating; Θ = diffraction angle; α = angular width of slit, or equivalent, projected onto the sky).

The luminosity resolution product is

$$\mathrm{L} \cdot R = 2DW \sin \Theta \tau \beta .$$

(τ = effective transmission factor of all optical components; β = acceptance angle along the slit on the sky). The maximum τ is obtained if Θ equals the blaze angle. In grating spectrometers the $L \cdot R$ product can be increased by larger grating dimensions and/or a larger blaze angle. If the source overfills the slit width of the spectrometer, only linear gain is possible with a larger telescope aperture.

Whenever feasible, grating spectrometers are attached to the Cassegrain focus of a telescope (Cassegrain spectrographs) to avoid light loss from unwanted reflections. They are designed for highest mechanical and thermal stability in order to minimize spectrum shifts during exposure time. Large grating spectrometers are installed in special rooms, temperature controlled, at the coudé focus or at an equivalent place (coudé spectrographs, solar spectrographs). Classical spectrographs employ low-order gratings, and only one order is measured at one time. Recently, high-order échelle spectrometers [4], employing échelle gratings of large blaze angle, have come into use. Many spectral orders, separated by means of a cross dispersing element, can be measured simultaneously using a two-dimensional detector. Performance parameters of typical grating spectrographs are summarized in Table 1.

Table 1. Performance parameters of typical grating spectrographs.

Spectrograph		Spectral resolution	Ruled width of grating	Blaze angle	Luminosity resolution product	Simultaneity gain	
						spectral	spatial
Classical	Cassegrain	$0.5 \cdots 5 \cdot 10^3$	$\lessapprox 15$ cm	$\lessapprox 25°$	moderate	high	high
(low-order)	Coudé	$5 \cdot 10^3 \cdots 10^5$	$\lessapprox 40$ cm	$\lessapprox 25°$	high	very high	high
Echelle	Cassegrain	$10^4 \cdots 10^5$	$\lessapprox 20$ cm	$\approx 63°$	high	very high	low
(high order)	Coudé	$\approx 10^5$	$\lessapprox 60$ cm	$\approx 63°$	very high	very high	low

An increasing variety or combinations of different detector types (e.g. photographic emulsion, electronographic devices, image intensifiers, television type cameras, diode arrays) are now in use or under development [5]. Equipped with a two-dimensional detector, most grating spectrographs exhibit a very large spectral as well as a large spatial simultaneity gain (along the slit).

1.5.3 Fabry-Perot spectrometers

(see also 1.8.5.4)

These instruments [1] exploit the very high $L \cdot R$ product obtained by the étalon, an interferometric cavity consisting of a pair of plane parallel reflecting layers. All-dielectric multilayer reflection coatings considerably improved the efficiency of the étalon, which now can reach 100 times the $L \cdot R$ product of a low-order grating. High spectral resolution ($\approx 5 \cdot 10^5$) is achieved by high-order interference, which leads to small inter-order separation, so that narrow filters are needed to isolate several useful orders.

Fabry-Perot spectrographs employ image detectors and are most effectively used in high resolution spectroscopy of extensive emission line sources for only a small number of spectral but a large number of spatial elements.

Fabry-Perot monochromators allow scanning through a certain spectral interval, e.g. by changing the pressure of the gas in a gas-spaced étalon. Whereas single-étalon scanners with a premonochromator can examine only reasonably narrow emission line features, the multi-étalon scanners (e.g. PEPSIOS) are designed to scan small regions of a continuum light source [6].

1.5.4 Fourier transform spectrometers

The two-beam interferometric Fourier transform spectrometer [1, 7] has found most useful applications in high resolution spectroscopy, when the receiver noise dominates other sources (e.g. in the infrared domain, see 1.8.5.3). The spectrum is derived from the interferogram obtained by scanning the path difference in a two-beam interference. The $L \cdot R$ product achieved with these instruments is comparable with that of Fabry-Perot spectrometers. However, Fourier transform spectroscopy is not restricted to line sources, since its multiplex advantage [8] results in a high spectral simultaneity gain comparable with grating spectrographs.

1.5.5 References for 1.5

1 Meaburn, J.: Detection and Spectrometry of Faint Light, Reidel Publ. Comp., Dordrecht (1976).
2 Jacquinot, P.: J. Opt. Soc. Am. **44** (1954) 761.
3 Bowen, I.S. in: Astronomical Techniques, Stars and Stellar Systems Vol. II., (Hiltner, W.A., ed.), Univ. of Chicago Press (1962) p. 34.
4 Chaffee, F.H., jr., Schroeder, D.J.: Annu. Rev. Astron. Astrophys. **14** (1976) 23.
5 Ford, W. Kent jr.: Annu. Rev. Astron. Astrophys. **17** (1979) 189.
6 Mack, J.E., McNutt, D.P., Roessler, F.L., Chabbal, R.: Appl. Opt. **2** (1963) 873.
7 Connes, P.: Annu. Rev. Astron. Astrophys. **8** (1970) 209.
8 Fellgett, P.B.: J. Phys. Radium **19** (1958) 187.
9 Schmahl, G., Rudolph, D.: Holographic Diffraction Gratings, in: Progress in Optics **XIV** (Wolf, E., ed.) (1975) 195.

1.6 Optical high resolution methods

The angular resolution of astronomical observations is limited by the turbulent atmosphere and optical aberrations of the telescopes. However, the resolution can be improved by interferometric methods, such as Michelson's stellar interferometry, intensity interferometry, and speckle interferometry and its modifications. Usually, these methods yield the modulus square of the Fourier transform of the object intensity distribution, also called "visibility" or modulus square of the complex degree of coherence.

These high resolution methods can be used to measure the angular diameter of stars, the separation of close binaries and multiple stars, star clusters, limb darkening of stellar surfaces, infrared objects, and galactic nuclei.

1.6.1 Michelson's stellar interferometry and related techniques

In Michelson interferometry one does not use the whole telescope aperture but only two small 10 cm-apertures. This size is equal to the size of typical turbulence cells. When the two beams interfere in the focal plane of the telescope, interference fringes, called a Michelson interferogram, appear. One can use such Michelson interferograms for high resolution measurements. This is due to the fact that the fringe width is usually much smaller than the seeing disk, as drawn in the example of Fig. 1. For example, the Michelson interferogram of a binary star consists of two identical fringe systems, shifted by an angle which is equal to the separation of the binary star. In the example of Fig. 1, the Michelson interferogram would have nearly no contrast if the separation of a binary star were be 0.5·0."015. The fringe contrast of a Michelson interferogram changes with increasing base line if the angular diameter of the object is larger than $1.22\,\lambda/D$ (D = base line, λ = wavelength). A detailed description of the method and experimental results is given in [1···4]. Modern investigations and modifications are contained in [5, 6].

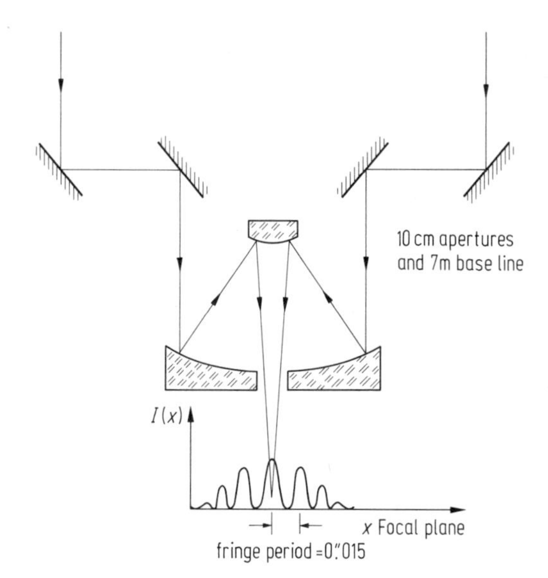

Fig. 1. Classical Michelson stellar interferometer. Example with 7 m base line and, therefore, 0."015 fringe period. $I(x)$ is the intensity distribution of the Michelson interferogram.

1.6.2 Intensity interferometry of Hanbury Brown and Twiss

In intensity interferometry, the light from the object is focused by two large mirrors onto two photomultipliers. The fluctuations of the output currents are correlated electronically. If one performs the measurement with different base lines, the modulus square of the Fourier transform of the object intensity distribution is obtained. A resolution of $6 \cdot 10^{-4}$ arc seconds was achieved with base lines up to 188 m. Stars hotter than type F8 and brighter than about $m_B = +2.5$ were measured. A detailed description of the method and experimental results is given in [7].

1.6.3 Labeyrie's speckle interferometry and related techniques

Astronomical short-exposure photographs consist of many small, randomly distributed interference maxima if one uses an exposure time of less than about 0.1 s. These interference maxima are called speckles. The size of individual speckles is equal to the size of the airy disk, i.e. $1.22\,\lambda/D$ or 0."02 in the case of a 5 m telescope and $\lambda = 400$ nm. Therefore, short-exposure photographs carry diffraction-limited information, although in encoded form.

Labeyrie's speckle interferometry [8] is a technique for extracting high-resolution object information from a sequence of N short-exposure photographs, called speckle interferograms. The intensity distribution $I_n(x, y)$ of a speckle interferogram can be described by the following space-invariant, incoherent imaging equation:

$$I_n(x, y) = O(x, y) * F_n(x, y), \quad n = 1, 2, \cdots, N,$$

where I_n describes the intensity distribution of the n-th recorded speckle interferogram, $O(x, y)$ denotes the object intensity distribution, $*$ denotes convolution and F_n is the point spread function of atmosphere and telescope during the exposure of the n-th photograph. In speckle interferometry, one determines the average power spectrum of $N=10^2 \cdots 10^6$ speckle interferograms. The whole procedure yields the modulus square of the Fourier transform of the object intensity distribution or after another Fourier transformation the autocorrelation of the object. This autocorrelation is obtained with diffraction-limited resolution. The image processing procedure is described in more detail in [8, 9, 24].

An advantage of speckle interferometry is that it can also be applied to faint objects [10···12, 25···29]. Magnitude 13^m was easily achieved [28, 29]. The theoretical limit is about 20^m.

Speckle interferometry yields a resolution of $1.22\ \lambda/D$, i.e. 0.″02 in the case of a 5 m telescope or even much more if one uses large multiple mirror telescopes or two separate telescopes with coherent light combination [14, 15, 30].

Speckle interferometry yields the high resolution object autocorrelation instead of actual images. Therefore, various authors proposed modifications for the reconstruction of actual images [16···22, 31]. The method described in [18] was applied to α Ori. The method described in [16] and [17] is called speckle holography. Speckle holography was used to reconstruct diffraction-limited images of binary stars and triple stars from speckle interferograms [23, 32].

1.6.4 References for 1.6

1 Michelson, A.A.: Astrophys. J. **51** (1920) 257.
2 Anderson, J.A.: Astrophys. J. **51** (1920) 263.
3 Michelson, A.A., Pease, F.G.: Astrophys. J. **53** (1921) 249.
4 Pease, F.G.: Ergebnisse der exakten Naturwissenschaften **10** (1931) 84.
5 Proc. Int. Astron. Union Coll. No. 50: High Angular Resolution Stellar Interferometry (1978).
6 Proc. ESO Conference: Optical Telescopes of the Future (Pacini, F., Richter, W., Wilson, R.N., eds.), Geneva (1978).
7 Hanbury Brown, R.: The Intensity Interferometer, its Application to Astronomy, Taylor & Francis Ltd., London (1974).
8 Labeyrie, A.: Astron. Astrophys. **6** (1970) 85.
9 Weigelt, G.: Optik **43** (1975) 111.
10 Ebersberger, J., Weigelt, G.: The Messenger (ESO) **18** (1979) 24.
11 Gezari, D.Y., Labeyrie, A., Stachnik, R.V.: Astrophys. J. **173** (1972) L 1.
12 McAlister, H.A.: Astrophys. J. **215** (1977) 159.
14 Labeyrie, A.: Astrophys. J. **196** (1975) L 71.
15 Labeyrie, A.: Annu. Rev. Astron. Astrophys. **16** (1978) 77.
16 Liu, C.Y.C., Lohmann, A.W.: Opt. Commun. **8** (1973) 372.
17 Bates, R.H.T., Gough, P.T., Napier, P.J.: Astron. Astrophys. **22** (1973) 319.
18 Lynds, C.R., Worden, S.P., Harvey, J.W.: Astrophys. J. **207** (1976) 174.
19 Knox, K.T., Thomson, B.J.: Astrophys. J. **193** (1974) L 45.
20 Ehn, D.C., Nisenson, P.: J. Opt. Soc. Am. **65** (1975) 1196.
21 Weigelt, G.: Opt. Commun. **21** (1977) 55.
22 von der Heide, K.: Astron. Astrophys. **70** (1978) 777.
23 Weigelt, G.: Appl. Opt. **17** (1978) 2660.
24 Dainty, J.C. in: Laser Speckle and Related Phenomena (Dainty, J.C., ed.), Springer Series Topics in Applied Physics, Vol. **9**, Springer, Berlin (1975) Ch. 7.1.2.
25 Blazit, A., Bonneau, D., Josse, M., Koechlin, L., Labeyrie, A.: Astrophys. J. **214** (1977) L 79.
26 McAlister, H.A.: Astrophys. J. **225** (1978) 932.
27 Weigelt, G.P.: Astron. Astrophys. **68** (1978) L5.
28 Arnold, S.J., Boksenberg, A., Sargent, W.L.W.: Astrophys. J. **234** (1979) L159.
29 Weigelt, G.P.: Soc. of Photo-Opt. Instr. Eng., proc. conf. "Speckle and related Phenomena", **243** (1980) 103.
30 Blazit, A., Bonneau, D., Josse, M., Koechlin, L., Labeyrie, A., Oneto, J.L.: Astrophys. J. **217** (1977) L55.
31 Fienup, J.R.: Optics Letters **3** (1978) 27.
32 Weigelt, G.P.: Optica Acta **26** (1979) 1351.

1.7 X-ray and γ-ray instruments

1.7.1 X-ray instruments

X-ray instruments are designed to measure the properties of single photons in the energy range 0.1···500 keV: energy, direction of arrival, time of arrival and polarization.

1.7.1.1 Non-focusing instruments

1.7.1.1.1 Non-focusing/non-imaging/non-dispersive instruments

The elementary interaction utilized by detectors to be discussed in this section is the photo-absorption process.

The main design features for these instruments are: large area, high efficiency, low background and high energy- as well as high time-resolution. The minimum source flux S_{min} in [photons cm^{-2} s^{-1} keV^{-1}] that can be detected with k standard deviations in an on/off observing mode (for $S \ll B$ and no systematic errors) is given by $S_{min} = \frac{k}{\varepsilon}\left(\frac{2B}{AT\Delta E}\right)^{1/2}$, where B is the background count rate in the same units as S, A is the detector area in [cm^2], T in [s] is the observation time spent on the source as well as on the background, ΔE is the energy range in [keV] and ε is the detector efficiency [57].

When observing pointlike X-ray sources there are two components contributing to the background: first, a diffuse X-ray background radiation from the sky and, second, the so-called detector background which is mainly due to particles or secondary photons produced by photon or particle interactions in the detector material itself. To minimize the detector background, various background reduction techniques like passive shielding, active anti-coincidence or rise time discrimination are applied [57, 6]. The diffuse sky background is reduced by limiting the field of view with mechanical collimators [25].

The simplest collimator is a set of parallel metallic plates. When a point source is scanned in the direction perpendicular to the direction defined by the plates the count rate in the photon detector follows a triangular response curve. A typical width for existing instruments is 1° (FWHM = full-width at half-maximum). Other collimators are made of tubes with rectangular, circular or hexagonal cross-section (e.g. honeycomb). For higher photon energies, also active borehole collimators have been used [36, 17].

A very special method is to observe sources during an occultation by the moon. Source positions and one-dimensional intensity profiles in the case of extended sources have been measured in this way [3, 65].

Four types of detectors are available to measure X-ray photons: gas proportional counters (PC), gas scintillation proportional counters (SPC), crystal scintillation counters (SC) and solid state detectors (SSD). These detectors are all comparable in time resolution; however, they differ however in efficiency, energy resolution, background properties and ease of constructing large areas. Their characteristics are compared in Table 1.

A common feature of all detectors is that the energy resolution scales with energy E as $E^{-1/2}$ due to the statistics of the interaction process.

The proportional counter has been the main instrument in X-ray astronomy. The principle of operation is that of a Geiger-Müller tube operated in the proportional mode. Modern detectors are built as multiwire proportional counters [57, 59]. They have the advantage of being easily constructed in large areas, but the efficiency decreases with increasing energy. The most commonly used filling gases are argon and xenon.

In gas scintillation proportional counters [58, 1] the primary electrons are accelerated only such that they can excite gas atoms. The de-excitation light is then detected by a photomultiplier, giving a signal proportional to the energy of the incident X-ray photon. The advantage over conventional proportional counters is a gain in energy resolution by a factor of 2. A disadvantage is the higher complexity of constructing large areas.

In scintillation counters crystal materials such as Na I (Tl) or Cs I (Na) are used as wavelength shifters from the X-ray range into the optical range [57]. They are mainly employed for hard (>10 keV) X-rays, providing good stopping power but only medium energy resolution. Large area detectors with low internal background can be constructed using the "phoswich"-technique [45, 47].

Solid state detectors [40, 51, 46] combine good stopping power with by far the best energy resolution. They can therefore be used in the whole energy range from 0.1···500 keV. Disadvantages are the relatively small area and technical difficulties due to the necessary cooling.

Table 1. Comparison of X-ray detectors.

Detector [area in cm^2]	Detector material	Energy range keV	Energy resolution ΔE(FWHM)/E	Typical efficiences
Proportional counter (PC) [several 1000]	C_3H_8, Ne, A, Kr, Xe + few % N_2, CO_2 or CH_4	0.1···60	0.18 at 6 keV 0.08 at 60 keV	depth 3.6 cm, pressure 1 atm: A: 0.80 at 6 keV; 0.31 at 10 keV; 0.28 at 20 keV Xe: 0.97 at 10 keV; 0.40 at 20 keV; 0.14 at 60 keV
Scintillation proportional counter (SPC) [few 100]	Xe	1···60	0.09 at 6 keV 0.04 at 60 keV	depth 3.6 cm, pressure 1 atm: see PC depth 3.6 cm, pressure 3 atm: 0.78 at 20 keV; 0.36 at 60 keV
Scintillation counter (SC) [few 1000]	Na I(TI) Cs I(TI) Cs I(Na)	10···500 (γ:···10 MeV,	0.20 at 60 keV 0.09 at 500 keV 0.03 at 5 MeV)	Na I(TI), depth 3 mm: 1.0 at 20 keV; 0.80 at 100 keV; 0.09 at 500 keV
Solid state detector (SSD) [several 10]	Ge(Li) Si(Li)	0.1···500 (γ:···5 MeV,	0.02 at 10 keV 0.006 at 500 keV 0.001 at 5 MeV)	Ge(Li), depth 7 mm: 1.0 at 10 keV; 0.80 at 100 keV; 0.03 at 500 keV

1.7.1.1.2 Non-focusing/imaging instruments

Conventionally, collimators restricting the field of view provide only a crude information on source positions in scanning observations (down to $\approx 0.°1$ for strong sources). Greatly improved precision on position and angular extent of sources can be reached with modulation collimators [8, 61]. The radiation passes through two or more carefully aligned grids of absorbing wires. A scanning movement – either by rotation [62] or linear translation [31] – modulates the intensity of the radiation received by the detector in a way which is characteristic for the position and the extent of the source. Resolution of 5″···10″ has been achieved. By applying sets of crossed grids even imaging of extended sources can be performed. However, this type of optics produces multiple images on the detector surface, and if several sources are in the field of view the problem of source confusion arises [48].

A pin hole camera can be built by a position-sensitive detector and a single-hole-mask. The sensitive area of such an instrument is quite small but the field of view is large (tens of degrees), so that it can be usefully employed as a sky monitor [35].

The pin hole camera principle can be extended to a multihole system, the so-called random mask or Dicke transform technique [18]. A quasi-random hole pattern and a position-sensitive detector are combined to a wide field camera. The sky image is reconstructed by Fourier deconvolution techniques.

1.7.1.2 Spectrometers and polarimeters

The detectors discussed in 1.7.1.1.1 provide a rather limited spectral resolution. High spectral resolution may be achieved by Bragg crystal spectrometers or diffraction grating spectrometers.

Bragg crystal spectrometers

A monocrystal which has a lattice spacing d will reflect X-rays of wavelength λ with high efficiency if the Bragg condition

$$n\lambda = 2d \sin\theta \quad (n = 1, 2, 3, ...)$$

is fulfilled (θ is the angle of incidence). To record a spectrum the crystal must be rotated with respect to the incident parallel beam. The spectral resolution is given by

$$n\Delta\lambda = 2d \cos\theta \, \Delta\theta \quad \text{or} \quad \frac{\lambda}{\Delta\lambda} = \frac{\tan\theta}{\Delta\theta}.$$

With good crystals, $\lambda/\Delta\lambda$ up to 10^4 may be achieved, however the integrated reflectivity is rather low (10^{-4}). Mosaic crystals provide a higher integrated reflectivity (10^{-2}) at the expense of spectral resolution ($\approx 10^2$).

A discussion of high resolution Bragg spectrometers is given in [53, 44].

Grating spectrometers

Free-standing curved gratings do not play a role because the fluxes in cosmic X-ray astronomy are too low. For transmission grating spectrometers in X-ray telescopes, see 1.7.1.3.

Polarimeters

To measure polarization two methods can be applied: at low X-ray energies (up to a few keV), Bragg reflection on crystalline surfaces is used [7]. The efficiency of reflection depends on the direction of the polarization vector. At higher energies, the dependence of the Thompson scattering angle on the direction of the polarization is used [54]. A general problem is the relatively low sensitivity of the present days instruments [55].

1.7.1.3 Focusing X-ray telescopes

These instruments consist of an X-ray mirror system and focal plane instruments like image detectors, spectrometers and polarimeters. Instruments of this type have been flown on rockets, on Skylab for solar studies [67, 66], and on the Einstein satellite [28].

1.7.1.3.1 X-ray mirror systems

X-ray mirror systems are based on the specular reflection of X-rays by polished surfaces under glancing angle conditions. This process can be considered as total external reflection of the incident photons by the atomic electron plasma of the mirror material. For photon energies far from absorption edges, the refractive index is given by

$$n^2 = 1 - \pi^{-1} r_0 \lambda^2 N_e \lessapprox 1 \ ,$$

where N_e is the electron density, λ is the wavelength and r_0 is the classical electron radius ($r_0 = 2.818 \cdot 10^{-13}$ cm). According to Snell's law, total external reflection occurs up to a critical glancing angle

$$\cos \alpha_c = n \ .$$

Typical glancing angles are 1⋯2° for wavelength 5⋯10 Å. Theoretical reflectivities as function of wavelength and glancing angle for different materials have been calculated by [26].

The main types of X-ray mirror system are:

paraboloidal mirrors,
Kirkpatrick-Baez mirrors,
Wolter type I systems.

Paraboloidal mirrors are not truely imaging; they may be used as X-ray collectors [24]. The flux from a cone with half-opening angle θ is focused into a circle with radius θf in the focal plane, when f is the focal distance. Several confocal paraboloids may be nested in order to give a larger collecting power.

The Kirkpatrick-Baez telescope [41] consists of a set of confocal one-dimensional parabolic mirrors followed by an orthogonal set of the same type of mirrors.

With such systems, imaging with an angular resolution of a few arc minutes has been achieved [29].

The standard instrument of modern X-ray astronomy is based on the Wolter-type I optics. It consists of a paraboloidal mirror followed by a confocal hyperboloidal mirror [69]. In contrast to Kirkpatrick-Baez telescopes, the telescope blurring disappears for on-axis radiation. A high resolution imaging is possible over a restricted field of view (0.5⋯2°).

These mirror systems can be nested in order to increase the collecting area. The on-axis resolution depends on the figuring errors of the reflecting surface and on the microroughness which gives rise to X-ray scattering. In practice, an on-axis angular resolution of 2″ has been achieved [28].

Wolter has shown that also other combinations of confocal conic sections can be used to build imaging optics [69].

1.7.1.3.2 X-ray imaging detectors

Photographic film may be used only for solar observations, where X-ray fluxes are large enough (Skylab, Apollo telescope mount (ATM) telescopes) [67, 66]. Imaging detectors used in cosmic X-ray telescopes are based on single photon counting.

The imaging proportional counter (IPC) contains crossed wire grids in order to locate impinging X-ray photons [11]. The electronic readout of X-ray positions may be performed by several methods, e.g. risetime or current division techniques. A linear resolution of the order 0.2···1 mm can be achieved. Pulse height measurements provide crude spectral information with a resolution of ≈60% FWHM at 1 keV.

The spectral response is determined by the transparency of the window material and the absorption probability in the filling gas. The efficiency may be close to one near 1 keV. Gas leakage through the thin (1 μm) front window must be compensated for by a gas flow system. Background may be reduced by anticoincidence and risetime discrimination methods. The imaging proportional counters of the Einstein- and EXOSAT-satellites are described in [28] and [60], respectively.

Channel plate arrays are characterized by high spatial resolution (12···50 μm), but do not give any spectral information. Their detection efficiency is ≈0.1 at 1 keV and increases towards lower photon energies. Readout systems are based on crossed wire grids [42] or resistive disks [56].

Future X-ray image detectors. At present this field is in rapid development: imaging gas scintillation proportional counter [15], CCD cameras [63], NEAD cameras [2]. (CCD = charged coupled device; NEAD = negative electron affinity X-ray detector.)

1.7.1.3.3 Focal plane spectrometers

Imaging proportional counters (see 1.7.1.3.2) combine medium spatial resolution with coarse spectral resolution (≈60% at 1 keV scaling with $E^{-1/2}$, energy range 0.1···10 keV).

Solid state detectors with liquid nitrogen or liquid helium cooling provide improved spectral resolution (≈10% at 1 keV, scaling with $E^{-1/2}$, energy range 0.8···10 keV). The spatial resolution depends on the detector size.

Bragg crystal spectrometers are placed behind a field stop located in the focal plane. By using curved Bragg crystals in a Rowland mounting a spectral resolution of $E/\Delta E \approx 10^4$ may be achieved.

Transmission gratings placed behind (or before) the mirror system of an imaging X-ray telescope can be used as objective grating spectrometers [30, 26]. The spectrum may be recorded simultaneously with an image detector in the focal plane. The spectral resolution is given approximately by

$$\Delta\lambda = g\,\delta\theta,$$

where g is the grating constant and $\delta\theta$ is the effective angular resolution of the X-ray telescope. With $g = 5\cdot 10^{-5}$ cm [9] and $\delta\theta = 2''$, a $\Delta\lambda \approx 0.05$ Å can be achieved. At long wavelengths ($\lambda \gtrsim 50\cdots 100$ Å), coma errors become dominant which can be corrected by using curved gratings [4].

1.7.2 γ-ray instruments

The general remarks made at the beginning of the X-ray instruments section apply also here. A description of γ-ray instruments as well as their early developments and modern realization, can be found in the following references: [10, 20, 21, 36, 37, 38, 50, 52, 68] and references therein.

In the low energy γ-ray range (up to a few MeV), non-organic scintillators (like Na I and Cs I) and solid state detectors are used (cf. Table 1). The detector depth is increased as compared to the X-ray detectors in order to provide sufficient stopping power. The high background caused by large volume detectors and the increasing problems of effectively shielding the central detectors from background radiation makes this energy range a very difficult one for astronomical observations. The principal photon interaction is the Compton scattering which a γ-ray photon must undergo several times before its total energy can be confined within the detector volume.

With increasing energy, the Compton scattered photon will escape from the first interaction detector and the Double Compton telescope may be applied [20, 52].

Also Compton scattering polarimeters have been proposed [52].

At still higher energies (>tens of MeV), the principal interaction process is pair production in high Z materials. The paths of the electron and the positron are followed by stacks of spark chambers or drift chambers and the energy may be measured in total absorption scintillation counters or by analysing the scattering characteristics of the electron and positron [16, 5, 20].

Angular and energy resolution are intercorrelated and a function of energy and angle of incidence. Typical values for the COS-B instrument for photon energies of 150 MeV and angles of incidence between 0° and 10° are $\Delta E/E \approx 50\%$ and $\Delta\psi \approx 8°$ (FWHM) [20]. Another approach has been used in balloon experiments: the Cerenkov radiation from electron-positron pairs in low pressure gas cylinders is detected. Spectral information can be gained by varying the gas pressure during the observation [37, 49].

For very high energy γ-rays (>100 GeV), the interaction of primary photons with the earth atmosphere and the resulting development of a cascade shower have been utilized. Two different ways have been followed: a search for muon-poor air showers, since primary photons should give rise to pure electromagnetic showers without a nuclear component [34] and the detection of Cerenkov light from air showers [20, 64].

1.7.3 X- and γ-ray satellites

In Table 2 the major X- and γ-ray satellites that have been flown to date are compiled. The Table is limited to satellites only, despite the fact that major discoveries in X- and γ-ray astronomy have been made, and continue to be made, by balloon and rocket experiments.

Table 2: next page.

1.7.4 References for 1.7

1 Andresen, R.D., Leimann, E.A., Peacock, A.: Nucl. Instrum. Methods **140** (1977) 371.
2 Bardas, D., Kellog, E., Murray, S., Enck, R., jr.: Rev. Sci. Instrum. **49** (1978) 1273.
3 Bowyer, C.S., Byram, E.T., Chupp, T.A., Friedmann, H.: Science **146** (1964) 912.
4 Beuermann, K.P., Lenzen, R., Bräuninger, H.: Appl. Opt. **16** (1977) 5.
5 Bignami, G.F., Boella, G., Burger, J.J., Keirle, P., Mayer-Hasselwander, H.A., Paul, J.A., Pfeffermann, E., Scarsi, L., Swanenburg, B.N., Taylor, B.G., Voges, W., Wills, R.D.: Space Sci. Instrum. **1** (1975) 245.
6 Bleeker, J.A.M., Overtoom, J.M.: Nucl. Instrum. Methods **167** (1979) 505.
7 Bragg, W.H.: Philos. Mag. **27** (1914) 881.
8 Bradt, H., Garmire, G., Oda, M., Spada, G., Sreekantan, B.V., Gorenstein, P., Gursky, H.: Space Sci. Rev. **8** (1968) 471.
9 Bräuninger, H., Kraus, H., Dangschat, H., Beuermann, K.P., Predehl, P., Trümper, J.: Appl. Opt. **18** (1979) 3502.
10 Chupp, E.L.: Gamma-ray Astronomy, Geophys. and Astrophys. Monographs **14**, Reidel, Dordrecht (1976).
11 Charpack, G., Petersen, G., Pollicarpo, A., Sauli, F.: Nucl. Instrum. Methods **148** (1978) 471.
12 Clark, G.W., Bradt, H.V., Lewin, W.H.G., Markert, T.H., Schnopper, H.W., Sprott, G.F.: Astrophys. J. **179** (1973) 263.
13 Cooke, B.A., Ricketts, M.J., Maccacaro, T., Pye, J.P., Elvis, M., Watson, M.G., Griffiths, R.E., Pounds, K.A., McHardy, I., Maccagni, D., Seward, F.D., Page, C.G., Turner, M.J.L.: Mon. Not. R. Astron. Soc. **182** (1978) 489.
14 Dailey, C., Parnell, T.: NASA TMX-73396 (1977).
15 Davelaar, J., Manzo, G., Peacock, A., Taylor, B.G., Bleeker, J.A.M.: preprint ESLAB/79/49, to be published in Inst. Electr. Electron. Eng. Trans. Nucl. Sci. NS-27 (1979).
16 Derdeyn, S.M., Ehrmann, C.H., Fichtel, C.E., Kniffen, D.A., Ross, R.W.: Nucl. Instrum. Methods **98** (1972) 557.
17 Dennis, B.R., Frost, K.J., Lencho, R.J., Orwig, L.E.: Space Sci. Instrum. **3** (1977) 325.
18 Dicke, R.H.: Astrophys. J. Lett. **153** (1968) L 101.
19 Sanford, R.: Proc. ESRO Symp., ESRO SP-87, Frascati (1972).
20 Wills, R.D., Battrick, B. (eds.): Proc. 12th ESLAB Symp.: Recent Advances in Gamma-ray Astronomy, ESA-SP 124 (1977).

continued p. 41

Table 2. X- and γ-ray satellites.
Code for the instrumentation:
1 = gas proportional counter
2 = scintillation counter
3 = solid state detector
4 = collecting mirror
5 = dispersive spectrometer
6 = modulation collimator
7 = pin hole camera
8 = imaging telescope
9 = focal plane image detector
10 = polarimeter
11 = spark chamber
12 = Cerenkov counter

Name	Launch	Major scientific objectives	Instrumentation	Ref.
OSO – 3	March 1967	solar and nonsolar X- and γ-ray measurements	2, 2	32, 43
"UHURU" (SAS – 1)	Dec. 1970	X-ray sky survey (339 sources in 4 U-catalog)	1	27, 22
OSO – 7	Sep. 1971	solar γ-ray and nonsolar X-ray measurements	1, 2, 2	12, 38, 33
"Copernicus" (OAO – 3)	Aug. 1972	UV- and X-ray observations of stars, supernova remnants and extragalactic objects	1, 4	19
SAS – 2	Nov. 1972	γ-ray sky survey	11, 12	16
ANS	Aug. 1974	UV- and X-ray observation of selected sources	1, 4, 5, 1	19
ARIEL V	Oct. 1974	X-ray deep sky survey, study of spectral and temporal characteristics of selected sources	1, 2, 5, 6, 7	19, 13
SAS – 3	May 1975	measurement of accurate positions, spectra and time variability of selected sources and sky survey	1, 4, 6	19
OSO – 8	June 1975	measurement of X-ray spectra of a wide energy range, time variability and polarimetry	1, 2, 10	19, 55, 77
COS – B	Aug. 1975	γ-ray sky survey	1, 2, 11	5, 20
HEAO – 1	Aug. 1977	advanced mission for non-imaging X-ray astronomy	1, 2, 6	23, 14
"Einstein" (HEAO – 2)	Nov. 1978	first X-ray satellite with an imaging telescope: imaging with arc sec resolution, detection of very faint sources, spectroscopy	1, 3, 5, 8, 9	28
ARIEL VI	June 1979	cosmic ray measurement and X-ray observations of selected sources in the range 0.1 ··· 50 keV	1, 4, 1	
Hakucho	Feb. 1979	observations of X-ray burst sources	1, 2, 6	39

References for 1.7, continued

21 Fazio, G.G. in: [38] p. 303.
22 Forman, W., Jones, C., Cominsky, L., Julien, P., Murray, S., Peters, G., Tananbaum, H., Giacconi, R.: Astrophys. J. Suppl. **38** (1978) 357.
23 Friedmann, H., Wood, K.S.: Sky Telesc. **56** (1978) 490.
24 Giacconi, R., Rossi, B.: J. Geophys. Res. **65** (1960) 773.
25 Giacconi, R., Gursky, H., van Speybroeck, L.P.: Observational Techniques in X-ray Astronomy, Annu. Rev. Astron. Astrophys. **6** (1968) 373.
26 Giacconi, R., Reidy, W.P., Vaiana, G.S., van Speybroeck, L.P., Zehnpfennig, T.F.: Space Sci. Rev. **9** (1969) 3.
27 Giacconi, R., Kellogg, E., Gorenstein, P., Gursky, H., Tananbaum, H.: Astrophys. J. Lett. **165** (1971) L 27.
28 Giacconi, R., Branduardi, G., Briel, U., Epstein, A., Fabricant, D., Feigelson, E., Forman, W., Gorenstein, P., Grindlay, J., Gursky, H., Harnden, F.R., jr., Henry, J.P., Jones, C., Kellogg, E., Koch, D., Murray, S., Schreier, E., Seward, F., Tananbaum, H., Topka, K., van Speybroeck, L., Holt, S.S., Becker, R.H., Boldt, E.A., Serlemitsos, P.J., Clark, G., Canizares, C., Markert, T., Novick, R., Helfand, D., Long, K.: Astrophys. J. **230** (1979) 540.
29 Gorenstein, P., Gursky, H., Harnden, F.R., jr., DeCaprio, A., Bjorkholm, P.: Inst. Electr. Electron. Eng. Trans. Nucl. Sci. NS-22 (1975) 616.
30 Gursky, H., Zehpfennig, T.F.: Appl. Opt. **5** (1966) 8.
31 Gursky, H., Bradt, H., Doxsey, R., Schwartz, D., Schwartz, J., Dower, R., Fabbiano, G., Griffiths, R.E., Johnston, M., Leach, R., Ramsey, A., Spada, G.: Astrophys. J. **223** (1978) 973.
32 Hicks, D.B., Reid, L., Peterson, L.E.: Inst. Electr. Electron. Eng. Trans. Nucl. Sci. NS-12 (1965) 54.
33 Higbie, P.R., Chupp, E.L., Forrest, D.J., Gleske, I.U.: Inst. Electr. Electron. Eng. Trans. Nucl. Sci. NS-19 (1972) 606, No. 1.
34 Hochart, J.P., Maze, R., Milleret, G., Zawadski, A., Gawin, J., Wdowczyk, J.: Proc. 14th Int. Cosmic Ray Conf., Munich **8** (1975) 2822.
35 Holt, S.S.: Astrophys. Space Sci. **42** (1976) 123.
36 Gratton, L., (ed.): Int. Astron. Union Symp. No. **37**, Non-Solar X- and Gamma-ray Astronomy, Reidel, Dordrecht (1970).
37 Labuhn, F., Lüst, R., (eds.): Int. Astron. Union Symp. No. **41**, New Techniques in Space Astronomy, Reidel, Dordrecht (1971).
38 Bradt, H., Giacconi, G., (eds.): Int. Astron. Union Symp. No. **55**, X- and Gamma-ray Astronomy, Reidel, Dordrecht (1973).
39 Inoue, H., Kuyama, K., Makishima, K., Matsuoka, M., Murakami, I., Oda, M., Ogawara, Y., Ohashi, I., Shibazaki, N., Tanaka, Y., Tawara, Y., Kondo, I., Hayakawa, S., Kunieda, H., Makino, F., Masai, K., Nagase, F., Yamashita, K., Miyamoto, S., Tsunemi, H., Yushimori, M.: Proc. 16th Int. Cosmic Ray Conf., Kyoto (1979) Vol. 1, 5.
40 Jacobsen, A.S., Bishop, R.J., Culp, G.W., Jung, L., Mahoney, W.A., Willett, J.B.: Nucl. Instrum. Methods **127** (1975) 115.
41 Kirkpatrick, P., Baez, A.V.: J. Opt. Soc. Am. **38** (1948) 766.
42 Kellog, E., Henry, P., Murray, S., van Speybroeck, L.: Rev. Sci. Instrum. **47** (1976) 282.
43 Kraushaar, W.L., Clark, G.W., Garmire, G.P., Borken, R., Higbie, P., Leong, C., Thorsos, T.: Astrophys. J. **177** (1972) 341.
44 Kestenbaum, H.L., Cohen, G.G., Long, K.S., Novick, R., Silver, E.H., Weißkopf, M.C., Wolff, R.S.: Astrophys. J. **210** (1976) 805.
45 Kurfess, J.D., Johnson, W.N.: Inst. Electr. Electron. Eng. Trans. Nucl. Sci. NS-**22** (1975) 626.
46 Leventhal, M., MacCullum, C., Watts, A.: Astrophys. J. **216** (1977) 491.
47 Matteson, J., Nolan, P., Paciesas, W.S., Pelling, R.M.: Univ. of California, San Diego, Report SP 76-07 (1976).
48 Makishima, K., Miyamoto, S., Murakami, T., Nishimura, J., Oda, M., Ogawara, Y., Tawara, Y. in: [70] p. 277.
49 McBreen, B., Ball, S.E., Campbell, M., Greisen, K., Koch, D.: Astrophys. J. **184** (1973) 571.
50 Ögelmann, H., Wayland, J.R. (eds.): Introduction to Experimental Techniques of High Energy Astrophysics, NASA SP-243 (1970).
51 Nakano, G.H., Imhof, W.L., Reagan, J.B.: Space Sci. Instrum. **2** (1976) 219.
52 Cline, T.L., Ramaty, R. (eds.): Proc. Symp. on γ-ray Spectroscopy in Astrophysics, NASA-TM 79619 (1978).
53 Novick, R. in: [38] p. 118.
54 Novick, R.: Space Sci. Rev. **18** (1975) 389.
55 Novick, R., Weißkopf, M.C., Silver, E.H., Kestenbaum, H.L., Long, K.S., Wolff, R.S. in: [70] p. 127.

56 Parkes, W., Evans, K.D., Mathieson, E.: Nucl. Instrum. Methods **121** (1974) 151.
57 Peterson, L.E.: Instrumental Technique in X-ray Astronomy, Annu. Rev. Astron. Astrophys. **13** (1975) 423.
58 Policarpo, A.J.P.L., Alves, M.A.F., Dos Santos, M.C.M., Carvalho, M.J.T.: Nucl. Instrum. Methods **102** (1972) 337.
59 Rothschild, R., Boldt, E., Holt, S., Serlemitsos, P., Garmire, G., Agrawal, P., Riegler, G., Bowyer, S., Lampton, M.: NASA-GSFC TM 79574 (1978).
60 Sanford, P.W. in: [70] p. 217.
61 Schnopper, H.W., Thompson, R.I.: Space Sci. Rev. **8** (1968) 534.
62 Schnopper, H.W., Delvaille, J.P., Epstein, A., Helmken, H., Murray, S.S., Clark, G., Jernigan, G., Doxsey, R.: Astrophys. J. Lett. **210** (1976) L 75.
63 Schwartz, D.A., Griffith, R.E., Murray, S.S., Zombeck, M.V., Barrett, J., Bradley, W.: Center for Astrophysics, Cambridge, Mass., preprint No. 1189 (1979).
64 Stephanian, A.A., Vladimirsky, B.M., Neshor, Y.L., Fomin, V.P.: Astrophys. Space Sci. **38** (1975) 267.
65 Staubert, R., Kendziorra, E., Trümper, J., Hoffmann, J.A., Pounds, K.A., Giles, A.B., Morrison, L.V.: Astrophys. J. Lett. **201** (1975) L 15.
66 Underwood, J.H., Milligan, J.E., de Loach, A.C., Hoover, R.B.: Appl. Opt. **16** (1977) 858.
67 Vaiana, G.S., van Speybroeck, L., Zombeck, M.V., Krieger, A.S., Silk, J.K., Timothy, A.: Space Sci. Instrum. **3** (1977) 19.
68 Vedrenne, G. in: [70] p. 151.
69 Wolter, H.: Ann. Phys. **10** (1952) 94.
70 Van der Hucht, K.A., Vaiana, G.,(eds.): New Instrumentation for Space Astronomy, Pergamon Press (1978).

1.8 Infrared techniques

The infrared (IR) region covers the wavelength range 0.8 μm ⋯1000 μm. According to different observational techniques this region can be divided into the near IR (0.8⋯1.2 μm), the middle IR (1.2⋯30 μm) and the far IR (30⋯1000 μm). The range $\lambda > 300$ μm is also designated the submillimeter region.

1.8.1 Infrared detectors (cf. also 1.3)

In the near IR basically the same types of detector are used as in the visible range. In the middle and far IR quantum (photoelectric) detectors and thermal detectors (bolometers) are used. Although both detector arrays and coherent detectors have great potential, they are not yet commonly used nor easily available [1, 2].

1.8.1.1 Detector types

Photographic plate (cf. 1.5)

In the near IR several emulsions (Kodak I–N, I–Z, IV–Z) are sensitive as far as 1.15 μm [3]. These plates have to be stored under refrigeration and are hypersensitized ($AgNO_3$-bath) shortly before use [4].

Photocathodes (cf. 1.3)

These are used in the near IR with image tubes and photomultipliers. The S1-cathode (AgOCs) extends to 1.1 μm with a quantum efficiency of 0.1% at 1 μm. The InGaAsP-cathode [5] has a quantum efficiency of 4% at 1 μm and 0.1% at 1.1 μm; this tube also has to be kept refrigerated when not in operation. While the latter cathode is an end-on type, the semitransparent S1-cathode has been applied with image tubes to survey dust-obscured regions [67, 68].

Silicon-photodiodes (cf. 1.3)

The sensitivity of silicon detectors, which are now used in large two dimensional arrays (CCD's), is high in the near IR, with quantum efficiency ≈10% at 1 μm. For CCD, see 1.3.1.

Photodetectors

These are semiconductors in which absorbed photons excite electrons from the valence to the conduction band (intrinsic detector) or excite impurity centers in germanium or silicon crystals (extrinsic detector), see Table 1. The carriers generated can be separated by an external electrical field (photoconductive detector), or by an internal field at a p−n junction (photovoltaic detector).

Table 1. Detector materials.

Type	Wavelength μm	Operating temperature K	Ref.
PbS	1···4	77 or 193	9
InSb	2···5.5	77 or 4.2	1, 2, 9
Si:Ga	2···17.5	6···18	10, 11
Si:As	2···25	6···12	10, 11
Si:Sb	2···29	6···10	10, 11
Ge:Cu	2···27	12	10, 11
Ge:Ga	2···120	3	10, 11
Ge-bolometer	1···1000	1.6 or 0.3	1, 2, 6, 7
HgCdTe	5···14	77	9
PbSnTe	2···13	77	9

Thermal detectors

The most commonly used is the germanium bolometer [6]. This is a black painted crystal, the resistance of which is strongly dependent on temperature ($R \propto 1/T^4$) and therefore on radiation absorbed. Composite bolometers have an additional absorbing disk of blackened material in order to increase the effective detector area [7, 8]. Helium-3 bolometers operating at 0.3 K, giving an order of magnitude increase in sensitivity, have been recently constructed [7, 8].

1.8.1.2 Detector parameters

Spectral range

Photodetectors have a limited sensitivity range with a steep long-wavelength cutoff caused by the band energy levels. The responsivity ([A/W] or [V/W]) peaks close to the cutoff. The bolometer has nearly constant sensitivity over a wide spectral range and is most effectively used in the range 5 to 300 μm.

Sensitivity

The Noise Equivalent Power (NEP) is a measure of detector sensitivity, describing the minimum incident flux giving a signal equal to the rms noise in a 1 Hz electrical bandwidth. The NEP is given in units of [W $\mathrm{Hz}^{-1/2}$]. A typical NEP of a detector operated with a ground-based telescope is in the range 10^{-14} to 10^{-16} W $\mathrm{Hz}^{-1/2}$. As a figure of merit for photodetectors the detectivity $D^* = \sqrt{A}/\mathrm{NEP}$ [cm $\mathrm{Hz}^{1/2}$ W^{-1}] is sometimes given, where A [cm^2] is the area of the element. When quoting the sensitivity of a detector the test conditions should be specified, i.e., chopper frequency, wavelength, blackbody temperature and background power.

Noise

IR detectors are subject to many noise sources, the most important being (I) Johnson noise, (II) $1/f$-noise, (III) phonon noise, (IV) generation-recombination noise [9]. In addition to this "detector noise" an IR detection system at a telescope also suffers "sky noise" and "background photon noise".

Time constant

This is a measure of the response speed of the detector. If too large it limits the chopping frequency, thus preventing the elimination of relatively large low frequency noise components ($1/f$-noise, sky-noise). Usually thermal detectors have larger time constants ($\approx 10^{-2}$ s) than photodetectors ($\approx 10^{-3} \cdots 10^{-6}$ s).

Operating temperature

All IR detectors used in astronomy require cooling, which is performed with dry ice (193 K), liquid nitrogen (77 K), liquid helium-4 (4.2 K), pumped helium-4 (1.6 K) or helium-3 (0.3 K) [7].

1.8.1.3 Low background detectors

When operated under low thermal background conditions, modern extrinsic photodetectors can achieve NEP-values as low as 10^{-17} W $Hz^{-1/2}$. A corresponding low background flux ($<10^8$ photons s^{-1} cm^{-2}) can be realized with cooled space telescopes [1, 10, 11]. So far low background detectors are only available for $\lambda < 120$ μm. Stressed Ge:Ga can possibly be used for $\lambda < 200$ μm [11a].

1.8.2 Atmospheric transmission and emission

Ground-based IR observations are possible only through a few atmospheric windows, in which the standard photometric bands are located (1.8.4). Other windows with transmittance of the order of 10% are at 34 μm [12], 350 μm, 450 μm, 800 μm [13]. Temporal fluctuation of atmospheric transmission and emission (sky noise) limits the performance of IR detection systems, especially when operated with large fields of view [13]. Because water vapor is the most important absorber, observational conditions improve rapidly with increasing altitude. Values of transmission and emission for high mountains (4.2 km), aircraft (12 km), and balloons (42 km) are given in [14]. Even from balloon altitudes, sky emission due to airglow phenomena (in particular the OH-airglow in the 0.8···4 μm region [15]) is strong.

1.8.3 Infrared telescopes

Optical telescopes can be used for IR observations. For effective use, however, they have to be modified as described in 1.8.3.3. An IR detection system at a telescope normally includes a chopper, a phase sensitive amplifier, the IR focal plane instrument and the data system (Fig. 1).

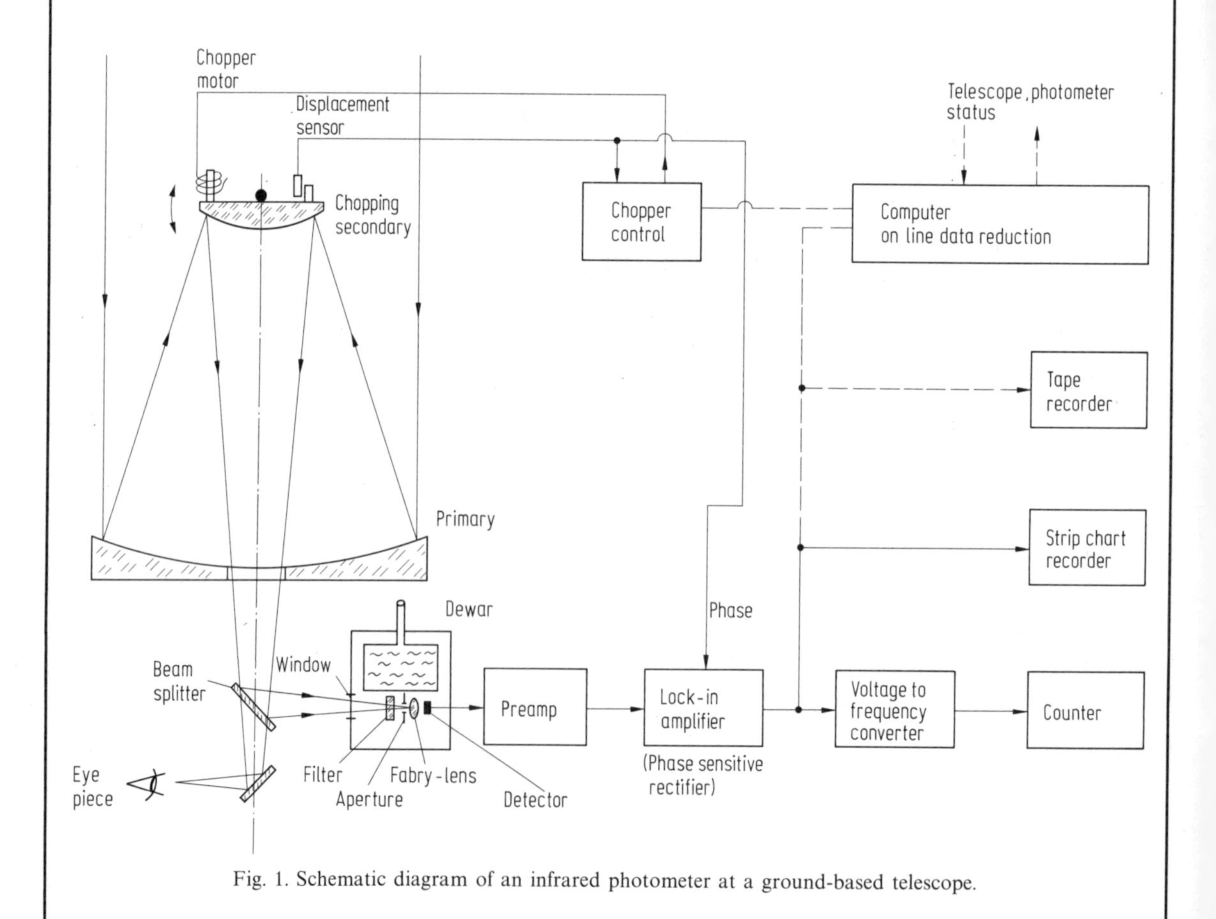

Fig. 1. Schematic diagram of an infrared photometer at a ground-based telescope.

1.8.3.1 Chopper

Since the combined thermal background radiation at the detector from a ground-based telescope and the sky is of the order of 10^{-8} W, beam switching or "chopping" has to be employed. Beam switching to an adjacent part of the sky allows measurements of fluxes several orders of magnitude less than the background. Beam switching (spatial modulation) is performed on small telescopes by rocking the primary and on larger telescopes by rocking the secondary or moving a third focal plane mirror [16, 17]. The modulation function of the chopper should preferably be square-wave. Chopper frequencies are typically about 10 Hz.

1.8.3.2 Optics of the IR telescope

An existing Cassegrain telescope is normally modified as described in [13] before it is used in the IR. All high emissivity parts close to the beam (baffles, rims, support, structure, central hole in the primary) should be avoided or minimized. The entrance aperture should preferably be defined by an undersized secondary. Mirrors should be coated with silver or gold, whose emissivity in the IR is considerably less than that of aluminum. The f ratio should be large ($f/20 \cdots f/100$) in order 1. to make the secondary as small as possible for easier chopping, 2. to make the solid angle at the detector small for closer cold baffling in the cryostat, 3. to reduce the diameter of the central hole in the primary.

1.8.3.3 Telescope platforms

In order to have access to the whole IR region, telescopes have to be sent above the atmosphere (1.8.2). Disadvantages of airborne telescopes, however, are limited observing time and telescope size, and cost.

Ground-based

Many of the major optical telescopes (cf. 1.1) are routinely used (after modification) for IR observations (e.g. the 5 m Palomar-, the 3.6 m ESO-, the 2.2 m Calar Alto-telescopes). Moreover there are several telescopes, on high mountains, designed specifically for IR observations (e.g. the 70 cm Mt. Lemmon-, the 1.3 m Kitt Peak-, the 1.5 m Teneriffe-, the 3.8 m UKIRT- and the 3.2 m NASA-telescopes on Mauna Kea, the Multi-Mirror-telescope (MMT) Mt. Hopkins).

Aircraft

Routine observations are made at altitudes of 12⋯15 km with the 0.9 m telescope of the NASA G. Kuiper Flying Observatory [18], and the 0.3 m telescope mounted on board a NASA Lear Jet [19].

Balloon-borne

Many IR telescopes have been flown on balloons in the altitude range 30⋯42 km. The largest one flown so far is 1 m in diameter [20], many others are listed in [1].

Rockets

A partial sky survey has been performed with small helium cooled rocket telescopes, the results are published as an AFGL-catalogue [21, 22]. IR measurements of the zodiacal light were obtained with a similar telescope [21a].

Satellites, Space Shuttle

Several liquid helium cooled telescopes are at present being planned [1]. They should realize the high potential of extrinsic low background detectors (1.8.1.3). Under construction are IRAS (= Infrared Astronomical Satellite; 60 cm telescope, sky survey [23]), Spacelab 2-IRT (= Infrared Telescope; 15 cm, sky survey, shuttle contamination [24]), GIRL (= German Infrared Laboratory; 40 cm, multi-purpose observatory [25, 25a]).

1.8.4 Infrared photometry

Photometric measurements in standard bands allow comparison of flux levels (or magnitudes) determined from different observations. As in the visible region the magnitude scale is applied and color indices can be used.

1.8.4.1 Photometric bands

The photometric bands are given in Table 2. The last column contains typical extinction values observed in these bands from high altitude observatories [13, 26]. The effective wavelengths depend on the extinction value.

Table 2. Infrared photometric bands (see also 4.2.5.12).
Range = cut-on⋯cut-off wavelength
λ_{eff} = effective wavelength
extinction = observed extinction range from high altitude observatories [13, 26]

Band	Range [μm]	λ_{eff} [μm]	Extinction
J	1.15⋯ 1.35	1.25	$0^m.06 \cdots 0^m.36$
H	1.45⋯ 1.8	1.63	0.06⋯0.25
K	1.9 ⋯ 2.5	2.22	0.05⋯0.18
L	3.05⋯ 4.1	3.6	0.07⋯0.28
M	4.5 ⋯ 5.5	5.0	0.27⋯0.70
N	7.9 ⋯13.2	10.6	0.14⋯0.55
Q	17 ⋯28	21	0.3 ⋯1.4

1.8.4.2 Absolute calibration

The zero points of the magnitude scale for the photometric bands are defined in analogy to the visible range by using the average A0 V star (cf. 4.2.5.12). The flux level calibration (Table 3) is defined by a 10000 K blackbody curve fitted to the zero point at 3.6 μm. Conversion of the flux units is given by $F(\lambda)=3\cdot 10^{10}\cdot F(\nu)/\lambda^2$, with units as in Table 3. The unit of flux density is Jansky [Jy], with $1\,\text{Jy} \mathrel{\hat=} 10^{-26}\,\text{W m}^{-2}\,\text{Hz}^{-1}$. Recent standard star calibrations are based on the comparison of the energy distribution of solar-like stars with the absolutely calibrated spectrum of the sun. A few IR standard stars are given in Table 4 [13]. In the far IR the standard stars are too weak, therefore planets are used whose fluxes can be calculated from their brightness temperatures: Jupiter 127 K, Saturn 85 K, Mars 235 K, Venus 240 K [13, 27].

Table 3. Flux level for 0.0 magnitude [13].

Band	λ_0 μm	$F(\lambda)$ W cm^{-2} μm^{-1}	$F(\nu)$ W m^{-2} Hz^{-1}
K	2.22	$4.14\cdot 10^{-14}$	$6.8\cdot 10^{-24}$
L	3.6	$6.38\cdot 10^{-15}$	$2.76\cdot 10^{-24}$
M	5.0	$1.82\cdot 10^{-15}$	$1.52\cdot 10^{-24}$
N	10.6	$9.7\cdot 10^{-17}$	$3.63\cdot 10^{-25}$
Q	21	$6.5\cdot 10^{-18}$	$9.56\cdot 10^{-26}$

Table 4. Magnitudes of standard stars [13, 28].
B. S. No. = number in Catalogue of Bright Stars, Yale University Observ. 1964.

Star	B. S. No.	Spectral class	*V*	*K*	*L*	*M*	*N*	*Q*
β And	337	M0 III	$2^m.03$	$-1^m.85$	$-2^m.10$	$-1^m.97$	$-2^m.06$	$-2^m.23$
α Tau	1457	K5 III	0.86	−2.89	−3.00	−2.89	−2.99	−3.12
α Aur	1708	G8 III	0.80	−1.78	−1.86	−1.92	−1.90	−1.93
α Boo	5340	K2 III	0.06	−2.99	−3.14	−2.98	−3.12	−3.30
γ Dra	6705	K5 III	2.22	−1.29	−1.50		−1.45	−1.52

1.8.4.3 Limiting magnitudes

The sensitivity of a photometer can be limited by system noise or sky noise. Sky noise can be reduced by reducing the beam and the chopper throw. The brightness of the faintest object measurable with a 3-sigma-value is considered the limiting magnitude. So far the best values reported are $K \approx 18$ mag in a 6″ aperture with a 2.2 m telescope [28, 28a] and $N \approx 8$ mag in a 6″ aperture with a 1.5 m telescope [29]. The minimum detectable flux can also be expressed in terms of noise equivalent flux density NEFD $[\text{W cm}^{-2}] = \text{NEP}\cdot\sqrt{f}/At$, where f [Hz] is the electrical bandwidth, A [cm^2] the telescope area, and t the transmission of the overall optical system.

1.8.4.4 Filters

Dielectric interference filters are available for $\lambda < 30\,\mu m$. They can be custom-made for a desired wavelength λ and passband $\Delta\lambda$ ($\lambda/\Delta\lambda < 200$), and achieve high transmission (0.5···0.9) inside and high rejection ($< 10^{-3}$) outside the band. In the far IR, interference filters made of two or more layers of metallic wire mesh have been used [30]. Wideband and lowpass filters for the far IR are made of cooled crystalline materials (BaF_2, sapphire etc.) where the cut-on wavelength can be selected by depositing different size diamond scattering particles on the filter [31, 32].

1.8.5 Infrared spectroscopy (see also 1.5)

1.8.5.1 Circular variable filter CVF

Effective wavelength and resolution $R = \lambda/\Delta\lambda$ depend on the angular position of the interference filter segment [33]. $R \leqq 100$ has been achieved in the middle IR. Only a single spectral element of a source can be measured at a given time. The NEFD of a cooled spectrometer increases only $\propto \sqrt{R}$. CFV's for the 8···13 µm region are cooled to 77 K [34, 35].

1.8.5.2 Prism, grating

Cooled prism spectrometers for the 8···13 µm region have been built with $R \approx 10$ using arrays with only a few detectors [36, 37]. Grating spectrometers with resolutions $R \approx 100$ to 2000 have been developed for the 10 and 20 µm-atmospheric window and for airborne applications [38···41]. Several spectrometers are cooled to 4 to 77 K and use arrays of up to 30 detectors.

1.8.5.3 Fourier spectrometer (cf. 1.5.4)

They record simultaneously all spectral elements of a given spectrum. However, this "multiplex advantage" diminished with the advent of low-background detector systems. Fourier spectrometers can operate with large throughputs (etendue) and high resolution R (Jaquinot advantage).

The Michelson spectrometer, with its derivatives, is the most common design. It is often used in the rapid scan mode in order to cancel atmospheric fluctuation and avoid spatial chopping [42, 43]. Michelsons are used at the Cassegrain focus or the Coudé focus of ground-based telescopes [45]. Numerous Michelsons have been flown on board aircraft [44, 44a], balloons, and satellites. References are listed in [1, 45].

A special Michelson modulates the polarization of the beam rather than its intensity [46]. These polarizing Michelsons have high beam splitter efficiency.

A few lamellar grating spectrometers have been built for ground-based and air-borne telescopes [47, 48]. In these instruments the beam splitter divides the wavefront, rather than the amplitude, and has high efficiency in the far infrared. In addition the beam is left unpolarized [49].

1.8.5.4 Fabry-Perot (cf. 1.5.3)

They are restricted to a very limited spectral range (single line profile), but offer high resolution R and large throughput. In the near and middle IR wavelength ranges scanning is performed by gas pressure change or piezo-electrically [50, 51, 52]. In the far IR Fabry-Perot mirrors consisting of wire mesh have been used [53a]. For $\lambda < 30\,\mu m$ cooled solid silicon Fabry-Perot's with $R \approx 3000$ have been tested; in this case wavelength scan is by tilting the etalon [53].

1.8.5.5 Heterodyne spectroscopy

The difference frequency between the source line and a laser oscillator line can be amplified with standard microwave techniques. Critical components are mixers: Schottky diodes have been applied usefully throughout the submillimeter region [54]. Very high resolution $R \approx 10^6$ can be achieved. Several test observations of bright objects (planets) have been made from the ground [55, 56].

1.8.6 Infrared polarimetry

Spectroscopic and photometric polarimeters have been used on ground-based [57, 57a] and air-borne telescopes [58]. For rotating polarizers polarization foils are available for $\lambda < 2.5\ \mu m$, while for $\lambda > 1.5\ \mu m$ wire grid polarizers are used [59, 60]. Photoelastic modulators have been applied in a polarimeter for the 1···8 µm region. This design has the advantage of only minor instrumental polarization [61].

1.8.7 Spatial resolution instrumentation

A spatial interferometry technique (Michelson, see 1.6.1) in which the baseline is varied with primary mirror masks has been used in the 5 and 10 µm region on large ground-based telescopes [62, 63]. Resolution of 0.''4 was possible at 10 µm. A similar system has been tested on the Multi-Mirror-Telescope in Arizona [64]. An angular resolution of 0.''05 at 2.2 µm can be achieved with its 6.5 m baseline. Lunar occultation measurements, giving resolution of about 0.''1, have been made with the 5 m Mt. Palomar telescope. Another recent development in speckle interferometry (cf. 1.6.3). In the future it may be possible to use two-dimensional detector arrays with this technique, thus enhancing its important advantage of multidirectional resolution [66].

1.8.8 References for 1.8

1 Soifer, B.T., Pipher, J.L.: Annu. Rev. Astron. Astrophys. **16** (1978) 335.
2 Gillett, F.C., Dereniak, E.L., Joyce, R.R.: Opt. Eng. **16** (1977) 544.
3 Kodak Plates and Films for Scientific Photography, Kodak Public. P-315, Rochester, New York.
4 Modern techniques in astronomical photography, European Southern Observatory Proceedings (West, R.M., Heudier, J.L., eds.) May (1978).
5 Data sheet, VPM-164 photomultiplier, Varian LSE, Palo Alto, Calif.
6 Low, F.J.: J. Opt. Soc. Am. **51** (1961) 1300.
7 Johnson, C., Low, F.J., Davidson, A.W.: Proc. Soc. Photo-Opt. Instrum. Eng. **172** (1979) 178.
8 Nishioka, N.S., Richards, P.L., Woody, D.P.: Appl. Opt. **17** (1978) 1562.
9 The Infrared Handbook (Woolfe, W.J., Zissis, G.J., eds.), Office of Naval Research, Dep. of the Navy, Washington, D.C. (1978).
10 Young, E.T., Low, F.J.: Proc. Soc. Photo-Opt. Instrum. Eng. **172** (1979) 184.
11 Bratt, P.R.: Impurity Germanium and Silicon Infrared Detectors, in: Semiconductors and Semimetals **12**, (Willardson, R.K., Beer, A.C., eds.), Academic Press, New York (1977).
11a Haller, E.E., Hueschen, M.R., Richards, P.L.: Appl. Phys. Lett. **34** (1979) 495.
12 Rieke, G.H., Low, F.J.: Astrophys. J. **200** (1975) L 67.
13 Low, F.J., Riecke, G.H.: The Instrumentation and Technique of Infrared Photometry, in: Methods of Exp. Physics (Carleton, N., ed.), Academic Press, New York (1974).
14 Traub, W.A., Stier, M.T.: Appl. Opt. **15** (1976) 364.
15 Hofmann, W., Lemke, D., Frey, A.: Astron. Astrophys. **70** (1978) 427.
16 Fahrbach, U., Haussecker, K., Lemke, D.: Astron. Astrophys. **33** (1974) 265.
17 Gautier, T.N., III, Hoffmann, W.F., Low, F.J., Reed, M.A., Rieke, G.H.: Proc. Soc. Photo-Opt. Instrum. Eng. **172** (1979) 54.
18 Cameron, R.M., Bader, M., Mobley, R.E.: Appl. Opt. **10** (1971) 2011.
19 Bader, M., Wagoner, C.B.: Appl. Opt. **9** (1970) 265.
20 Fazio, G.G., Kleinmann, D.E., Low, F.J.: Far infrared observations with a 1-m balloon-borne telescope, in: Far Infrared Astronomy, (Rowan-Robinson, M., ed.), Pergamon Press (1976).
21 Price, S.D., Walker, S.P.: The AFGL four color infrared sky survey, AFGL-TR 76-0208 (1976).
21a Price, S.D., Murdock, T.L., Marcotte, L.P.: Astron. J. **85** (1980) 765.
22 Gosnell, T.R., Hudson, H., Puetter, R.C.: Astron. J. **84** (1979) 538.
23 Low, F.J.: Proc. Soc. Photo-Opt. Instrum. Eng. **183** (1979) 11.
24 Koch, D.: Proc. Soc. Photo-Opt. Instrum. Eng. **183** (1979) 16.
25 Lemke, D., Klipping, G., Grewing, M., Trinks, H., Drapatz, S., Proetel, K.: Proc. Soc. Photo-Opt. Instrum. Eng. **183** (1979) 31.
25a Lemke, D., Proetel, K., Dahl, F.: Proc. Soc. Photo-Opt. Instrum. Eng. **265** (1981) 366.
26 Lemke, D., Frey, A., Hefele, H., Schulte in den Bäumen, J.: Mitt. Astron. Ges. **43** (1978) 98.
27 Wright, E.L.: Astrophys. J. **210** (1976) 250.
28 Rieke, G.H.: private communication.
28a Lebofsky, M.J.: Int. Astron. Union Symp. **92** (1980) 257.

29 Rieke, G.H.: Astrophys. J. **226** (1978) 550.
30 Ulrich, R.: Infrared Phys. **7** (1967) 37.
31 Armstrong, K.R., Low, F.J.: Appl. Opt. **13** (1974) 425.
32 Wijnbergen, J.J., Moolenaar, W.H., de Groob, G.: Filters for far infrared astronomy, in: Infrared detection techniques for space research (Manno, V., Ring, J., eds.), Reidel, Dordrecht, Holland (1972).
33 The Infrared Handbook, published by OCLI, Santa Rosa, California.
34 Gillett, F.C., Forrest, W.J.: Astrophys. J. **179** (1973) 483.
35 Willner, S.P.: Astrophys. J. **206** (1976) 728.
36 Gehrz, R.D., Hackwell, J.A., Smith, J.R.: Publ. Astron. Soc. Pacific **88** (1976) 971.
37 Hackwell, J.A., Gehrz, R.D., Smith, J.R., Briotta, D.A.: Astrophys. J. **221** (1978) 797.
38 Houck, J.R., Ward, D.: Publ. Astron. Soc. Pacific **91** (1979) 140.
39 Rank, D.M., Dinerstein, H.L., Lester, D.F., Bregman, J.D., Aitken, D.K., Jones, B.: Mon. Not. R. Astron. Soc. **185** (1978) 179.
40 Beckwith, S., Persson, S.E., Neugebauer, G., Becklin, E.E.: Astrophys. J. **223** (1978) 464.
41 McCarthy, J.F., Forrest, W.J., Houck, J.R.: Astrophys. J. **231** (1979) 711.
42 Larson, H.P., Fink, U.: Appl. Opt. **14** (1975) 2085.
43 Mertz, L.: Transformations in optics, Wiley, New York (1965).
44 Baluteau, J.P., Bussoletti, E., Anderegg, M., Moorwood, A.F.M., Coron, N.: Astrophys. J. **210** (1976) L 45.
44a Davis, D.S., Larson, H.P., Williams, M., Michel, G., Connes, P.: Appl. Opt. **19** (1980) 4138.
45 Maillard, J.P.: Status and prospects of Fourier transform spectroscopy in astronomy, in: Proc. High resolution spectro-photometry (Hack, M., ed.), Trieste July (1978).
46 Martin, D.H., Puplett, E.: Infrared Phys. **10** (1969) 105.
47 Hofmann, R., Drapatz, S., Michel, K.W.: Infrared Phys. **17** (1977) 451.
48 Pipher, J.L., Savedoff, M.P., Duthie, J.G.: Appl. Opt. **16** (1977) 223.
49 Bell, R.J.: Introductory Fourier transform spectroscopy, Academic Press, New York, London (1972).
50 Selby, M.J., Jorden, P.R., MacGregor, A.D.: Infrared Phys. **16** (1976) 317.
51 Van der Wal, P.B., Slingerland, J.: A high-resolution Fabry-Perot interferometer for the Ne II line at 12.8 μ, in: Proc. High resolution spectrophotometry (Hack, M., ed.), Trieste July (1978).
52 Geballe, T.: Ph.D. thesis, Univ. Calif., Berkeley (1974).
53 Auer, R.D., Michel, K.W.: Fabry-Perot line spectroscopy in the 30 μ range by tilt and in the angle scan mode, in: Proc. High resolution spectrophotometry (Hack, M., ed.), Trieste July (1978).
53a Storey, J.W.V., Watson, D.M., Townes, C.H.: Astrophys. J. **233** (1979) 109.
54 Mezger, P.G.: Beobachtungen im Millimeter- und Submillimeter-Bereich, Preprint No. 21, Max-Planck-Institut Radioastronomie, Bonn, Aug. (1978).
55 de Graauw, T., van de Staadt, H.: Nature Phys. Sci. **246** (1973) 73.
56 Townes, C.H.: Infrared heterodyne spectroscopy for astronomical purposes, in: Proc. 2. Intern. Conf. Infrared Phys. (Affolter, E., Kneubühl, F., eds.), ETH Zürich, March (1979).
57 Dyck, H.M., Forbes, F.F., Shawl, S.J.: Astron. J. **76** (1971) 901.
57a Knacke, R.F., Capps, R.W.: Astron. J. **84** (1979) 1705.
58 Coletti, A., Melchiorri, F., Natall, V.: Measurements of the polarized sky background in the far infrared, in: Far infrared astronomy (Rowan-Robinson, M., ed.), Pergamon Press (1976).
59 Auton, J.P., Hutley, M.C.: Infrared Physics **12** (1972) 99.
60 Auton, J.P.: Appl. Opt. **6** (1967) 1023.
61 Kemp, J.C., Rieke, G.H., Lebofsky, M.J., Coyne, G.V.: Astrophys. J. **215** (1977) L 107.
62 McCarthy, D.W., Low, F.J., Howell, R.: Astrophys. J. **214** (1977) L 85.
63 McCarthy, D.W., Low, F.J., Howell, R.: Opt. Eng. **16** (1977) 569.
64 Low, F.J.: The MMT and the infrared, in preliminary Proc.: The MMT and the future of ground-based astronomy (Weekes, T.C., ed.), Smithsonian Special Report (1979).
65 Zappala, R.R., Becklin, E.E., Matthews, K., Neugebauer, G.: Astrophys. J. **192** (1974) 109.
66 Sibille, F., Chelli, A., Léna, P.: Astron. Astrophys. **79** (1979) 315.
67 Beetz, M., Elsässer, H., Weinberger, R.: Astron. Astrophys. **34** (1974) 335.
68 Eiroa, C., Elsässer, H., Lahulla, J.F.: Astron. Astrophys. **74** (1979) 89.

1.9 Radio astronomical receiver systems

1.9.1 Radiometers (receivers)

The fundamental configuration and properties of a radiometer were briefly described in LB, NS, Vol. VI/1 (1965) p. 27. A comprehensive treatment of radiometers and of the newer correlation receivers and spectrometers may be found in [a].

Developments in radiometer technology since 1965 have made particular progress in the mm and cm wavelength ranges toward reducing the effective receiver noise temperature T_E. The presently attainable noise temperatures of various types of receivers are shown in Fig. 1 as a function of frequency.

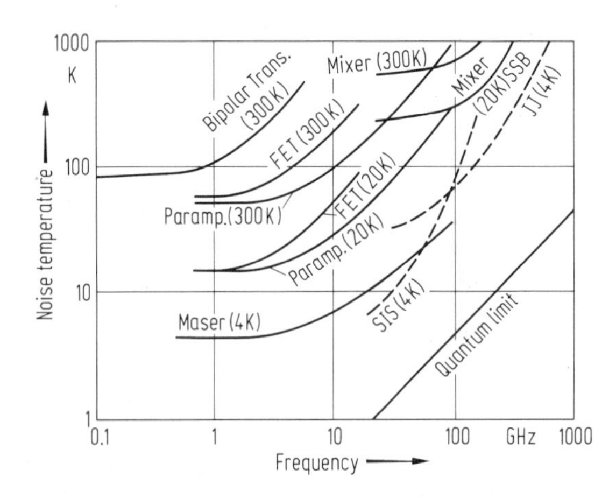

Fig. 1. Effective noise temperature of different amplifiers.
Maser 4 K
= maser amplifier cooled to 4 K [2]
FET 20 K
= field-effect transistor amplifier cooled to 20 K [b, 6]
Paramp. 20 K
= parametric amplifier cooled to 20 K
JJ 4 K
= Josephson-junction mixer [1, 4, 5, 7]
SIS 4 K
= superconductor-insulator-superconductor mixer [3]
Mixer SSB
= single side band mixer
Bipolar Trans.
= bipolar transistor amplifier

The receiver bandwidth B could also be significantly increased. A bandwidth $B \cong 200$ MHz has been achieved with masers at $f = 5$ GHz [2]. With parametric amplifiers one can attain bandwidths that are 10% of the receiving frequency, and this ratio can be raised to 20% with FET amplifiers [b, 6].

The threshold sensitivity of radiometers has thereby been increased by about a factor of ten. Similarly, the mean fluctuation of the system temperature, ΔT_a, could be lowered to just a few mK for certain continuum measurements:

$$\Delta T_a = c \frac{T_E}{\sqrt{B\tau}} \tag{1}$$

with the constant $c \cong 1$ and τ = integration time (e.g. LB, NS, Vol. VI/1 p. 28, Fig. 5).

References for 1.9.1

General references

a Evans, A., McLeish, G.M.: RF Radiometer Handbook, Artech House Inc., Massachusetts (1977).
b Liechti, C.A.: Inst. Electr. Electron. Eng., Trans. Microwave Theory Tech. **24** (1976) 279.

Special references

1 Josephson, B.D.: Phys. Lett. **1** (1962) 251.
2 Moore, C.R., Clauss, R.C.: Inst. Electr. Electron. Eng., Trans. Microwave Theory Tech. **27** (1979) 249.
3 Richards, P.L., Shen, T.M.: Appl. Phys. Lett. **34** (1979) 141.
4 Taur, Y., Claasen, J.H., Richards, P.L.: Inst. Electr. Electron. Eng., Trans. Microwave Theory Tech. **22** (1974) 1005.
5 Taur, Y., Kerr, A.R.: Appl. Phys. Lett. **32** (1978) 775.
6 Vowinkel, B.: Electron. Lett. **16** (1980) 730.
7 Vowinkel, B.: Nachrichtentech. Z. Archiv **2** (1980) 151.

1.9.2 The radio telescopes

The physical characteristics of various types of antennas were presented in LB, NS, Vol. VI/1 p. 23.

The range of radio waves accessible to ground-based stations for astronomical observations extends from wavelengths $\lambda \cong 30$ m down to $\lambda \cong 0.3$ mm, i.e. over about five decades. Segments of almost this entire wavelength interval are presently being used for radio astronomical investigations.

This large range of wavelengths necessitates the development of conceptually different receiver systems, which are specially designed for observations at long, medium and short wavelengths, respectively. Radio telescopes may therefore be roughly divided into three categories:

(1) Radio telescopes for m wavelengths ($\lambda > 0.5$ m)
(2) Radio telescopes for the lower dm and cm ranges
(3) Millimeter radio telescopes

1.9.2.1 Radio telescopes for m wavelengths

It is difficult even today to attain the generally desired angular resolution of one arc minute (1′) at wavelengths greater than 0.5 m. The half power beam width (HPBW) of the main antenna lobe is given by the formula

$$\mathrm{HPBW} \cong \frac{3800\,\lambda}{D}\,[\text{arc min}], \qquad (2)$$

where D is the diameter of the antenna. In order to obtain a HPBW less than 4′, a diameter $D \cong 1000\,\lambda$ is necessary. At m wavelengths that resolution would require a gigantic antenna with a diameter of up to 10 km or more.

Two specific methods have been devised to reduce the enormous constructional and financial constraints imposed on such antennas.

These are:
(1) The Mills cross antenna
(2) The method of aperture synthesis

Both of these concepts have been briefly described in LB, NS, Vol. VI/1 p. 33. For a more lengthy discussion, see [a].

The Mills cross uses a two-phase switching cycle to combine the two signals of the two arms of a cross shaped antenna (perpendicular arms of length D, usually oriented north/south and east/west) first directly and then with an additional half wave delay. The difference in the two signals during these two switching phases is then formed to produce the desired receiver output. In this way, one can obtain an antenna beam with a half width appropriate to a full antenna aperture of total diameter D.

The method of aperture synthesis employs a limited number of individual displaceable antenna elements, each of which measures the amplitude and phase of the incident radiation. The antennas are strategically moved in an array that adequately covers a desired larger aperture area. The distribution of radio intensity of an observed celestial object is obtained by combining the inputs of all antennas in a Fourier synthesis process, usually with the help of an on-line computer. The angular resolution attained in this way is equal to that of a single antenna with an aperture the size of the entire area swept out by the smaller antenna elements.

In order to shorten the time required for relocation of the antenna elements during an observation, one arranges the elements in a simple pattern such as a cross, a "T" or a "Y" (Fig. 2), and then exploits the diurnal rotation of the earth to rotate the system with respect to the celestial sphere. A complete synthesis map can be constructed in this manner in less than 12 hours.

The most important radio telescopes for m wavelengths (Table 1) employ one or the other of these two observing techniques. The ordering in Table 1 is random. Descriptive data for other smaller meter telescopes may be found in [b].

Fig. 2: p. 53.

Table 1. Radio telescopes for m wavelengths.
Type: M.C. = Mills cross; A.S. = aperture synthesis
A = collecting area; ν = frequency

No.	Operating Institute	Location	Type	Characteristics, size	A [m²]	ν [MHz]	Programs	Ref.
1	Cornell-Sydney University Astronomy Centre, University of Sydney, Australia	Molonglo Radio Obs. Hoskintown, N.S.W. 149°25′ E 35°22′ S	M.C.	cylindrical paraboloid cross; each arm 12 × 1600 m meridian transit	17000	408 111	galactic and extragalactic sources	6, 7
2	University of Bologna, Bologna, Italy	Univ. of Bologna, Obs. Medicina 11°39′ E 44°31′ N	M.C.	cylindrical paraboloid cross: each arm 30 m × 1200 m; meridian transit	35000	408	galactic and extragalactic sources	1
3	Clark Lake Radio Observatory, Borrego Springs, Calif., USA	Borrego Springs, Calif. 116°16.′8 W 33°20.′3 N	A.S.	720 spiral helix antennas in T-shaped array: EW arm 480 elements, 3000 m; NS arm 1800 m	5000 at 60 MHz	10⋯130	solar pictures every second; galactic and extragalactic sources	3, 4
4	Department of Physics and Astronomy, The University of Iowa, Iowa City, USA	Borrego Springs, Calif. 116°16.′8 W 33°20.′3 N	M.C.	colinear coaxial antenna elements; EW arm 1184 m; NS arm 832 m	76000	34.3	interplanetary scintillation observations; also radio-astronomy	b
5	Lebedev Physical Institute, Moscow, USSR	Serpukhov Radio-physical Station, Serpukhov	M.C.	cylindrical paraboloids: two arms 40 m × 1000 m	80000	600⋯300	galactic and extragalactic sources	b 9, 5
6	Institute of Radio-physics and Electronics, Ukrainian Academy of Sciences, Kharkhov, USSR	Kharkhov	A.S.	dipole array in T-shape cross, NS arm 1860 m, EW arm 900 m	14000 at 10 MHz	10⋯25	galactic and extragalactic sources	2
7	CSIRO Divison of Radiophysics, Epping, N.S.W., Australia	CSIRO Solar Radio Observatory, Culgoora, N.S.W. 149°34.′3 E 30°19.′3 S	A.S.	circular arrays 3 km in diameter, 96 steerable paraboloids each 15 m ∅	6000	80 160 327	solar pictures with 3000 points in 1 s	10 8

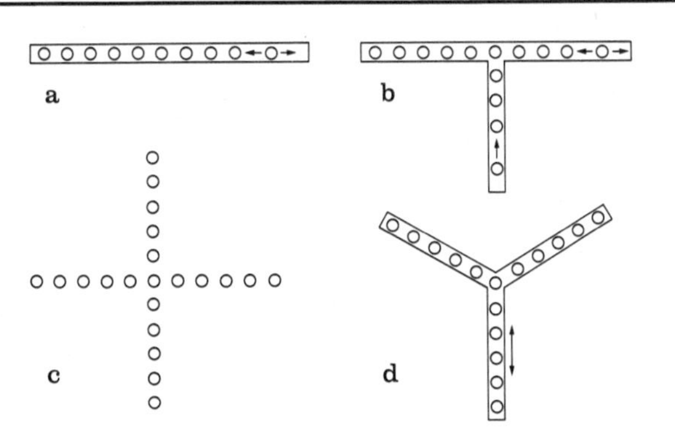

Fig. 2. Typical arrangements of the antenna elements in earth rotation synthesis radio telescopes. (The elements are partly movable on rails.)
a) east-west line arrays
b) T-shaped arrays
c) cross
d) 120° Y-shaped array

References for 1.9.2.1

General references

a Christiansen, W. N., Högbom, J.A.: Radiotelescopes, Cambridge University Press (1969).
b List of Radio and Radar Astronomy Observatories, ed. Committee on Radio Frequencies, National Academy of Sciences, Washington, D.C. 20418 (1979).

Special references

1 Braccesi, A., Ceccarelli, M., Colla, G., Ficarra, A., Gelato, G., Grueff, G., Sinigaglia, G.: Nuovo Cimento **62** B (1969) 13.
2 Braude, S.Ya., Megn, A.V., Ryabov, B.P., Sharykin, N.K., Zhouck, J.N.: Astrophys. Space Sci. **54** (1978) 3.
3 Erickson, W.C.: Proc. Inst. Electr. Electron. Eng. **61** (1973) 1276.
4 Erickson, W.C., Fisher, J.R.: Radio Sci. **9** (1974) 387.
5 Ilyasov, Yu.P., Kuzmin, A.D.: Radio Telescopes (ed. Skobeltsyn Consultants Bureau), Plenum Pub. Corpor., New York (1966) 7.
6 Mills, B.Y., Aitchison, R.E., Little, A.G., McAdam, W.B.: Proc. Inst. Radio Eng. Australia **24** (1963) 156.
7 Monro, R.E.B., Murdoch, H.S., Large, M.I.: Proc. Inst. Radio Eng. Australia **31** (1970) 19.
8 Sheridan, K.V., Labrum, N.R., Payten, W.J.: Proc. Inst. Electr. Electron. Eng. **61** (1973) 1312.
9 Vitkevich, V.V., Kalachev, P.D.: Radio Telescopes (ed. Skobeltsyn Consultants Bureau), Plenum Pub. Corpor., New York (1966) 1.
10 Wild, J.P.: Proc. Inst. Radio Eng. Australia **28** (1967) 277.

1.9.2.2 Radio telescopes for the lower dm and cm ranges

Large single telescopes with parabolic reflectors as well as aperture synthesis telescopes are in use today for observations in the lower dm and cm wavelength ranges. The large parabolic telescopes, particularly in the lower cm range, are used for scanning larger objects or larger regions in the sky and for polarization observations. They are also well suited for spectroscopic investigations with high frequency resolution of the interstellar medium. Table 2 lists the major telescopes of this type with parabolic reflectors $D > 30$ m. A directory of the numerous smaller telescopes may be found in the list of radio and radar observatories in [b]. It should only be mentioned that more than 30 telescopes with $20\,\text{m} < D < 30\,\text{m}$ in different countries are presently in use for radio astronomical observations.

The earth rotation synthesis telescopes have a much higher angular resolution than the large parabolic telescopes [30, 5, 11, a]. Very detailed radio maps of isolated celestial objects have been constructed. The synthesis telescopes of the Mullard Radio Astronomy Observatory in Cambridge and the new VLA (Very Large Array) of the National Radio Astronomy Observatory in New Mexico can obtain an angular resolution of about 1″ or better, which is comparable with the resolution of the larger optical telescopes. Some of the observation stations in Table 3 are also equipped with multi-channel receivers for spectroscopy, thus enabling spectroscopic line studies combined with high angular resolution.

Table 2. Large parabolic reflector telescopes ($D > 30$ m).
Mt = mounting; az = azimuthal; tr = transit; equ = equatorial; VLBI = very-long-baseline interferometry (see 1.9.2.4)

No.	Operating institute	Location	D m	Surface	Mt	Frequency range [MHz]	Programs	Ref.
1	Max-Planck-Institut f. Radioastronomie, Bonn, Germany	Effelsberg 6°53′ 1″.5 E 50°31′28″.6 N	100	$D < 85$ m Al-plates; $D > 85$ m wire-mesh	az	600···15000; inner part $D < 85$ m 15000···50000	galactic radiation continuum and lines, extragalactic sources, VLBI, solar radiation	12, 13
2	National Radio Astronomy Observatory, Charlottesville, USA	Green Bank W. Va. 79°50′.2 W 38°26′17″ N	91.5	wire-mesh (new surface)	tr	0.3···5000	galactic radiation continuum and lines, extragalactic sources	b
3	Nuffield Radio Astron. Laboratory, University of Manchester, England	Jodrell Bank Cheshire 2°18′25″ W 53°14′13″ N	76.2	steel plates (new surface)	az	0.3···4000	galactic radiation continuum and lines, extragalactic sources, interferometer VLBI	26
4	CSIRO-Division of Radiophysics, Epping, 2121, N.S.W. Australia	Parkes N.S.W. Australia 148°15′.7 E 33°00′.0 S	64	$D < 34$ m Al-plates; $D > 34$ m galvanized wire	az	500···5000; inner part $D < 17$ m 90000	galactic radiation continuum and lines, extragalactic sources, interferometer	4, 19 27
5	Jet Propulsion Lab. 4800 Oak Grove Drive, Pasadena, Calif. 91103, USA	Goldstone Calif. 116°53′20″ W 35°25′30″ N	64	$D < 25$ m Al-plates; $D > 25$ m perforated plates (50% porous)	az	600···10000	primary: deep space probe tracking; secondary use in radio astronomy, VLBI	24
6	National Institute of Aerospace Technology (INTA), Spain	Robledo (Spain) 4°14′49″.2 W 40°26′ 2″ N	64	$D < 25$ m Al-plates; $D > 25$ m perforated plates	az	600···10000	primary use in deep space probe tracking; secondary use in radio astronomy, VLBI	24
7	Jet Propulsion Lab. 4800 Oak Grove Drive, Pasadena, Calif. 91103, USA	JPL Deep Space Station Tidbinbilla, Australia 148°58′.8 E 35°24′.1 S	64	$D < 25$ m Al-plates; $D > 25$ m perforated plate	az	600···10000	primary use in deep space probe tracking; secondary use in radio astronomy, VLBI	24

8	National Research Council of Canada, Ottawa, Ontario, Canada	Algonquin Radio Observatory, Lake Traverse, Ont. 78°4′.4 W, 45°57′.3 N	45.7	$D<36$ m Fe-plates; $D>36$ m mesh	az	400···10000	galactic radiation continuum and lines, extragalactic sources, VLBI	20
9	AFCRL L.G. Hanscom Field, Bedford, Mass. 01730, USA	Sagamore Hill Radio Obs. Hamilton Mass. 70°49′ W 42°38′ N	45.7	wire-mesh	az	400···2000	radar, sun and planets, pulsars	b
10	Center for Radar Astron. Durand Building, Stanford University, Stanford, Calif. 94305, USA	Stanford Center for Radar Astron. 122°10′42″ W 37°24′31″ N	45.7	wire-mesh	az	259.7···2000	used both for radio and radar astronomy	b
11	National Radio Astronomy Observatory, Charlottesville W. Va., USA	Green Bank W. Va. 79°49′42″ W 38°26′08″ N	42.7	Al-plates	equ	1400···22000	galactic radiation continuum and lines, extragalactic sources, VLBI	b
12	Owens Valley Radio Observatory, Big Pine, Calif. 93513, USA	Big Pine, Calif. 118°16′.9 W 37°13′.9 N	39.6	Al-plates	az	500···12000 18000···24000	galactic continuum and spectral lines, extragalactic sources, VLBI	b
13	University of Illinois, 60 Electrical Engeneering Building, Urbana, Illinois 61801, USA	Vermilion River Obs. Danville, Ill. 87°33′49″ W 40°03′38″ N	36.6	mesh	equ	600···1720	galactic continuum and spectral lines, extragalactic sources	b
14	NEROC-Northeast Radio Observatory Corporation, Haystack Observatory, Westford, Mass. 01886, USA	Haystack Obs. Tyngsboro, Mass. 71°29′19″ W 42°37′23″ N	36	Al-honeycomb plate, antenna enclosed in radome	az	1.600···11200	continuum mapping, various spectral lines, and VLBI	34, 32
15	Nuffield Radio Astron. Laboratory, University of Manchester, England	Jodrell Bank Macclesfield, Cheshire 2°18′.4 W, 53°14′.2 N	38 m × 25.9 m elliptical shape	Fe-plates	az	2700···10000	galactic radiation continuum and lines, extragalactic sources, interferometer with No. 16	
16	Nuffield Radio Astron. Laboratory, University of Manchester, England	Wardle, Cheshire 2°24′.2 W 53° 6′.8 N	38 m × 25.9 m elliptical shape	Fe-plates	az	2700···5000	interferometer with No. 3 and No. 15	

Table 3. Earth rotation synthesis radio telescopes for cm wavelengths

D = diameter
n_{tot} = number of total elements
n_{mov} = number of movable elements
t_i = imaging time
ν = frequency
Beam = beam of elements
HPBW = half-power beam width

No.	Operating Institute	Location	Elements			ν[MHz]	Beam	Type of Array (s. Fig. 2)	Spacing [m]		HPBW	t_i[h]	Ref.
			D[m]	n_{tot}	n_{mov}				max	min			
1	National Radio Astronomy Observatory, Charlottesville, W. Va., USA	NRAO Very Large Array Socorro, New Mexico 107°37′04″ W 34°04′43″ N	25	27	27	1420 1700 5000 15000 24010	36′ 30′ 9′ 3′.7	120° "Y" with legs of 21; 21 and 18.9 km	21000	36	2″ 0″.5	6	16
2	Netherlands Foundation for Radio Astronomy, Westerbork, Post Hooghalen	Westerbork Radio Observatory Hooghalen 6°36′.25 E 52°55′.0 N	25 equatorial paraboloids	14	2	610 1415 4995	83′ 36′ 11′	east-west baseline	1600	36	56″ 24″ 6″.8	12	2, 3, 7
3	Mullard Radio Astronomy Observatory, Cambridge, England	Mullard Radio Astron. Observatory Cambridge 0°02′.4 E 52°10′.0 N	13	8	4	2695 5000 15375	6′	cross	4560	40	2″	192	31
4	Owens Valley Radio Observatory, Big Pine, Calif. 93513, USA	Big Pine, Calif. 118°17′.6 W 37°13′.9 N	27 39.6	2 1	2	1420	33′	cross	1080	30.5	7″		
5	Radio Astron. Institute, Stanford University, Stanford, Calif., USA	Stanford Calif. 122°11′.3 W 37°23′.9 N	18	5	0	10690	7′	east-west baseline	206	23	19″	10	6
6	School of Electrical Eng., University of Sydney, Sydney, N.S.W., Australia	Radio Astron. Observatory Fleurs, N.S.W. 150°46′ E 33°51′ S	5.7 14.0	64 4	0	1415	3°	cross	800	12.2	40″	12	9, 10

Fixed spherical or parabolic reflectors (see LB VI/1 p. 31)

Only the stationary 305 m telescope of the Arecibo Observatory in Puerto Rico should be given special recognition alongside the large steerable parabolic reflectors of Table 2 and the synthesis telescopes of Table 3. The recently refurbished reflecting surface at Arecibo is now operable down to a wavelength $\lambda = 6$ cm ($\triangleq$ 5000 MHz). The telescope is primarily used for pulsar observations, line spectroscopy and active radar investigations.

Fixed radio telescopes with fan beams (see LB VI/1 p. 32)

A large telescope of this type was brought into operation at Zelenchukskaya (Stavropolsky District) in the Soviet Union. It is named RATAN-600 and consists of individual flat (or slightly cylindrically curved) reflecting panels arranged in a circle 576 m in diameter. For observing purposes the panels of one segment (about 1/4 of the total circle) are adjusted in such a way as to form a paraboloid whose focus is within the circle at about $1/2\,R$. The reflector can be used over a wavelength range of 20 to 0.8 cm. At 8 mm wavelength the fanshaped main beam has a half-power width of $\approx 5''$. By observations in various azimuth directions it is possible to derive a two-dimensional image from the different linear brightness distributions of a source [8, 28]. Panels $\approx 2 \times 7.4\ \mathrm{m}^2$.

References for 1.9.2.2: see 1.9.2.3.

1.9.2.3 Radio telescopes for mm wavelengths

Telescopes of the mm wavelength range must have a high precision reflecting surface. The root-mean-square deviation of the reflector from an ideal paraboloid σ_{rms} is related to the minimum observable wavelength $\lambda_{\min}$ by the rule-of-thumb formula:

$$\lambda_{\min} = 15\,\sigma_{\mathrm{rms}}\,.$$

If a telescope is to be usable at $\lambda = 1$ mm, the surface must be constructed to a precision of $\sigma_{\mathrm{rms}} \cong 0.07$ mm. This high accuracy must be maintained taking into consideration the gravitational distortion of the telescope support structure and reflector elements, as well as the thermal expansion and the fabrication and adjustment errors of the surface panels. The thermal deformation of the reflecting surface, a result of solar radiative heating during daytime and ground radiation at night, is particularly difficult to control. One is practically required to either protect the telescope with a radome or to wrap the support structure with an insulating cover in order to keep it at a constant temperature [14].

Transparency of the earth's atmosphere

Strong absorption lines of H_2O and O_2 can occasionally disrupt the transparency of the terrestrial atmosphere at wavelengths $\lambda < 13$ cm. The off-line radiation can also be strongly attenuated by the wings of these absorption lines. The atmospheric absorption loss by this process is primarily dependent upon the columnar water vapor content along the ray path above the observing station. The zenith transmission for 1, 2, 4, and 8 mm of precipitated water is displayed in Fig. 3. This atmospheric absorption restricts the site selection for a millimeter telescope to dry regions at very high elevation. Table 4 is a list of the more important millimeter telescopes.

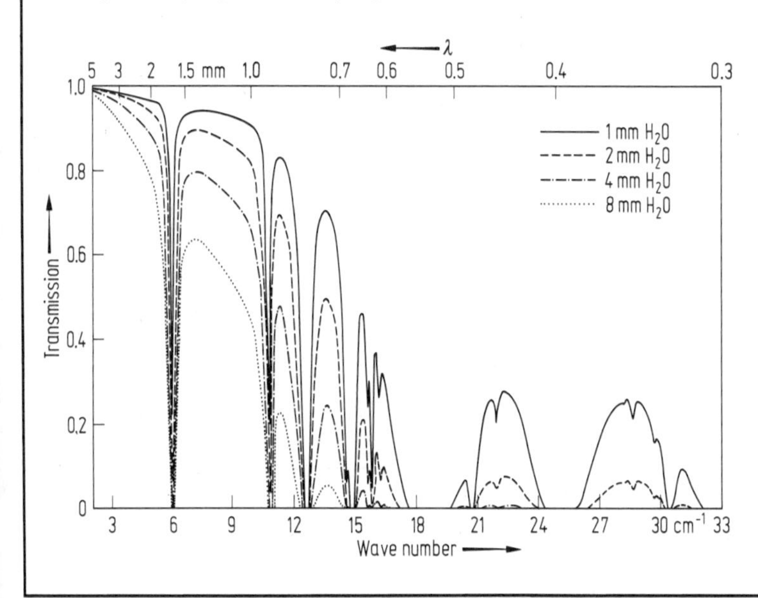

Fig. 3. Zenith transmission for 1, 2, 4, and 8 mm of precipitated H_2O [1].

Table 4. Radio telescopes for mm wavelengths. (No. 11···14 at present under construction).
All telescopes are mounted azimuthally.
σ, λ_{min}: see text
Protection:
o.a.: openair
rad.: radome
a.d.: astro-dome (with slit)

No.	Operating Institute	Location	Altitude m	D m	σ mm	λ_{min} mm	Protection	Programs	Ref.
1	Crimean Astrophys. Observatory, USSR	Simeis Crimea 34°01′.0 E 44°34′.7 N	550	22	0.25	2	o.a.	sun, planets, continuum, line spectroscopy, VLBI	23, 18
1a	the same type of telescope at Lebedev Physical Institute, Moscow, USSR			22	0.25	3	o.a.	sun, planets, continuum, line spectroscopy	21, 22
2	Onsala Space Observatory, 34 Onsala, Sweden	Onsala, Sweden 11°55′.2 E 57°23′.6 N	12	20	0.18	2.9	rad.	galactic continuum, molecular lines	29
3	Centro de Radio Astronomia e Astrophysica, Universidade Mackenzie, Sao Paulo, Brazil	Atibaia, S.P. 46°33′.8 W 23°11′.1 S	650	13.7	0.3		rad.	galactic continuum and lines, sun	
4	Helsinki University of Technology, Radio Laboratory, Otaniemi, Finland	Metsähovi 24°23′47″ E 60°13′04″ N	60	13.7	0.25	3	rad.	galactic continuum and lines	33
5	Observatorio Astronomica Nacional, Alfonso XII 3, Madrid 7, Spain	Yebes 3° 5′21″.6 W 40°31′24″.6 N	930	13.7	0.2	3	rad.	molecular lines, continuum	
6	Hasbrouck Laboratory, University of Massachusetts, Amherst, Mass., USA	Quabbin Reservation New Salem Mass. 72°20′40″.4 W 42°23′33″.2 N	310	13.7	0.15	2.4	rad.	molecular lines, continuum	
7	National Radio Astronomy Observatory, Charlottesville, W. Va., USA	Tucson, Arizona 111°36′50″ W 31°57′11″ N	1930	11	0.14	2.2	a.d.	molecular lines, continuum, extragalactic sources	17

8	California Institute of Technology, USA	Big Pine Cal. 118°17′.6 W 37°13′.9 N	1216	10.4	0.035	0.5	o.a.	molecular lines, continuum	25
8a	A second telescope of the same type							interferometer with No. 8	
9	Bell Laboratories, Crawford Hill, Holmdel New Jersey, USA	Crawford Hill Lab. 74°11′15″ W 40°23′31″ N	114	7	0.1	1.6		molecular lines	
10	Astronomy Department, University of California, Berkeley Calif. 94720, USA	Hat Creek Radio Astron. Station 121°28′.4 W 40°49′.0 N	1012	6.1 two telescopes	0.13	2.0	o.a.	two telescopes interferometer T-baseline, 302 m E–W 152 m N–S	35
11	SRC (Science Research Council) Japan	Nobeyama Nagano, Japan 138° E 36° N	1400	45	0.3	3	o.a.	continuum, lines, works as single dish telescope, as well as aperture synthesis with 11a	
11a	SRC (Science Research Council) Japan	same place	1400	10	0.3	3	o.a.	five 10 m telescopes movable along EW and NS base lines each 600 m long	
12	IRAM (Institut de Radio Astronomie Millimetrique), French-German cooperation, 53 Avenue des Martyrs, F-38026 Grenoble, France	Lomo de Dilar, Granada, Spain, – 3°.20 W 37°.03 N	2950	30	0.09	1.4	o.a.	begin of operation 1983; molecular lines, continuum, extragalactic sources, sun	
13	National Radio Astronomy Observatory, Charlottes-ville, W.Va., USA	Mauna Kea Hawaii, USA 155°.4 W 19°.9 N	4080	25	0.07	1.1	a.d.*)		
14	SRC (Science Research Council) England	Mouchachos La Palma, Spain 17°.45 W 28°.41 N	2400	15	0.05	0.8	a.d.	begin of operation 1985	

*) With "transparent" door.

References for 1.9.2.2 and 1.9.2.3

General references

a Christiansen, W. N., Högbom, J.A.: Radiotelescopes, Cambridge University Press (1969).
b List of Radio and Radar Astronomy Observatories, National Academy of Sciences, Washington, D.C. 20418 (1979).
c Structures Technology for Large Radio and Radar Telescope Systems (Mar, J.M., Liebowitz, H., eds.), The MIT Press (1969).

Special references

1 Arnold, E.M.: Thesis Universität Bonn (1979).
2 Baars, J.W.M., Van der Brugge, J.F., Casse, J.L., Hamaker, J.P., Sondaar, L.H., Visser, J.J., Wellington, K.J.: Proc. Inst. Electr. Electron. Eng. **61** (1973) 1258.
3 Baars, J.W.M., Hooghoudt, B.G.: Astron. Astrophys. **31** (1974) 323.
4 Bowen, E.G., Minett, H.C.: Proc. Inst. Radio Eng. Australia **24** (1963) 98.
5 Bracewell, R.N.: Australian J. Phys. **9** (1956) 198.
6 Bracewell, R.N., Colvin, R.S., D'Addario, L.R., Grebenkemper, C.J., Price, K.M., Thompson, A.R.: Proc. Inst. Electr. Electron. Eng. **61** (1973) 1249.
7 Casse, J.L., Muller, C.A.: Astron. Astrophys. **31** (1974) 333.
8 Chaikin, S.E., Kajdanowskij, N.L., Parijskij, Ju.N., Esepkina, N.A.: Rep. Pulkovo Obs. No. 188 (1972).
9 Christiansen, W.N.: Proc. Inst. Electr. Electron. Eng. **61** (1973) 1266.
10 Christiansen, W.N. et al.: Proc. Inst. Radio Eng. Australia **34** (1973) No. 8 (included 10 different papers on the Fleurs Synth. Telescope).
11 Fomalont, E.B.: Proc. Inst. Electr. Electron. Eng. **61** (1973) 1211.
12 Hachenberg, O.: Sky Telesc. **40** (1970) 338.
13 Hachenberg, O., Grahl, B.H., Wielebinski, R.: Proc. Inst. Electr. Electron. Eng. **61** (1973) 1288.
14 Hachenberg, O.: Techn. Report of the MPIfR, Bonn, Nr. 3 (1973).
15 Hills, R.E., Janssen, M.A., Thornton, D.D., Welch, W.J.: Proc. Inst. Electr. Electron. Eng. **61** (1973) 1278.
16 Hjellming, R.M.: An introduction to the NRAO Very Large Array, NRAO Socorro, New Mexico (1978).
17 Hvatum, H.: Inst. Electr. Electron. Eng. Trans. Antennas Propag. **18** (1970) 523.
18 Ivanov, V.N., Moiseev, I.G., Monin, Y.G.: Ivz. Krymskoj Astrofiz. Obs. **38** (1967) 141.
19 Jeffery, M.H.: Ann. N.Y. Acad. Sci. **116** (1964) 62.
20 Jeffery, M.H. in: [c] p. 219.
21 Kalachev, P.D.: Proc. Lebedev Phys. Inst. **28** (1965) 183.
22 Kalachev, P.D. in: Radio Telescopes (ed. Skobeltsyn Consultant Bureau) Plenum Pub. Corpor. New York (1966) 35, 143.
23 Kalachev, P.D., Salomonovich, A.E., Moiseev, I.G.: Inst. Electr. Electron. Eng. Trans. Antennas Propag. **18** (1970) 516.
24 Katow, M.S. in: [c] p. 185.
25 Leighton, R.B.: California Institute of Technology, Technical Report (1978).
26 Lovell, A.C.B.: Nature **180** (1957) 60.
27 Minett, H.C., Yabsley, D.E., Puttock, M.J. in: [c] p. 135.
28 Parijskij, Yu.N., Schivris, O.N.: Rep. Pulkovo Obs. No. 188 (1972).
29 Rydbeck, O.E.H.: Kosmos **52** (1975) (swedish), edited by Svenska Fysikersamfundet.
30 Ryle, M.: Nature **194** (1962) 517.
31 Ryle, M.: Nature **239** (1972) 435.
32 Stuart, D.G.: Mass. Inst. Technol. Technical Note 7 (1968).
33 Urpo, S.: Helsinki Univ. of Technology Radio Lab. Report S 73 (1975).
34 Weiss, H.G., Fanning, W.R., Folino, F.A., Muldoon, R.A. in: [c] p. 151.
35 Welch, W.J.: Inst. Electr. Electron. Eng. Trans. Antennas Propag. **18** (1970) 526.

1.9.2.4 Very-long-baseline interferometer

The interferometers described in LB VI/1 (1965) p. 33 could be expanded in the late 1960's to baselines (distance between the observing telescopes) of 120 km. The coherence of the oscillator frequency at the two telescopes and the transmission of the intermediate frequency to the central receiver were maintained with a station-to-station radio link.

The desire to measure the sizes of the emission regions of quasars and the diameters of galactic nuclei was the driving force behind the eventual expansions of the baseline to several thousand kilometers. For tandem observations over these intercontinental baselines, however, other methods had to be developed for producing oscillator coherence at the two telescopes [2].

This coherence was achieved, for example, either with quartz oscillators at both telescopes that are tuned to a constant frequency with rubidium clocks, or by using hydrogen masers to derive the oscillator frequency. The coherence of the oscillators could thus be guaranteed for the required observational intervals of several hours [1, 6].

The transmission of the intermediate frequency (IF) of both telescopes to a central correlation receiver was accomplished by first making a digital recording of the IF signal on a video magnetic tape. The data on these two tapes then serve as input for a specialized correlator (e.g. the Mark II system of NRAO) [4], which derives the correlation function and interference frings. This Very Long Baseline Interferometry (VLBI) technique can attain an angular resolution of 0.″1···0.″0001, depending on the observing wavelength. Review articles [6, 8].

VLBI therefore enables a very precise determination of the position of extraterrestrial radio sources, an attribute with many astronomical and geophysical applications [7, 3]. Furthermore, it is also possible in many cases to determine the diameter of the source, thus providing a first rough idea of the intensity distribution of an emitting region.

The distribution of brightness of a celestial object may be more fully determined by extending the observation time to several hours and using the daily rotation of the baseline with respect to the sky in a manner similar to aperture synthesis. A further step is to observe simultaneously with not just two, but an entire array of widely separated telescopes. Although it is still not possible in this case to derive the brightness distribution directly from a Fourier transform, one can compose an adequate map of the true distribution of intensity using model calculations or iteration processes.

The telescopes that have participated in VLBI observations up to 1979 are listed in Table 5. Correlators of the Mark II type are located at National Radio Astronomy Observatory, Charlottesville, USA; Owens Valley Radio Observatory, Big Pine, California, USA; Massachusetts Institute of Technology, Cambridge, USA; Max-Planck-Institut für Radioastronomie, Bonn, Germany.

Table 5. Telescopes that have participated in VLBI observations.

No.	Location operating institute	D m	λ_{min} cm	Longitude	Latitude
1	Tidbinbilla, Austral. NASA	64	3.8	−148.°980	−35.°404
2	Simeis, Krim, USSR Crimean Astrophys, Observ.	22	1.3	− 34.017	+44.728
3	Hartebeesthoek, South Africa CSIR, National Inst. for Telecommunications Research	26	3.8	− 27.685	−25.887
4	Onsala, Sweden Onsala Space Observatory	26	6	− 11.920	+57.393
5	Effelsberg, Germany MPI for Radioastronomy	100	1.3	− 6.884	+50.525
6	Westerbork, Netherlands	(93)	6	− 6.604	+52.917
7	Dwingeloo, Netherlands	25	18	− 6.397	+52.813
8	Cambridge, England Mullard Radio Astronomy Observatory	(32)	2	− 0.040	+52.163

continued

Table 5, continued

No.	Location operating institute	D m	λ_{min} cm	Longitude	Latitude
9	Chilbolton, England Appleton Laboratory	25	1.3	+ 1.437	+51.145
10	Jodrell Bank, England Nuffield Radio Astronomy Laboratories	76	18	+ 2.307	+53.237
11	Robledo, Spain INTA-NASA	64	2	+ 4.247	+40.434
12	Arecibo, Puerto Rico Cornell University, USA	300	13	+ 66.753	+18.344
13	Haystack, Mass., USA Haystack Observatory NEROC	37	1.3	+ 71.489	+42.623
14	Maryland Point, Maryland USA, NRL	26	1.3	+ 77.231	+38.374
15	Algonquin Park, Canada Algonquin Radio Observatory, NRCC	46	2.8	+ 78.073	+45.956
16	Green Bank, W. Va., USA NRAO	43	2.8	+ 79.836	+38.438
17	Vermilion River, Illinois University of Illinois, USA	37	13	+ 87.557	+40.065
18	Fort Davis, Texas, USA Harvard Radio Astronomy Station	26	2.8	+103.947	+30.636
19	Goldstone, California, USA NASA/JPL	64	2	+116.889	+35.426
20	Big Pine Owens Valley Radio Observatory/ CALTECH	40	2.8	+118.282	+37.232
21	Hat Creek, California, USA	26	3.8	+121.473	+40.628

References for 1.9.2.4

1 Bare, C. et al.: Science **157** (1967) 189.
2 Broten, N.W. et al.: Science **156** (1967) 1592.
3 Broten, N.W. et al.: Mon. Not. R. Astron. Soc. **146** (1969) 313.
4 Clark, B.G.: Proc. Inst. Electr. Electron. Eng. **61** (1973) 1242.
5 Cohen, M.H.: Annu. Rev. Astron. Astrophys. **7** (1969) 619.
6 Cohen, M.H.: Proc. Inst. Electr. Electron. Eng. **61** (1973) 1192.
7 Counselman III, C.C.: Proc. Inst. Electr. Electron. Eng. **61** (1973) 1225.
8 Moran, J.M. in: Methods of Experimental Physics (Meeks, M.L., ed.) **12** C (1976) 174, 228.
9 Preuss, E., Kellermann, K.I., Pauliny-Toth, I.I.K., Witzel, A., Shaffer, D.B.: Astron. Astrophys. **79** (1979) 268.
10 Readhead, A.C.S., Wilkinson, P.N.: Astrophys. J. **223** (1978) 25.
11 Shapiro, I.I. in: Methods of Experimental Physics (Meeks, M.L., ed.) **12** C (1976) 261.

2 Positions and time determination, astronomical constants

2.1 Determination of astronomical latitude and longitude

2.1.1 Introduction

Textbooks [a, b, c, d]

Positions on the earth may be designated by giving the astronomical latitude and longitude. These coordinates express the direction of the astronomical vertical (direction of local gravity) with respect to a terrestrial reference system. Astronomical and geophysical observations are referred to this system; geodetic reference systems are attached to it.

Due to the existence of gravity anomalies, astronomical latitudes and longitudes cannot be used directly to furnish the geometric position of a station with respect to the earth's centre of mass. The angle between the astronomical vertical at a station, S, and the normal at S to an earth-ellipsoid is up to about 10″ in the plains and hills [10]; in the mountains it can be as much as 20″···50″ in exceptional cases [10, 11, 12] with maximum gradients >10″/km [10].

2.1.2 Polar motion

The earth's axis of rotation, R, continually changes direction slightly within the earth. The term polar motion, according to present use, refers to the motion of the point where, near the "north pole", R intersects the surface of the earth (pole of rotation), with respect to a point quasi-fixed on the earth (origin of the coordinates of the pole).

Polar motion is almost entirely geophysical in origin [13, 14, 52] and must be determined by observation. An externally-forced nearly diurnal motion of R in space with $0 \cdots < 0\rlap{.}''02$ radius [15] will be included in the computation of nutation from 1984 onwards [16]; then that diurnal motion will no longer be assumed to be part of polar motion.

2.1.3 Definition and observation of astronomical latitude and longitude

The instantaneous latitude φ_S of a station S is the acute angle between the direction of the astronomical vertical at S and any plane perpendicular to R. Latitudes are positive north.

If M_S is that plane which contains the astronomical vertical at S and is parallel to R, and L the corresponding plane at a designated origin of longitudes, then the angle from L to M_S, measured westward 0°···360° or $0^h \cdots 24^h$, is the instantaneous longitude λ_s of S.

Conventional latitudes and longitudes, $\varphi_{O,S}$ and $\lambda_{O,S}$ refer to that axis and those planes for which the pole is the adopted origin of the polar coordinates. The conventional origin of longitudes, O, is a fictitious point of 0° conventional latitude; it is thus not affected by polar motion. Due to the choice of O, the mean instantaneous longitude of the Greenwich transit circle, old site, is nearly zero; its conventional longitude is about $+0\rlap{.}^s02$ [2a, b; 17].

Measurements of star directions with regard to the local vertical, together with appropriate timing, provide the instantaneous latitude and/or local astronomical time.

Table 1. Instrument types applied to latitude, L, and longitude (time), T, observations at permanent (P) and field (F) stations.

Instrument type	Application	Station	Ref.
Danjon astrolabe (A)	$L+T$	P	a, b, c, 18b
Photographic zenith tube (PZT)	$L+T$	P	a, b, 18a
Floating zenith telescope	L	P	a, 19
Zenith telescope (ZT)	L	P, F	a, 13
Circumzenithal	$L+T$	P, F	a
Visual transit instrument (TIv)	T, L	P, F	a, b, c
Photoelectrical transit instrument (TIp)	T	P	20, 21
Portable astrolabe	$L+T$	F	a, b, c
Geodetic theodolite	L, T $L+T$	F	a, b, c
Photographic zenith camera	$L+T$	F	22

The instantaneous longitude is obtained from the local astronomical time by comparison with astronomical time observed nominally at O (2.1.4.2). The coordinates observed directly, φ_S and λ_S, are related to $\varphi_{O,S}$ and $\lambda_{O,S}$ through

$$\varphi_{O,S}-\varphi_S=-(x\cos\lambda_S+y\sin\lambda_S),$$
$$\lambda_{O,S}-\lambda_S=-(x\sin\lambda_S-y\cos\lambda_S)\tan\varphi_S.$$

x, y = coordinates of the pole, generally expressed as an angular displacement (0.″01 equivalent to 0.31 m in linear displacement); $+x$ directed to O, $+y$ to 90° W. Neglecting geophysical effects, observational and other errors, changes of systems and constants, the values of $\varphi_{O,S}$ and $\lambda_{O,S}$ are time independent.

2.1.4 Definition and realization of the terrestrial reference system

2.1.4.1 Origin of the coordinates of the pole

By international agreement, the coordinates of the pole are referred to the Conventional International Origin (CIO) defined by adopted values of the latitudes of the present five stations of the International Latitude Service (ILS), which was established in 1899.

Table 2. Station latitudes defining the CIO [23, 24].

Station	Longitude	Latitude
Mizusawa	−141°08′	39°08′03.″602
Kitab	− 66 53	01.850
Carloforte	− 8 19	08.941
Gaithersburg	+ 77 12	13.202
Ukiah	+123 13	12.096

The regular latitude observations taken at the stations with zenith telescopes are evaluated by a central bureau to obtain current values of x and y. Since 1962, the bureau has also determined polar motion from all adequate data available, this under the designation International Polar Motion Service (IPMS) [1]).

x and y from IPMS are nominally on CIO through the adoption of initial coordinates of the collaborating stations (50···60 instruments, generally identical with those enumerated in Table 3; but only the present ILS is strictly on CIO): ILS and IPMS results are published monthly [3] and, with more details, annually [4]. For revised ILS polar coordinates 1900.0···1978.9, see 2.1.7.

2.1.4.2 Origin of longitudes

The origin of longitudes, O, and the adhering pole, empirically attached to CIO, are defined by the values of φ_O and/or λ_O and weights of a number of instruments, as determined by the Bureau International de l'Heure (BIH) [2]) in the "1968 BIH System" [2].

Table 3. Instruments defining the 1968 BIH system.

Type instrument type: see Table 1
Weight adopted weights of one instrument
L = latitude
T = time (longitude)
N = number of instruments

Type	Weight L	Weight T
A	4···100	4···49
PZT	25···100	25···49
ZT	1···100	
TIv		1··· 9
TIp		1···49

Geographical distribution continent	N
Africa	2
America N	5
America S	5
Australia	1
Eurasia	51
Japan	4
	68

[1]) IPMS, International Latitude Observatory, Mizusawa-shi, 023 Japan.
[2]) BIH, 61, avenue de l'Observatoire, 75014 Paris.

The evaluation of observed data results in astronomical time on O and the operational coordinates of the pole. In practice, this time, denoted UT1 (2.2.2.2), is referred to UTC (2.2.4.3) which is available worldwide through time signals. Values of UT1 – UTC, x, y are published monthly [1] and annually [2]; [2] also contains basic data of computations, and, from the 1978 Volume onwards, results from new techniques which operate independently of the vertical, such as laser ranging to the moon and artificial satellites, and radio interferometry [25, 26, 27, 52, 53].

Since 1972, the BIH solution includes results from Doppler satellite measurements [2c] taken at about 20 tracking stations [28, 29]. The apparent diurnal variations of their geocentric coordinates as deduced from these measurements, provide current values of x and y (BIH weights: 400 each).

Corrections to relate results in the 1968 System to the improved 1979 BIH System [1a, 2k]:

$$C_{\mathrm{UT1}} = 0.7 \sin 2\pi (t-0.477) + 0.7 \sin 4\pi (t-0.397) \text{ in [ms]},$$
$$C_x = 0.024 \sin 2\pi (t-0.158) + 0.007 \sin 4\pi (t-0.289) \text{ in [arc sec]},$$
$$C_y = 0,$$

with t in fractions of the Besselian year (2.2.3.2).

ILS, IPMS, BIH, and Doppler [7] polar coordinates deviate [29, 30] to some extent from each other and may be chosen according to the application for which they are required.

Standard errors of BIH 5-day values: $\sigma_x = 0''.007$, $\sigma_y = 0''.007$, $\sigma_{\mathrm{UT1}} = 0.8$ ms [26]. Estimated systematic errors of results in the 1979 BIH System: 0''.01, 1 ms [1a].

The International Astronomical Union Colloquium No. 56 held in September 1980 recommended that a proposal should be prepared "for the establishment and maintenance of a Conventional Terrestrial Reference System ... for the replacement of the presently used terrestrial reference system ... providing continuity".

2.1.5 Errors in latitude and longitude or time

Table 4. Estimated errors associated with various instrument types.

Type [1])	Standard deviation for one night [2])		Long-term stability, standard deviation [3])	
	latitude	longitude [4])	latitude	longitude [4])
A	0''.08	0ˢ.007	0''.04	0ˢ.003
PZT	0.06	0.005	0.04	0.003
ZT	0.10		0.03	
TIv		0.012		0.006
TIp		0.009		0.004

[1]) See Table 1.
[2]) From differences between night results and a curve fitted to a series of observations (computed from data given in [3, 31]).
[3]) Standard deviation of annual average values of φ_O (λ_O) in the BIH System versus 10-year means (computed from data given in [2]).
[4]) Referred to equator.

The values given are considered to be representative; errors of individual instruments may vary (see Table 3).

2.1.6 Coordinates of observatories

Precise conventional coordinates of time and/or latitude stations [2, 4a]. Coordinates of astronomical observatories [5, 32]; [5] contains separate lists from optical and radio observatories.

2.1.7 Polar coordinates

Table 5. Revised ILS polar coordinates 1900.0···1978.9 in a uniform system [35]. Units: 0″001.

	1900		1901		1902		1903		1904		1905		
	x	*y*	*x*	*y*	*x*	*y*	*x*	*y*	*x*	*y*	*x*	*y*	
.0	+ 8	+ 18	− 18	+ 20	−114	− 33	−147	− 72	− 49	−157	+ 86	−125	.0
.1	+ 57	− 44	+ 20	+ 29	− 95	+ 53	−187	+ 4	−152	−103	− 26	−170	.1
.2	+ 31	− 66	+ 52	+ 34	− 51	+127	−153	+ 92	−189	− 28	−121	−139	.2
.3	− 11	− 85	+ 85	+ 9	+ 33	+171	− 77	+170	−177	+ 58	−154	− 68	.3
.4	− 38	−106	+104	− 56	+127	+159	+ 24	+215	−104	+160	−146	+ 10	.4
.5	− 36	−104	+125	−114	+203	+ 85	+111	+213	− 4	+191	− 94	+110	.5
.6	− 54	− 73	+ 85	−160	+204	− 8	+186	+146	+ 85	+164	− 8	+163	.6
.7	− 72	− 29	+ 29	−151	+140	− 90	+208	+ 25	+147	+104	+ 76	+160	.7
.8	− 76	+ 14	− 35	−126	+ 44	−144	+165	−101	+165	+ 32	+121	+114	.8
.9	− 65	+ 27	− 95	− 97	− 59	−129	+ 78	−169	+144	− 41	+139	+ 36	.9

	1906		1907		1908		1909		1910		1911		
	x	*y*	*x*	*y*	*x*	*y*	*x*	*y*	*x*	*y*	*x*	*y*	
.0	+ 98	− 36	+ 37	+111	− 73	+169	−272	− 11	−201	−251	+ 19	−290	.0
.1	+ 16	− 72	+ 52	+ 46	+ 6	+193	−265	+123	−269	−112	−135	−252	.1
.2	− 37	− 75	+ 57	− 3	+101	+176	−178	+240	−308	+ 59	−220	−134	.2
.3	− 88	− 53	+ 63	− 43	+184	+107	− 3	+297	−248	+227	−249	+ 30	.3
.4	−116	− 2	+ 18	− 80	+222	− 2	+178	+271	− 85	+315	−192	+191	.4
.5	−108	+ 61	− 7	−101	+219	− 91	+296	+164	+108	+339	− 92	+317	.5
.6	− 77	+ 96	− 31	− 83	+147	−167	+324	+ 23	+257	+269	+ 56	+342	.6
.7	− 35	+107	− 95	− 43	+ 28	−199	+265	−130	+320	+ 93	+181	+284	.7
.8	− 6	+105	−154	+ 19	−115	−191	+115	−249	+289	− 89	+257	+159	.8
.9	+ 15	+116	−130	+101	−225	−128	− 75	−293	+182	−220	+275	+ 7	.9

	1912		1913		1914		1915		1916		1917		
	x	*y*	*x*	*y*	*x*	*y*	*x*	*y*	*x*	*y*	*x*	*y*	
.0	+211	− 98	+104	+123	− 87	+129	−194	+ 3	−123	−192	+ 65	−244	.0
.1	+104	−168	+122	+ 69	− 50	+145	−193	+ 91	−190	−115	− 53	−234	.1
.2	− 16	−173	+126	+ 12	+ 64	+148	−134	+214	−205	+ 55	−137	−134	.2
.3	−107	−114	+124	− 43	+162	+131	− 17	+296	−156	+181	−145	+ 2	.3
.4	−153	− 31	+118	− 64	+183	+ 93	+104	+262	− 27	+279	− 86	+133	.4
.5	−136	+ 60	+ 73	− 56	+189	+ 19	+220	+194	+ 98	+286	− 18	+238	.5
.6	−100	+149	− 2	− 79	+149	− 61	+276	+ 93	+223	+172	+ 82	+233	.6
.7	− 61	+177	− 67	− 68	+ 65	−137	+257	− 52	+305	+ 32	+151	+173	.7
.8	− 10	+195	−105	− 5	− 44	−171	+159	−206	+291	− 81	+161	+ 75	.8
.9	+ 69	+186	− 91	+ 71	−145	− 92	+ 23	−234	+198	−196	+159	− 8	.9

continued

Table 5, continued

	1918		1919		1920		1921		1922		1923		
	x	y	x	y	x	y	x	y	x	y	x	y	
.0	+122	-103	+98	+94	-17	+139	-82	-53	-19	-27	-9	-85	.0
.1	+18	-112	+73	+55	+28	+171	-105	+47	-106	+46	-84	-36	.1
.2	-47	-26	+61	+63	+51	+172	-57	+90	-97	+124	-121	+49	.2
.3	-68	+44	+74	+27	+132	+151	+45	+125	-26	+199	-86	+161	.3
.4	-85	+120	+74	+20	+194	+68	+155	+134	+82	+214	-23	+244	.4
.5	-67	+170	+56	-21	+208	-7	+220	+53	+177	+143	+73	+247	.5
.6	+10	+155	+19	-35	+191	-83	+245	-2	+249	+49	+188	+234	.6
.7	+64	+132	-32	+14	+141	-129	+206	-75	+259	-12	+248	+174	.7
.8	+83	+95	-63	+42	+63	-141	+126	-132	+212	-79	+239	+97	.8
.9	+96	+95	-63	+62	-11	-137	+60	-98	+117	-81	+184	+30	.9

	1924		1925		1926		1927		1928		1929		
	x	y	x	y	x	y	x	y	x	y	x	y	
.0	+81	-12	+96	+21	+42	-6	+57	+80	-6	+26	-27	+118	.0
.1	-6	-13	+26	-38	+7	-43	+68	+71	-34	+74	-36	+125	.1
.2	-57	+59	-13	-27	-47	+11	+58	+51	-29	+103	-14	+152	.2
.3	-53	+113	-8	+56	-40	+88	+61	+71	+4	+113	+63	+189	.3
.4	-40	+156	+27	+90	-26	+116	+73	+107	+29	+151	+118	+197	.4
.5	+3	+182	+33	+90	+11	+145	+84	+114	+45	+113	+155	+136	.5
.6	+92	+195	+74	+88	+61	+159	+94	+80	+73	+90	+162	+53	.6
.7	+145	+167	+105	+57	+100	+138	+88	+37	+49	+63	+121	-4	.7
.8	+143	+121	+113	+47	+88	+103	+61	+43	-6	+70	+33	-19	.8
.9	+130	+88	+84	+23	+68	+117	+36	+18	-19	+101	-33	-29	.9

	1930		1931		1932		1933		1934		1935		
	x	y	x	y	x	y	x	y	x	y	x	y	
.0	-80	+19	-91	+8	-42	-35	+48	+31	+34	+39	+39	+115	.0
.1	-116	+124	-121	+110	-140	+28	-59	+16	-30	+48	+12	+97	.1
.2	-82	+192	-114	+173	-151	+125	-99	+65	-82	+84	+26	+104	.2
.3	-10	+202	-64	+246	-85	+226	-100	+134	-109	+114	+32	+121	.3
.4	+74	+236	+28	+280	-62	+254	-74	+196	-84	+154	+36	+141	.4
.5	+157	+210	+155	+260	+18	+251	-18	+223	-18	+215	+56	+129	.5
.6	+207	+156	+241	+205	+118	+234	+58	+201	+70	+215	+66	+113	.6
.7	+180	+61	+252	+131	+183	+174	+100	+166	+96	+202	+62	+140	.7
.8	+107	+21	+142	+13	+183	+144	+98	+130	+83	+173	+26	+120	.8
.9	+5	+20	+44	-29	+143	+79	+81	+77	+59	+125	+2	+112	.9

	1936		1937		1938		1939		1940		1941		
	x	y	x	y	x	y	x	y	x	y	x	y	
.0	-9	+128	-47	+108	-78	+27	-100	-3	+43	+6	+91	+25	.0
.1	-37	+169	-109	+157	-123	+95	-62	+63	-58	+27	+37	+38	.1
.2	-42	+181	-83	+200	-97	+190	-61	+159	-93	+86	-33	+88	.2
.3	+4	+195	-6	+223	-40	+251	-17	+209	-91	+180	-42	+122	.3
.4	+64	+197	+68	+246	+2	+269	+5	+247	-59	+247	0	+136	.4
.5	+122	+175	+132	+213	+116	+256	+62	+255	+39	+267	+18	+190	.5
.6	+108	+116	+155	+136	+241	+209	+163	+222	+133	+234	+78	+216	.6
.7	+63	+53	+121	+71	+208	+136	+200	+143	+181	+185	+115	+194	.7
.8	+19	+30	+94	+52	+90	+40	+177	+87	+165	+103	+126	+165	.8
.9	-16	+41	+41	+43	-28	-7	+130	+43	+120	+42	+117	+113	.9

continued

Table 5, continued

	1942		1943		1944		1945		1946		1947		
	x	y	x	y	x	y	x	y	x	y	x	y	
.0	+86	+66	+29	+211	−59	+156	−101	+40	−58	−105	+186	−72	.0
.1	+28	+24	+52	+205	−33	+198	−150	+117	−187	−53	+33	−97	.1
.2	+40	+46	+81	+184	−8	+227	−129	+212	−227	+53	−75	−52	.2
.3	+62	+101	+111	+143	+81	+229	−56	+298	−166	+196	−148	+74	.3
.4	+44	+107	+160	+128	+212	+209	+80	+322	−58	+302	−149	+180	.4
.5	+32	+118	+168	+69	+281	+139	+217	+278	+53	+366	−63	+263	.5
.6	+23	+114	+143	+33	+292	+32	+290	+161	+179	+358	+70	+321	.6
.7	+20	+143	+74	+26	+223	−59	+312	+34	+293	+253	+175	+322	.7
.8	+1	+166	−29	+38	+35	−89	+265	−85	+346	+116	+259	+282	.8
.9	+13	+184	−83	+97	−56	−42	+100	−142	+299	0	+291	+229	.9

	1948		1949		1950		1951		1952		1953		
	x	y	x	y	x	y	x	y	x	y	x	y	
.0	+273	+159	+116	+325	−125	+279	−229	+54	−91	−166	+143	−194	.0
.1	+240	+64	+218	+284	−44	+361	−266	+235	−243	−3	−22	−182	.1
.2	+209	−21	+278	+214	+101	+395	−169	+395	−299	+206	−188	−88	.2
.3	+124	−49	+282	+117	+248	+332	+13	+496	−187	+378	−253	+72	.3
.4	+11	−12	+261	+24	+346	+235	+200	+451	−53	+492	−251	+254	.4
.5	−38	+28	+220	−35	+391	+103	+344	+331	+140	+510	−117	+400	.5
.6	−44	+88	+102	−76	+360	−27	+454	+147	+326	+445	+64	+461	.6
.7	−51	+170	−31	−87	+221	−118	+425	−40	+426	+290	+238	+436	.7
.8	−8	+244	−116	+4	+21	−181	+279	−182	+411	+78	+308	+272	.8
.9	+47	+313	−146	+163	−139	−131	+105	−219	+285	−104	+301	+97	.9

	1954		1955		1956		1957		1958		1959		
	x	y	x	y	x	y	x	y	x	y	x	y	
.0	+231	+6	+146	+254	+18	+311	−229	+208	−157	+23	+78	−56	.0
.1	+118	−90	+190	+171	+120	+345	−174	+364	−232	+168	−58	−70	.1
.2	−11	−116	+214	+114	+188	+309	−73	+467	−210	+310	−136	+19	.2
.3	−95	−50	+168	+49	+241	+244	+78	+496	−97	+427	−154	+199	.3
.4	−152	+28	+108	+6	+271	+127	+219	+445	+32	+480	−84	+335	.4
.5	−106	+145	+32	−3	+237	+12	+331	+310	+188	+445	+24	+392	.5
.6	−43	+254	−45	+5	+189	−62	+382	+140	+346	+369	+161	+411	.6
.7	−4	+316	−99	+66	+91	−59	+296	−18	+386	+241	+269	+378	.7
.8	+36	+339	−116	+167	−57	+5	+139	−94	+335	+102	+306	+272	.8
.9	+107	+326	−100	+240	−188	+80	−14	−85	+232	−5	+255	+149	.9

	1960		1961		1962		1963		1964		1965		
	x	y	x	y	x	y	x	y	x	y	x	y	
.0	+138	+33	+94	+224	+1	+300	−125	+249	−171	+114	−10	+50	.0
.1	+53	−16	+59	+157	+19	+302	−107	+340	−204	+227	−135	+96	.1
.2	+7	+14	+59	+148	+65	+282	−42	+374	−162	+347	−199	+210	.2
.3	−74	+77	+42	+162	+119	+267	+70	+384	−74	+456	−210	+320	.3
.4	−104	+154	+33	+150	+163	+217	+193	+346	+69	+454	−131	+397	.4
.5	−72	+208	+39	+151	+171	+129	+282	+245	+219	+395	+5	+451	.5
.6	+8	+265	+47	+151	+137	+66	+295	+135	+260	+283	+137	+409	.6
.7	+71	+301	+48	+169	+62	+70	+196	+37	+272	+181	+214	+320	.7
.8	+91	+284	+7	+180	−17	+95	+62	−22	+234	+94	+248	+233	.8
.9	+118	+259	−30	+241	−88	+152	−59	+39	+106	+52	+202	+161	.9

continued

Table 5, continued

	1966		1967		1968		1969		1970		1971		
	x	y	x	y	x	y	x	y	x	y	x	y	
.0	+ 62	+109	+103	+237	− 51	+299	−114	+282	−133	+161	− 58	+ 23	.0
.1	+ 5	+ 96	+ 71	+194	− 10	+303	− 95	+296	−171	+253	−184	+ 95	.1
.2	− 74	+ 86	+ 27	+156	+ 35	+285	− 33	+351	−141	+359	−223	+213	.2
.3	−119	+152	+ 15	+172	+ 68	+257	+ 50	+391	− 68	+439	−136	+340	.3
.4	−127	+211	+ 4	+152	+ 62	+230	+134	+356	+ 28	+457	− 37	+448	.4
.5	−102	+289	+ 41	+161	+ 80	+201	+177	+299	+164	+394	+ 84	+489	.5
.6	− 50	+351	+ 50	+183	+100	+166	+187	+210	+251	+296	+220	+449	.6
.7	+ 63	+343	+ 13	+198	+ 17	+165	+121	+139	+244	+180	+280	+335	.7
.8	+139	+316	− 33	+214	− 44	+168	+ 44	+110	+172	+102	+220	+232	.8
.9	+136	+274	− 67	+255	−109	+239	− 60	+115	+ 64	+ 40	+160	+160	.9

	1972		1973		1974		1975		1976		1977		
	x	y	x	y	x	y	x	y	x	y	x	y	
.0	+ 75	+ 61	+207	+155	+132	+239	+ 10	+261	− 44	+202	− 63	+ 82	.0
.1	− 27	+ 21	+147	+117	+133	+172	+ 5	+283	− 65	+239	−134	+158	.1
.2	− 89	+101	+ 71	+113	+ 95	+147	+ 67	+300	− 31	+328	−145	+269	.2
.3	−119	+239	+ 6	+135	+ 56	+171	+103	+323	+ 23	+419	− 77	+399	.3
.4	− 67	+354	− 40	+218	+ 65	+172	+146	+290	+ 52	+419	+ 8	+477	.4
.5	+ 38	+402	− 37	+289	+ 67	+179	+181	+252	+125	+360	+120	+454	.5
.6	+ 90	+411	+ 9	+336	+ 76	+205	+176	+206	+179	+297	+253	+376	.6
.7	+140	+379	+ 52	+352	+ 88	+221	+131	+141	+154	+227	+286	+270	.7
.8	+208	+322	+107	+320	+ 75	+259	+ 68	+112	+168	+153	+244	+151	.8
.9	+221	+259	+113	+285	+ 40	+270	+ 3	+152	+ 72	+ 89	+126	+ 71	.9

	1978		1979 [1]		1980 [1]	
	x	y	x	y	x	y
.0	− 16	+ 26	+143	+ 90	+175	+207
.1	− 94	+ 49	+ 26	+ 82	+145	+195
.2	−124	+173	− 67	+133	+183	+152
.3	−165	+314	−107	+226	+119	+191
.4	− 95	+436	− 94	+277	+111	+244
.5	+ 31	+489	− 17	+336	+ 79	+242
.6	+189	+466	+106	+385	+ 87	+292
.7	+256	+414	+254	+424	+ 87	+350
.8	+262	+310	+309	+425	+107	+359
.9	+232	+188	+264	+311	+228	+416

[1]) Preliminary values [3].

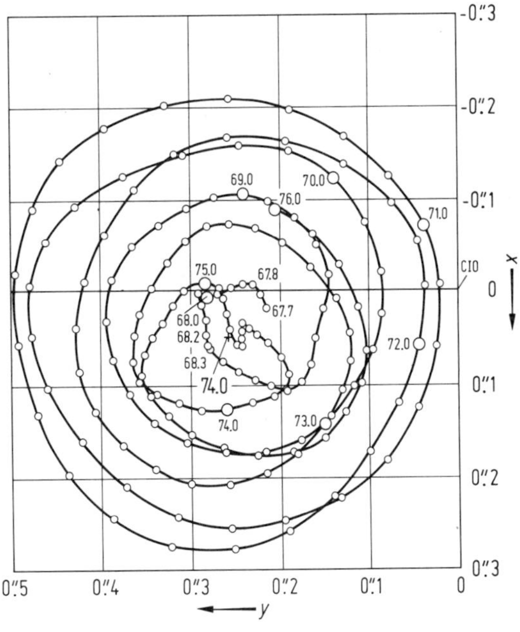

Fig. 1. IPMS polar path 1967.7···1976.0. Cross shows position of mean pole at 1974.0. The mean pole at about 1903 was near CIO.

Polar motion includes two semi-periodic, rotating (counter-clockwise at north) components, the forced annual and the free, 14.2-months Chandler.Excitations are caused by seasonal meteorological and by unknown geophysical effects, respectively. Normal paths are elliptical and circular, respectively, but vary randomly; therefore a rigorous separation is impossible and approximations must be used which are affected by observational errors.

The components produce a beat effect and the pole spirals in and out. Fig. 1 shows observed IPMS polar motion, 1967.7···1976.0, when both components were essentially constant and circular [33].

Recent investigations on long term and short term polar motions from pre-ILS (since 1840), revised ILS, IPMS, and independent Doppler data led to the following results [33, 34]:

The Chandler semi-amplitude had two sharp maxima: 0.″24 at 1911 and 0.″30 at 1953. The three minima were essentially flat for about 15 a each; two at about 0.″13 and one 0.″07. Increases and decreases were smooth, as were changes in phase. The apparent period, varying 1.12···1.20 a, is affected by indeterminate phase changes. The true free period probably lies between 1.18 a and 1.19 a.

The annual component had significant changes: in semi-amplitude (0.″07···0.″11), in phase (±10°), and in eccentricity.

Chandler inelastic damping has often been considered to be strong [13, 14], but long periods of constant amplitude indicates otherwise.

The mean pole has a somewhat irregular secular motion, the average of which is about 0.″004/a toward 75° W.

References for 2.1, see 2.2.7 .

2.2 Time determination

2.2.0 Notations used in 2.2

The designations of the time units (s, min, h, d) are differentiated, if necessary or appropriate, according to the following rules:

s_U = second of mean solar time; s^* = second of mean sidereal time; s_E = second of ephemeris time. s normally represents the atomic time second.

2.2.1 Systems of time measurement

Periodic phenomena applied to time measurements in astronomy, and time scales derived from these phenomena:

(i) The rotation of the earth about its axis: sidereal time and universal time both rigorously connected through a numerical formula; non-uniform due to the variations of the earth's rotation.

(ii) The orbital motions of earth, moon, and planets in the solar system: ephemeris time (concept to be dropped from 1984 onwards); theoretically uniform by definition.

(iii) The electromagnetic oscillation of the cesium 133 atom: international atomic time, coordinated universal time (the latter approximated to universal time in a prescribed manner), dynamical time for ephemerides; theoretically uniform by definition.

Systems (i), (ii) treated in [a, b, d, e], in full detail [e]; atomic time (iii) in [a, b, e] according to the state of development at that time.

2.2.2 Sidereal, solar, and universal time

2.2.2.1 Definitions of sidereal and solar time, relations between their units

Table 6. Definitions and units.

Definition	Unit
mean sidereal time = hour angle of the mean (vernal) equinox [1]	1 mean sidereal day (d*) = time interval between two succeeding transits of the mean equinox through the upper meridian
relation: 1 d* = 23 h_U 56 min_U 04.0905 s_U [2] [3]	
apparent solar time = hour angle of the true sun, counted from midnight	1 apparent solar day = time interval between two succeeding transits of the true sun through the lower meridian
apparent solar time – mean solar time = equation of time (2.2.2.4)	
mean solar time = hour angle of the fictitious mean sun counted from midnight [4]	1 mean solar day (d_U) = time interval between two succeeding transits of the fictitious mean sun through the lower meridian
relation: 1 d_U = 24 h* 03 min* 56.5554 s*	

[1]) Apparent sidereal time = hour angle of the true equinox = mean sidereal time + equation of the equinoxes (due to nutation). Daily values of the latter [5].

[2]) Owing to precession, 1 d* is about 0.0084 s* shorter than the period of the earth's rotation (fixed star to fixed star).

[3]) Precise relation, valid until the end of 1983 [16]: $1\,d^* = 0.997\,269\,566\,414\, d_U - 0.586\, T_U \cdot 10^{-10}\, d_U$, where T_U is the number of Julian centuries of 36 525 d_U elapsed since Greenwich noon on 1900 January 0.

[4]) The fictitious mean sun (sometimes called "universal mean sun") is defined by an expression for the right ascension of an abstract fiducial point on the celestial equator. The hour angle of that point is determined, through the intermediary of sidereal time (see 2.2.2.2), by the observation of stars. The position of the fictitious mean sun is very close to the mean position of the true sun projected onto the equator, but it is not rigorously related to it [e]. Additional departures will occur from 1984 onwards [16].

2.2.2.2 Universal time (UT)

The time forms defined in 2.2.2.1 refer to the local meridian. Mean sidereal and mean solar time are called Greenwich mean sidereal time (GMST) and UT, respectively, when they are referred to the origin of longitudes (2.1.4.2) by adding the longitude of the location to local time (for counting of longitudes, see 2.1.3). Relation: GMST of 0^h UT:

$$6^{h*}\, 38^{m*}\, 45.836^{s*} + 8\,640\,184.542\, T_U\, s^* + 0.0929\, T_U^2\, s^* .$$

From 1984 onwards, new values will be substituted. Preliminary estimate of new minus old: $-0.0075\, s^* + 0.085\, T_U\, s^*$ [54]. Daily values of GMST of 0^hUT (strictly, of 0^hUT1; see below) are given in [5].

In astronomical practice before 1925, solar time was reckoned from noon, but the term Greenwich mean (solar) time (GMT) has also been used for the midnight time scale from 1925 onwards [36]. The designation UT always refers to the latter (1924 December 31, 12^h GMT (old) = 1925 January 1, 0^h UT).

Different forms of UT were introduced in 1956 [37]:

UT0(S) is the mean solar time of the origin of longitudes obtained from direct astronomical observation at station S and derived on the basis of the conventional longitude (2.1.3);

UT1 (S) is UT0 (S) corrected for the effect of polar motion:

$$\mathrm{UT1\,(S)} = \mathrm{UT0\,(S)} - \tfrac{1}{15}(\lambda_{\mathrm{O,S}} - \lambda_{\mathrm{S}})\ (2.1.3);$$

UT2 (S) is UT1 (S) corrected for the effect of the seasonal variation in the earth's rate of rotation, as published by the Bureau International de l'Heure (BIH).

UT1 represents the true angular motion of the earth about its axis. The comparison with a uniform time scale reveals the irregularities in the earth's rotation speed (2.2.6). UT1 (BIH) is regarded as the best result of UT1 (2.1.4.2). Relevant publications [1, 2].

BIH correction for seasonal variation as valid from 1962 onwards [2]:

$$\mathrm{UT2} - \mathrm{UT1} = 22\sin 2\pi t - 12\cos 2\pi t - 6\sin 4\pi t + 7\cos 4\pi t \text{ in [ms]},$$

where t is expressed in fractions of the Besselian year (2.2.3.2).

2.2.2.3 Standard times

Standard or zone times chosen for civil timekeeping correspond basically to the mean solar time of a particular standard meridian which is uniformly used within certain limits of longitude, or geographical, or political area. Hence,

standard time = universal time − longitude of standard meridian.

Generally, the longitudes of the standard meridians are integers of 1 h (1 h $\hat{=}$ 15°) and so, in principle, the standard times differ from UT1 by integers of 1 h_U. In practice, civil timekeeping is based upon UTC (2.2.4.3), and the standard times differ from it by integers of 1 hour of atomic time. Lists of standard times [8, 9, 32], diagrams [a, 8a, 9].

2.2.2.4 Equation of time

Daily values (precision at least 1 s) of

apparent solar time − mean solar time,

are tabulated in most almanacs for navigation; they are obtained with an error <0.2 s by substracting the time of the sun's "ephemeris transit" [5] from 12^h. Diagrams [a, 6], rough values [32].

Approximate dates of turning points in the seasonal variation of the equation of time: mid-February (−14 min), mid-May (+4 min), end of July (−6 min), beginning of November (+16 min).

2.2.2.5 Julian date, modified Julian date; Greenwich sidereal date

The number assigned to a day in a continuous day count starting at 12^h UT on B.C. 4713 January 1, Julian proleptic calendar, is the Julian day number. The Julian date (JD) is the Julian day number followed by the fraction of the mean solar day elapsed since the preceding UT noon. The terminology Julian ephemeris date (JED) may be used when it is necessary to distinguish the Julian date in ephemeris time (2.2.3) from the Julian date in universal time.

Julian day number of 1900, January 0: 2415 020. The Julian date of January 0, 12^h UT (or ET) for every leap year 1904···2096, JD (L), can be computed by

$$\mathrm{JD\,(L)} = 2415\,019.0 + 1461.0\,l,$$

where l is the number of the leap year after 1900.

Tables covering B.C. 1600···A.D. 1999 [e, 5]; table 1920···1999 [32].

The modified Julian date (MJD) is a continuous day count originating on 1858 November 17, 0^h UT and is related to JD by [38]:

$$\mathrm{MJD} = \mathrm{JD} - 2400\,000.5\,.$$

The Greenwich sidereal date (GSD) is the interval in sidereal days, determined by the equinox of date that has elapsed on the Greenwich meridian since the beginning of the sidereal day which was in progress at JD 0.0. An approximation adequate to give the corresponding day number [e], is:

$$\mathrm{GSD} \hat{=} +0.671 + 1.00273\,79093 \cdot \mathrm{JD}.$$

Table covering B.C.2000···A.D.1999 [e], daily data of respective year [5].

2.2.3 Ephemeris time (ET)

ET is the independent variable in the dynamical theories of the sun, moon, and planets. It has been the argument of the fundamental ephemerides in the ephemeris from 1960 onwards. New dynamical time scale from 1984 onwards, see 2.2.4.4.

2.2.3.1 Definition of epoch and unit, and determination of ephemeris time

"Ephemeris time is reckoned from the instant, near the beginning of the calendar year A.D. 1900, when the geometric mean longitude of the sun was 279° 41′ 48″.04, at which instant the measure of ephemeris time was 1900 January 0, 12 hours precisely" [37].

"The second is the fraction 1/31556 925.9747 of the tropical year for 1900 January 0 at 12 hours of ephemeris time" [39].

The ET second was the time unit of the International System of Units (SI) from 1960 to 1967 (see 2.2.4.1). Its length is in approximate agreement with the average length of the UT second during the 18th and 19th century. On 1900 January 0, $\Delta T = \mathrm{ET} - \mathrm{UT} \approx -4.5$ s.

ET is obtained by comparison of observed positions of the sun, moon, and planets with their corresponding ephemerides. Observations of the moon [18a, 40] are the most effective since its motion between the stars is much greater than those of other bodies. Actually, a distinction must be made between the various ET time scales according to the ephemerides used (j) [24]: ET (j) may differ systematically from ET as defined by reference to the sun's mean longitude. Relation between ET and TAI, the latter being available direct through UTC (2.2.4.3): ET = TAI + 32.184 s.

2.2.3.2 Lengths of the year

Lengths of the principal years as derived from the expressions of the true sun's mean motion [e], with T_E = number of Julian (ephemeris) centuries of 36525 d_E elapsed since 1900 January $0^d.5$ ET: [1])

tropical year (equinox to equinox)	$365.24219\,878\,d_E - 0.00000\,614\,T_E\,d_E$
sidereal year (fixed star to fixed star) [2])	$365.25636\,590\,d_E + 0.00000\,011\,T_E\,d_E$
anomalistic year (perihelion to perihelion) [3])	$365.25962\,642\,d_E + 0.00000\,316\,T_E\,d_E$
eclipse year (moon's node to moon's node)	$346.62003\,1\,d_E + 0.00003\,2\,T_E\,d_E$

[1]) From 1984 onwards, the number of Julian centuries will be expressed in terms of the new dynamical time scale for ephemerides (2.2.4.4) with D as notation for the unit. Since 1 D = 1 d_E, no change in the numerical values in the expressions for the lengths of the year will occur.

[2]) Computed on the basis of new constant for general precession in longitude (2.3.2).

[3]) Computed [55] with new values for the masses of the planets (3.2.1.2) and the moon (3.2.1.5), and with new constant for general precession in longitude (2.3.2).

The Besselian (fictitious) solar year begins at the instant when the right ascension of the ephemeris mean sun (mean position of the true sun projected onto the equator), affected by aberration and measured from the mean equinox, is 280°. This instant always occurs near the beginning of the calendar year and is identified by .0 after the year; for example, the beginning of the Besselian year 1980 is January $1^d.189$ ET = 1980.0. Tables 1900 ··· 1999 [e], of current year [5]. The Besselian year is shorter than the tropical year by the amount of 0.148 T_E s.

The lengths in units of a uniform time scale of the average calender years, Julian (365.25 d_U) and Gregorian (365.2425 d_U), depend upon variations in the earth's orbit and upon the variable speed of the earth's rotation and are not basically related to those of the years already given. The calendar year lengths were historically chosen so as to approximate the tropical year length which is in accordance with the change of seasons.

2.2.3.3 Lengths of the month

Lengths of the mean months on 1900 January $0^d_.5$ ET [e]:

synodic month (new moon to new moon)	29.530589 d_E
tropical month (equinox to equinox)	27.321582 d_E
sidereal month (fixed star to fixed star)	27.321661 d_E
anomalistic month (perigee to perigee)	27.554551 d_E
draconic month (node to node)	27.212220 d_E

The lengths of the mean months undergo only slight secular variations ($\ll 0.1$ s/cent) which partly depend upon the variations in the earth's rotation [40, 41]. However, due to periodic perturbation of the lunar motion, the actual lengths of the months can vary by several hours (up to 13 h in the case of the synodic month).

As to notations of units, and length of the sidereal month, see 2.2.3.2, notes 1), 2).

2.2.4 Atomic time

2.2.4.1 Definition of the second of the International System of Units (SI)

"The second is the duration of 9192631770 periods of the radiation corresponding to the transition between the two hyperfine levels of the fundamental state of the atom of cesium 133" [42]. This definition made the new SI (atomic) second equal to the old SI (ephemeris) second, within the error of measurement.

2.2.4.2 International atomic time (TAI)

Integration of seconds and their dating, by means of the operation of a cesium atomic clock, results in an independent atomic time scale. The internationally accepted scientific time reference is TAI, as established by the BIH on the basis of atomic clock data supplied by cooperating institutions [2e]. The nominal value of the TAI scale unit is the SI second referred to sea level. TAI began in 1955 July (under a different name); on 1958 January 1, 0^h TAI, it was in agreement with the same date in UT2 (BIH). On 1980 January 1: TAI − UT2 = +18.36 s.

Deviation of the duration of the TAI second from nominal (since 1977): $<1\cdot10^{-13}$ s [2f]. Operational procedures of TAI [2g, h, i].

Generation of atomic time scales [43]; relativistic effects on time scales [44, 45]; performance of cesium frequency standards [45].

2.2.4.3 Coordinated universal time (UTC)

Since 1972, almost all time signals have been transmitted on the basis of the time scale UTC (new) which has exactly the same rate as TAI and differs from it by an integral number of seconds. By the introduction of leap seconds UTC is kept in agreement with UT1 within ± 0.9 s. In 1975, the General Conference of Weights and Measures recommended UTC as the basis of civil time.

Operational rules on UTC [46, 2 pt. C]. Relationship between UTC (old) and TAI 1961⋯1971 [2].

2.2.4.4 Dynamical time for ephemerides (TD)

The concept of the time scale to be used in the ephemerides for the years 1984 onwards and in all other relevant astronomical work, TD, is expressed in the following terms [25]:
(a) On 1977 January 1, 0^h TAI (exactly), the value of the time scale for apparent geocentric ephemerides be 1977 January $1^d_.0003725$ exactly;
(b) the unit of this time scale be a day (D) of 86400 SI seconds at mean sea level.

Continuity of TD with ET has been achieved because 1 s = 1 s_E, and because the chosen offset between TD and TAI was the current estimate of ET − TAI (expressed in [s] in 2.2.3.1).

If it is necessary to differentiate between the time scale for apparent geocentric ephemerides (which is a proper time scale) and the time scale referred to the barycentre of the solar system (which is a coordinate time scale), then the former is called Terrestrial Dynamical Time (TDT) and the latter Barycentric Dynamical Time (TDB) [16]. TDT and TDB differ only by periodic terms. The largest is an annual one with coefficient 1.658 ms. Transformation from TDT to TDB [47].

2.2.5 Time signals

Lists of time signal emissions: [2] pt. C, [8, 9, 45, 46]. [45, 46] also provide information about methods of high precision time comparisons over long distances.

2.2.6 Long term fluctuations of the earth's rotation speed

The changes in the earth's rate of rotation include the following parts:
(i) A secular deceleration in the region of 2.5 ms/d cent (recent estimate; see below) due to tidal friction.
(ii) Irregular fluctuations covering the range of roughly 5···30 a. Mainly attributed to changes in the electromagnetic coupling between the earth's mantle and core; long term variations in the distribution of water, ice, and air could contribute to some extent. Changes of some 3···4 ms/d within 20 a have taken since 1850.
(iii) Periodic variations. Fairly regular annual and – although intermittent – biennial changes, almost entirely caused by the variable strength of the zonal wind circulation, amount to ± 0.4 ms/d and ± 0.1 ms/d, respectively; semi-annual changes, of which about 50% are due to winds and 50% are due to solar zonal tide, amount to ± 0.3 ms/d. Lunar tides with periods near 14 d and 27 d result in variations of ± 0.4 ms/d and ± 0.2 ms/d, respectively. Tidal perturbations occurring with periods of 8.8 a and 18.6 a cannot be separated from noise of data available at present. For the integrated effect of the seasonal variation expressed by $-(\mathrm{UT2}-\mathrm{UT1})$, see 2.2.2.2.

Detailed treatment of the whole subject of the rotation of the earth, which includes both the rate of rotation and polar motion [14, 52].

Table 7 gives corrections of universal time versus a uniform time scale for the interval 1627···1980. The pre-1956 data – obtained from lunar observations – are based upon the value of $-26''\pm 2''/\mathrm{cent}^2$ for the tidal acceleration of the moon, as deduced in an analysis of the transits of Mercury 1677···1943 [50]. The estimated fractional increase in the length of the mean solar day due to tidal friction, derived from those figures, is [41]:

$$\dot{\omega}/\omega = (29\pm 2)\cdot 10^{-9}/\mathrm{cent.} \mathrel{\hat{=}} (2.5\pm 0.2)\ \mathrm{ms/d\ cent.}$$

Fig. 2 shows the changes in the length of day for 1956···1979 and the acceleration of earth rotation.

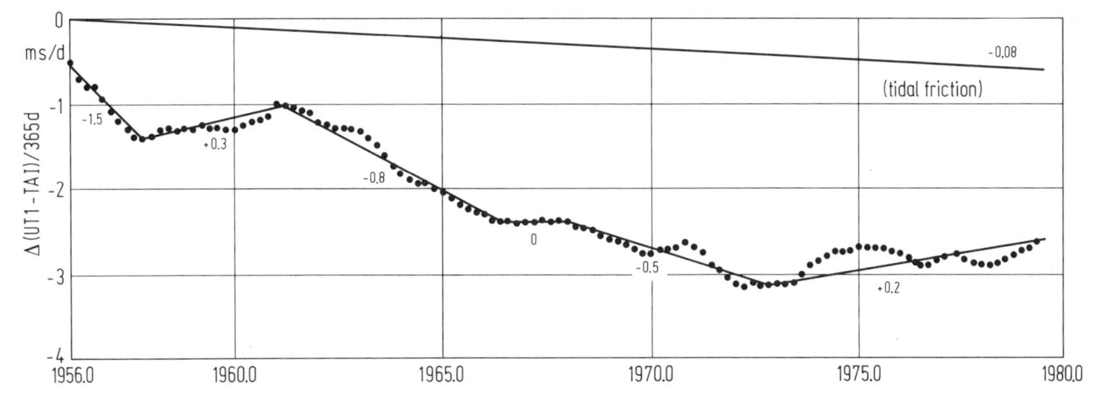

Fig. 2. Changes in the length of day, deduced from yearly differences of UT1 – TAI. The values are drawn for every 0.2 a, and can be represented fairly well by segments of straight lines, the slope of which is a measure for the acceleration of the rate of earth rotation. The figures characterizing the acceleration, according to the straight lines, are given as relative change of the length of two succeeding days in units of 10^{-11}. The line "tidal friction" shows the expected acceleration if only the tidal friction would be effective.

Table 7. Time correction ΔT. ΔT in [s] stands for TD − UT (1627···1955) and for 32.184s + (TAI − UT1) (1956···1981), respectively. The data 1861···1978 are taken from [48]; the data 1627···1860.5 are those published in [49] after the correction [51]

$$C_{\Delta T} = 1.25 - 1.821\,(-1.54 + 2.33\,T - 1.78\,T^2) \text{ in [s]}$$

(T in Julian centuries from 1900.0) has been applied. Thereby, all values of ΔT are referred to a common system.

Date	ΔT [s]	Date	ΔT [s]	Date	ΔT [s]	Date	ΔT [s]	Date	ΔT [s]
1627	+74.93	1739	+13.80	1805	+12.13	1852.5	+ 7.68	1895.0	− 6.47
37	79.80	46	14.05	07	11.03	53.5	8.70	96.0	6.09
38	36.15	47	15.35	08	14.21	54.5	8.01	97.0	5.76
41	0.18	53	16.71	09	13.25	55.5	9.22	98.0	4.66
44	53.86	54	15.38	10.5	13.27	56.6	8.55	99.0	3.74
45	+39.47	1757	+ 9.52	1811.5	+12.58	57.5	+ 8.12	1900.0	− 2.72
56	42.26	58	15.54	12.5	12.34	58.5	8.23	1.0	1.54
58	77.62	64	17.50	13.5	12.59	59.5	8.41	2.0	− 0.02
63	37.74	65	17.02	14.5	11.98	60.5	7.84	3.0	+ 1.24
64	23.13	66	18.03	16.5	17.25			4.0	2.64
1671	+24.53	67	+16.75	17.5	+10.44	1861.0	+ 7.82	5.0	+ 3.86
72	37.80	68	15.82	18.5	13.36	62.0	7.54	6.0	5.37
73	26.74	69	18.54	19.5	14.56	63.0	6.97	7.0	6.14
74	27.51	70	16.77	20.5	10.65	64.0	6.40	8.0	7.75
76	24.83	71	17.23	21.5	11.72			9.0	9.13
78	+22.13	1772	+16.65	1822.5	+11.88	65.0	+ 6.02	1910.0	+10.46
79	21.41	73	16.19	23.5	11.34	66.0	5.41	11.0	11.53
80	20.75	74	16.96	24.5	10.26	67.0	4.10	12.0	13.36
82	14.95	75	17.37	25.5	10.58	68.0	2.92	13.0	14.65
83	19.72	76	18.32	26.5	11.91	69.0	1.82	14.0	16.01
1684	+25.19	77	+17.84	27.5	+10.54	1870.0	+ 1.61	15.0	+17.20
85	18.20	78	17.18	28.5	10.97	71.0	+ 0.10	16.0	18.24
86	17.39	79	15.30	29.5	8.85	72.0	− 1.02	17.0	19.06
99	20.52	83	14.12	30.5	9.01	73.0	1.28	18.0	20.25
1701	21.44	84	17.00	31.5	8.35	74.0	2.69	19.0	20.95
1705	+16.40	1785	+16.85	1832.5	+ 7.95	75.0	− 3.24	1920.0	+21.16
06	18.91	86	17.07	33.5	9.84	76.0	3.64	21.0	22.25
07	19.21	87	16.05	34.5	7.33	77.0	4.54	22.0	22.41
09	19.51	88	15.14	35.5	7.56	78.0	4.71	23.0	23.03
10	17.14	89	14.27	36.5	7.60	79.0	5.11	24.0	23.49
1711	+23.39	1790	+13.53	37.5	+ 7.48	1880.0	− 5.40	25.0	+23.62
12	21.04	91	14.34	38.5	7.69	81.0	5.42	26.0	23.86
13	18.75	92	14.58	39.5	7.22	82.0	5.20	27.0	24.49
14	12.81	93	12.29	40.5	7.56	83.0	5.46	28.0	24.34
15	14.55	94	13.15	41.5	7.46	84.0	5.46	29.0	24.08
17	+12.41	95	+14.85	1842.5	+ 7.78	85.0	− 5.79	1930.0	+24.02
18	17.90	96	17.16	43.5	8.08	86.0	5.63	31.1	24.00
19	12.01	97	14.37	44.5	6.99	87.0	5.64	32.0	23.87
20	20.66	98	16.80	45.5	7.48	88.0	5.80	33.0	23.95
27	19.86	99	15.16	46.5	8.04	89.0	5.66	34.0	23.86
1729	+15.62	1800	+12.59	47.5	+ 7.97	1890.0	− 5.87	35.0	+23.93
33	14.77	01	13.77	48.5	7.74	91.0	6.01	36.0	23.73
36	16.84	02	11.64	49.5	7.30	92.0	6.19	37.0	23.92
37	14.14	03	9.20	50.5	7.90	93.0	6.64	38.0	23.96
38	13.74	04	13.64	51.5	8.58	94.0	6.44	39.0	24.02

continued

Table 7, continued

Date	ΔT [s]	Date	ΔT [s]	Date	ΔT [s]	Date	ΔT [s]	Date	ΔT [s]
1940.0	+24.33	1950.0	+29.15	1960.0	+33.15	1970.0	+40.18	1980.0	+50.54
41.0	24.83	51.0	29.57	61.0	33.59	71.0	41.17	81.0	+51.38
42.0	25.30	52.0	29.97	62.0	34.00	72.0	42.22		
43.0	25.70	53.0	30.36	63.0	34.47	73.0	43.37		
44.0	26.24	54.0	30.72	64.0	35.03	74.0	44.48		
45.0	+26.77	55.0	+31.07	65.0	+35.73	75.0	+45.47		
46.0	27.28	56.0	31.35	66.0	36.54	76.0	46.46		
47.0	27.78	57.0	31.68	67.0	37.43	77.0	47.52		
48.0	28.25	58.0	32.18	68.0	38.29	78.0	48.52		
49.0	28.71	59.0	32.68	69.0	39.20	79.0	49.59		

2.2.7 References for 2.1 and 2.2

General references

a Mueller, I.I.: Spherical and practical astronomy as applied to geodesy. Ungar, New York (1969).
b Ramsayer, K.: Geodätische Astronomie, Jordan/Eggert/Kneissl, Handbuch der Vermessungskunde, Bd. IIa. Metzler, Stuttgart (1970).
c Tardi, P., Laclavère, G.: Traité de géodésie, Tome II, Astronomie géodésique de position. Gauthier-Villars, Paris (1955).
d Woolard, E.W., Clemence, G.M.: Spherical astronomy. Academic Press, New York and London (1966).
e Nautical Almanac Offices of the United Kingdom and the United States of America: Explanatory Supplement to the Astronomical Ephemeris and the American Ephemeris and Nautical Almanac, third impression (with amendments). H. M. Stationary Office, London (1974).

Circulars and annual publications

1 Bureau International de l'Heure: Circular D (monthly), No. 1 ..., Paris (1966 ...). a: No. 150 (1979).
2 Guinot, B. et al.: Bureau International de l'Heure, Rapport Annuel pour 1967 ..., Paris (1968 ...), published alternatively in French and English.
a: (1968) 13. b: (1969) 11. c: (1973) A–2. d: (1979) and subsequent Volumes. e: (1972) A–9. f: (1979) B–53. g: (1974) A–8. h: (1975) A–7. i: (1978) A–6. k: (1980) A–1.
3 International Polar Motion Service: Monthly Notes, No. 1 ..., Mizusawa (1962 ...).
4 Yumi, S.: Annual Report of the International Polar Motion Service for the year 1962 ..., Mizusawa (1964 ...). a: (1977, 1978).
5 H. M. Nautical Almanac Office: The astronomical ephemeris for the year ... H. M. Stationary Office, London. (The same publication is also issued by U.S. Naval Observatory, Nautical Almanac Office, under title "The American Ephemeris and Nautical Almanac", U.S. Government Printing Office, Washington, New name for both publications from 1981 onwards: "Astronomical Almanac".)
6 Bureau des Longitudes: Annuaire pour l'an ... Gauthier-Villars, Paris.
7 U.S. Naval Observatory: Time Service Publ., Series 7 (weekly), No. 161 ..., Washington (1971 ...).

Publications with supplements

8 Hydrographer of the Navy: The Admiralty List of Radio Signals, Vol. 5, London. a: Vol. 5a, Diagrams.
9 Deutsches Hydrographisches Institut: Nautischer Funkdienst, Bd. I, D, Hamburg.

Special references

10 Straub, G.: Veröff. Deutsche Geod. Komm., Reihe C, Heft 65 (1963).
11 Kaula, M.: Trans. Am. Geophys. Union **38** (1957) 578.
12 Fischer, I.: Intern. Hydrographic Review **53** (1976) No. 1.
13 Melchior, P.J.: Commun. Obs. R. Belgique **130** (1957).
14 Munk, W.H., McDonald, G.J.F.: The rotation of the Earth, University Press, Cambridge (1960).
15 Woolard, E.W.: Astron. Papers Washington XV, pt. I (1953) 161.

16 Trans. Int. Astron. Union **17B**, Report Comm. 4 (1980) 63.
17 O'Hora, N.P.J.: Observatory **91** (1971) 155.
18 Kuiper, G., Middlehurst, B. (eds.): Stars and stellar systems, Vol. I, Telescopes, University of Chicago Press (1969).
18a Markowitz, W.: The Photographic Zenith Tube and the Dual-Rate Moon-Position Camera, in [18] p. 88.
18b Danjon, A.: The Impersonal Astrolabe, in [18] p. 115.
19 Hattori, T.: Publ. Intern. Latitude Obs. Mizusawa **I**, 1 (1951).
20 Pavlov, N.N.: Trudy 15. Astrometr. Konf. SSSR, Moskva-Leningrad (1963) 246.
21 Potthoff, H.: Wiss. Z. Techn. Univ. Dresden **17** (1968) 1477.
22 Gessler, J.: Entwicklung und Erprobung einer transportablen Zenitkamera für astronomisch-geodätische Ortsbestimmungen, Dr.-Ing. thesis, Universität Hannover (1975).
23 C.R. XIVe Assembl. Gén. Union Géodésique et Géophysique Int. (1967) 135.
24 Trans. Int. Astron. Union **13B** (1968) 49, 111.
25 Trans. Int. Astron. Union **16B** (1977) 60, 155.
26 Trans. Int. Astron. Union **17A** pt. 1 (1979) 127, 130.
27 Proc. Int. Astron. Union Symp. **82** "Time and the Earth's rotation" (McCarthy, D.D., Pilkington, J., eds.), Reidel, Dordrecht (1979).
28 Anderle, R.J.: Naval Weapons Lab. Techn. Rep., TR-2432, Dahlgren (1970).
29 Anderle, R.J.: Bull. Géodésique **50** (1976) 377.
30 Markowitz, W.: Comparisons of ILS, IPMS, BIH, and Doppler polar motions with theoretical. Report to Comm. 19 and 31, XVIth Int. Astron. Union General Assembly, Grenoble (1976).
31 Bureau International de l'Heure: Rotation de la terre. Observations traitées et résidus en 1974···1976, Paris (1975···1977).
32 Larink, J. in: Landolt-Börnstein, NS, Vol. VI/1 (1965) 71, 73.
33 Markowitz, W.: Independent polar motions, optical and Doppler; Chandler uncertainties. Report to Comm. 19 and 31, XVIIth Int. Astron. Union General Assembly, Montreal (1979).
34 Markowitz, W.: Article in preparation
35 Yumi, S., Yokoyama, K.: Results of the International Latitude Service in a homogeneous system 1899.9–1979.0, Mizusawa (1980).
36 Sadler, D.H.: Quart. J. R. Astron. Soc. **19** (1978) 290.
37 Trans. Int. Astron. Union **10** (1960) 72, 489.
38 Trans. Int. Astron. Union **15B** (1974) 55.
39 Comité International des Poids et Mesures: Procès verbaux des séances **25** (1957) 77.
40 Gondolatsch, F.: Veröff. Astron. Recheninst. Heidelberg Nr. 5 (1953).
41 Morrison, L.V.: Tidal Friction and the Earth's rotation (Brosche, P., Sündermann, J., eds.), Springer, Berlin (1978) 22.
42 Conférence Générale des Poids et Mesures: C. R. des Séances de la 13e conférence générale des poids et mesures. Gauthier-Villars, Paris (1968).
43 Becker, G.: Radio Science **14** (1979) 593.
44 Becker, G.: Proceedings second Cagliari international meeting on time, Anastatiche, Cagliari (1975) 63.
45 International Radio Consultative Committee: Recommendations and reports of the CCIR, 1978, Vol. 7, Geneva (1978) 14, 62, 78, 86.
46 International Radio Consultative Committee: 13th plenary assembly, Vol. 7, Geneva (1975) 18, 25, 45.
47 Moyer, T.D.: Jet Propulsion Laboratory, Technical Memorandum 33-786, Pasadena (1976).
48 Morrison, L.V.: Geophys. J. R. Astron. Soc. **58** (1979) 349.
49 Martin, C.F.: A study of the rate of rotation of the Earth from occultations of stars by the Moon 1627–1860, PhD thesis, Yale University (1969).
50 Morrison, L.V.: Mon. Not. R. Astron. Soc. **173** (1975) 183.
51 Morrison, L.V.: Private communication (1979).
52 Lambeck, K.: The Earth's variable rotation, University Press, Cambridge (1980).
53 Wilkins, G.A. (ed.): Project Merit, R. Greenwich Obs., Hailsham, and Inst. Angew. Geodäsie, Frankfurt (1980).
54 Fricke, W.: Proc. Int. Astron. Union Coll. **56** "Reference coordinate systems for Earth dynamics" (Gaposchkin, E.M., Kolaczek, B., eds.), in press (1981).
55 Lederle, T.: Private communication (1980).

2.3 The system of astronomical constants

2.3.1 Introduction

A certain number of constants are required for the interpretation of observations of positions of celestial objects and for the construction of ephemerides. Such constants are often called "fundamental constants", and it may be noted that some of them vary with time. In the case of variation with time one gives the numerical value of the constant for an adopted standard epoch, and it is understood that the value for any other epoch can be computed by means of well-known formulae.

It might appear attractive to use at any time the best determined values available, but the comparison of observations with theories is greatly facilitated if all authorities of ephemerides and all observers use an internationally agreed set of numerical values, which form a consistent system of constants. The first agreement on a system was reached by a conference held in Paris in 1896 [a]. Most of the proposed constants were determined by Newcomb [i, 7]. They were in common use from about 1900 to 1964. On recommendation by an IAU Symposium held in Paris in 1963 [d] a first revision of "Newcomb's system" was carried out. As a result the IAU adopted a revised system in 1964 [e], later called the "IAU (1964) System of Astronomical Constants", which was formulated by Fricke et al. [c].

It was known at that time that another revision would be necessary as soon as improved values of planetary masses and of some important constants could be included. Due to rapid progress made within a few years, an IAU Colloquium held in Heidelberg in 1970 [f] suggested the rediscussion of changes in precession, planetary ephemerides, and units and timescales. As a result important changes of the IAU (1964) system were proposed, and the "IAU (1976) System of Astronomical Constants" was formulated and presented in a report by Duncombe et al. [b]. This system was adopted by the IAU in 1976 [g] with the recommendation "that it shall be used in the preparation of the fundamental catalogue FK5 and of the national and international ephemerides for the years 1984 onwards, and in all other relevant astronomical work". The numerical values of the constants in the IAU (1976) system are given in 2.3.3 in units specified in 2.3.2. The values reported by Böhme and Fricke [1] are now of historical interest only.

2.3.2 Units

The meter (m), kilogram (kg) and second (s) are the units of length, mass and time in the International System of Units (SI); the astronomical units are defined as follows:

astronomical unit of time: one day, D, of 86400 s (cf. 2.2.4.4); the interval of 36525 days is one Julian century,

astronomical unit of mass: the mass of the sun, S,

astronomical unit of length: the length, A, for which the Gaussian constant, k, takes the value 0.01720209895 when the units of measurement are the astronomical units of length, mass, and time.

The dimensions of k^2 are those of the constant of gravitation, G. One may use the term "unit distance" for the length A.

2.3.3 The IAU (1976) system of astronomical constants

The last column of the following Table indicates the limits (in units of the last decimal) between which the true value of the constant is believed to lie.

Defining constant		
1. Gaussian gravitational constant	$k = 0.01720209895$	± 0
Primary constants		
2. Speed of light	$c = 299792458\ \mathrm{m\,s^{-1}}$	± 1.2
3. Light-time for unit distance	$\tau_A = 499.004782\ \mathrm{s}$	± 6
4. Equatorial radius for earth	$a_e = 6378140\ \mathrm{m}$	± 5
5. Dynamical form-factor for earth	$J_2 = 0.00108263$	± 1
6. Geocentric gravitational constant	$GE = 3.986005 \cdot 10^{14}\ \mathrm{m^3\,s^{-2}}$	± 3
7. Constant of gravitation	$G = 6.672 \cdot 10^{-11}\ \mathrm{m^3\,kg^{-1}\,s^{-2}}$	± 4
8. Ratio of mass of moon to that of earth	$\mu = 0.01230002$	± 5
9. General precession in longitude per Julian century, at standard epoch 2000	$p = 5029''.0966$	± 1534
10. Obliquity of the ecliptic, at standard epoch 2000	$\varepsilon = 23°26'21''.448$	± 102
11. Constant of nutation, at standard epoch 2000	$N = 9''.2055$	± 3
Derived constants		
12. Unit distance	$c\tau_A = A = 1.49597870 \cdot 10^{11}\ \mathrm{m}$	± 2
13. Solar parallax	$\arcsin(a_e/A) = \pi_\odot = 8''.794148$	± 7
14. Constant of aberration, at standard epoch 2000	$\kappa = 20''.49552$	± 0.2
15. Flattening factor for the earth	$f = 0.00335281$	± 2
16. Heliocentric gravitational constant	$GS = 1.32712438 \cdot 10^{20}\ \mathrm{m^3\,s^{-2}}$	± 5
17. Ratio of mass of sun to that of earth	$S/E = 332946.0$	± 3
18. Ratio of mass of sun to that of earth + moon	$(S/E)/(1+\mu) = 328900.5$	± 3
19. Mass of the sun	$S = 1.9891 \cdot 10^{30}\ \mathrm{kg}$	± 12

2.3.4 Notes

The system contains primary and derived constants where the latter ones follow from the numerical values of the primary constants by well-known formulae. In some cases one had to decide whether a constant should be considered a primary one or be derived from anotherquantity. For instance, since recent measurements of light-time have yielded the numerical value of the unit distance, one decided to include the light-time for unit distance among the primary constants.

The system also contains new values of the planetary masses in ratios of mass of sun to those of the planets and their satellites. These values are given in 3.2.1 .

According to the IAU resolution on the IAU (1976) system, the numerical values of the time-depending constants (general precession, obliquity, nutation, aberration) are given at the new standard epoch 2000 which is 2000 January $1^d.5$ = JD 2451545.0; the new standard equinox will correspond to this instant. The Julian century of 36525 days is applied as the unit of time in the fundamental formulae for precession (see 3.2.1.4.2).

The following notes give brief information on the authorities of the numerical values of primary constants. More information can be found in [b] and in a review by Lederle [5].

Speed of light: as recommended by the XV General Conference on Weights and Measures in 1975.

Light-time for unit distance: based on radar measurements of planetary distances.

Equatorial radius and dynamical form-factor for earth, geocentric gravitational constant: as recommended by the International Association of Geodesy in 1975.

Constant of gravitation: as given in the CODATA system of physical constants of 1973.

Mass ratio moon/earth: based on data from lunar and planetary spacecraft.

General precession in longitude: based on the value of a correction to Newcomb's lunisolar precession determined from FK4 proper motions by Fricke [2, 4] and a correction to planetary precession determined from the new planetary masses by Lieske et al. [6]. The expressions for the precession quantities were developed by Lieske et al. [6]. The reasons for the change were explained by Fricke [3].

Obliquity of the ecliptic: determined by Lieske et al. [6] in applying secular terms computed with the new planetary masses to the current value for 1900.

Constant of nutation: based on the IAU (1979) theory of nutation [h].

2.3.5 References for 2.3

General references

a Conférence internationale des étoiles fondamentales de 1896. Gauthier-Villars, Paris (1896).
b Duncombe, R.L., Fricke, W., Seidelmann, P.K., Wilkins, G.A.: Int. Astron. Union Trans. **16B** (1977) 56.
c Fricke, W., Brouwer, D., Kovalevsky, J., Mikhailov, A.A., Wilkins, G.A.: Int. Astron. Union Trans. **12B** (1966) 593.
d Int. Astron. Union Symp. **21** "The System of Astronomical Constants" held in Paris 1963. Bull. Astron. l'Obs. de Paris, Vol. XXV, fasc. 1–3. Gauthier-Villars, Paris (1965).
e Int. Astron. Union Trans. **12B** (1966) 94.
f Int. Astron. Union Coll. **9** "The System of Astronomical Constants" (Emerson, B., Wilkins, G.A., eds.), Celestial Mechanics **4** (1971) 128.
g Int. Astron. Union Trans. **16B** (1977) 31.
h Int. Astron. Union Trans. **17B** (1980) 80.
i Newcomb, S.: The elements of the four inner planets and the fundamental constants of astronomy (Suppl. American Ephem. and Naut. Almanac for 1897), Government Printing Office, Washington (1895).

Special references

1 Böhme, S., Fricke, W.: Landolt-Börnstein, NS, Vol. VI/1 (1965) 76.
2 Fricke, W.: Astron. J. **72** (1967) 1368.
3 Fricke, W.: Astron. Astrophys. **54** (1977) 363.
4 Fricke, W.: Veröff. Astron. Rechen-Institut, Heidelberg (1977) No. 28.
5 Lederle, T.: Mitt. Astron. Gesellschaft **48** (1980) 59.
6 Lieske, J.H., Lederle, T., Fricke, W., Morando, B.: Astron. Astrophys. **58** (1977) 1.
7 Newcomb, S.: Astron. Papers Washington **8** (1897) 1.

3. The solar system

3.1 The sun

3.1.1 The quiet sun

3.1.1.1 Solar global parameters

Parallax	$\pi_\odot$	$= 8\overset{''}{.}79415$
mean distance (astronomical unit)	$r_\odot$	$= 1.495979 \cdot 10^{13}$ cm
minimum distance (perihelion)	$(r_{min})_\odot$	$= 1.4710 \cdot 10^{13}$ cm
maximum distance (aphelion)	$(r_{max})_\odot$	$= 1.5210 \cdot 10^{13}$ cm
radius	$R_\odot$	$= 6.960 \cdot 10^{10}$ cm
semi-angle-diameter (mean distance)	$\frac{1}{2}\bar{d}_\odot$	$= 959\overset{''}{.}63$
minimum semi-angle-diameter (aphelion)	$\frac{1}{2}(d_{min})_\odot$	$= 943\overset{''}{.}84$
maximum semi-angle-diameter (perihelion)	$\frac{1}{2}(d_{max})_\odot$	$= 975\overset{''}{.}93$
mass	$\mathfrak{M}_\odot$	$= 1.989 \cdot 10^{33}$ g
mean density	$\bar{\varrho}_\odot$	$= 1.409$ g cm^{-3} *)
central density (see 3.1.1.2)	$(\varrho_\odot)_{central}$	$= 140 \cdots 180$ g cm^{-3} *)
central temperature (see 3.1.1.2)	$T_{central}$	$= (14.9 \cdots 15.7) \cdot 10^6$ K *)
effective temperature	T_{eff}	$= 5780$ K *)

Absolute magnitudes

M_V	$= 4\overset{M}{.}87$ *)
M_B	$= 5.54$ *)
M_U	$= 5.72$ *)
M_{bol}	$= 4.74$ *)

Apparent magnitudes

$m_V = V$	$= -26.70$
B	$= -26.03$ *)
U	$= -25.85$ *)
m_{bol}	$= -26.83$ *)

Colour indices (symbols: see 4.2.5.12)

$B-V$	$= 0\overset{m}{.}67$
$U-B$	$= 0.18$
$V-R$	$= 0.52$
$V-I$	$= 0.81$
$V-K$	$= 1.42$
$V-M$	$= 1.53$

1″ } geocentric, at mean solar distance	725.3 km *)
1′ } geocentric, at mean solar distance	43518 km *)
1° heliocentric	12147 km *)
surface area	$6.087 \cdot 10^{22}$ cm^2 *)
volume	$1.412 \cdot 10^{33}$ cm^3 *)
gravitational acceleration at surface	$2.740 \cdot 10^4$ cm s^{-2} *)
moment of inertia	$5.7 \cdot 10^{53}$ g cm^2 *)
angular momentum based on surface rotation	$1.63 \cdot 10^{48}$ g cm^2 s^{-1} *)
rotational energy based on surface rotation	$2.4 \cdot 10^{42}$ erg *)
escape velocity at surface	$6.177 \cdot 10^7$ cm s^{-1} *)
luminosity	$3.853 \cdot 10^{33}$ erg s^{-1} *)
specific surface emission	$6.329 \cdot 10^{10}$ erg cm^{-2} s^{-1} *)
specific mean energy production	1.937 erg g^{-1} s^{-1} *)
solar constant S (see 3.1.1.3)	$1.370 \cdot 10^6$ erg cm^{-2} s^{-1}

age $t_\odot \simeq$ solar system age

composition: see Ross and Aller (1976) [4]

*) Indicates a derived quantity not directly measured.

References for 3.1.1.1

1 Allen, C.W.: Astrophysical Quantities, 3rd ed., Athlone Press, London (1973).
2 Barry, D.C., Cromwell, R.H., Schoolman, S.A.: Astrophys. J. **222** (1978) 1032.
3 Gallouët, L.: Ann. Astrophys. **27** (1964) 423.
4 Ross, J.E., Aller, L.H.: Science **191** (1976) 1223.
5 Waldmeier, M.: Landolt-Börnstein, NS, Vol. VI/1 (1965) 95.

3.1.1.2 Solar interior

3.1.1.2.1 Standard models

Produced by evolving a $1\mathfrak{M}_\odot$ main-sequence model with originally homogeneous chemical composition assuming stationarity, and no mixing, rotation nor magnetic fields. Initial composition fixes either present photospheric metals/hydrogen ratio (Z/X=0.019) or cosmic helium abundance (Y=0.25), then is adjusted to produce solar luminosity after $(4.5 \cdots 5.0) \cdot 10^9$ a. The mixing length parameter (see convection zone) is adjusted to reproduce the solar radius.

Models 1964···1976: [63, 10, 48, 12, 78, 8, 2, 18, 49], see Table 1.

Table 1. Solar standard model with initial X=0.732, Y=0.253, Z=0.015 [2].
$\mathfrak{M}/\mathfrak{M}_\odot$ = mass fraction $L/L_\odot$ = scaled luminosity
P = pressure $R/R_\odot$ = scaled radius
T = temperature ϱ = density

$\mathfrak{M}/\mathfrak{M}_\odot$	P 10^{15} dyn cm^{-2}	T 10^6 K	$L/L_\odot$	$R/R_\odot$	ϱ g cm^{-3}
0.000	254.3	15.38	0.000	0.000	161.3
0.102	129.4	12.47	0.545	0.111	84.03
0.201	83.7	10.98	0.793	0.150	58.28
0.299	55.2	9.777	0.912	0.184	42.24
0.402	35.1	9.687	0.968	0.218	30.72
0.503	21.6	7.732	0.990	0.252	20.91
0.599	12.7	6.868	0.998	0.289	13.79
0.702	6.32	5.954	1.000	0.337	7.940
0.801	2.57	4.998	1.000	0.401	3.840
0.902	0.591	3.821	1.000	0.511	1.160
0.990	0.0153	1.580	1.000	0.819	0.077

In radiative equilibrium interior $\varrho/T^3 \simeq$ constant.
Evolutionary tracks in HR diagram [48].
Change of luminosity with time [38]:

$$\frac{L}{L_\odot} \simeq \frac{1}{1+0.4(1-t/t_\odot)}$$

($L_\odot$ and $t_\odot$ are present solar luminosity and age, see 3.5.)

3.1.1.2.2 Non-standard models

Produced by relaxing one or more assumptions of standard model.

Table 2. Non-standard models.
M ≡ model details
N ≡ neutrino flux
O ≡ oscillatory modes

Type of model	M	N	O	Author	Ref.
Non-stationary	×	×		Rood (1972)	56
	×	×		Ezer and Cameron (1972)	33
		×		Dilke and Gough (1972)	32
		×		Ulrich and Rood (1973)	83
			×	Scuflaire et al. (1975)	62
		×		Gabriel et al. (1976)	35
Continuously mixed	×	×		Shaviv and Beaudet (1968)	66
	×	×		Bahcall et al. (1968)	11
		×		Shaviv and Salpeter (1971)	67
Rapidly rotating		×		Ulrich (1969)	80
	×	×		Demarque et al. (1973)	27
Strong internal magnetic field	×	×		Abraham and Iben (1971)	2
	×	×		Bahcall and Ulrich (1971)	9
	×	×		Chitre et al. (1973)	24
Rotating magnetic	×	×		Snell et al. (1976)	69
Non-standard abundances:					
low Z	×	×		Abraham and Iben (1971)	2
	×	×		Bahcall and Ulrich (1971)	9
	×	×		Bahcall et al. (1973)	13
	×	×		Bhavsar and Härm (1977)	16
high Z in core	×	×		Prentice (1976)	54
high He^3	×	×		Abraham and Iben (1970)	1
high Y in core	×	×		Faulkner et al. (1975)	34
	×			Rouse (1975)	57
	×	×		Wheeler and Cameron (1975)	87
			×	Rouse (1977)	58
low Y in core	×	×		Faulkner et al. (1975)	34
	×	×		Hoyle (1975)	47
	×	×		Bhavsar and Härm (1977)	16
Time varying gravity	×	×		Shaviv and Bahcall (1969)	65

3.1.1.2.3 Solar neutrinos

Solar neutrino measurement is based on detection of radioactive argon produced by

$$\nu + {}^{37}\mathrm{Cl} \rightarrow {}^{37}\mathrm{Ar} + e^-$$

with threshold energy 0.814 MeV [7].

Flux is measured in solar neutrino units:
1 SNU $= 10^{-36}$ neutrino absorptions per target atom per second.

Measured flux: (1.8 ± 0.4) SNU [26].

Solar neutrino producing reactions and energy spectrum of standard solar model [7].

Model neutrino fluxes: see Table 3.

Table 3. Approximate lower bounds to neutrino fluxes for various solar models. Published bounds are scaled to the standard flux of 9 SNU.
1 SNU $=10^{-36}$ neutrino absorptions per target atom per second
B = magnetic field

Model	Neutrino flux SNU	Model requirement	Author	Ref.
Standard	9		Bahcall and Sears (1972)	7
	4.7		Bahcall et al. (1973)	13
Non-standard				
non-stationary	1.0	large luminosity changes	Gabriel et al. (1976)	35
mixed	3.0	100% mixing	Shaviv and Salpeter (1971)	67
rotating	1.0	quadrupole moment $>3\cdot10^{-5}$	Demarque et al. (1973)	27
magnetic	< 1.5	$B>10^9$ G at centre	Chitre et al. (1973)	24
rotating, magnetic	0.64	$B>2\cdot10^9$ G at centre, core period 30 min	Snell et al. (1976)	69
low Z	2.0	$Z \ll 0.01$, $Y \simeq 0.1$	Bahcall and Ulrich (1971)	9
high core Z	2.0	metal-rich core of 0.03 $\mathfrak{M}_\odot$	Prentice (1976)	54
high He^3	0.3	primordial $He^3/He^4 \simeq 0.3$	Abraham and Iben (1970)	1
high core Y	2.0	fractionated core 0.1···0.2 $\mathfrak{M}_\odot$	Wheeler and Cameron (1975)	87
low core Y	0.5	$Y \simeq 0.15$, iron-peak high	Hoyle (1975)	47
variable G	18		Shaviv and Bahcall (1969)	65

3.1.1.2.4 Global oscillations

Reported solar oscillatory periods: see Table 4. Observations probably do not refer to single modes but to a mixture of many high order non-radial modes.

Table 4. Solar oscillations and eigenmodes of a standard solar model.

Observed period min	Mode	Ref.	Model eigenperiod min	Ref.	Mode [1])
(12.2d)	diameter	30			
160.0	velocity, luminosity, magnetic field	19, 20, 50, 64	158.9	49	g_{10}
148.4 }	velocity	50, 64	145.9		g_9
134.5 }	velocity	50, 64	132.9		g_8
67.2 }	diameter	42	62.3		p_1
45.5 }	diameter	42	45.9		f
40.3 }	diameter	42	40.9		p_2
36.0 }	diameter	42			
30.9 }	diameter	42	30.9		p_3
27.5 }	diameter	42			
24.8···6.5 discrete	diameter	21	24.5···6.3		$p_4 \cdots p_{18}$
6.5···3.6 } continuum	velocity { photosphere	28, 55	7···3	55, 82	$p_1 \cdots p_9$ ($l=150 \cdots 800$)
6.5···2.8 } continuum	velocity { chromosphere	28, 55	7···3	55, 82	$p_1 \cdots p_9$ ($l=150 \cdots 800$)

[1]) Pressure modes (p) are radial except where otherwise noted; gravity (g) and fundamental (f) are quadrupolar ($l=2$).

Model oscillatory modes [3, 62, 25, 49, 58, 82].

3.1.1.2.5 Convection zone

All solar convection zone models are based on mixing-length theories. The local form has usually one free parameter, the ratio of the mixing-length to the local pressure scale height, α; non-local forms have more. Local forms due to Böhm-Vitense (BV) and Öpik (Ö) yield almost identical models, except in uppermost 2000 km, when properly calibrated [39, 71].

Mixing-length parameter: $\alpha = 1.1$ (BV), 2.4 (Ö) [39].
Depth of convection zone: $D = 149\,000$ km (BV), 152000 km (Ö) [39].
In non-standard solar models the extent can be up to twice as deep or much shallower.
Modified local model: see Table 5.

Table 5. Local mixing-length model of the solar convection zone [70].
z = depth, T = temperature, P = pressure, ϱ = density, $\Delta\nabla$ = superadiabatic gradient

z km	T 10^3 K	P 10^7 dyn cm^{-2}	ϱ 10^{-5} g cm^{-3}	$\Delta\nabla$
33	7.14	0.0154	0.0335	0.93
54	8.28	0.0173	0.0322	0.72
78	9.00	0.0194	0.0329	0.36
170	10.3	0.0287	0.0411	0.10
329	11.5	0.0510	0.0625	0.05
513	12.6	0.0907	0.0973	0.03
720	13.7	0.161	0.153	0.02
954	14.7	0.287	0.243	0.01
1215	15.9	0.510	0.386	0.007
3200	24.9	10.2	4.08	0.001
24300	179.0	10200.0	433.0	0.00
198200	2200.0	5810000.0	19700.0	0.00

Throughout most of the convection zone, stratification is adiabatic with constant specific entropy.

Non-local theories allow for overshooting a) at lower boundary: [17], b) at upper boundary: see 3.1.1.4. and [81, 79].

Acoustic energy flux from dipole emission [73].
$F = 7 \cdot 10^6 \cdots 10^8$ erg cm^{-2} s^{-1} [72] (dependent on assumed turbulence spectrum).

Period of maximum acoustic energy production
$P_{max} = 30 \cdots 60$ s [72].

3.1.1.2.6 Solar rotation

Internal rotation is measured by solar oblateness

$$\varepsilon = 2 \cdot \frac{R_e - R_p}{R_e + R_p}$$ (R_e and R_p are the equatorial and polar radii.)

$\varepsilon = (4.51 \pm 0.34) \cdot 10^{-5}$ [31].

$\varepsilon = (0.96 \pm 0.65) \cdot 10^{-5}$ [43] (at minimum excess equatorial brightness).

The former value requires a rapidly rotating core (period $\simeq 1$ day), the latter is consistent with uniform rotation, period = 25 days.

Rotation of convection zone appears to increase with depth ($\partial\omega/\partial z > 0$, $z < 15\,000$ km) [29] (from splitting of oscillatory modes).

Surface rotation is measured by Doppler shifts (plasma) or proper motions of surface features (tracers). The rotation law depends on latitude, longitude and feature. The rate and degree of rigidity varies within the activity cycle and from cycle to cycle [36, 14].

Mean rates, angular velocity, period and linear equatorial velocity: see Table 6.

Table 6. Mean equatorial rotation rates of various features.
ω = angular velocity [degrees/d]
$T = \frac{360°}{\omega} = 50.51/v$ = sidereal period [d]
v = linear equatorial velocity [km s^{-1}]

Feature	ω degrees/d	T d	v km s^{-1}	Ref.
photospheric plasma (1966...68)	13.76	26.16	1.93	45
Spots (1905...54) (all)	14.52	24.79	2.04	86
(1878...1944) (longlived)	14.38	25.03	2.02	53
Faculae (1967...74) (large)	14.0	25.7	1.97	15
(small)	14.5	24.8	2.04	
Weak magnetic field (1959...70)	14.4	25.0	2.02	74
Corona (1947...70)	14.2	25.4	1.99	5
Carrington's co-ordinate system	14.18	25.38	1.99	85

Rotation laws:	Ref.
a) plasma	45, 46, 51, 50a
b) spots	86, 23
c) faculae/plage	52, 61, 6, 15
d) polar filaments	22
e) weak magnetic fields	88, 74, 75
f) EUV patterns	68, 41, 84
g) corona	40, 5, 4, 37

The more rapid rotation of tracers is consistent with their reflecting subsurface plasma rotation.
Conventional rotating co-ordinate system due to Carrington (1863) based on mean sunspot rotation rate at latitude $\simeq 16°$ [85].

3.1.1.2.7 General magnetic field

There is no permanent low-order multipolar field with surface average field strength
>1 G (total) [59, 32a]
>4 G (latitude 60···80°) [44].
Persistent field pattern in polar regions, flux beyond 55° latitude:
$\Phi = 3 \cdot 10^{22}$ Mx (solar minimum) [77].

Mean field in integrated light:
$\bar{B} = 0.15...0.5$ G (probably dependent on cycle) [59, 60].
Structure: essentially identical to interplanetary sector structure.
Random rms field strength: $B_{rms} \lesssim 110$ G;
outside network regions: $B_{rms} \lesssim 90$ G [76].

3.1.1.2.8 References for 3.1.1.2

1 Abraham, Z., Iben, I.: Astrophys. J. Lett. **162** (1970) L125.
2 Abraham, Z., Iben, I.: Astrophys. J. **170** (1971) 157.
3 Ando, H., Osaki, Y.: Publ. Astron. Soc. Japan **27** (1975) 581.
4 Antonucci, E., Dodero, M.A.: Sol. Phys. **53** (1977) 179, **62** (1979) 107.
5 Antonucci, E., Svalgaard, L.: Sol. Phys. **34** (1974) 3.
6 Antonucci, E., Azzarelli, L., Casalini, P., Cerri, S.: Sol. Phys. **53** (1977) 519, **63** (1979) 17, **61** (1979) 9.
7 Bahcall, J.N., Sears, R.L.: Annu. Rev. Astron. Astrophys. **10** (1972) 25.
8 Bahcall, J.N., Ulrich, R.K.: Astrophys. J. Lett. **160** (1970) L57.
9 Bahcall, J.N., Ulrich, R.K.: Astrophys. J. **170** (1971) 593.
10 Bahcall, J.N., Bahcall, N.A., Shaviv, G.: Phys. Rev. Lett. **20** (1968) 1209.
11 Bahcall, J.N., Bahcall, N.A., Ulrich, R.K.: Astrophys. Lett. **2** (1968) 91.
12 Bahcall, J.N., Bahcall, N.A., Ulrich, R.K.: Astrophys. J. **156** (1969) 559.
13 Bahcall, J.N., Huebner, W.F., Magee, N.H., Merts, A.L., Ulrich, R.K.: Astrophys. J. **184** (1973) 1.
14 Belvedere, G., Paternò, L. (eds.): Proc. Workshop on Solar Rotation, Oss. Astrofiz. Catania Publ. 162 (1979).
15 Belvedere, G., Godoli, G., Motta, S., Paternò, L., Zappalà, R.A.: Astrophys. J. Lett. **214** (1977) L91.
16 Bhavsar, S.P., Härm, R.: Astrophys. J. **216** (1977) 138.
17 Böhm, K.H., Stückl, E.: Z. Astrophys. **66** (1967) 487.
18 Boury, A., Gabriel, M., Noels, A., Scuflaire, R., Ledoux, P.: Astron. Astrophys. **41** (1975) 279.
19 Brookes, J.R., Isaak, G.R., van der Raay, H.B.: Nature (London) **259** (1976) 92.
20 Brookes, J.R., Isaak, G.R., McLeod, C.P., van der Raay, H.B., Roca Cortes, T.: Mon. Not. R. Astron. Soc. **184** (1978) 759.
21 Brown, T.M., Stebbins, R.T., Hill, H.A.: Astrophys. J. **223** (1978) 324.
22 Bruzek, A.: Z. Astrophys. **51** (1961) 75.
23 Chistyakov, V.F.: Bull. Astron. Inst. Czech. **27** (1976) 84.
24 Chitre, S.M., Ezer, D., Stothers, R.: Astrophys. Lett. **14** (1973) 37.
25 Christensen-Dalsgaard, J., Gough, D.O.: Nature (London) **259** (1976) 89.

26 Davis, R., Evans, J.C., Cleveland, B.T.: Proc. Neutrino 78 Conf. (Fowler, ed.) Purdue Univ. (1978) 53.
27 Demarque, P., Mengel, J.G., Sweigart, A.V.: Astrophys. J. **183** (1973) 997.
28 Deubner, F.L.: Proc. Symp. on Large-Scale Motions on the Sun, Sacramento Peak Observatory, Sunspot (1977) 77.
29 Deubner, F.L., Ulrich, R.K., Rhodes, E.J.: Astron. Astrophys. **72** (1979) 177.
30 Dicke, R.H.: Sol. Phys. **47** (1976) 475.
31 Dicke, R.H., Goldenberg, H.M.: Astrophys. J. Suppl. **27** (1974) 131.
32 Dilke, F.W.W., Gough, D.O.: Nature (London) **240** (1972) 262, 293.
32a Duvall, T.L., Scherrer, P.H., Svalgaard, L., Wilcox, J.M.: Sol. Phys. **61** (1979) 233.
33 Ezer, D., Cameron, A.G.W.: Nature (London) **240** (1972) 180.
34 Faulkner, D.J., Da Costa, G.S., Prentice, A.J.R.: Mon. Not. R. Astron. Soc. **170** (1975) 589.
35 Gabriel, M., Noels, A., Scuflaire, R., Boury, A.: Astron. Astrophys. **47** (1976) 137.
36 Gilman, P.A.: Annu. Rev. Astron. Astrophys. **12** (1974) 47.
37 Golub, L., Vaiana, G.S.: Astrophys. J. Lett. **219** (1978) L55.
38 Gough, D.O.: The Solar Output and its Variation (White, ed.), Colorado Assoc. Univ. Press, Boulder (1977).
39 Gough, D.O., Weiss, N.O.: Mon. Not. R. Astron. Soc. **176** (1976) 589.
40 Hansen, R.T., Hansen, S.F., Loomis, H.G.: Sol. Phys. **10** (1969) 135.
41 Henze, W., Dupree, A.K.: Sol. Phys. **33** (1973) 425.
42 Hill, H.A., Caudell, T.P.: Mon. Not. R. Astron. Soc. **186** (1979) 327.
43 Hill, H.A., Stebbins, R.T.: Astrophys. J. **200** (1975) 471.
44 Howard, R.: Sol. Phys. **52** (1977) 243.
45 Howard, R., Harvey, J.: Sol. Phys. **12** (1970) 23.
46 Howard, R., Yoshimura, H.: Int. Astron. Union Symp. **71** (1976) 19.
47 Hoyle, F.: Astrophys. J. Lett. **197** (1975) L127.
48 Iben, I.: Ann. Phys. N.Y. **54** (1969) 164.
49 Iben, I., Mahaffy, J.: Astrophys. J. Lett. **209** (1976) L39.
50 Kotov, V.A., Severny, A.B., Tsap, T.T.: Mon. Not. R. Astron. Soc. **183** (1978) 61.
50a Livingston, W., Duvall, T.L.: Sol. Phys. **61** (1979) 219.
51 Livingston, W., Milkey, R.: Sol. Phys. **25** (1972) 267.
52 Müller, R.: Z. Astrophys. **35** (1954) 61.
53 Newton, H.W., Nunn, M.L.: Mon. Not. R. Astron. Soc. **111** (1951) 413.
54 Prentice, A.J.R.: Astron. Astrophys. **50** (1976) 59.
55 Rhodes, E.J., Ulrich, R.K., Simon, G.W.: Proc. OSO-8 Workshop, LASP, Univ. Colorado, Boulder (1977) 365.
56 Rood, R.T.: Nature (London) **240** (1972) 178.
57 Rouse, C.A.: Astron. Astrophys. **44** (1975) 237.
58 Rouse, C.A.: Astron. Astrophys. **55** (1977) 477.
59 Scherrer, P.H., Wilcox, J.M., Kotov, V., Severny, A.B., Howard, R.: Sol. Phys. **52** (1977) 3.
60 Scherrer, P.H., Wilcox, J.M., Svalgaard, L., Duvall, T.L., Dittmer, P.H., Gustafson, E.K.: Sol. Phys. **54** (1977) 353.
61 Schröter, E.H., Wöhl, H.: Sol. Phys. **49** (1976) 19.
62 Scuflaire, R., Gabriel, M., Noels, A., Boury, A.: Astron. Astrophys. **45** (1975) 15.
63 Sears, R.L.: Astrophys. J. **140** (1964) 477.
64 Severny, A.B., Kotov, V.A., Tsap, T.T.: Proc. 2nd European Solar Physics Meeting, CNRS, Paris (1978) 123.
65 Shaviv, G., Bahcall, J.N.: Astrophys. J. **155** (1969) 135.
66 Shaviv, G., Beaudet, G.: Astrophys. Lett. **2** (1968) 17.
67 Shaviv, G., Salpeter, E.E.: Astrophys. J. **165** (1971) 171.
68 Simon, G.W., Noyes, R.W.: Sol. Phys. **26** (1972) 8.
69 Snell, R.L., Wheeler, J.C., Wilson, J.R.: Astrophys. Lett. **17** (1976) 157.
70 Spruit, H.C.: Sol. Phys. **34** (1974) 277.
71 Staude, J.: Bull. Astron. Inst. Czech. **27** (1976) 365.
72 Stein, R.F.: Astrophys. J. **154** (1968) 297.
73 Stein, R.F., Leibacher, J.: Annu. Rev. Astron. Astrophys. **12** (1974) 407.
74 Stenflo, J.O.: Sol. Phys. **36** (1974) 495.
75 Stenflo, J.O.: Astron. Astrophys. **61** (1977) 797.
76 Stenflo, J.O., Lindgren, L.: Astron. Astrophys. **59** (1977) 367.
77 Svalgaard, L., Duvall, T.L., Scherrer, P.H.: Sol. Phys. **58** (1978) 225.
78 Torres-Peimbert, S., Simpson, E., Ulrich, R.K.: Astrophys. J. **155** (1969) 957.
79 Travis, L.D., Matsushima, S.: Astrophys. J. **180** (1973) 975.

Durrant

80 Ulrich, R.K.: Astrophys. J. **158** (1969) 427.
81 Ulrich, R.K.: Astrophys. Space Sci. **9** (1970) 80.
82 Ulrich, R.K., Rhodes, E.J.: Astrophys. J. **218** (1977) 521.
83 Ulrich, R.K., Rood, R.T.: Nature (London) **241** (1973) 111.
84 Wagner, W.F.: Astrophys. J. Lett. **198** (1975) L141.
85 Waldmeier, M.: Landolt-Börnstein, NS, Vol. VI/1 (1965) 95.
86 Ward, F.: Astrophys. J. **145** (1966) 416.
87 Wheeler, J.C., Cameron, A.G.W.: Astrophys. J. **196** (1975) 601.
88 Wilcox, J.M., Schatten, K.H., Tanenbaum, A.S., Howard, R.: Sol. Phys. **14** (1970) 255.

3.1.1.3 Solar energy spectrum

3.1.1.3.1 Absolute energy distribution

Varies with position on the disk; spatially averaged to give values at a given heliocentric angle θ, ($\cos\theta=\mu$). Commonly given in terms of the spectral radiance (intensity) $I_\lambda(\mu)$ or the average radiance (mean intensity of the disk or flux) F_λ. Also given as brightness temperatures [u]:

$$T_B=\frac{14388}{\lambda \ln(11909/\lambda^5 I_\lambda+1)}$$ (I_λ in [W cm^{-2} µm^{-1} sr^{-1}], λ in [µm], T_B in [K]),

and as spectral irradiance S_λ, the flux F_λ received outside the earth's atmosphere at the mean sun-earth distance:

$$S_\lambda=6.799\cdot10^{-5}\,F_\lambda=\left(\frac{R_\odot}{\bar{r}_\odot}\right)^2\pi F_\lambda$$ ($R_\odot\equiv$solar radius, $\bar{r}_\odot\equiv$mean radius of earth's orbit).

Total irradiance, the solar constant:

$$S=\int_0^\infty S_\lambda \mathrm{d}\lambda$$ $S=(1370\pm1)\,\mathrm{W\,m^{-2}}$ [u] (see also [45]).

Systematic variation during activity cycle <0.75% [u].

Effective temperature defined by

$$\sigma T_{\rm eff}^4=\pi\int_0^\infty F_\lambda \mathrm{d}\lambda$$ with $\sigma=(5.6692\pm0.0007)\cdot10^{-5}$ erg cm^{-2} deg^{-4} s^{-1}:

$T_{\rm eff}=5780$ K.

Spectral energy distribution: see Tables 1 and 2.

Table 1. Sources of absolute data.
T_B=mean brightness temperature of disk
$T_B(0)$=central brightness temperature
S_λ=solar irradiance
$I_\lambda(\mu)$=intensity for given μ (centre of solar disk: $\mu=1$)
F_λ=flux

Wavelength range µm	Author	Ref.	Data given	Form
1000 ···10	White (1977)	u	$T_B(0)$, T_B, S_λ [1]	Table (compilation)
300 ··· 0.1	Vernazza et al. (1976)	75	$T_B(0)$, T_B, $I_\lambda(1)$, S_λ	Table, Atlas (compilation)
10 ··· 0.1	Thekaekara (1974)	73 [2]	S_λ	Table
3 ··· 0.3	Labs and Neckel (1973)	44	$I_\lambda(1)$, F_λ, S_λ	Table
0.33 ··· 0.017	Delaboudinière et al. (1978)	f	S_λ	Table, Atlas (compilation)
0.32 ··· 0.225	Kohl et al. (1978)	j	$I_\lambda(1)$, $I_\lambda(0.23)$	Table, Atlas
0.299··· 0.176	Brinkman et al. (1966)	d	S_λ	Table
0.21 ··· 0.175	Brückner et al. (1976)	15	$I_\lambda(1)$, F_λ, S_λ	Table, Atlas
0.21 ··· 0.14	Samain (1979)	69	$T_B(0)$, T_B, $I_\lambda(1)$, S_λ	Table
0.194··· 0.025	Heroux and Higgins (1977)	35	F_λ	Table
0.155··· 0.088	Brinkman et al. (1966)	d	S_λ	Table
0.03 ··· 0.001	White (1977)	u	S_λ	Table (compilation)

[1]) See Table 2 and footnote [1]).
[2]) Standard for American Society for Testing and Materials and NASA Space Vehicle Design Criteria. See also [43].

Table 2. Solar irradiances S_λ and central brightness temperatures $T_B(0)$ for $\lambda > 1500$ Å, and irradiances integrated over finite bandwidths ΔS for $\lambda < 1500$ Å. Dependence on activity and solar cycle is indicated in column A: blank ≡ no variation, × × × ≡ very strong variation. In cases of strong variation the figures given refer to weak to moderate activity. Major source of the radiation is indicated in the last column. For $1 > \lambda > 0.3$ µm the values are averaged over 200 Å bands, for $0.3 > \lambda$, over 100 Å bands.

λ µm	S_λ mW cm^{-2} µm^{-1}	Ref. [1])	$T_B(0)$ K	Ref.	A	Spectrum
1000	$3.37 \cdot 10^{-10}$	u	6000	u	×	low chromospheric H continuum
750	$1.01 \cdot 10^{-9}$	u	5612	u	×	
500	$4.76 \cdot 10^{-9}$	u	5100	u		
300	$3.33 \cdot 10^{-8}$	u	4600	u		
200	$1.58 \cdot 10^{-7}$	u	4417	u		
150	$4.88 \cdot 10^{-7}$ [2])	u	4394	u		
100	$2.45 \cdot 10^{-6}$	u	4440	u		
75	$7.75 \cdot 10^{-6}$	u	4478	u		photospheric H continuum
50	$3.92 \cdot 10^{-5}$	u	4530	u		
30	$3.01 \cdot 10^{-4}$	u	4610	u		
20	$1.51 \cdot 10^{-3}$	u	4709	u		
15	$4.75 \cdot 10^{-3}$	u	4841	u		
10	$2.40 \cdot 10^{-2}$	u	5100	u		
7	$9.90 \cdot 10^{-2}$	73				
5	$3.79 \cdot 10^{-1}$	73				
3	$3.1 \cdot 10^{0}$	73				
2	$1.17 \cdot 10^{1}$	44	6493	44		
1	$7.38 \cdot 10^{1}$	44	6080	44		
0.9	$9.09 \cdot 10^{1}$	44	6028	44		+molecular bands
0.8	$1.15 \cdot 10^{2}$	44	6050	44		
0.7	$1.43 \cdot 10^{2}$	44	6086	44		+photospheric absorption lines
0.6	$1.75 \cdot 10^{2}$	44	6147	44		
0.5	$1.93 \cdot 10^{2}$	44	6178	44		
0.4	$1.38 \cdot 10^{2}$	44	5993	44		photospheric absorption dense
0.35	$9.21 \cdot 10^{1}$	44	5819	44		
0.29	$4.41 \cdot 10^{1}$	f	5820	j		neutral metal continua +dense photospheric line absorption
0.27	$2.41 \cdot 10^{1}$	f	5588	j		
0.25	$6.0 \cdot 10^{0}$	f	5049	j		
0.23	$5.7 \cdot 10^{0}$	f	5172	j		
0.21	$2.4 \cdot 10^{0}$	f			×	
0.19	$2.9 \cdot 10^{-1}$	f			×	
0.17	$4.3 \cdot 10^{-2}$	f			×	
0.15	$7.5 \cdot 10^{-3}$	f			×	chromospheric emission lines

continued

Table 2 (continued)

Wavelength interval Å	ΔS mW cm^{-2}	Ref.	A	Spectrum
1400···1260	$4.0\cdot10^{-5}$	f	×	chromospheric emission lines
1260···1160	$4.0\cdot10^{-4}$	f	× [3])	Lα
1160···1027				
1027··· 911	$2.7\cdot10^{-5}$	f	×	+transition region lines
911··· 800	$2.7\cdot10^{-5}$	f	×	Lyman continuum
800··· 630	$8.6\cdot10^{-6}$	f	×	
630··· 460	$3.1\cdot10^{-5}$	f	×	
460··· 370	$3.3\cdot10^{-6}$	f	×	+coronal lines
370··· 280	$9.6\cdot10^{-5}$	f	×	
280··· 205	$5.5\cdot10^{-5}$	f	× ×	coronal lines
205··· 155	$9.0\cdot10^{-5}$	f	× ×	
155··· 31	$5.0\cdot10^{-5}$	u	× ×	
31··· 10	$(8\cdots13)\cdot10^{-7}$	u	× × ×	

[1]) Figures corrected for error in original [u].
[2]) Minimum mean brightness temperature of disk: $T_B=4423$ K, later value $T_B=4530^{+100}_{-150}$ K[65].
[3]) Variation of Lα with cycle ≃40···60% [f, u, 75a].

3.1.1.3.2 Relative energy distribution

Table 3. Atlases of the solar spectrum.
Symbols: see Table 1; furthermore: Id ≡ identification; E ≡ integrated intensity in emission lines; LSS ≡ quiet sun loops; CH ≡ coronal holes

Wavelength interval μm	Author	Ref.	Data
23.7 ···2.8	Migeotte et al. (1956)	l	$I_\lambda(1)$, Id
3.4 ···1.3	Benner et al. (1972)	7	I_λ
2.52 ···0.846	Mohler et al. (1950)	p	$I_\lambda(1)$
2.47 ···1.96; 1.80 ···1.49; 1.35 ···1.14	Hall (1973)	i	$I_\lambda(1)$, Id
1.20 ···0.750	Delbouille and Roland (1963)	g	$I_\lambda(1)$
1.00 ···0.300	Delbouille et al. (1973)	h	$I_\lambda(1)$
1.00 ···0.294	Brault and Testerman (1972)	c	$I_\lambda(1)$, $I_\lambda(0.20)$
0.877···0.333	Minnaert et al. (1940)	m	$I_\lambda(1)$
0.70 ···0.38	Beckers et al. (1976)	b	S_λ
0.363···0.299	Brückner (1960)	e	$I_\lambda(1)$, $I_\lambda(0)$
0.320···0.225	Kohl et al. (1978)	j	$I_\lambda(1)$, $I_\lambda(0.23)$, Id
0.299···0.223	Tousey et al. (1974)	t	S_λ
0.293···0.268	Allen et al. (1978)	a	$I_\lambda(1)$
0.21 ···0.117	Moe et al. (1976)	n	$I_\lambda(0.73)$, $I_\lambda(0.32)$
0.135···0.028	Vernazza and Reeves (1978)	74	$I_\lambda(1)$, $I_\lambda(0-)$, Id, E, (LSS, CH)
0.030···0.005	Malinovsky and Heroux (1973)	52	F_λ

Table 4. Line wavelengths, identifications and strengths. See also Table 3.

λ ≡ wavelength
W_λ ≡ equivalent width
I ≡ intensity (minimum for absorption lines, maximum for emission lines)
E ≡ integrated intensity in emission lines
Id ≡ identification
LSS ≡ quiet sun loops
CH ≡ coronal holes

Wavelength interval μm	Author	Ref.	Data
10.0 ···1.0	Biémont (1976)	9	λ, Id, W_λ (iron group neutrals)
8.0 ···3.0	Biémont and Zander (1977)	10	λ, Id, W_λ
3.0 ···1.0	Biémont (1973)	8	λ, Id, W_λ (CI)
2.56 ···1.20	Mohler (1955)	o	λ, Id, W_λ
1.20 ···0.75	Swensson et al. (1970)	r	λ, Id, I
0.927···0.304	Pierce (1968)	60	λ, Id (off-limb)
0.910···0.32	Dunn et al. (1968)	26	λ, Id, E (off-limb)
0.900···0.292	Pierce and Breckinridge (1973)	61	λ, Id
0.877···0.2935	Moore et al. (1966)	q	λ, Id, W_λ
0.760···0.610	Boyer et al. (1975···1978)	13	λ, Id, W_λ (diatomic molecules)
0.644···0.490	Sotirovski (1972)	71	λ, Id, W_λ (diatomic molecules)
0.353···0.340	Canfield et al. (1978)	17	λ, Id (emission lines) (limb)
0.32 ···0.20	Doschek et al. (1977)	24	λ, Id, E (limb)
0.30 ···0.12	Moore et al. (1977)	55	λ, Id (SiI)
0.283···0.276	Greve and McKeith (1977)	31	λ, Id
0.280···0.030	Burton and Ridgeley (1970)	16	λ, Id, E (limb)
0.194···0.1175	Doschek et al. (1976) Feldman et al. (1976)	25 29	λ, Id, I (limb) (LSS, CH)
0.191···0.1175	Moe and Nicolas (1977)	54	λ, Id, E (limb) (LSS)
0.194···0.100	Cohen et al. (1978)	18	λ, Id, I (flare)
0.150···0.080	Fawcett (1974)	27	λ, Id (highly ionized species)
0.080···0.0001	Fawcett (1974)	27	λ, Id
0.077···0.016	Behring et al. (1976)	6	λ, Id
0.063···0.0171	Dere (1978) Sandlin et al. (1976)	22 70	λ, Id (flare)
0.0171···0.0066	Kastner et al. (1974)	40	λ, Id (flare)

Table 5. General limb darkening measurements.
For W_λ and E, see Table 4.

Wavelength interval μm	Author	Ref.	Data
115.0 ···8.6	White (1977)	u	continuum (compilation)
20.0 ···0.148	Makarova and Kharitonov (1976)	50	continuum (compilation)
4.0 ···1.0	Koutchmy et al. (1977)	42	continuum
2.4 ···0.74	Pierce et al. (1977)	63	continuum
1.75	Wöhl (1975)	85	continuum, lines
0.73 ···0.30	Pierce and Slaughter (1977)	62	"continuum" [1])
0.7 ···0.4	Müller and Mutschlecner (1964)	56	Ca, Ti, V, Cr, Mn, Fe, Co lines (W_λ)
0.56 ···0.32	Makarova and Kharitonov (1977)	51	continuum plus lines (compilation)
0.52 ···0.37	Withbroe (1968)	83	C_2, CH, CN, CO, MgH lines (W_λ)
0.28 ···0.195	Bonnet (1968)	11	"continuum" [1]), lines
0.21 ···0.14	Samain (1979)	69	"continuum" [1])
0.140···0.050	Withbroe (1970) Reeves and Parkinson (1970)	84 67	lines (E)

[1]) Intensity extrema, subject to line haze.

Tables of strongest lines in the solar spectrum: see LB, NS, VI/1 [k] p. 100ff.
True central intensities (3083···7699 Å): [14].
Intensity and flux are related by the limb-darkening (or brightening) function $R_{\lambda\mu}$:

$$F_\lambda = I_\lambda(1) \int_0^1 R_{\lambda\mu} \mu \mathrm{d}\mu.$$

Limb darkening, broad band: see Table 5; in individual lines: see Table 6.
For the extreme limb and beyond in visible (eclipse observations): lines, see Table 4; continuum, see [72, 39].

Table 6. Individual line profile limb darkening measurements.

Species	Lines; λ in [Å]	Ref.	Species	Lines; λ in [Å]	Ref.
H	Hα···Hγ	78	Al I	1932	49
	Hα···Hδ	21	Si II	5 multiplets (1190···1818)	58
	Hα···H 16	37	Si III	1207, 1299, 1892	58
	Lα, Lβ	12	Ca I	4227	48
He I	10830	38		6573	4
	584, 537	53	Ca II	H, K	79
He II	304, 256	53, 20		8498, 8542, 8662	47, 70a
O I	9 lines (5577···8446)	57		[7324]	4
			Ti I	3641, 3685, 3913	48
Na I	D	76	Fe I	18 lines (3440···6430)	48
Mg I	b	77	Ba II	4554, 4131	68
	4571	80			
Mg II	h, k	41, 23	CO	2−0, 3−1 bands	3

Atlases of full-disk images:
a) visible — Title (1966) [s]
b) EUV — Reeves and Parkinson (1970) [67]
c) X-ray — Zombeck et al. (1978) [87]

Spatial variation of chromospheric lines:
Lα [11a]
Hα [32, 19]
Hβ [33]
Ca II K [34, 64]
Ca II K, 8542 Å [19]
Na D [19]
Fe I 3930 Å [19]
UV lines 1200···1560 Å [28]
UV lines 250···1350 Å [66]
Mg II h, k [23]

Line haze [36].

Line blocking coefficients η
2.5 ···0.3 μm averaged over 100 Å intervals [86]
1.25···0.33 μm averaged over 20 Å intervals [44]
0.62···0.43 μm averaged over 10 Å intervals [2].

Fraunhofer line statistics (counts per frequency and per equivalent width ranges) [1,31].

The sun as a star
a) spectrum ≡ G2V
b) colours *), $B-V=0.67\pm0.01$, $U-B=0.18\pm0.02$, ≡ G3V···G4V [5]
β-index = $2^{\mathrm{m}}.5955$, $b-y=0.39$ [59].

*) For definitions, see 4.2.5

3.1.1.3.3 Limb polarization

Measurements of linear polarization, P:
a) continuum: visible and near IR: [46]; UV: [30].
b) lines [81, 82].

Wavelength variation at fixed heliocentric angle: 2000···10 000 Å at $\mu=0.3$ [30].

Variation with heliocentric angle at fixed wavelength [46]:

$$\log 10^6 P = 2.35 - 1.93\,\mu + \frac{0.00106}{0.0004+\mu^3} - \frac{0.00017}{(1-\mu)^3};\quad (0.10<\mu<0.80 \text{ at } \lambda=5900\,\text{Å}).$$

3.1.1.3.4 References for 3.1.1.3

Catalogues and monographs

a Allen, M.S., McAllister, H.C., Jefferies, J.T.: High Resolution Atlas of the Solar Spectrum 2678···2913 Å, Institute for Astronomy, Univ. Hawaii (1978).
b Beckers, J.M., Bridges, C.A., Gilliam, L.B.: A High Resolution Spectral Atlas of the Solar Irradiance from 380 to 700 nm, AFGL-TR-76-0126 (I and II), Air Force Geophysics Lab., Hanscom AFB, Mass. (1976).
c Brault, J., Testerman, L.: Preliminary Kitt Peak Photoelectric Atlas, Kitt Peak National Observatory, Tucson (1972).
d Brinkman, R.T., Green, A.S., Barth, C.A.: Jet Propulsion Lab. Tech. Rept. 32–951 (1966).
e Brückner, G.E.: Photometric Atlas of the Near Ultraviolet Solar Spectrum 2988···3629 Å, Vandenhoeck and Ruprecht, Göttingen (1960).
f Delaboudinière, J.P., Donnelly, R.F., Hinteregger, H.E., Schmidtke, G., Simon, P.C.: Intercomparison/Compilation of Relevant Solar Flux Data related to Aeronomy, COSPAR Technique Manual 7, COSPAR, Paris (1978).
g Delbouille, L., Roland, G.: Photometric Atlas of the Solar Spectrum from λ 7498 to λ 12016, Mem. Soc. R. Sci. Liège, Special Vol. 4 (1963).
h Delbouille, L., Roland, G., Neven, L.: Photometric Atlas of the Solar Spectrum from 3000 Å to 10 000 Å, Inst. Astrophys., Univ. Liège, Liège (1973).
i Hall, D.N.B.: An Atlas of the Infrared Spectra of the Solar Photosphere and of Sunspot Umbrae, Kitt Peak National Observatory, Tucson (1973).
j Kohl, J.L., Parkinson, W.H., Kurucz, R.L.: Center and Limb Solar Spectrum in High Resolution, 225.2···319.6 nm, Center for Astrophysics, Cambridge, Mass. (1978).
k Landolt-Börnstein, NS, Vol. VI/1, Astronomy and Astrophysics (Voigt, H.H., ed.), Berlin, Heidelberg, New York: Springer (1965).
l Migeotte, M., Neven, L., Swensson, J.: The Solar Spectrum from 2.8 to 23.7 μm, Mem. Soc. R. Sci. Liège, Special Vols. 1, 2 (1956, 1957).
m Minnaert, M., Mulders, G.F.W., Houtgast, J.: A Photometric Atlas of the Solar Spectrum, de Schnabel, Amsterdam (1940).
n Moe, O.K., van Hoosier, M.E., Bartoe, J.D.F., Brückner, G.E.: A Spectral Atlas of the Sun between 1175 Å and 2100 Å, Naval Research Laboratory Rept. 8057, Washington (1976).
o Mohler, O.C.: Table of Solar Spectrum Wavelengths, Univ. Michigan Press, Ann. Arbor (1955).
p Mohler, O.C., Pierce, A.K., McMath. R.R., Goldberg, L.: Photometric Atlas of the Near Infrared Solar Spectrum λ 8465 to λ 25242, Univ. Michigan Press, Ann Arbor (1950).
q Moore, C.E., Minnaert, M.G.J., Houtgast, J.: The Solar Spectrum 2935 Å to 8770 Å, NBS Monograph 61, Washington (1966).
r Swensson, J.W., Benedict, W.S., Delbouille, L., Roland, G.: Solar Spectrum 7498 Å···12 016 Å: Table of Measures and Identifications, Mem. Soc. R. Sci. Liège, Special Vol. 5 (1970).
s Title, A.: Selected Spectroheliograms, California Inst. of Technology, Pasadena (1966).
t Tousey, R., Milone, E.F., Purcell, J.D., Palm Schneider, W., Tilford, S.G.: An Atlas of the Solar Ultraviolet Spectrum between 2226 and 2992 Å, Naval Research Laboratory Rept. 7788, Washington (1974).
u White, O.R. (ed.): The Solar Output and its Variation, Colorado Associated Univ. Press, Boulder (1977).

Special references

1 Allen, C.W.: Mon. Not. R. Astron. Soc. **148** (1970) 435.
2 Ardeberg, A., Virdefors, B.: Astron. Astrophys. **45** (1975) 19.
3 Ayres, T.R.: Astrophys. J. **225** (1978) 665.
4 Ayres, T.R., Testerman, L.: Sol. Phys. **60** (1978) 19.
5 Barry, D.C., Cromwell, R.H., Schoolman, S.A.: Astrophys. J. **222** (1978) 1032.
6 Behring, W.E., Cohen, L., Feldman, U., Doschek, G.A.: Astrophys. J. **203** (1976) 521.
7 Benner, D.C., Kuiper, G.P., Randić, L., Thomson, A.B.: Comm. Lunar Planet. Lab., Tucson **9** (1972) 155 and earlier.
8 Biémont, E.: Sol. Phys. **32** (1973) 117.
9 Biémont, E.: Astron. Astrophys. Suppl. **26** (1976) 89.
10 Biémont, E., Zander, R.: Astron. Astrophys. **56** (1977) 315.
11 Bonnet, R.M.: Ann. Astrophys. **31** (1968) 597.
11a Bonnet, R.M., Bruner, E.C., Acton, L.W., Brown, W.A., Decaudin, M.: Astrophys. J. **237** (1980) L47.
12 Bonnet, R.M., Lemaire, P., Vial, J.C., Artzner, G., Gouttebroze, P., Jouchoux, A., Leibacher, J.W., Skumanich, A., Vidal-Madjar, A.: Astrophys. J. **221** (1978) 1032.
13 Boyer, R., Sotirovski, P., Harvey, J.W.: Astron. Astrophys. Suppl. **19** (1975) 359; **24** (1976) 111; **33** (1978) 145.
14 Brault, J.W., Slaughter, C.D., Pierce, A.K., Aikens, R.S.: Sol. Phys. **18** (1971) 366.
15 Brückner, G.E., Bartoe, J.D.F., Moe, O.K., van Hoosier, M.E.: Astrophys. J. **209** (1976) 935.
16 Burton, W.M., Ridgeley, A.: Sol. Phys. **14** (1970) 3.
17 Canfield, R.C., Pasachoff, J.M., Stencel, R.E., Beckers, J.M.: Sol. Phys. **58** (1978) 263.
18 Cohen, L., Feldman, U., Doschek, G.A.: Astrophys. J. Suppl. **37** (1978) 393.
19 Cram, L.E., Brown, D.R., Beckers, J.M.: Astron. Astrophys. **57** (1977) 211.
20 Cushman, G.W., Rense, W.A.: Sol. Phys. **58** (1978) 299.
21 David, K.H.: Z. Astrophys. **53** (1961) 37.
22 Dere, K.P.: Astrophys. J. **221** (1978) 1062.
23 Doschek, G.A., Feldman, U.: Astrophys. J. Suppl. **35** (1977) 471.
24 Doschek, G.A., Feldman, U., Cohen, L.: Astrophys. J. Suppl. **33** (1977) 101.
25 Doschek, G.A., Feldman, U., van Hoosier, M.E., Bartoe, J.D.F.: Astrophys. J. Suppl. **31** (1976) 417.
26 Dunn, R.B., Evans, J.W., Jefferies, J.T., Orrall, F.Q., White, O.R., Zirker, J.B.: Astrophys. J. Suppl. **15** (1968) 275.
27 Fawcett, B.C.: Adv. At. Mol. Phys. **10** (1974) 223.
28 Feldman, U., Doschek, G.A., Patterson, N.P.: Astrophys. J. **209** (1976) 270.
29 Feldman, U., Doschek, G.A., van Hoosier, M.E., Purcell, J.D.: Astrophys. J. Suppl. **31** (1976) 445.
30 Goutail, F.: Astron. Astrophys. **64** (1978) 73.
31 Greve, A., McKeith, C.D.: Astron. Astrophys. Suppl. **30** (1977) 387.
32 Grossmann-Doerth, U., von Uexküll, M.: Sol. Phys. **28** (1973) 319.
33 Grossmann-Doerth, U., von Uexküll, M.: Sol. Phys. **42** (1975) 303.
34 Grossmann-Doerth, U., Kneer, F., von Uexküll, M.: Sol. Phys. **37** (1974) 85.
35 Heroux, L., Higgins, J.E.: J. Geophys. Res. **82** (1977) 3307.
36 Holweger, H.: Astron. Astrophys. **4** (1970) 11.
37 de Jager, C.: Rech. Utrecht **13** (1952) 1.
38 de Jager, C., Namba, O., Neven, L.: Bull. Astron. Inst. Neth. **18** (1966) 128.
39 Jordan, C., Ridgeley, A.: Mon. Not. R. Astron. Soc. **168** (1974) 533.
40 Kastner, S.O., Neupert, W.M., Swartz, M.: Astrophys. J. **191** (1974) 261.
41 Kohl, J.L.. Parkinson, W.H.: Astrophys. J. **205** (1976) 599.
42 Koutchmy, S., Koutchmy, O., Kotov, V.: Astron. Astrophys. **59** (1977) 189.
43 Labs, D.: Problems in Stellar Atmospheres and Envelopes (Baschek, Kegel and Traving, eds.), Berlin, Heidelberg, New York: Springer (1975) 1.
44 Labs, D., Neckel, H.: Proc. Symp. on Solar Radiation (Klein and Hickey, eds.), Smithson. Inst. Radiation Biology Lab., Rockville (1973) 269; Sol. Phys. **15** (1970) 79; Z. Astrophys. **69** (1968) 1.
45 Labs, D., Neckel, H.: Sol. Phys. **19** (1971) 3.
46 Leroy, J.L.: Lund Obs. Rept. **12** (1977) 161; Sol. Phys. **36** (1974) 81; Astron. Astrophys. **19** (1972) 287.
47 Linsky, J.L., Teske, R.G., Wilkinson, C.W.: Sol. Phys. **11** (1970) 374.
48 Lites, B.W., Brault, J.W.: Sol. Phys. **30** (1973) 283.
49 McAllister, H.C.: Sol. Phys. **35** (1974) 3.
50 Makarova, E.A., Kharitonov, A.V.: Astron. Zh. **53** (1976) 1234; Sov. Astron. AJ (English Transl.) **20** (1976) 698.

51 Makarova, E.A., Kharitonov, A.V.: Astron. Zh. **54** (1977) 115; Sov. Astron. AJ (English Transl.) **21** (1977) 65.
52 Malinovsky, M., Heroux, L.: Astrophys. J. **181** (1973) 1009.
53 Mango, S.A., Bohlin, J.D., Glackin, D.L., Linsky, J.L.: Astrophys. J. **220** (1978) 683.
54 Moe, O.K., Nicolas, K.R.: Astrophys. J. **211** (1977) 579.
55 Moore, C.E., Brown, C.M., Sandlin, G.D., Tilford, S.G., Tousey, R.: Astrophys. J. Suppl. **33** (1977) 393.
56 Müller, E.A., Mutschlecner, J.P.: Astrophys. J. Suppl. **9** (1964) 1.
57 Müller, E.A., Baschek, B., Holweger, H.: Sol. Phys. **3** (1968) 125.
58 Nicolas, K.R., Brückner, G.E., Tousey, R., Tripp, D.A., White, O.R., Athay, R.G.: Sol. Phys. **55** (1977) 305.
59 Olsen, E.H.: Astron. Astrophys. **50** (1976) 117.
60 Pierce, A.K.: Astrophys. J. Suppl. **17** (1968) 1.
61 Pierce, A.K., Breckinridge, J.B.: Kitt Peak National Obs. Contrib. 559 (1973).
62 Pierce, A.K., Slaughter, C.D.: Sol. Phys. **51** (1977) 25.
63 Pierce, A.K., Slaughter, C.D., Weinberger, D.: Sol. Phys. **52** (1977) 179.
64 Punetha, L.M.: Bull. Astron. Inst. Czech. **25** (1974) 207.
65 Rast, J., Kneubühl, F.K., Müller, E.A.: Astron. Astrophys. **68** (1978) 229.
66 Reeves, E.M.: Sol. Phys. **46** (1976) 53.
67 Reeves, E.M., Parkinson, W.H.: Astrophys. J. Suppl. **21** (1970) 1.
68 Rutten, R.J.: Sol. Phys. **56** (1978) 237.
69 Samain, D.: Astron. Astrophys. **74** (1979) 225.
70 Sandlin, G.D., Brückner, G.E., Scherrer, V.E., Tousey, R.: Astrophys. J. Lett. **205** (1976) L47.
70a Shine, R.A., Linsky, J.L.: Sol. Phys. **25** (1972) 357.
71 Sotirovski, P.: Astron. Astrophys. Suppl. **6** (1972) 85.
72 Tanaka, K., Hiei, E.: Publ. Astron. Soc. Japan **24** (1972) 323.
73 Thekaekara, M.P.: Appl. Opt. **13** (1974) 518.
74 Vernazza, J.E., Reeves, E.M.: Astrophys. J. Suppl. **37** (1978) 485.
75 Vernazza, J.E., Avrett, E.H., Loeser, R.: Astrophys. J. Suppl. **30** (1976) 1.
75a Vidal-Madjar, A., Phissamay, B.: Sol. Phys. **66** (1980) 259.
76 Waddell, J.: Astrophys. J. **136** (1962) 223.
77 Waddell, J.: Astrophys. J. **137** (1963) 1210.
78 White, O.R.: Astrophys. J. Suppl. **7** (1962) 333.
79 White, O.R., Suemoto, Z.: Sol. Phys. **3** (1968) 523.
80 White, O.R., Altrock, R.C., Brault, J.W., Slaughter, C.D.: Sol. Phys. **23** (1972) 18.
81 Wiehr, E.: Astron. Astrophys. **38** (1975) 303.
82 Wiehr, E.: Astron. Astrophys. **67** (1978) 257.
83 Withbroe, G.L.: Sol. Phys. **3** (1968) 146.
84 Withbroe, G.L.: Sol. Phys. **11** (1970) 42, 208.
85 Wöhl, H.: Sol. Phys. **43** (1975) 285.
86 Wöhl, H.: Astron. Astrophys. **40** (1975) 343.
87 Zombeck, M.V., Vaiana, G.S., Haggerty, R., Krieger, A.S., Silk, J.K., Timothy, A.: Astrophys. J. Suppl. **38** (1978) 69.

3.1.1.4 Solar photosphere and chromosphere

3.1.1.4.1 Models

The emergent intensity from a plane-parallel homogeneous atmosphere is

$$I_\lambda(\mu) = \int_0^\infty S_\lambda(\tau_\lambda)\, e^{-\tau_\lambda/\mu} \frac{d\tau_\lambda}{\mu},$$

where τ_λ is the optical depth at wavelength λ, S_λ is the source function and $\mu = \cos\theta$, θ being the inclination of the line of sight to the vertical ($\simeq$ heliocentric angle). Direct inversion of frequency or centre-to-limb intensity variations generally yields limited information [63, 22, 38]. Most models are semi-empirical in which a $T(\tau)$ relationship is adjusted until a chosen set of observed properties is reproduced. Hydrostatic balance is assumed since observed velocity fields are subsonic. In the low photosphere the source function is given to a good approximation by the Planck function and ionization/excitation balance by the Saha-Boltzmann equation (local thermodynamic equilibrium, LTE). Higher in the atmosphere, the dependence of the radiation field on atmospheric parameters requires a full non-LTE (NLTE) treatment (e.g. [e]).

Semi-empirical models: see Table 1.

Table 1. Semi-empirical solar atmosphere models.

Author	Ref.	Scope	Observational basis
Gingerich and de Jager (1968) (Bilderberg)	30	photosphere – middle chromosphere	continua
Gingerich et al. (1971) (HSRA)	31	photosphere – middle chromosphere	continua
Vernazza et al. (1973) (VAL)	78	photosphere – high chromosphere	continua and lines
Holweger and Müller (1974)	39	photosphere	continuum and line strengths
Vernazza et al. (1976) (VAL)	79	photosphere – low chromosphere	continua
Tanaka and Hiei (1972)	73	upper photosphere – low chromosphere	continuum at limb
Linsky and Avrett (1970)	48	upper photosphere – low chromosphere	Ca II lines
Lites (1973)	50	upper photosphere – low chromosphere	Fe I lines
Mount and Linsky (1974)	58	upper photosphere	CN lines
Ayres and Linsky (1976)	9	upper photosphere – low chromosphere	Ca II, Mg II lines
Basri et al. (1979)	10	middle chromosphere	Lα line
Athay and Canfield (1970)	5	upper chromosphere	O I lines
Linsky et al. (1976)	49	upper chromosphere	He II
Lites et al. (1978)	52	upper chromosphere	C II lines

Deviations from plane-parallel atmospheres: see below.

Theoretical models predict the temperature structure from physical premises:

a) Photospheric models are based on radiative/convective equilibrium; LTE [44, 13]; NLTE [4].
b) Chromospheric models introduce mechanical heating [76].

Run of thermodynamic parameters in the solar photosphere-chromosphere: see Table 2.

Table 2. Photospheric and chromospheric models.

h = height
τ_{5000} = optical depth at 5000 Å
T = temperature
T(VAL) taken from [78];
T(AL) and T(K) = temperatures at the same column mass density interpolated from the models AL and K, respectively
P_e = electron pressure (VAL)
P = total pressure (VAL)
ξ_v = vertical component of microturbulence [12]
ξ_h = horizontal component of microturbulence [12]

VAL model: Vernazza, Avrett, Loeser [78]
AL model: Ayres, Linsky [9]
K model: Kurucz [44]

h km	τ_{5000}	T(VAL) K	T(AL) K	T(K) K	P_e dyn cm^{-2}	P dyn cm^{-2}	ξ_v km s^{-1}	ξ_h km s^{-1}
800	$2.20\cdot10^{-5}$	5360	5280	3970	$7.37\cdot10^{-2}$	$1.08\cdot10^{2}$	1.7	1.7
700	$3.53\cdot10^{-5}$	4890	4980	4070	$4.93\cdot10^{-2}$	$2.41\cdot10^{2}$		
600	$7.35\cdot10^{-5}$	4350	4590	4200	$6.30\cdot10^{-2}$	$5.94\cdot10^{2}$	1.0	1.0
500	$3.36\cdot10^{-4}$	4150	4450	4350	$1.29\cdot10^{-1}$	$1.60\cdot10^{3}$	0.7	0.8
400	$1.92\cdot10^{-3}$	4330	4570	4510	$3.44\cdot10^{-1}$	$4.28\cdot10^{3}$	0.6	0.8
300	$1.01\cdot10^{-2}$	4600	4775	4640	$9.40\cdot10^{-1}$	$1.09\cdot10^{4}$	0.7	1.0
200	$4.79\cdot10^{-2}$	4920	4980	4820	$2.48\cdot10^{0}$	$2.61\cdot10^{4}$	0.9	1.4
100	$2.04\cdot10^{-1}$	5445	5445	5150	$7.59\cdot10^{0}$	$5.87\cdot10^{4}$	1.3	2.0
0	1.00	6423	6423	6160	$5.85\cdot10^{1}$	$1.20\cdot10^{5}$	1.8	2.7
− 30	2.00	7040	7040	7030	$1.92\cdot10^{2}$	$1.45\cdot10^{5}$		
− 60	4.94	7880	7880	7900	$7.66\cdot10^{2}$	$1.71\cdot10^{5}$		

Temperature plateau in upper chromosphere: 20 000 K (VAL model) [78],
16 500 K [52].

Unresolved velocities, known as microturbulence, determined by line fitting. Centre-to-limb variation analysed with mean atmospheric models requires anisotropic broadening:

a) photosphere: see Table 2 [12, 14, 41],

b) chromosphere: isotropic, increasing rapidly to 12 km s^{-1} at 2500 km [12].
Perhaps identifiable with short period wave broadening [24].

Total velocity field (microturbulence plus resolved fields) [12, 36].

The sun-as-a-star [34]:
microturbulence $V_t = 0.5$ km s^{-1},
radial-tangential macroturbulence $V_{RT} = 3.8 \cdots 3.1$ km s^{-1} from low to high photosphere.

3.1.1.4.2 NLTE studies

Ionization and excitation balances depart from the local thermodynamic equilibrium in the upper photosphere and higher layers. In Fe I, ionization balance goes into NLTE at $\tau_{5000} \approx 0.05$ and excitation of low lying levels goes into NLTE around the temperature minimum [50]. Profile analysis requires full NLTE treatment; equivalent widths of photospheric lines do not depart significantly from LTE values.

Atomic species analysed in NLTE: see Table 3.

Table 3. NLTE studies in the sun. Listed are the atomic species and the principal lines and continua considered.

Species	Major continua and lines in [Å]	Author	Ref.
H I	Hα···γ, Pα, β, Bα	Zelenka (1977)	83
	911 continuum, Lα	Basri et al. (1979)	10
	911 continuum, Lα, β	Gouttebroze et al. (1978)	33
He I	504 continuum, 584, 537, 10 830	Milkey et al. (1973)	57
	504 continuum	Avrett et al. (1976)	8
He II	227 continuum, 304, 256, 243, 237, 234	Linsky et al. (1976)	49
	227 continuum, 304	Avrett et al. (1976)	8
C I	1100, 1239, 1444 continua	Vernazza (1972)	77
	1560, 1657	Shine et al. (1978)	67
C II	1335	Lites et al. (1978)	52
C III	1176	Chipman (1971)	18
O I	1304, 1357	Chipman (1971)	18 [1])
	7773, 8446	Sedlmayr (1974)	66
Na I	D, 5683, 6161, 8183, 11 404	Gehren (1975)	29
	D	Schleicher (1976)	64a
Mg I	b, 2852, 4571	Altrock and Canfield (1974)	2
	2852	Canfield and Cram (1977)	17
		Heasley and Allen (1980)	38a
	b, 4571	Schleicher (1976)	64a
Mg II	h, k	Ayres and Linsky (1976)	9
		Heasley and Allen (1980)	38a
Al I	1932, 1936	Finn and Jefferies (1974)	28
Si I	1525, 1682 continua	Vernazza et al. (1976)	79
Si II	1265, 1533, 1817	Tripp et al. (1978)	74
Si III	1207	Tripp et al. (1978)	74
S I	1807, 1900	Doherty and McAllister (1978)	26
Ca I	4227	Lites (1974)	51
Ca II	H, K, 8542	Shine et al. (1975)	68
	K, 8542	Ayres and Linsky (1976)	9
	H, K, 8498, 8542, 8662	Schleicher (1976)	64a
Fe I	18 lines	Athay and Lites (1972)	6
		Lites (1973)	50
Ba II	4554	Rutten (1978)	64
V II, Fe I, Zr II		Tanaka (1971)	72
Rare earths		Canfield (1971)	16

[1]) See also non-solar work, Haisch et al. (1977) [37].

3.1.1.4.3 Morphology of the solar photosphere and chromosphere

3.1.1.4.3.1 Granulation

See general reference Bray and Loughhead (1967) [b].

rms intensity fluctuations:
6.7···15.0 in continuum at 5500···6000 Å at disk centre [80],
decreasing with increasing heliocentric angle and wavelength [1, 65],
in lines increasing with height [43].

Mean granule/intergranule contrast [15].
Structural size distribution [42, 25].
Mean horizontal wavelength $\lesssim$1040 km [25].
Mean cell size 1380 km [15].

Granular evolution and fragmentation [55, 41a].
Life times: number decay time 7.4···8.3 min,
correlation decay time 5.9 min.

rms velocity fluctuations [27, 42a, 43]:
$\simeq$0.5 km s^{-1} vertical, $\simeq$2.5 km s^{-1} horizontal at h=150 km;
both decrease with height, h, in the photosphere.

Models
a) multicomponent without horizontal exchange [54]
b) multicomponent with exchange [75, 61]
c) two-dimensional [19a, 47, 60, 59].

rms temperature fluctuations [60]:
$\simeq$ 370 K at h=0
decreasing to zero at h=100 km
small values in high photosphere anticorrelated with low photosphere.

Measured total temperature fluctuations in upper photosphere [56, 3].

3.1.1.4.3.2 Supergranulation and network

A cellular flow structure in the low photosphere and a brightness pattern in the upper photosphere and chromosphere [69].
Horizontal velocity 0.3...0.4 km s^{-1} in low photosphere decreasing with height [12].
Vertical velocities [82]:
50 m s^{-1} upwards at cell centres
200 m s^{-1} downwards at some boundary points.

Mean cell size: 34000 km [71].
Life times, network pattern $\approx$36 h,
velocity correlation decay time 19···21 h [82].
Temperature gradient across cell undetectable [81].
Steady chromospheric flow 0.8···3.0 km s^{-1}, increasing with height, almost isotropic [62].

3.1.1.4.3.3 Oscillations

rms velocity:
0.5 km s^{-1} at $h\simeq$100 km
increasing slightly outwards to low chromosphere [43],
increasing rapidly in chromosphere [21],
predominantly vertical [70].

Power distribution in frequency and wavenumber space [23].
Variation with height [20].

Maximum power at longest wavelengths $\gtrsim$10000 km and periods $\simeq$300 s in photosphere,
$\simeq$300···180 s in chromosphere
showing strong spatial variation associated with large brightness variations in both network and cell interiors [32, 19, 7].
Temperature fluctuation $\simeq$20 K at h=300 km [40].

Models: see 3.1.1.2 Global oscillations, also [46].

Other wave modes [20].

3.1.1.4.3.4 Network elements

Appear in the chromosphere as mottles seen on the solar disk and as spicules seen beyond the limb.

Properties of chromospheric structures: see Table 4.

Interspicular region is not well determined but certainly contains a coronal component [11].

Height distribution of spicules [11, 45, 53].
Physical parameters in spicules [11]:

h km	T_e K	N_e 10^{10} cm^{-3}
3 400	9 000	18.9
7 000	15 000	7.5
10 800	16 500	3.0

Durrant

Table 4. Properties of chromospheric mottles and spicules [c].

Feature	Height at mid point km	Extension km		Flow $km\,s^{-1}$	Lifetime min
		vertical	horizontal		
Bright mottle	725···3 300	1 000··· 2 200	1 450···4 000	4 [1])	11···12
Dark mottle	5 000···7 600	8 000···12 000	1 000···8 000	4 [1])	5
Spicule	4 900	9 800	800··· 900	25	5

[1]) From [35].

3.1.1.4.4 References for 3.1.1.4

Monographs

a Athay, R.G.: The Solar Chromosphere and Corona: The Quiet Sun, Reidel, Dordrecht (1976).
b Bray, R.J., Loughhead, R.E.: The Solar Granulation, Chapman and Hall, London (1967).
c Bray, R.J., Loughhead, R.E.: The Solar Chromosphere, Chapman and Hall, London (1974).
d Gibson, E.A.: The Quiet Sun, NASA SP-303, Washington (1973).
e Mihalas, D.: Stellar Atmospheres, Freeman, San Francisco (1978).

Special references

1 Albregtsen, F., Hansen, T.L.: Sol. Phys. **54** (1977) 31.
2 Altrock, R.C., Canfield, R.C.: Astrophys. J. **194** (1974) 733.
3 Altrock, R.C., Keil, S.L.: Astron. Astrophys. **57** (1977) 159.
4 Athay, R.G.: Astrophys. J. **161** (1970) 713.
5 Athay, R.G., Canfield, R.C.: Spectrum Formation in Stars with Steady-State Extended Atmospheres (Groth and Wellmann, eds.), NBS Special Pub. 332 (1970) 65.
6 Athay, R.G., Lites, B.W.: Astrophys. J. **176** (1972) 809.
7 Athay, R.G., White, O.R.: Astrophys. J. Suppl. **39** (1979) 333.
8 Avrett, E.H., Vernazza, J.E., Linsky, J.L.: Astrophys. J. Lett. **207** (1976) L199.
9 Ayres, T.R., Linsky, J.L.: Astrophys. J. **205** (1976) 874.
10 Basri, G.S., Linsky, J.L., Bartoe, J.D.F., Brückner, G., Van Hoosier, M.E.: Astrophys. J. **230** (1979) 924.
11 Beckers, J.M.: Annu. Rev. Astron. Astrophys. **10** (1972) 73.
12 Beckers, J.M., Canfield, R.C.: Motions in the Solar Atmosphere, AFCRL-TR-75-0592, Air Force Cambridge Hanscom, Mass. (1975).
13 Bell, R.A., Ericksson, K., Gustafsson, B., Nordlund, Å.: Astron. Astrophys. Suppl. **23** (1976) 37.
14 Blackwell, D.E., Ibbetson, P.A., Petford, A.D., Willis, R.B.: Mon. Not. R. Astron. Soc. **177** (1976) 227.
15 Bray, R.J., Loughhead, R.E.: Sol. Phys. **54** (1977) 319.
16 Canfield, R.C.: Astron. Astrophys. **10** (1971) 54, 64.
17 Canfield, R.C., Cram, L.E.: Astrophys. J. **216** (1977) 654.
18 Chipman, E.G.: Ph. D. Thesis, Harvard Univ. (1971).
19 Chipman, E.G.: Sol. Phys. **55** (1977) 277.
19a Cloutman, L.D.: Astrophys. J. **227** (1979) 614.
20 Cram, L.E.: Astron. Astrophys. **70** (1978) 345.
21 Cram, L.E., Brown, D.R., Beckers, J.M.: Astron. Astrophys. **57** (1977) 211.
22 Curtis, G.W., Jefferies, J.T.: Astrophys. J. **150** (1967) 1061.
23 Deubner, F.L.: Astron. Astrophys. **44** (1975) 371.
24 Deubner, F.L.: Astron. Astrophys. **51** (1976) 189.
25 Deubner, F.L., Mattig, W.: Astron. Astrophys. **45** (1975) 167.
26 Doherty, L.R., McAllister, H.C.: Astrophys. J. **222** (1978) 716.
27 Durrant, C.J., Mattig, W., Nesis, A., Reiss, G., Schmidt, W.: Sol. Phys. **61** (1979) 251.
28 Finn, G.D., Jefferies, J.T.: Sol. Phys. **34** (1974) 57.
29 Gehren, T.: Astron. Astrophys. **38** (1975) 289.
30 Gingerich, O., de Jager, C.: Sol. Phys. **3** (1968) 5.
31 Gingerich, O., Noyes, R.W., Kalkofen, W., Cuny, Y.: Sol. Phys. **18** (1971) 347.
32 Giovanelli, R.G.: Int. Astron. Union Symp. **56** (1974) 137.

33 Gouttebroze, P., Lemaire, P., Vial, J.C., Artzner, G.: Astrophys. J. **225** (1978) 655.
34 Gray, D.F.: Astrophys. J. **218** (1977) 530.
35 Grossmann-Doerth, U., von Uexküll, M.: Sol. Phys. **20** (1971) 31.
36 Gurtovenko, E.A.: Sol. Phys. **45** (1975) 25.
37 Haisch, B.M., Linsky, J.L., Weinstein, A., Shine, R.A.: Astrophys. J. **214** (1977) 785.
38 Hearn, A.G., Holt, J.N.: Astron. Astrophys. **23** (1973) 347.
38a Heasley, J.N., Allen, M.S.: Astrophys. J. **237** (1980) 255.
39 Holweger, H., Müller, E.A.: Sol. Phys. **39** (1974) 19.
40 Holweger, H., Testerman, L.: Sol. Phys. **43** (1975) 271.
41 Holweger, H., Gehlsen, M., Ruland, F.: Astron. Astrophys. **70** (1978) 537.
41a* Ichiro, K.: Sol. Phys. **65** (1980) 207.
42 Karpinsky, V.N., Mekhanikov, V.V.: Sol. Phys. **54** (1977) 25.
42a Keil, S.L.: Astrophys. J. **237** (1980) 1024.
43 Keil, S.L., Canfield, R.C.: Astron. Astrophys. **70** (1978) 169.
44 Kurucz, R.L.: Sol. Phys. **34** (1974) 17.
45 Lantos, P., Kundu, M.R.: Astron. Astrophys. **21** (1972) 119.
46 Leibacher, J.W.: Ph. D. Thesis, Harvard Univ. (1971).
47 Levy, M.L.: Astron. Astrophys. **31** (1974) 451.
48 Linsky, J.L., Avrett, E.H.: Publ. Astron. Soc. Pac. **82** (1970) 169.
49 Linsky, J.L., Glackin, D.L., Chapman, R.D., Neupert, W.M., Thomas, R.J.: Astrophys. J. **203** (1976) 509.
50 Lites, B.W.: Sol. Phys. **32** (1973) 283.
51 Lites, B.W.: Astron. Astrophys. **30** (1974) 297.
52 Lites, B.W., Shine, R.A., Chipman, E.G.: Astrophys. J. **222** (1978) 333.
53 Lynch, D.K., Beckers, J.M., Dunn, R.B.: Sol. Phys. **30** (1973) 63.
54 Margrave, T.E., Swihart, T.L.: Sol. Phys. **6** (1969) 12.
55 Mehltretter, J.P.: Astron. Astrophys. **62** (1978) 311.
56 Mein, P.: Sol. Phys. **20** (1971) 3.
57 Milkey, R.W., Heasley, J.N., Beebe, H.A.: Astrophys. J. **186** (1973) 1043.
58 Mount, G.H., Linsky, J.L.: Sol. Phys. **35** (1974) 259.
59 Nelson, G.D.: Sol. Phys. **60** (1978) 5.
60 Nelson, G.D., Musman, S.: Astrophys. J. **214** (1977) 912.
61 Nordlund, Å.: Astron. Astrophys. **50** (1976) 23.
62 November, L.J., Toomre, J., Gebbie, K.B., Simon, G.W.: Astrophys. J. **227** (1979) 600.
63 Pierce, A.K., Waddell, J.: Mem. R. Astron. Soc. **68** (1961) 89.
64 Rutten, R.J.: Sol. Phys. **56** (1978) 237.
64a Schleicher, H.: Dissertation, Göttingen (1976).
65 Schmidt, W., Deubner, F.L., Mattig, W., Mehltretter, J.P.: Astron. Astrophys. **75** (1979) 223.
66 Sedlmayr, E.: Astron. Astrophys. **31** (1974) 23.
67 Shine, R.A., Lites, B.W., Chipman, E.G.: Astrophys. J. **224** (1978) 247.
68 Shine, R.A., Milkey, R.W., Mihalas, D.: Astrophys. J. **199** (1975) 724.
69 Skumanich, A., Smythe, C., Frazier, E.N.: Astrophys. J. **200** (1975) 747.
70 Stix, M., Wöhl, H.: Sol. Phys. **37** (1974) 63.
71 Sykora, J.: Sol. Phys. **13** (1970) 292.
72 Tanaka, K.: Publ. Astron. Soc. Japan **23** (1971) 217.
73 Tanaka, K., Hiei, E.: Publ. Astron. Soc. Japan **24** (1972) 323.
74 Tripp, D.A., Athay, R.G., Peterson, V.L.: Astrophys. J. **220** (1978) 314.
75 Turon, P.: Sol. Phys. **41** (1975) 271.
76 Ulmschneider, P., Schmitz, F., Kalkofen, W., Bohn, H.U.: Astron. Astrophys. **70** (1978) 487.
77 Vernazza, J.E.: Ph. D. Thesis, Harvard Univ. (1972).
78 Vernazza, J.E., Avrett, E.H., Loeser, R.: Astrophys. J. **184** (1973) 605.
79 Vernazza, J.E., Avrett, E.H., Loeser, R.: Astrophys. J. Suppl. **30** (1976) 1.
80 Wittmann, A., Mehltretter, J.P.: Astron. Astrophys. **61** (1977) 75.
81 Worden, S.P.: Sol. Phys. **45** (1975) 521.
82 Worden, S.P., Simon, G.W.: Sol. Phys. **46** (1976) 73.
83 Zelenka, A.: Astron. Astrophys. **48** (1976) 75.

* **Durrant**

p. 101 ref. 41a: *instead of* Ichiro, K. *read* Kawaguchi, I.

3.1.1.5 Solar transition region and quiet corona

Seen as line emission (transition region and L-corona) or continuum radiation of photospheric origin scattered by free electrons (K-corona) or interplanetary dust (F-corona = scattered Fraunhofer spectrum; see also 3.3.4).

TZ = transition region or transition zone is the atmospheric plasma with temperatures between chromospheric ($\simeq 2\cdot 10^4$ K) and coronal ($\simeq 10^6$ K).

QC = quiet corona has two morphological forms:

a) LSS = large scale structure, in which the magnetic field lines are closed over large distances but generally low heights of a few tenths of $R_\odot$ (quiet sun loops),

b) CH = coronal holes, in which the magnetic field lines are open over large areas.

Both overlie normal quiet photosphere/chromosphere [a, 56, 59].

Spectrum: see 3.1.1.4, Table 4.
Prominent coronal line list [d].

3.1.1.5.1 Models

The plasma is highly heterogeneous throughout, so recent work has emphasized mapping structural features rather than constructing mean models.

Empirical models:

a) one-dimensional

TZ	Dupree (1972) [15], Withbroe and Gurman (1973) [57], Burton et al. (1973) [8], Withbroe (1975) [55]
CH	Munro and Withbroe (1972) [40], Dulk et al. (1977) [14], Drago (1980) [13 b]
CH, LSS	Trottet and Lantos (1978) [49],

b) two-dimensional

QC	Waldmeier (1965) [d], Newkirk (1967) [41], Saito (1970) [45]
CH	Munro and Jackson (1977) [39].

Semi-empirical based on demanding energy balance:

a) one-dimensional, static

TZ	McWhirter et al. (1975) [34], Flower and Pineau des Forêts (1976) [21],

b) one-dimensional, dynamic

TZ	Pneuman and Kopp (1978) [42],

c) two-dimensional, static

TZ	Gabriel (1976) [23]
CH	Rosner and Vaiana (1977) [44], Mariska (1978) [35],

d) two-dimensional, dynamic

LSS	Kopp and Orrall (1976) [28]
CH	Kopp and Orrall (1976) [28], Rosner and Vaiana (1977) [44], McWhirter and Kopp (1979) [33].

Analysis in terms of small-scale structures: Brückner and Nicolas (1973) [7], Withbroe and Mariska (1976) [58], Mariska et al. (1978) [36], Feldman et al. (1979) [18].

Contributions of spicule-type structures in TZ range from 20 ··· 100 %.

3.1.1.5.2 Physical parameters

Electron density: $N_e \simeq 2\cdot 10^{10}\ \mathrm{cm}^{-3}$ at $T \simeq 6\cdot 10^4$ K [13]
$\simeq 3\cdot 10^9\ \mathrm{cm}^{-3}$ at $T \simeq 10^6$ K [25]
$\simeq 10^9\ \mathrm{cm}^{-3}$ at $T \simeq 1.3\cdot 10^6$ K [20].

$T(N_e)$ relation varies little between LSS and CH, network and cell interior [13, 52, 16].

Pressure is almost constant with height for 15 000 km above limb [18].

Height differences between emission levels in LSS and CH [19];
scale height is much greater in CH than LSS [52].

Maximum temperature in CH: $T \lesssim 1\cdot 10^6$ K [49, 11].

Velocity fields
a) vertical flows: $-10 \cdots +20$ km s^{-1}, predominantly downward in network (10^5 K), up to -10 km s^{-1} (upward) in cell interiors [54, 12];
average outflow at 10^5 K, $\lesssim 4 \cdots 5$ km s^{-1} in LSS and CH [54, 22];
16 km s^{-1} outflow in CH at 10^6 K, [10], but see [11].
b) oscillations: some quasi-periods ≈ 7 min, mainly aperiodic transients at 10^5 K [4].
c) non-thermal, unresolved velocities: increase with T to 25 km s^{-1} at $2 \cdot 10^5$ K [36, 37];
similar in CH and LSS, network and cell interior [19];
$6 \cdots 16$ km s^{-1} in corona [50, 9].

3.1.1.5.3 Diagnostics

Lines are optically thin so coronal conditions apply in which collisional excitation and ionization processes are balanced by radiative de-excitation and recombination. Ionization balance can be complicated by autoionization and dielectronic recombination.
Ionization balance (steady state balance has been seriously questioned) [26, 3, 31, 46, 53].

Electron density diagnostics:
a) Most ions emit only at sharply defined temperatures so the lines can be used to derive $N_e(T)$ from density dependent line ratios [17]:
He I sequence [13a]
Be I sequence [4a, 13c, 16, 32, 27]
N I sequence [20]
Si III [37]
Fe IX [25]
different ions formed at same T [13].
b) Continuum scattering by electrons [b, 45, 30].
Temperatures from line ratios are not reliable in the presence of inhomogeneities [17].

3.1.1.5.4 Morphology

Transition region network
(not related to low coronal fine structure) [6, 18].
Coronal holes, structure and evolution [47,5].
Loops [48, 51, 43].
Polar plumes [2, 1].
Macrospicules/coronal spikes [60, 38, 29].
X-ray bright points (XBP) [24].

3.1.1.5.5 References for 3.1.1.5

Monographs

a Athay, R.G.: The Solar Chromosphere and Corona. The Quiet Sun, Reidel, Dordrecht (1976).
b Billings, D.E.: A Guide to the Solar Corona, Academic Press, New York (1966).
c Macris, C.J. (ed.): Physics of the Solar Corona, Reidel, Dordrecht (1971).
d Waldmeier, M.: Landolt-Börnstein, NS, Vol. VI/1 (1965) p. 112.
e Zirker, J.B. (ed.): Coronal Holes and High Speed Wind Streams, Colorado Associated Univ. Press. Boulder (1977).

Special references

1 Ahmad, I.A., Webb, D.F.: Sol. Phys. **58** (1978) 323.
2 Ahmad, I.A., Withbroe, G.L.: Sol. Phys. **53** (1977) 397.
3 Allen, J.W., Dupree, A.K.: Astrophys. J. **155** (1969) 27.
4 Athay, R.G., White, O.R.: Astrophys. J. **229** (1979) 1147.
4a Berrington, K.A., Burke, P.G., Dufton, P.L., Kingston, A.E.: J. Phys. B **10** (1977) 1465.
5 Bohlin, J.D.: Sol. Phys. **51** (1977) 377.
6 Brückner, G.E., Bartoe, J.D.F.: Sol. Phys. **38** (1974) 133.
7 Brückner, G.E., Nicolas, K.R.: Sol. Phys. **29** (1973) 301.
8 Burton, W.M., Jordan, C., Ridgeley, A., Wilson, R.: Astron. Astrophys. **27** (1973) 101.
9 Cheng, C.C., Doschek, G.A., Feldman, U.: Astrophys. J. **227** (1979) 1037.
10 Cushman, G.W., Rense, W.A.: Astrophys. J. Lett. **207** (1976) L61.
11 Doschek, G.A., Feldman, U.: Astrophys. J. Lett. **212** (1977) L143.
12 Doschek, G.A., Feldman, U., Bohlin, J.D.: Astrophys. J. Lett. **205** (1976) L177.
13 Doschek, G.A., Feldman, U., Bhatia, A.K., Mason, H.E.: Astrophys. J. **226** (1978) 1129.

13a Doyle, J.G.: Astron. Astrophys. **87** (1980) 183.
13b Drago, F.C.: Sol. Phys. **65** (1980) 237.
13c Dufton, P.L., Doyle, J.G., Kingston, A.E.: Astron. Astrophys. **78** (1979) 318.
14 Dulk, G.A., Sheridan, K.V., Smerd, S.F., Withbroe, G.L.: Sol. Phys. **52** (1977) 349.
15 Dupree, A.K.: Astrophys. J. **178** (1972) 527.
16 Dupree, A.K., Foukal, P.V., Jordan, C.: Astrophys. J. **209** (1976) 621.
17 Feldman, U., Doschek, G.A., Behring, W.E.: Space Sci. Rev. **22** (1978) 191.
18 Feldman, U., Doschek, G.A., Mariska, J.T.: Astrophys. J. **229** (1979) 369.
19 Feldman, U., Doschek, G.A., Patterson, N.P.: Astrophys. J. **209** (1976) 270.
20 Feldman, U., Doschek, G.A., Mariska, J.T., Bhatia, A.K., Mason, H.E.: Astrophys. J. **226** (1978) 674.
21 Flower, D.R., Pineau des Forêts, G.: Astron. Astrophys. **52** (1976) 191.
22 Francis, M.H., Roussel-Dupré, R.: Sol. Phys. **53** (1977) 465.
23 Gabriel, A.H.: Phil. Trans. Roy. Soc. London, A **281** (1976) 339.
24 Golub, L., Krieger, A.S., Silk, J.K., Timothy, A.F., Vaiana, G.S.: Astrophys. J. Lett. **189** (1974) L93.
25 Haug, E.: Astrophys. J. **228** (1979) 903.
26 Jordan, C.: Mon. Not. R. Astron. Soc. **142** (1969) 501.
27 Jordan, C.: Astron. Astrophys. **34** (1974) 69.
28 Kopp, R.A., Orrall, F.Q.: Astron. Astrophys. **53** (1976) 363.
29 Koutchmy, S., Stellmacher, G.: Sol. Phys. **49** (1976) 253.
30 Koutchmy, S., Picat, J.P., Dantel, M.: Astron. Astrophys. **59** (1977) 349.
31 Landini, M., Monsignori Fossi, B.C.: Astron. Astrophys. Suppl. **7** (1972) 291.
32 Loulergue, M., Nussbaumer, H.: Astron. Astrophys. **51** (1976) 163.
33 McWhirter, R.W.P., Kopp, R.A.: Mon. Not. R. Astron. Soc. **188** (1979) 871.
34 McWhirter, R.W.P., Thonemann, P.C., Wilson, R.: Astron. Astrophys. **40** (1975) 63; **61** (1977) 859.
35 Mariska, J.T.: Astrophys. J. **225** (1978) 252.
36 Mariska, J.T., Feldman, U., Doscheck, G.A.: Astrophys. J. **226** (1978) 698.
37 Moe, O.K., Nicolas, K.R.: Astrophys. J. **211** (1977) 579.
38 Moore, R.L., Tang, F., Bohlin, J.D., Golub, L.: Astrophys. J. **218** (1977) 286.
39 Munro, R.H., Jackson, B.V.: Astrophys. J. **213** (1977) 874.
40 Munro, R.H., Withbroe, G.L.: Astrophys. J. **176** (1972) 511.
41 Newkirk, G.: Annu. Rev. Astron. Astrophys. **5** (1967) 213.
42 Pneuman, G.W., Kopp, R.A.: Sol. Phys. **57** (1978) 49.
43 Priest, E.R.: Sol. Phys. **58** (1978) 57.
44 Rosner, R., Vaiana, G.S.: Astrophys. J. **216** (1977) 141.
45 Saito, K.: Ann. Tokyo Ser. 2 **12** (1970) 53.
46 Summers, H.P.: Culham Lab. Internal Memo 367 (1974).
47 Timothy, A.F., Krieger, A.S., Vaiana, G.S.: Sol. Phys. **42** (1975) 135.
48 Tousey, R., Bartoe, J.D.F., Bohlin, J.D., Brückner, G.E., Purcell, J.D., Scherrer, V.E., Sheeley, N.R., Schumacher, R.J., Van Hoosier, M.E.: Sol. Phys. **33** (1973) 265.
49 Trottet, G., Lantos, P.: Astron. Astrophys. **70** (1978) 245.
50 Tsubaki, T.: Sol. Phys. **43** (1975) 147.
51 Vaiana, G.S., Krieger, A.S., Timothy, A.F.: Sol. Phys. **32** (1973) 81.
52 Vernazza, J.E., Reeves, E.M.: Astrophys. J. Suppl. **37** (1978) 485.
53 Vernazza, J.E., Raymond, J.C.: Astrophys. J. Lett. **228** (1979) L89.
54 White, O.R.: The Energy Balance and Hydrodynamics of the Solar Chromosphere and Corona (Bonnet and Delache, eds.), de Bussac, Clermont-Ferrand (1976) 75.
55 Withbroe, G.L.: Sol. Phys. **45** (1975) 301.
56 Withbroe, G.L.: The Energy Balance and Hydrodynamics of the Solar Chromosphere and Corona (Bonnet and Delache, eds.), de Bussac, Clermont-Ferrand (1976) 263.
57 Withbroe, G.L., Gurman, J.B.: Astrophys. J. **183** (1973) 279.
58 Withbroe, G.L., Mariska, J.T.: Sol. Phys. **48** (1976) 21.
59 Withbroe, G.L., Noyes, R.W.: Annu. Rev. Astron. Astrophys. **15** (1977) 363.
60 Withbroe, G.L., Jaffe, D.T., Foukal, P.V., Huber, M.C.E., Noyes, R.W., Reeves, E.M., Schmahl, E.J., Timothy, J.G., Vernazza, J.E.: Astrophys. J. **203** (1976) 528.

3.1.1.6 Radio emission of the quiet sun

Daily measurements of the solar radio flux have been made at different observatories at various fixed frequencies for more than two sunspot cycles. It is evident from these flux measurements as well as from synoptic maps of the solar disk that in addition to the radio radiation emitted by the entire solar disk intensive radiation is occasionally generated in confined local areas on the solar surface. These are known as active regions.

We may thus classify solar radio radiation according to two components:
A. radiation of the quiet sun,
B. variable radiation of active regions (see 3.1.2.8).

The flux level of the quiet sun can be satisfactorily determined without extrapolation only when the sun has been free of spots for some weeks. The radiation flux of the quiet sun thus refers to the condition of the sun at sunspot minimum. The solar atmosphere can be described in this case as follows: The solar poles are enveloped by extensive regions with a predominantly open magnetic field topology, often referred to as "coronal holes". On the other hand, the equatorial zone up to 50° latitude has a predominantly closed magnetic field configuration as evidenced by the extended weak arcs in EUV-images of the lower solar corona. Both regions are devoid of noteworthy solar activity. For the quiet corona, see also 3.1.1.5.

3.1.1.6.1 Flux density of the quiet sun

The values for radiation flux density, $F_\odot$, in Table 1 are based on measurements performed during the sunspot minimum years 1964 and 1976. Part of the data at various frequencies has been taken from the Quarterly Bulletin of Solar Activity [a] and from the Solar Geophysical Data [b]. The remainder has been taken directly from the publications of the observatories, e.g. [c, d, e, f].

Table 1. Flux density, $F_\odot$, and radiation temperature, T_{rad}, of the quiet sun.
For references and more details, see text.
$F_\odot$ = flux density of the quiet sun during sunspot minimum
T_{rad} = radiation temperature of an optical solar disk with $d = 32'$
T_c = radiation temperature at the center of the solar disk
T_H = radiation temperature in a coronal hole. The values of T_H marked with a colon have been interpolated assuming a λ^2 wavelength dependence.

f MHz	λ cm	$F_\odot$ 10^{-22} W m^{-2} Hz^{-1}	T_{rad} K	T_c K	T_H K
30	1000	0.17	$9.03 \cdot 10^5$	$5.1 \cdot 10^5$	$5.1 \cdot 10^5$:
50	600	0.54	$10.3 \cdot 10^5$	$6.2 \cdot 10^5$	$6.2 \cdot 10^5$
100	300	2.40	$11.5 \cdot 10^5$	$8.9 \cdot 10^5$	$6.7 \cdot 10^5$:
150	200	5.10	$11.0 \cdot 10^5$	$9.5 \cdot 10^5$	$6.4 \cdot 10^5$
200	150	8.1	$9.7 \cdot 10^5$	$8.6 \cdot 10^5$	$6.1 \cdot 10^5$:
300	100	14.9	$7.9 \cdot 10^5$	$7.03 \cdot 10^5$	$5.4 \cdot 10^5$:
400	75	21.7	$6.5 \cdot 10^5$	$5.6 \cdot 10^5$	$4.4 \cdot 10^5$
600	50	32.1	427000	363000	265000:
1000	30	41.3	197600	162000	108000:
1500	20	48.0	102200	83800	57800
3000	10	69	36680	31180	25100:
3750	8	82	27900	24550	20770:
5000	6	107	20670	18810	16650:
10000	3	275	13160	12240	11700
15000	2	574	12210	11600	11360:
20000	1.5	890	11560	11100	
30000	1.0	1862	10480	10110	
37500	0.8	2816	9580	9290	
50000	0.6	4503	8619	8450	
100000	0.3	14709	7038	6900	
300000	0.1	113199	6018	5900	

Reliable measurements of absolute flux density are available over the range of wavelengths $1\ \mathrm{cm} < \lambda < 100\ \mathrm{cm}$. The values for 1964 were used after correction according to [14] and [6]. The values for 1976 were compiled in a similar way. These values were used to derive graphically a mean variation of the flux density with wavelength. The deviations of the measurements about this mean curve are about $\pm 5\%$.

The flux densities of Table 1 at long wavelengths ($\lambda > 1$ m) are based on the measurements [10, 9, 7, 2] following the discussions in [1] and [15]. The absolute values of the radiation flux in this range of wavelengths are usually calibrated with intense galactic and extragalactic radio sources. The residuals about a mean curve amount to about $\pm 8\%$.

The absolute determination of solar radiation flux in the millimeter range is impeded by the difficult measurement technique. A more serious problem, however, is the variable influence of the earth's atmosphere. Results of previous measurements compiled e.g. in the summaries [13, 3, 4] show a considerable scatter. Measurements relating the solar radiation flux to that of the new moon seem to be more reliable. Using the moon for calibration, it is not necessary to know accurately the gain and beam pattern of the receiving antenna due to the almost identical angular diameters of the two objects. The atmospheric influence can also be reduced considerably. The values of Table 1 in the mm-range are mainly based on measurements using the moon as a calibrator [12, 11, 8]. Comparable measurements by [16, 17] employed an artificial calibration source in the far field of the receiving antenna.

Radiation flux density, $F_\odot$, and the corresponding radiation temperature, T_{rad}, are related by:

$$F_\odot = 2.09 \cdot 10^{-44} \cdot f^2 \cdot T_{rad},$$

$F_\odot$ in $[\mathrm{W\,m^{-2}\,Hz^{-1}}]$ with f = frequency in [Hz], T_{rad} = radiation temperature in [K].

The temperature T_c at disk center was derived from $F_\odot$ using the known brightness distribution over the solar surface (see also 3.1.1.6.2). Some values for T_c in the mm- and m-ranges were obtained from direct measurements followed by a graphical adjustment.

In general, T_c rises monotonically with wavelength from 5900 K at $\lambda = 1$ mm to $9 \cdot 10^5$ K at $\lambda = 1.5$ m. The gradient of the increase is distinctly lower from $\lambda = 8$ mm to $\lambda = 3$ cm; in this range a slight jump in the gradient is present which may be associated with the upper chromosphere.

According to the definition used here, the radiation temperature T_c in the center of the solar disk refers to a surface without active regions but showing arc structure in the corona. In regions of open magnetic field configuration, i.e. coronal holes, the radiation temperature T_H is reduced. The radiation temperature differences indicated in [5, 1, 15] were applied to T_c.

References for 3.1.1.6.1

General references, current data series

a Quart. Bull. Solar Activity, Zürich.
b Solar-geophys. Data, Boulder, Col.
c Covington, E.: A Working Collection of Daily 2800 MHz Solar Flux Values, Natl. Res. Coun. Canada, Report No. ARO-5 (1977).
d Solar Radio Flux Measurements of the Sagamore Hill Radio Observatory.
e Complete Summary of Daily Solar Radio Flux, Toyokawa Observatory, Nagoya University, Toyokawa (1975).
f Heinrich-Hertz-Institut Solar Data, Akad. Wiss. DDR, Berlin.

Special references

1 Dulk, G.A., Sheridan, K.V., Smerd, S.F., Withbroe, G.L.: Sol. Phys. **52** (1977) 349.
2 Erickson, W.C., Gergely, T.E., Kundu, M.R., Mahoney, M.J.: Sol. Phys. **54** (1977) 57.
3 Felli, M., Tofani, G.: Physics of the Solar Corona (Macris, C.J., ed.), D. Reidel Publ., Dordrecht (1971) 267.
4 Fürst, E.: Int. Astron. Union Symp. **86** (1980) 25 (cf. [a] Ref. p. 111).
5 Fürst, E., Hirth, W.: Sol. Phys. **42** (1975) 157.
6 Krüger, A., Michel, H.-St.: Nature **206** (1965) 601.
7 Kundu, M.R., Gergely, T.E., Erickson, W.C.: Sol. Phys. **53** (1977) 489.
8 Kuseski, R.A., Swanson, P.N.: Sol. Phys. **48** (1976) 41.
9 Lantos, P., Avignon, Y.: Astron. Astrophys. **41** (1975) 137.

10 Leblanc, Y., Le Squeren, A.M.: Astron. Astrophys. **1** (1969) 239.
11 Linsky, J.L.: Sol. Phys. **28** (1973) 409.
12 Reber, E.E.: Sol. Phys. **16** (1971) 75.
13 Shimabukuro, F.I., Stacey, J.M.: Astrophys. J. **152** (1968) 777.
14 Tanaka, H., Castelli, J.P., Covington, A.E., Krüger, A., Landecker, T.L., Tlamicha, A.: Sol. Phys. **29** (1973) 243.
15 Trottet, G., Lantos, P.: Astron. Astrophys. **70** (1978) 245.
16 Wrixon, G.T., Hogg, D.C.: Astron. Astrophys. **10** (1971) 193.
17 Wrixon, G.T., Schneider, M.V.: Nature **250** (1974) 314.

3.1.1.6.2 The brightness distribution across the solar disk

Relatively good solar radio maps are available today in the wavelength range 2 mm < λ < 5 cm. Although most of the measurements were performed using large single dishes, especially the 100 m radio telescope in Effelsberg, some good measurements of brightness distribution were also obtained with multi-element interferometers. The angular resolution of the telescopes in this range of wavelength runs from 30″ to 120″, which is still insufficient to produce truly reliable measurements of the intensity distribution at the solar limb. Consequently it is still an open question whether or not a theoretically predicted bright ring exists at the limb for λ < 3 cm.

For wavelengths λ > 10 cm we must still resort to measurements made at synthesis telescopes or at Christiansen cross arrays. The high-resolution synthesis telescopes, which employ the earth's rotation for construction of complete, very detailed maps, have been used up to now for only a few special projects in solar physics.

Radio maps of the sun at times of high solar activity are characterized by the daily varying location and intensity of active regions on the solar disk. At these times it is difficult to ascribe a mean solar brightness distribution to the sun. The following description of the brightness distribution of the undisturbed solar surface thus refers to conditions at sunspot minimum (review papers [4, 7, 18]).

a) Observed brightness distribution at λ = 1 mm

At λ = 1.2 mm, a limb darkening seems to exist, which is a well-known effect in the visible. By correcting for the antenna pattern, one can recognize a faint bright ring superimposed over the limb darkening [11, 12].

b) Brightness distribution at 3 mm < λ < 4 mm

The brightness distribution of the solar disk is a constant for $R < 0.8\,R_\odot$ [23, 26]. According to [16], there seems to be a weak decrease at the solar limb. Comparing these measurements with a constant brightness solar model with $R = R_\odot$ which has been convolved with the antenna diagram, one can see a bright ring at the limb (dashed outline in Fig. 1). On the other hand, the measurements may also be represented by a model with the disk enlarged by 1%. The dashed outline of Fig. 1 should thus be considered as an upper bound.

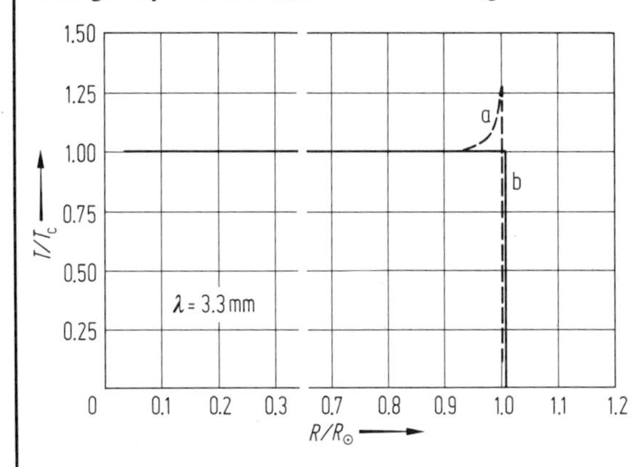

Fig. 1. Observed brightness distribution at λ = 3.3 mm. After convolution with the antenna pattern both profiles: a) with a spike at the rim (dashed line), and b) a constant brightness distribution of a disk with $R = 1.005\,R_\odot$, represent the measurements.
R = distance from the center of solar disk
T_c = radiation temperature at the center of the solar disk.

c) Brightness distribution at 6 mm < λ < 10 mm

At λ = 8.5 mm (Fig. 2), the brightness distribution is constant within 2% over the solar disk for $R \leqq 0.9\,R_\odot$ [9]. Measurements taken during a solar eclipse have shown that the abrupt cut-off at the limb occurs at $R = 1.02\,R_\odot$ [24]. A weak bright ring at the limb was found [13, 16] after deconvolution with the antenna diagram. The variation at the limb was also derived from multi-element interferometer [10, 25]. The bright ring (indicated in Fig. 2 by the dotted line) contributes less than 1% of the total solar radiation flux.

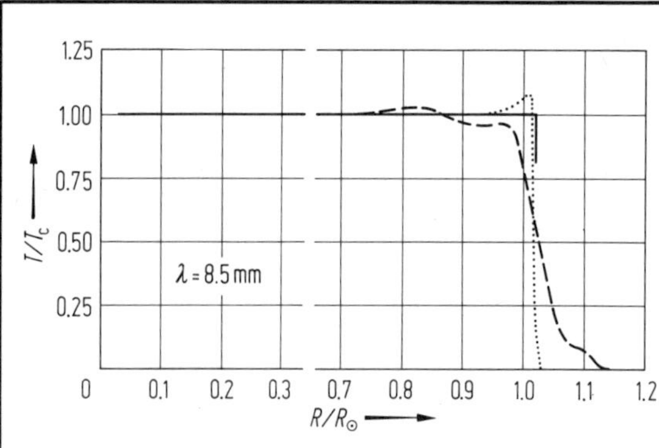

Fig. 2. Observed brightness distribution at λ=8.5 mm. Full line: measurements [9]; dashed line: measurements [10]; dotted line: after deconvolution with the antenna pattern.

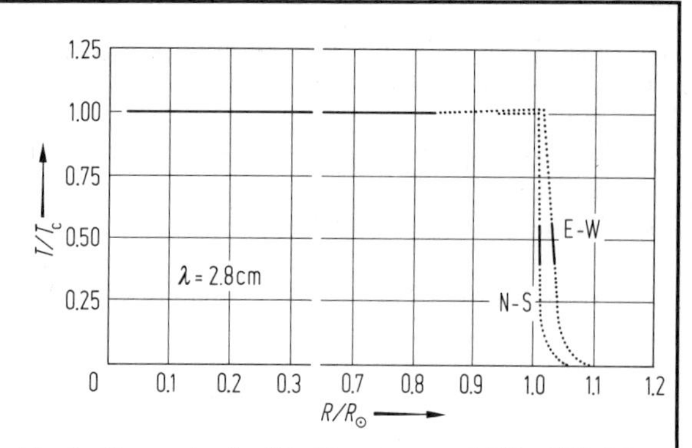

Fig. 3. Observed polar (N – S) and equatorial (E – W) brightness distribution at λ=2.8 cm.
Full line: measurements [5]; dotted line: theoretical models [17].

d) Brightness distribution at 1 cm < λ < 3 cm

At λ=2.8 cm (Fig. 3) the solar disk brightness distribution is again constant within 1% for $R \leqq 0.85\, R_\odot$ [5]. The solar disk, however, is already slightly elliptical at these wavelengths, measuring 1.04 $R_\odot$ in diameter along the equator and 1.01 $R_\odot$ along the polar direction [6]. The theoretical variation in brightness at the limb derived from model calculations [17] is represented by the dotted lines in Fig. 3.

e) Brightness distribution at λ=6 cm

Moving up to a wavelength λ=6 cm, the measurements of [5] have shown that an increase in brightness is now observed by travelling from center disk to the limb along the equator. On the other hand, a slight decrease in intensity is apparent when scanning out along the polar direction. The weakly elliptic solar disk measures 1.07 $R_\odot$ and 1.03 $R_\odot$ along the major and minor axes respectively [6]. The theoretically determined variation in brightness at the limb is given in Fig. 4 by a dotted line [17].

f) Brightness distribution at $\lambda \cong$ 10 cm

New measurements at λ=9.1 cm have recently been published [22]. These values for the increase in brightness along the equator were used for Fig. 5. The limb variation from the same model calculations applied above [17] is represented here again by a dotted line.

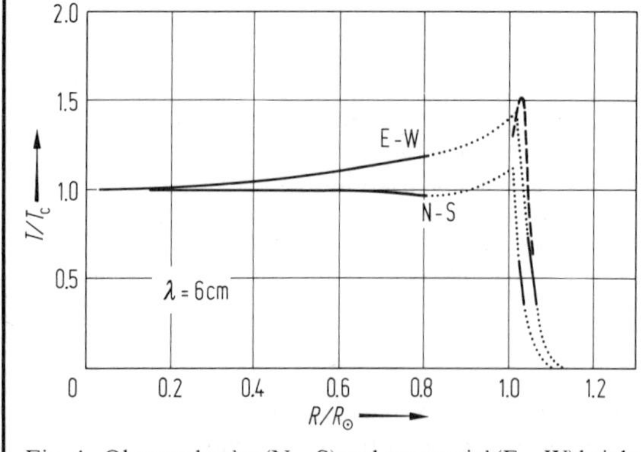

Fig. 4. Observed polar (N – S) and equatorial (E – W) brightness distribution at λ=6 cm. Full line: measurements [5]; dashed line: measurements [15] deconvolved; dotted line: theoretical models [17].

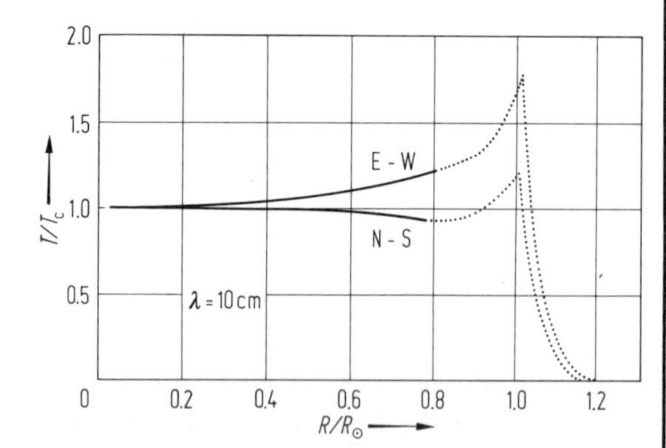

Fig. 5. Observed polar (N – S) and equatorial (E – W) brightness distribution at λ=10 cm. Full line: measurements [22]; dotted line: theoretical models [17].

g) Brightness distribution at decimeter wavelengths

The brightness distribution at $\lambda = 20$ cm must still be taken either from interferometer measurements [2] or from eclipse observations [8], both of which were used for LB, NS, Vol. VI/1 (1965) p. 119. A mean brightness distribution was derived from both sets of measurements, giving higher weight to the eclipse observations at the limb because of their superior resolving power. This resulting brightness distribution is shown in Fig. 6.

For increasing decimeter wavelengths, the maximum in intensity apparently shifts along the equator from the solar limb toward the center [21]. A theoretical explanation of this effect has been given by [20].

h) Brightness distribution at meter wavelengths

The solar radiation at meter wavelengths originates primarily in the corona, so that radio maps at these frequencies reveal primarily coronal structure. Observations with the large synthesis telescope at Culgoora, Australia [3] yielded an approximately constant intensity in front of the solar disk with positive and negative deviations of $\pm 30\%$. Active regions, which do appear brighter at decimetric wavelengths, are not uniquely associated with higher intensities in the meter range. A more likely correlation seems to exist between filaments and meter wavelength brightness enhancements. The depressions in intensity, on the other hand, are connected with coronal holes. Thus it is difficult to derive a general description of the quiescent sun from the observations.

The coronal brightness distribution at sunspot minimum was formerly derived from one-dimensional solar scans using an antenna with a typical lobed pattern. A large number of scans, taken on various days during solar minimum, were used for a determination of a minimum intensity envelope. This, in turn, was defined as the one-dimensional brightness distribution of the quiet sun. In view of this definition, special consideration is given to the distribution of coronal holes over the solar disk.

Fig. 7 shows the brightness distribution at $\lambda = 1.78$ m (169 MHz) in both the equatorial and polar directions as derived in this way by [19] from 900 measurements. The solar disk is elliptical, measuring $38' \pm 1'$ along the major (equatorial) axis and $32' \pm 1'$ along the minor (polar) axis. Using a similar technique, the one-dimensional brightness distribution was derived by [14] at $\lambda = 2.47$ m, 4.06, and 11.4 m. The results for $\lambda = 4.06$ m are given in Fig. 8.

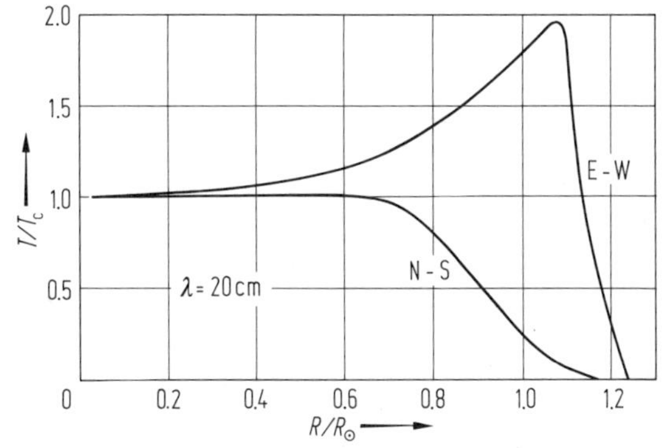

Fig. 6. Observed polar (N – S) and equatorial (E – W) brightness distribution at $\lambda = 20$ cm.

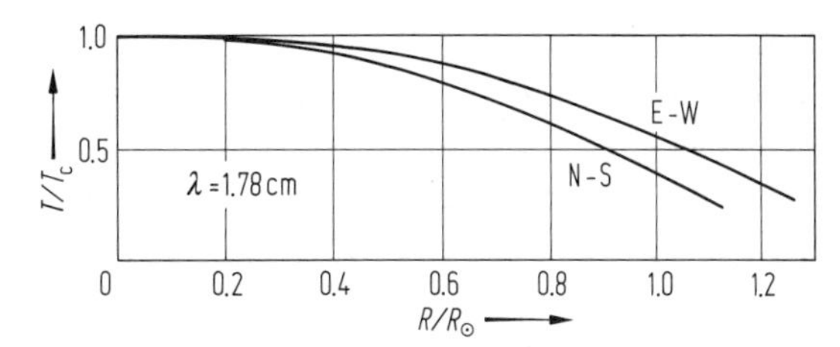

Fig. 7. One-dimensional brightness distribution in equatorial (E – W) and polar (N – S) directions at $\lambda = 1.78$ m [19].

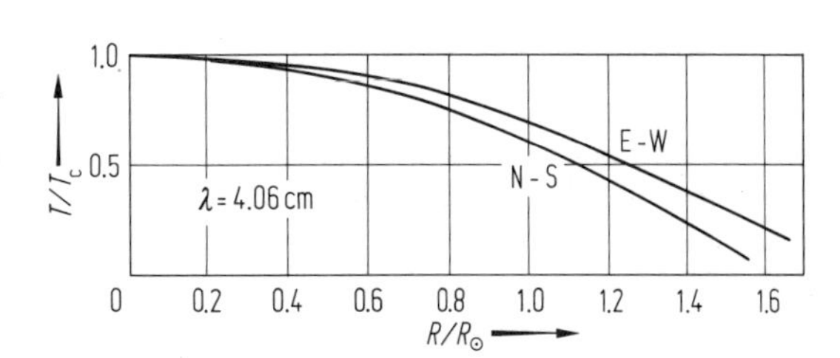

Fig. 8. One-dimensional brightness distribution in equatorial (E – W) and polar (N – S) directions at $\lambda = 4.06$ m [14].

The radio radius of the sun

The equatorial and polar radii of the sun are characteristic for its changing radio image at different wavelengths. They are useful standards for comparison of observational results with models of coronal composition and structure.

The radius of the sun's radio image has been defined [6] to be the distance of the isophote $T=0.5\ T_c$ from the disk center. Another possibility, used by [5, 24], is to derive the radius from the times of first and fourth contact during a solar eclipse. These two definitions are not equivalent. The radii derived from the contact points at eclipse agree only approximately with the isophote $T=0.5\ T_c$. It is true that the contact points are easily determined and yield exact values for the solar radius. Their disadvantage, however, is that they are strongly dependent on the momentary state of the corona at the time of contact and thus are of only limited use in a determination of "mean" values.

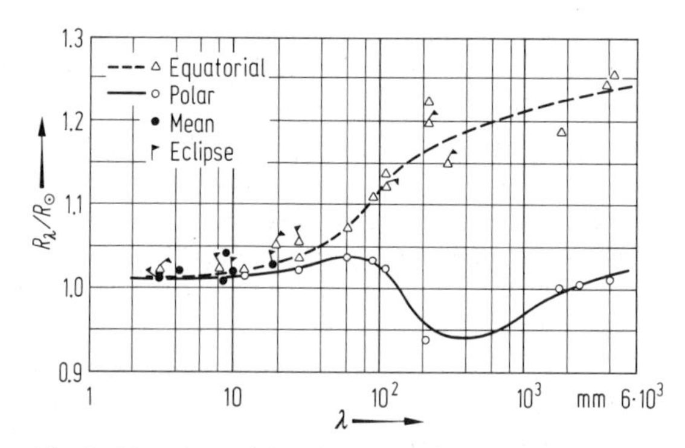

Fig. 9. The equatorial and polar radio radius of the sun as function of wavelength.

The values from [5, 6, 14, 19, 24] were used in Fig. 9.

Since long wavelength radiation ($\lambda > 15$ cm) originates primarily in the transition region and lower corona, it is understandable that the radio images reflect the brightness distribution of the corona. Particularly at polar latitudes, where the solar corona is often weaker due to the open magnetic field configuration, it is possible by the above definition to obtain a radius at decimeter wavelengths that is less than 1 $R_\odot$.

Fine structure in quiet areas of the solar surface

A weak fine structure can be detected in radio maps even in regions on the solar surface completely devoid of activity. The characteristic size of this structure at millimeter wavelengths is approximately that of the supergranules (ca. 30″···50″ in diameter). However, a possible connection of this fine structure with supergranules, which are distinguishable in optical spectroheliograms, has not yet been firmly established. The variation of the radiation temperature in these structures is $\pm 2\%$ at $\lambda = 8$ mm [9]. The median value of this variation defines the radiation temperature $T_\odot$ of the "quiet solar surface". Regions with radiation temperature $T < 0.98\ T_\odot$ very often contain either coronal holes or filaments in various stages of evolution. Regions with $T > 1.02\ T_\odot$ coincide usually with areas where either a longitudinally oriented magnetic field or increasing solar activity are present. The magnetic field appears to be of particular importance for the fine structure [9].

A coarser structure is observed in radio maps at upper decimeter and meter wavelengths [3], which seems to be associated with the arch-like features in the corona.

Interpretation of the brightness distribution

Models designed to explain the brightness distribution across the solar disk have been applied practically since the beginning of solar radio observations. All of the calculations assume that the radiation of the quiet solar atmosphere is of thermal origin. In order to account for the variations in radio maps at meter and upper decimeter wavelengths with a spherically symmetric model, different coronal electron density distributions along the equatorial and polar directions must be introduced. Maps at lower cm- and mm-wavelengths, which show a constant center-to-limb brightness, additionally require the consideration of chromospheric structure including the supergranules and spicules. Summaries of the various models have been given by [6, 7, 18, 1a].

References for 3.1.1.6.2

General references

a Radiophysics of the Sun; Int. Astron. Union Symp. **86** (Kundu, M.R., Gergely, T.E., eds.), D. Reidel, Dordrecht (1980).

Special references

1 Aubier, M., Leblanc, Y., Boischot, A.: Astron. Astrophys. **12** (1971) 435.
1a Ahmad, I.A., Kundu, M.R.: Sol. Phys. **69** (1981) 273.
2 Christiansen, W.N., Warburton, J.A.: Austr. J. Phys. **8** (1955) 474.
3 Dulk, G.A., Sheridan, K.V.: Sol. Phys. **36** (1974) 191.
4 Felli, M., Tofani, G.: Physics of the Solar Corona (Macris, C.J., ed.), D. Reidel, Dordrecht (1971) 267.
5 Fürst, E., Hachenberg, O., Hirth, W.: Astron. Astrophys. **36** (1974) 129.
6 Fürst, E., Hirth, W., Lantos, P.: Sol. Phys. **63** (1979) 257.
7 Fürst, E. in: [a] p. 25.
8 Hachenberg, O., Fürstenberg, F., Prinzler, H.: Z. Astrophys. **39** (1956) 232.
9 Hachenberg, O., Steffen, P., Harth, W.: Sol. Phys. **60** (1978) 105.
10 Kawabata, K., Fujishita, M., Kato, T., Ogawa, H., Omodaka, T.: Sol. Phys. **65** (1980) 221.
11 Kundu, M.R.: Sol. Phys. **21** (1971) 130.
12 Kundu, M.R., Sou-Yang Liu: Sol. Phys. **44** (1975) 361.
13 Kundu, M.R., Sou-Yang Liu, McCulbough, T.P.: Sol. Phys. **51** (1977) 321.
14 Kundu, M.R., Gergely, T.E., Erickson, W.C.: Sol. Phys. **53** (1977) 489.
15 Kundu, M.R., Rao, A.P., Erskine, F.T., Bregman, J.D.: Astrophys. J. **234** (1979) 1122.
16 Lantos, P., Kundu, M.R.: Astron. Astrophys. **21** (1972) 119.
17 Lantos, P., Fürst, E., Hirth, W.: Sol. Phys. **63** (1979) 271.
18 Lantos, P. in: [a] p. 41.
19 Leblanc, Y., LeSqueren, A.M.: Astron. Astrophys. **1** (1969) 239.
20 Molchanov, A.P.: Physics of the Solar System (Mikhailov, A.A., ed.) (1964) 206.
21 O'Brien, P.A., Tandberg-Hansen, E.: Observatory **39** (1956) 232.
22 Riddle, A.C.: Sol. Phys. **7** (1969) 434.
23 Shimabukuro, F.I.: Sol. Phys. **12** (1970) 438.
24 Swanson, P.N.: Sol. Phys. **32** (1973) 77.
25 Suzuki, I., Kawabata, K., Ogawa, H.: Sol. Phys. **46** (1976) 205.
26 Tlamicha, A.: Sol. Phys. **10** (1969) 150.

3.1.2 Solar activity

3.1.2.1 Active regions

3.1.2.1.1 Features of active regions

Enhanced magnetic flux, sunspots, faculae and plages, filaments (prominences), active coronal regions (condensations, enhancements), flares [j].

Typical characteristics

$3\cdot10^{16}$ cm^2 = 10^{-6}·area of solar hemisphere
EFR = ephemeral flux region [1]

	EFR	Medium	Large
total magnetic flux [10^{20} Mx]	1	100	500
size of K-plage ≡ magnetic region [10^{18} cm^2]	≦7	100	400
max. number of spots	–	20	80
max. total spot area [10^{18} cm^2]	–	6	25
lifetime spot group [solar rotations]	–	1···2	2···3
lifetime plage (rot. = solar rotations)	$1^d\cdots2^d$	2···3 rot.	4 rot.

3.1.2.1.2 Active region development

Spotgroups [a, b, c, e, g]; filaments, K-plages [d]; Hα-, EUV-, X-ray-, radio-heliograms [a]; magnetograms [a].

3.1.2.1.3 Spotgroups [j, k]

a) Classification: according to phase of development and geometry: Zürich Classification; according to magnetic configuration: Mt. Wilson Classification.

b) Development:

growth rate:	up to $200\cdot10^{-6}$ solar hemisphere/day
decay rate after maximum:	up to $100\cdot10^{-6}$ solar hemisphere/day
late decay of remnant leader spot:	$6\cdot10^{-6}$ solar hemisphere/day independent of size
lifetime:	1···100 days.

3.1.2.1.4 Activity indices, global data (daily values)

Zürich Sunspot Relative Number (Wolf's Number)
$R = k\cdot(10g + s)$ [a, h, f]
(g = number of spotgroups; s = number of individual spots; k = calibration constant).

Total spot area [b, e, h].
Total facula area [b, f].
K-plage index (McMath) [a].
Solar radio flux 2800 MHz (Ottawa) [a, h].

3.1.2.1.5 References for 3.1.2.1

General references, current data series

a Solar Geophysical Data, US Department of Commerce, NOAA Boulder, Col.
b Photoheliographic Results, until 1952: Greenwich Photoheliographic Results, Royal Greenwich Observatory; 1956–1961: in Royal Greenwich Observatory Bulletins, London; since 1962: in Royal Observatory Annals, Herstmonceux.
c Solnechnye Dannye (Solar Data), Academy of Science, Leningrad.
d Cartes Synoptiques de la Chromosphere Solaire, Observatoire de Paris – Meudon.

e Solar Phenomena, Osservatorio Astronomico di Roma.
f Astronomische Mitteilungen der Eidgenössischen Sternwarte, Zürich.
g Heliographic Maps of the Photosphere (until 1965: Heliographische Karten der Photosphäre), Publikationen der Eidgenössischen Sternwarte, Zürich.
h Quarterly Bulletin on Solar Activity, Zürich.
j Waldmeier, M.: Landolt-Börnstein, NS, Vol. VI/1 (1965) p. 119.
k Bruzek, A., Durrant, C.J. (eds.): Illustrated Glossary for Solar and Solar-Terrestrial Physics, D. Reidel, Dordrecht 1977.

Special references

1 Harvey, K.L., Martin, S.F.: Sol. Phys. **32** (1973) 389.

3.1.2.2 11-year solar cycle

Epochs of sunspot minima and maxima [b, 1, 2].

Cycle No.	Min	Max	Cycle No.	Min	Max	Cycle No.	Min	Max	Cycle No.	Min	Max
1	1755.2	1761.5	6	1810.6	1816.4	11	1867.2	1870.6	16	1923.6	1928.4
2	1766.5	1769.7	7	1823.3	1829.9	12	1878.9	1883.9	17	1933.8	1937.4
3	1775.5	1778.4	8	1833.9	1837.2	13	1889.6	1894.0	18	1944.2	1947.5
4	1784.7	1788.1	9	1843.5	1848.1	14	1901.7	1907.1	19	1954.3	1957.9
5	1798.3	1805.2	10	1856.0	1860.1	15	1913.6	1917.6	20	1964.7	1968.9
									21	1976.5	1980.0

Mean monthly and annual sunspot numbers R for 1749···1963 and annual mean sunspot areas 1874···1958: [b] pp. 125···129; values up to date, see refs. [a, e, f, h] of 3.1.2.1.

Solar cycle variation of facula area, prominence area and coronal λ 5303 Å intensities: [b] p. 130.

Mean solar characteristics during a sunspot cycle: [a].

Positions and migration of activity zones during the solar cycle: [b] p.131.

References for 3.1.2.2

General references

a Allen, C.W.: Astrophysical Quantities, 3rd ed., The Athlone Press, London 1973.
b Waldmeier, M.: in Landolt-Börnstein, NS, Vol. VI/1, 1965.

Special references

1 Waldmeier, M.: The Sunspot Activity in the Years 1610–1960, Zürich 1961.
2 Waldmeier, M.: Astron. Mitt. Zürich **335** (1977) 6.

3.1.2.3 Sunspots

3.1.2.3.1 General characteristics

Table 1. Characteristics of spots [70].

A = spot area A_U = umbral area
R = spot radius R_U = umbral radius
Φ = total magnetic flux B = magnetic field strength

	Large spots	Small spots	Pores
R [10^3 km]	28	5	0.5 ···3
R_U [10^3 km]	11	2	–
Φ [10^{20} Mx]	300	5	0.25···5
B (center) [G]	3000±400		≈2000
T_{eff} [K]	4000±100 [70] 3700 [62]		

$A_U/A = 0.17 \pm 0.03$; $R_U/R = 0.41 \pm 0.035$ for $A > 50 \cdot 10^{-6}$ solar hemispheres

References: [a, b, c, e, 7, 8, 26, 50, 62, 67, 70].

Bruzek

3.1.2.3.2 Magnetic field [50]

a) Radial variation of magnetic field strength B: $B(\varrho)$ for $0 \leqq \varrho \leqq 1$: see Fig. 1.

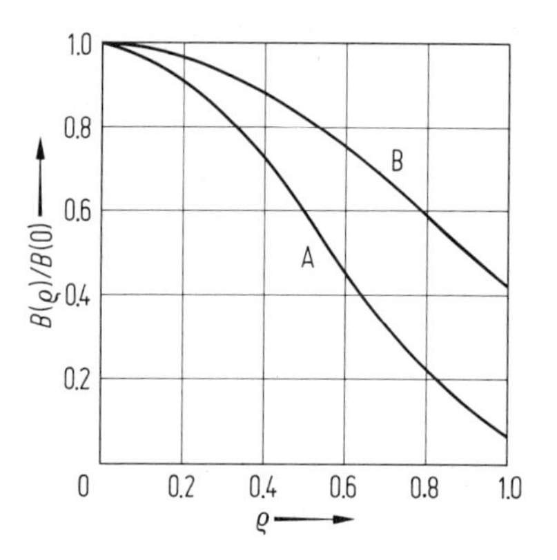

Fig. 1. Radial variation of spot magnetic field strength B: B vs. ϱ ($0 \leqq \varrho \leqq 1$) with ϱ = radial distance in units of spot radius.
Curve A: average of measurements of [2, 19, 40, 46]
curve B: average of measurements of [9, 48].

b) Inclination of field vector
zenith angle $\alpha(\varrho) \approx \varrho \cdot \pi/2$ [9] and refs. therein [20].

c) Vertical gradient [50] and references therein:

spot center	1.0···0.5 G km^{-1}
umbra-penumbra border	0.3···0.2 G km^{-1}
penumbra	0.2···0.05 G km^{-1}.

3.1.2.3.3 Spot umbra

a) Spectrum, continuum intensities $I(\lambda)$: [a, e] and refs. therein; in addition see:
compilation of measurements 1968···1973 in the range 0.4···4.0 μm [62] Fig. 3; [68] Table and Fig.
continuum intensities λ 0.387···1.67 μm: [21, 30]; [62] Table V
computed infrared opacities 1···10 μm [28a]
center-limb variation of continuum intensities for three wavelengths: [60] Table IV
line spectrum: Atlas 3900···8000 Å [62]
infrared spectrum [d]
molecular spectrum [58, 61].

b) Turbulent (non-thermal) velocities [6] Table III and refs. therein:

mean velocity from curve of growth measurements	1.8 km s^{-1}
mean velocity from line widths	1.5 km s^{-1}
mean velocity from $g(0)$ lines (without Zeeman splitting)	0.8 km s^{-1}.

c) Wilson effect shown by 84% of all spots [16].
Wilson depression (independent of spot area for $A > 25 \cdot 10^{-6}$ solar hemispheres and $0.1 \leqq \tau \leqq 1$): (600 ± 200) km.
References: [26, 41, 50, 56, 59, 60].

d) Spot umbra models
empirical (hydrostatic): [30] Table I; [28, 31, 32a, 54, 55]; [58] Table IV (from molecular lines); [60] Table VI; [64] Table I and compilation of previous models in Fig. 2; [68] Table II; [69] Table I.
magnetostatic models: [18, 26, 52, 53]; [65] Table 1.
thermal models: [16a] refs.

e) Fine structure and dynamics

Table 2. Fine structure.
t = lifetime, v = velocity, d = diameter

	d [km]	T [K]	t [s]	v [km s^{-1}]	Ref.
umbral dots	150	6300	1500		7, 8, 9, 33, 13b, 35a
umbral flashes	1500		150	40 (horizontal) 1.5 (vertical)	10

Table 3. Umbral oscillations.

Observed periods	Region	Amplitudes	Horiz. size	Ref.
300···470 s	photosphere			11, 51
145···196 s	photosphere	0.2 km s^{-1}	2000 km	51, 6 (refs.)
	chromosphere	1···6 km s^{-1}		
110···123 s	photosphere			51

f) Chromosphere of spot umbra models: for profiles of Hα, Ca II K, Na I D_2 [3], equivalent widths of weak metal lines [3a], Ca II lines [57] Figs. 3···5, Ca II lines [32] Figs. 2, 3.

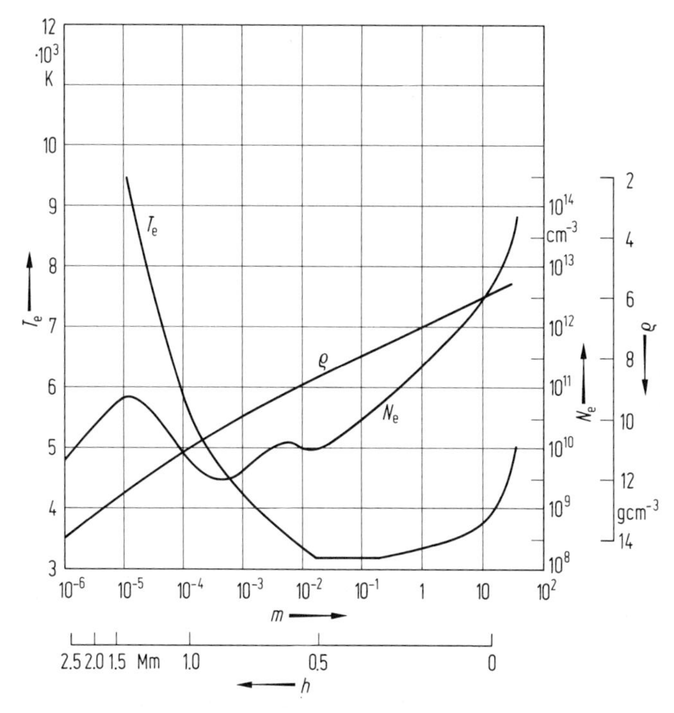

Fig. 2. Umbra model (chromosphere, photosphere): T, N_e, ϱ vs. m (model C of [32]).
T_e = electron temperature
N_e = electron density
ϱ = density
m = mass in g per cm^2-column
h = height

g) Transition region of spot umbra
temperature region $4 \cdot 10^4 \cdots 2 \cdot 10^5$ K ($d \approx 90$ km): spectrum 1200···1817 Å [14], model [15] Table II;
temperature region $2 \cdot 10^5 \cdots 8 \cdot 10^5$ K: spectrum 600···1050 Å [24, 47], emission heights and density scale heights [23].

3.1.2.3.4 Spot penumbra

a) Intensity: mean intensity (averaged over bright and dark penumbral filaments) $3870 < \lambda < 38000$ Å: [22, 37],[39] Table I, [62] Fig. 4, Table V.

Table 4. Penumbral filaments.
I_p = penumbral intensity
I_{phot} = photospheric intensity

	Bright	Dark	Ref.
I_p/I_{phot} ($\lambda = 4600$ Å)	0.78	0.62	33
I_p/I_{phot} ($\lambda = 5280$ Å)	0.93	0.56	43, 44
relative total area	0.43	0.57	44
mean magnetic field [G]	950	1200	1

b) Evershed effect, reviews [35, 49]
velocities:
averaged: 2.0 km s^{-1} outflow [e];
dark filaments: < 6.0 km s^{-1} outflow [5, 13a];
bright grains: ≦0.5 km s^{-1} inflow [43, 45];
transition region photosphere-chromosphere: [12, 29];
chromosphere, Hα [4, 38] and refs. therein.

c) Penumbral models:
homogenous hydrostatic [36];
two-component model [29, 35];
convective rolls [17, 25].

d) Penumbral waves:
v (horizontal, outwards) 10···20 km s^{-1} [25a, 66].

3.1.2.3.5 Sunspot energy

Table 5. Energetics for large spots (diameter $3.5 \cdot 10^4$ km) [6].

	erg cm^{-2} s^{-1}	Total erg s^{-1}
missing flux, umbra	$4.7 \cdot 10^{10}$	$7 \cdot 10^{28}$
missing flux, penumbra	$1.2 \cdot 10^{10}$	$1 \cdot 10^{29}$
Evershed effect		$7 \cdot 10^{27}$
bright ring	$2 \cdot 10^9$	$2 \cdot 10^{28}$
Alfvén waves (umbra)	10^{11}	$1 \cdot 10^{29}$
running penumbral waves (Hα)	$3 \cdot 10^8$	$3 \cdot 10^{27}$
Hα umbral oscillations	$8 \cdot 10^8$	$1 \cdot 10^{27}$

3.1.2.3.6 References for 3.1.2.3

General references

a Allen, C.W.: Astrophysical Quantities, 3rd ed., The Athlone Press, London 1973.
b Bray, R.J., Loughhead, R.E.: Sunspots, Chapman and Hall, London 1964.
c Bruzek, A., Durrant, C.J. (eds.): Illustrated Glossary for Solar and Solar-Terrestrial Physics, D. Reidel, Dordrecht 1977.
d Hall, D.N.B.: An Atlas of the Infrared Spectra of the Solar Photosphere and of Sunspot Umbrae, Kitt Peak National Observatory, Tucson 1973.
e Waldmeier, M.: Landolt-Börnstein, NS, Vol. VI/1, 1965.

Special references

1 Abdusamatov, H.I.: Sol. Phys. **48** (1976) 117.
2 Adam, M.G.: Mon. Not. R. Astron. Soc. **136** (1967) 71.
3 Baranovsky, E.A.: Izv. Krymskoj Astrofiz. Obs. **49** (1974) 25.
3a Baranovsky, E.A.: Izv. Krymskoj Astrofiz. Obs. **51** (1974) 56.
4 Beckers, J.M.: Australian J. Phys. **15** (1962) 327.
5 Beckers, J.M.: Sol. Phys. **3** (1968) 258.
6 Beckers, J.M.: New View of Sunspots, Sacramento Peak Obs. Contr. No. 249 = AFCRL-TR-75-0089 (1975).
7 Beckers, J.M., Schröter, E.H.: Sol. Phys. **4** (1968) 303.
8 Beckers, J.M., Schröter, E.H.: Sol. Phys. **7** (1969) 22.
9 Beckers, J.M., Schröter, E.H.: Sol. Phys. **10** (1969) 384.
10 Beckers, J.M., Tallant, P.E.: Sol. Phys. **7** (1969) 351.

11 Bhatnagar, A., Livingston, W.C., Harvey, J.W.: Sol. Phys. **27** (1972) 80.
12 Bønes, J., Maltby, P.: Sol. Phys. **57** (1978) 65.
13 Bumba, V.: Izv. Krymskoj Astrofiz. Obs. **23** (1960) 212.
13a Bumba, V.: Izv. Krymskoj Astrofiz. Obs. **23** (1960) 252.
13b Bumba, V., Suda, J.: Bull. Astron. Inst. Czech. **31** (1980) 101.
14 Cheng, C.C., Doschek, G.A., Feldman, U.: Astrophys. J. **201** (1976) 836.
15 Cheng, C.C., Kjeldseth Moe, O.: Sol. Phys. **52** (1977) 327.
16 Chystiakov, V.F.: Soviet Astron.-A.J. **6** (1962) 363.
16a Clark, A., jr.: Sol. Phys. **62** (1979) 305.
17 Danielson, R.E.: Astrophys. J. **134** (1961) 289.
18 Deinzer, W.: Astrophys. J. **141** (1965) 281.
19 Deubner, F.L.: Sol. Phys. **7** (1969) 87.
20 Deubner, F.L., Göhring, R.: Sol. Phys. **13** (1970) 118.
21 Ekman, G., Maltby, P.: Sol. Phys. **35** (1974) 317.
22 Eriksen, G., Maltby, P.: Proc. First Europ. Astron. Meeting **1** (1973) 87.
23 Foukal, P.V.: Astrophys. J. **210** (1976) 575.
24 Foukal, P.V., Huber, M.C.E., Noyes, R.W., Reeves, E.M., Schmahl, E.J., Timothy, J.G., Vernazza, J.E., Withbroe, G.L.: Astrophys. J. **193** (1974) L143.
25 Galloway, D.J.: Sol. Phys. **44** (1975) 409.
25a Giovanelli, R.G.: Sol. Phys. **27** (1972) 71.
26 Gokhale, M.H., Zwaan, C.: Sol. Phys. **26** (1972) 52.
27 Henoux, J.C.: Ann. Astrophys. **31** (1968) 511.
28 Henoux, J.C.: Astron. Astrophys. **2** (1969) 288.
28a Joshi, G.C., Punetha, L.M., Pande, M.C.: Sol. Phys. **64** (1979) 255.
29 Kjeldseth Moe, O., Maltby, P.: Sol. Phys. **36** (1974) 101.
30 Kjeldseth Moe, O., Maltby, P.: Sol. Phys. **36** (1974) 109.
31 Kneer, F.: Astron. Astrophys. **18** (1972) 39.
32 Kneer, F., Mattig, W.: Astron. Astrophys. **65** (1978) 17.
32a Kollatschny, W., Stellmacher, G., Wiehr, E., Falipou, M.A.: Astron. Astrophys. **86** (1980) 245.
33 Krat, V.A., Karpinsky, V.N., Prandjuk, L.M.: Sol. Phys. **26** (1972) 305.
34 Krat, V.A., Vyalshin, G.F.: Sol. Phys. **60** (1978) 47.
35 Lamb, S.A.: Mon. Not. R. Astron. Soc. **172** (1975) 205.
35a Loughhead, R.E., Bray, R.J., Tappere, E.J.: Astron. Astrophys. **79** (1979) 128.
36 Maltby, P.: Sol. Phys. **8** (1969) 275.
37 Maltby, P.: Sol. Phys. **26** (1972) 76.
38 Maltby, P.: Sol. Phys. **43** (1975) 91.
39 Maltby, P., Mykland, N.: Sol. Phys. **8** (1969) 23.
40 Mattig, W.: Z. Astrophys. **31** (1953) 273.
41 Mattig, W.: Sol. Phys. **8** (1969) 291.
42 Mattig, W.: Sol. Phys. **36** (1974) 275.
43 Muller, R.: Sol. Phys. **29** (1973) 55.
44 Muller, R.: Sol. Phys. **32** (1973) 409.
45 Muller, R.: Sol. Phys. **48** (1976) 101.
46 Nishi, K.: Publ. Astron. Soc. Japan **14** (1962) 325.
47 Noyes, R.W., Foukal, P.V., Huber, M.C.E., Reeves, E.M., Schmahl, E.J., Timothy, J.G., Vernazza, J.E., Withbroe, G.L.: Int. Astron. Union Symp. **68** (1975) 3.
48 Rayrole, J.: Ann. Astrophys. **30** (1967) 257.
49 Schröter, E.H.: in Xanthakis (ed.): Solar Physics, Interscience Publishers, London 1967, p 325.
50 Schröter, E.H.: Int. Astron. Union Symp. **43** (1971) 167.
51 Soltau, D., Schröter, E.H., Wöhl, H.: Astron. Astrophys. **50** (1976) 367.
52 Spruit, H.C.: Sol. Phys. **50** (1976) 269.
53 Staude, J.: Bull. Astron. Inst. Czech. **29** (1978) 71.
54 Stellmacher, G., Wiehr, E.: Astron. Astrophys. **7** (1970) 432.
55 Stellmacher, G., Wiehr, E.: Astron. Astrophys. **45** (1975) 69.
56 Suzuki, Y.: Publ. Astron. Soc. Japan **19** (1967) 220.
57 Teplitskaja, R.B., Grigoryeva, S.A., Skochilov, V.G.: Sol. Phys. **56** (1978) 293.
58 Webber, J.C.: Sol. Phys. **16** (1971) 340.
59 Wilson, P.R., McIntosh, P.S.: Sol. Phys. **10** (1969) 370.

60 Wittmann, A., Schröter, E.H.: Sol. Phys. **10** (1969) 357.
61 Wöhl, H.: Sol. Phys. **16** (1971) 362.
62 Wöhl, H., Wittmann, A., Schröter, E.H.: Sol. Phys. **13** (1970) 104.
63 Yun, H.S.: Astrophys. J. **162** (1970) 975.
64 Yun, H.S.: Sol. Phys. **16** (1971) 379.
65 Yun, H.S.: Sol. Phys. **16** (1971) 398.
66 Zirin, H., Stein, A.: Astrophys. J. **178** (1972) L85.
67 Zwaan, C.: Annu. Rev. Astron. Astrophys. **6** (1968) 135.
68 Zwaan, C.: Sol. Phys. **37** (1974) 99.
69 Zwaan, C.: Sol. Phys. **45** (1975) 115.
70 Zwaan, C.: Sol. Phys. **60** (1978) 213.

3.1.2.4 Faculae and plages

3.1.2.4.1 Continuum

a) Structure
Faculae appear resolved into facular granules: diameter 1″, lifetime 1···2 h, [a, 12].
Facular granules are a composite of facular points (filigrees): diameter 100···200 km, lifetime 5···15 min [7, 9, 10, 11].

b) Continuum contrast, center-limb variation (CLV):
unresolved facular regions: [2, 5, 14]; [19] Fig. 1, Table 1;
facular granules: [1] Table 1; [12] Fig. 3;
facular points, filigrees: [9, 11] $\Delta I/I_{phot} \gtrsim 100\%$.
Dependence of continuum contrast on measured magnetic field: [5] Figs. 2, 4, 6.

Table 1. CLV continuum contrast $I_{fac}/I_{phot}(\lambda \approx 5200$ Å).

cos θ	1.0	0.7	0.6	0.5	0.4	0.3	0.2	0.1
Ref.	I_{fac}/I_{phot}							
12 (Fig. 3)					1.35	1.405	1.385	1.23
1 (Table 1)		1.072	1.095	1.129	1.180	1.194	1.145	
19 (Table 1)	1.015	1.04				1.11		
5 (Fig. 4) (B=150 G)	1.014	1.018		1.048		1.132		

3.1.2.4.2 Facula models

a) For CLV of continuum contrast [1] Table 2; [2] Table I; [8]; [12] Figs. 4, 5; [13]; [23] Fig. 3:
$\Delta T \sim 400 \cdots 1000$ K for $\tau_{5000} < 1$
$\Delta T \sim -(100 \cdots 400)$ K for $\tau_{5000} > 1$.

b) For weak lines and wings of Ca II lines [4, 13]; [15] Table 1, Fig. 3; [18, 19]:
$\Delta T \approx 100 \cdots 300$ K.

c) For filigree continuum and line intensities [10]; [20] Fig. 1:
$\Delta T \approx 1000$ K.

d) Magnetic elements (B=1000···2000 G):
models for weak lines [3] Table 3; [22] Fig. 3, Table I: $\Delta T \approx 800$ K;
magnetic, velocity and brightness structure: [6] Fig. 6;
multi-dimensional NLTE calculations for magnetic tubes [17, 21]:
Wilson depression: ≈100 km, ΔT=0.

3.1.2.4.3 Chromospheric plage

Model for Ca II line profiles [16] Figs. 3, 4.

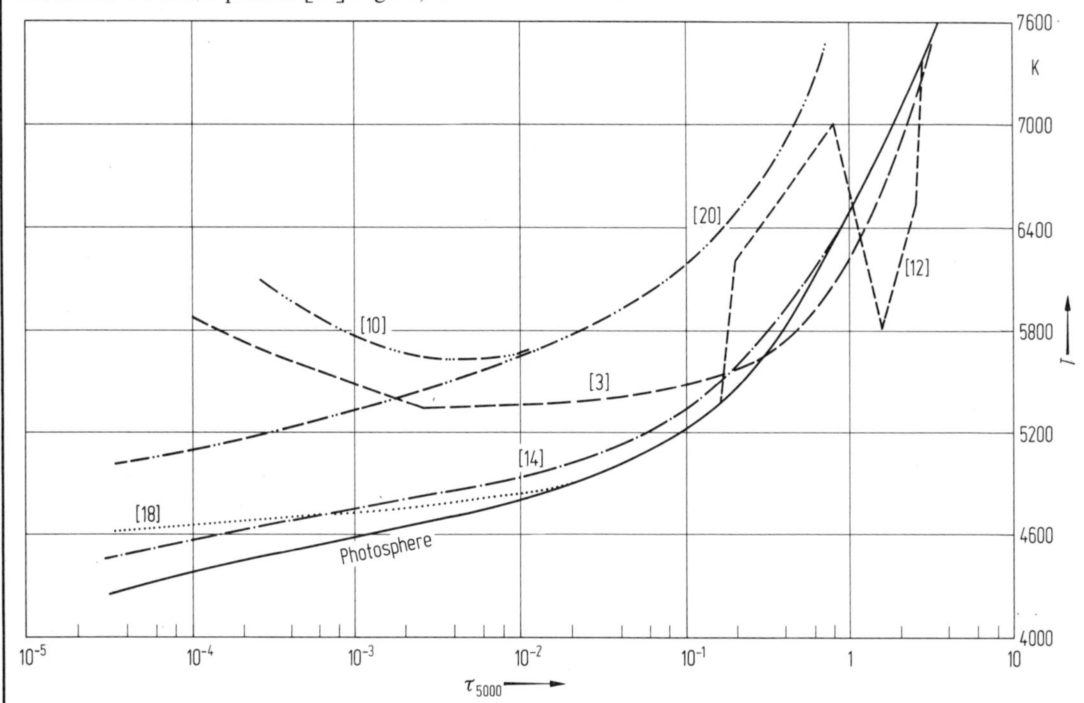

Fig. 1. Temperature models of faculae and filigrees from [20, (Fig. 1)]; numbers in the figure are reference numbers.

3.1.2.4.4 References for 3.1.2.4

General references

a Bray, R.J., Loughhead, R.E.: Sunspots, Chapman and Hall, London 1964.

Special references

1 Badalyan, O.G., Prudkovskii, A.G.: Sov. Astron.- A.J. **17** (1973) 356.
2 Chapman, G.A.: Sol. Phys. **14** (1970) 315.
3 Chapman, G.A.: Astrophys. J. **232** (1979) 923.
4 Chapman, G.A., Sheeley, N.R.: Sol. Phys. **5** (1968) 442.
5 Frazier, E.N.: Sol. Phys. **21** (1971) 42.
6 Frazier, E.N., Stenflo, J.O.: Astron. Astrophys. **70** (1978) 789.
7 Harvey, J.W., Breckinridge, J.B.: Astrophys. J. **182** (1973) L137.
8 Hirayama, T.: Publ. Astron. Soc. Japan **30** (1978) 330.
9 Koutchmy, S.: Astron. Astrophys. **61** (1977) 397.
10 Koutchmy, S., Stellmacher, G.: Astron. Astrophys. **67** (1978) 93.
11 Mehltretter, P.: Sol. Phys. **38** (1974) 43.
12 Muller, R.: Sol. Phys. **45** (1975) 105.
13 Rees, D.E.: Int. Astron. Union Symp. **56** (1974) 177.
14 Schmahl, G.: Z. Astrophys. **66** (1967) 81.
15 Shine, R.A., Linsky, J.L.: Sol. Phys. **37** (1974) 145.
16 Shine, R.A., Linsky, J.L.: Sol. Phys. **39** (1974) 49.
17 Spruit, H.C.: Sol. Phys. **50** (1976) 209.
18 Stellmacher, G., Wiehr, E.: Sol. Phys. **18** (1971) 220.
19 Stellmacher, G., Wiehr, E.: Astron. Astrophys. **29** (1973) 13.
20 Stellmacher, G., Wiehr, E.: Astron. Astrophys. **75** (1979) 263.
21 Stenholm, L.G., Stenflo, J.O.: Astron. Astrophys. **58** (1977) 273.
22 Stenflo, J.O.: Sol. Phys. **42** (1975) 79.
23 Wilson, P.R.: Sol. Phys. **21** (1971) 101.

3.1.2.5 Prominences and ejecta

3.1.2.5.1 General characteristics

Table 1. Classifications of prominences [a, b].

Gross phenomenological	Detailed morphological
quiescent prominences	Pettit
active prominences: loops	Newton
	Menzel-Evans
active prominences: ejections	Severny
	de Jager
	Zirin

Table 2. Dimensions of prominences [a, b, c].
in ( ): mean values

$[10^3$ km]	Quiescent prominences (filaments)	Prominence threads	Ref.
length	60···1000 (200)	5	33
width	4··· 15 (7)	0.2···0.4	2, 13
height	15··· 120 (45)		

Table 3. Lifetime of quiescent prominences [c].

Lifetime (solar rotations)	≦1	1···2	2···3	3···4	4···5	5···6	6···7	7···8
% of all quiescents	8	26	26	19	8	5	4	2

Inclination of the filament axis towards the parallels of latitude as a function of latitude [c] p. 124, Table 2.

3.1.2.5.2 Prominence spectrum

Compilation [b]; recent spectral investigations, see Table 4.

Table 4. Spectral investigations.

Lines or region	Ref.
H Balmer lines	3, 14, 21, 22, 23, 43
H Ly α	45
H Ly continuum	30, 38
He I lines	3, 12, 14, 20, 22, 23, 43
Ca II H, K	3, 7, 21, 23, 43, 45
Ca II infrared triplet	7, 21, 22
Mg II h, k	45
EUV lines (1175···1930 Å)	15a (Table II), 27

Characteristic differences between spectra of the quiet chromosphere and of various types of prominences, and spectral classification of prominences:
Zirin-Tandberg-Hanssen classification [b] Table II.3,
Waldmeier classification [b] Table II.2,
Hirayama classification [12] Table 2.

3.1.2.5.3 Physical characteristics of quiescent prominences

Table 5. Temperature.

Method	T [K]	Ref.
from line widths (H, He, metals)	4500···8500, central parts	8, 9, 10, 22, 28, 42, 46
	8000···12000, outer parts	3, 9, 10, 30, 38
colour temperature of H Ly continuum	8000	30, 38
excitation temperature Balmer lines	7170±310	23
excitation temperature He triplet	4460± 60	23
other possible methods		12 (Table 1)

Table 6. Internal motions.

Type (method)	Region	v [km s^{-1}]	Ref.
turbulent velocities (widths of weak metal lines)	inner parts	3··· 8	3, 9, 10, 11, 22, 28
	outer parts	10···20	
	interface	2··· 7	4
internal motion (Doppler shift)	threads	15···30	2

Table 7. Magnetic fields B [b, 26].

Method	$\bar{B}$ [G]	Ref.
longitudinal Zeeman effect	3···8	b (Table II.4, lines used Table II.5)
Hanle effect [1]	5.5	24, 25, 26 (Fig. 9)

Table 8. Number densities N.

Method	N [10^{10} cm^{-3}]	Ref.
Stark effect of H Balmer lines	$N_e = 1 \cdots 25$	11, 15, 29, 47
He triplet/ singlet intensity ratio	$N_H = 1.6 \cdots 64$ (mean 27)	22

3.1.2.5.4 Ejections

Table 9. Physical characteristics of mass ejecta [35] Table 7.1.

p = gas pressure
T = temperature
h = height
v = characteristic velocity
Δ = change from surroundings
B = magnetic field strength
E_{kin} = kinetic energy
t = lifetime

Phenomenon	$\Delta p/p$	T 10^4 K	B G	h 10^9 cm	v 10^7 cm s^{-1}	Mass 10^{14} g	E_{kin} erg	t 10^3 s
eruptive prominences	2	10	10 at 1.1 $R_\odot$	5	2	10	$2 \cdot 10^{29}$	1···5
sprays	10	1···10	100 at 1.02 $R_\odot$	5	6	10	$2 \cdot 10^{30}$	1
surges	5	0.8	30 at 1.02 $R_\odot$	1	1	1···10	$5 \cdot 10^{28}$	1
coronal transients	10···100	100	2 at 2.0 $R_\odot$	30	4	10···100	$10^{29} \cdots 10^{30}$	10

3.1.2.5.5 Interface prominence — corona [39]

EUV line spectrum ($T = 3 \cdot 10^4 \cdots 5 \cdot 10^5$ K): [4]; [27] Table I; [30, 31]; [32] Table II; [38].

Table 10. Characteristics of interface.

		Ref.
width: range	10···100 km	38, 48
C III region ($T = 9 \cdot 10^4$ K)	10 km	38
O VI region ($T = 3 \cdot 10^5$ K)	41 km	38
$\Delta T/h$ ($3 \cdot 10^4 < T < 10^5$ K)	10···100 K m^{-1}	48
N_e (intensity ratio C III 1176/977)	$1.3 \cdot 10^9$ cm^{-3}	32 (Table IV) *)
P_e	0.01···0.02 dyn cm^{-2}	20, 31, 32, 38 *)

*) [27] Table 3 and [48] Table I give one order of magnitude larger values of electron number density N_e and electron pressure P_e.

Cavity surrounding prominence [40] $T = 1 \cdot 10^6$ ($\cdots 2 \cdot 10^6$) K
$N_e = 3 \cdot 10^8$ cm^{-3}.

Total mass deficiency compared to quiet corona (white light observations): $10^{37} \cdots 10^{38}$ electrons [36, 37].
X-ray cavity [40], radio cavity [17, 18].
Filaments at radio wavelengths [17] Table 1; [33a]; [34] Fig. 4.

3.1.2.5.6 References for 3.1.2.5

General references

a Bruzek, A., Durrant, C.J.: Illustrated Glossary for Solar and Solar-Terrestrial Physics, D. Reidel, Dordrecht 1977.
b Tandberg-Hanssen, E.: Solar Prominences, D. Reidel, Dordrecht 1974.
c Waldmeier, M.: Landolt-Börnstein, NS, Vol. VI/1, 1965.

Special references

1 Bommier, V., Sahal-Bréchot, S.: Int. Astron. Union Coll. **44** (1979) 87.
2 Engvold, O.: Sol. Phys. **49** (1976) 289.
3 Engvold, O.: Sol. Phys. **56** (1978) 87.
4 Feldman, U., Doschek, G.A.: Astrophys. J. **216** (1977) L119.
5 Heasley, J.N., Milkey, R.W.: Astrophys. J. **210** (1976) 827.
6 Heasley, J.N., Milkey, R.W.: Astrophys. J. **221** (1978) 677.
7 Heasley, J.N., Milkey, R.W., Engvold, O.: Sol. Phys. **51** (1977) 315.
8 Hirayama, T.: Publ. Astron. Soc. Japan **15** (1963) 122.
9 Hirayama, T.: Publ. Astron. Soc. Japan **16** (1964) 104.
10 Hirayama, T.: Sol. Phys. **17** (1971) 50.
11 Hirayama, T.: Sol. Phys. **24** (1972) 310.
12 Hirayama, T.: Int. Astron. Union Coll. **44** (1979) 4.
13 Hirayama, T., Nakagomi, Y., Okamoto, T.: Int. Astron. Union Coll. **44** (1979) 48.
14 Illing, R.M.E., Landman, D.A., Mickey, D.L.: Sol. Phys. **45** (1975) 339.
15 Jefferies, J.T., Orrall, J.Q.: Astrophys. J. **137** (1963) 1232.
15a Kjeldseth Moe, O., Cook, J.W., Margo, S.A.: Sol. Phys. **61** (1979) 319.
16 Kotov, V.A.: Izv. Krymskoj Astrofiz. Obs. **39** (1969) 276.
17 Kundu, M.R.: Int. Astron. Union Coll. **44** (1979) 122.
18 Kundu, M.R., Fürst, E., Hirth, W., Butz, M.: Astron. Astrophys. **62** (1978) 431.
19 Landman, D.A.: Sol. Phys. **50** (1976) 383.
20 Landman, D.A., Illing, R.M.E.: Astron. Astrophys. **49** (1976) 277.
21 Landman, D.A., Illing, R.M.E.: Astron. Astrophys. **55** (1977) 103.
22 Landman, D.A., Edberg, S.J., Laney, C.D.: Astrophys. J. **218** (1977) 888.
23 Landman, D.A., Mongillo, M.: Astrophys. J. **230** (1979) 581.
24 Leroy, J.L.: Astron. Astrophys. **60** (1977) 79.
25 Leroy, J.L.: Astron. Astrophys. **64** (1978) 247.
26 Leroy, J.L.: Int. Astron. Union Coll. **44** (1979) 56.
27 Mariska, J.T., Doschek, G.A., Feldman, U.: Astrophys. J. **232** (1979) 929.
28 Mouradian, A., Leroy, J.L.: Sol. Phys. **51** (1977) 103.
29 Nikolsky, G.M., Gulyaev, R.A., Nikolskaya, K.I.: Sol. Phys. **21** (1971) 332.
30 Noyes, R.W., Dupree, A.K., Huber, M.C.E., Parkinson, W.H., Reeves, E.M., Withbroe, G.L.: Astrophys. J. **178** (1972) 515.
31 Orrall, F.Q., Speer, R.: Int. Astron. Union Symp. **56** (1974) 193.
32 Orrall, F.Q., Schmahl, E.: Sol. Phys. **50** (1976) 365.
33 Ramsey, L.W.: Sol. Phys. **51** (1977) 307.
33a Rao, A.P., Kundu, M.R.: Astron. Astrophys. **86** (1980) 373.
34 Raoult, A., Lantos, P., Fürst, E.: Sol. Phys. **61** (1979) 335.
35 Rust, D.M., Hildner, E., Dryer, M., Hansen, R.T., McClymont, A.N., McKenna-Lawlor, S.M.P., McLean, D.J., Schmal, E.J., Steinolfson, R.S., Tandberg-Hanssen, E., Tousey, R., Webb, D.F., Wu, S.T.: in Sturrock (ed.): Solar Flares, Boulder, Col. (1979) 273.
36 Saito, K., Tandberg-Hanssen, E.: Sol. Phys. **31** (1973) 105.
37 Saito, K., Hyder, C.L.: Sol. Phys. **5** (1968) 61.
38 Schmahl, E.J., Foukal, P.V., Huber, M.C.E., Noyes, R.W., Reeves, E.M., Timothy, J.G., Vernazza, J.E., Withbroe, G.L.: Sol. Phys. **39** (1974) 337.
39 Schmahl, E.J.: Int. Astron. Union Coll. **44** (1979) 102.
40 Serio, S., Vaiana, S., Godoli, G., Motta, S., Pirronello, V., Zappala, R.A.: Sol. Phys. **59** (1978) 65.
41 Smolkov, G.Y.: Int. Astron. Union Symp. **43** (1971) 710.
42 Stellmacher, G.: Astron. Astrophys. **1** (1969) 62.

43 Stellmacher, G.: Sol. Phys. **61** (1979) 61.
44 Tsubaki, T.: Sol. Phys. **43** (1975) 147.
45 Vial, I.C., Gouttebroze, P., Artzner, G., Lemaire, P.: Sol. Phys. **61** (1979) 39.
46 Yakovkin, N.A., Zeldina, M.Y.: Soviet Astron.-A.J. **7** (1964) 643.
47 Yakovkin, N.A., Zeldina, M.Y., Rakhbovsky, A.S.: Soviet Astron.-A.J. **19** (1975) 66.
48 Yang, C.Y., Nicholls, R.W., Morgan, F.J.: Sol. Phys. **45** (1975) 351.

3.1.2.6 Coronal active region

3.1.2.6.1 Visible

a) White light coronal condensation [23],[33] Fig. 6,
model: $N_e = 5 \cdot 10^9\ \text{cm}^{-3}$ in the center,
$N_e = 4 \cdot 10^8\ \text{cm}^{-3}$ in outer shells [33].

b) Monochromatic (forbidden line) condensation:
list of observed lines [10, 11, 28],
two-component model [25, 33]:

T 10^6 K	N_e $10^9\ \text{cm}^{-3}$
2.5···5	3.6
1.0···1.5	1.1

3.1.2.6.2 EUV line coronal enhancement

Review [18],
observed lines [17, 18, 24],
enhancement over quiet corona: from 3 times for transition region lines to 40 times for coronal lines [18] Fig. 10.
EUV model for transition region: [2]; [33] Fig. 2; typical $N_e \approx 10^{11}\ \text{cm}^{-3}$ [7].

3.1.2.6.3 X-ray corona

Reviews [12, 29, 30].

a) Continuous spectrum: observed [30] Fig. 5; computed (radiative recombination, bremsstrahlung and two-photon decay of metastable levels of H-like and He-like ions) [5, 26]; [29] Fig. 24; [30] Fig. 4.

b) Line spectrum: observed [9] Table I, [20] Table I, [32] Table V; computed $\lambda < 100$ Å [13, 16, 26].

c) Emission measure $M = \int N_e^2 \mathrm{d}V\ [\text{cm}^{-3}]$

differential emission measure $M(T) = \dfrac{\mathrm{d}M}{\mathrm{d}T}\ [\text{cm}^{-3}\,(10^6\ \text{K})^{-1}]$

exponential model [1, 3, 4, 31, 12 (Fig. 2, Table VII), 30 (Figs. 7, 8)]:

$$M(T) = C \cdot 10^{-T/T_0}\ [\text{cm}^{-3}\,(10^6\ \text{K})^{-1}]$$

multilayer model [19] Table II.

d) $N_e(T)$ model: hot core models [2, 8, 14, 27], average:

	T_e 10^6 K	N_e $10^9\ \text{cm}^{-3}$	h 10^4 km
hot core	4	10	3
cool shells	2	2	10

multilayer model [19] Fig. 4 and Table II.

e) Coronal loop model [15a] refs.

3.1.2.6.4 Total radiation loss [6, 22]

Typical coronal active region: $\mathrm{d}E/\mathrm{d}t\ (T > 10^4\ \text{K}) = (1 \cdots 4) \cdot 10^{26}\ \text{erg s}^{-1}$.

3.1.2.6.5 References for 3.1.2.6

1 Acton, L.W., Catura, R.C., Meyerott, A.I., Wolfson, C.J., Culhane, J.L.: Sol. Phys. **26** (1972) 183.
2 Boardman, W.J., Billings, D.E.: Astrophys. J. **156** (1969) 731.
3 Bonnelle, C., Senemaud, C., Senemaud, G., Chambe, G., Guionnet, M., Henoux, J.C., Michard, R.: Sol. Phys. **29** (1973) 341.
4 Chambe, G.: Astron. Astrophys. **12** (1971) 210.
5 Culhane, J.L.: Mon. Not. R. Astron. Soc. **144** (1969) 375.
6 Evans, K.D., Pye, J.P., Hutcheon, R.J., Gerassimenko, M., Krieger, A.S., Davis, J.M., Vesecky, J.F.: Sol. Phys. **55** (1977) 387.
7 Feldman, U., Doschek, G.A.: Astron. Astrophys. **65** (1978) 215.
8 Gabriel, A.H., Jordan, C.: Mon. Not. R. Astron. Soc. **173** (1975) 397.
9 Hutcheon, R.J.: Int. Astron. Union Symp. **68** (1975) 69.
10 Jefferies, J.T., Orrall, F.Q., Zirker, J.B.: Sol. Phys. **16** (1971) 103.
11 Jefferies, J.T.: Mem. Soc. R. Sci. Liege 5th Series XVII (1969) 213.
12 Jordan, C.: Int. Astron. Union Symp. **68** (1975) 109.
13 Landini, M., Monsignori Fossi, B.C.: Astron. Astrophys. **6** (1970) 468.
14 Landini, M., Monsignori Fossi, B.C.: Sol. Phys. **17** (1971) 379.
15 Landini, M., Monsignori Fossi, B.C., Krieger, A.S., Vaiana, G.S.: Sol. Phys. **44** (1975) 69.
15a Levine, R.H., Pye, J.P.: Sol. Phys. **66** (1980) 39.
16 Mewe, R.: Sol. Phys. **22** (1972) 459.
17 Noyes, R.W., Withbroe, G.L., Kirschner, R.P.: Sol. Phys. **11** (1970) 388.
18 Noyes, R.W.: in Macris (ed.): Physics of the Solar Corona, Reidel, Dordrecht 1971, p.192.
19 Parkinson, J.H.: Sol. Phys. **28** (1973) 487.
20 Parkinson, J.H.: Sol. Phys. **42** (1975) 183.
21 Parkinson, J.H.: Int. Astron. Union Symp. **68** (1975) 45.
22 Pye, J.P., Evans, K.D., Hutcheon, R.J., Gerassimenko, M., Davis, J.M., Krieger, A.S., Vesecky, J.F.: Astron. Astrophys. **65** (1977) 123.
23 Saito, K., Billings, D.E.: Astrophys. J. **140** (1964) 760.
24 Sheeley, N.R. jr., Bohlin, J.D., Brueckner, G.E., Purcell, J.D., Scherrer, V., Tousey, R.: Sol. Phys. **40** (1975) 103.
25 Suzuki, T., Hirayama, T.: Publ. Astron. Soc. Japan **16** (1964) 58.
26 Tucker, W.H., Koren, M.: Astrophys. J. **168** (1971) 283.
27 Vaiana, G.S., Krieger, A.S., Timothy, A.F.: Sol. Phys. **32** (1974) 81.
28 Wagner, W.J., House, L.L.: Sol. Phys. **5** (1968) 55.
29 Walker, A.B.C.: Space Sci. Rev. **13** (1972) 672.
30 Walker, A.B.C.: Int. Astron. Union Symp. **68** (1975) 73.
31 Walker, A.B.C., Rugge, H.R., Weiss, K.: Astrophys. J. **188** (1974) 423; Astrophys. J. **192** (1974) 169.
32 Withbroe, G.L.: Sol. Phys. **21** (1971) 272.
33 Zirker, J.B.: in Macris (ed.): Physics of the Solar Corona, Reidel, Dordrecht 1971, p. 140.

3.1.2.7 Flares

3.1.2.7.1 General

a) Flare classifications [a, b, g, i]

Table 1. Flare importance.

Importance (column 1) is defined by the area (column 2) in Hα at intensity maximum; letters F, N, B (for faint, normal, bright) are added in order to indicate flare brightness;

S ≡ subflare

t = lifetime;

$I_{H\alpha}$ = Hα intensity of flare;

I_{cont} = continuum intensity of undisturbed surroundings;

E = total energy.

Importance	Area 10^{18} cm^2	t 10^3 s	$I_{H\alpha}/I_{cont}$ *)	E erg *)
S	< 3.0	0.1 ··· 1		10^{28}
1	3.0 ··· 7.5	0.5 ··· 2	0.6	10^{29}
2	7.5 ··· 18	1 ··· 5	1.0	10^{30}
3	18 ··· 36	2 ··· 10	1.5	10^{31}
4	> 36	> 5	2.0	10^{32}

*) Hα intensity and total energy E are order of magnitude estimates, detailed estimates for E: see [2, 8, 23]; [f] appendix A, B.

Table 2. X-ray importance.
X-ray importance is defined by the peak intensity I_{max} measured at the earth in the 1···8 Å band.

Importance	I_{max} erg cm^{-2} s^{-1}
C1 ···C9	(1··· 9)·10^{-3}
M1···M9	(1··· 9)·10^{-2}
X1 ···X10	(1···10)·10^{-1}

b) Frequency of flares: [a, b, g], compilation of flares [b, d, e, h].

Comprehensive Flare Index CFI [b] is a combination of indices for the most characteristic flare emissions:

CFI = A + B + C + D + E,

A: importance of ionizing radiation as indicated by the importance of the associated SID, scale 1···3; *)
B: importance of Hα flare, scale 1···3 (3 stands for classes 3 and 4);
C: log of 2800 MHz flux in units [10^{-22} W m^{-2} Hz^{-1}];
D: radio bursts: type II = 1, continuum = 2, type IV = 3;
E: log of 200 MHz flux in the same units as C.

For radio flux and bursts, see also 3.1.2.8

*) SID: Sudden Ionospheric Disturbance [a] p. 173.

3.1.2.7.2 Flare spectrum

a) Thermal emission ($10^4 < T < 5 \cdot 10^7$ K):
Visible and UV radiation is essentially enhanced chromospheric line emission; most important lines [g] Tables V and VI. Visible continuum is extremely rare ("White light flares"). Limb flares show Balmer lines up to H 30, Balmer continuum and forbidden coronal lines [g].

EUV and XUV spectrum includes chromospheric, transition region and coronal lines and continuum; lists with intensities, identifications, line widths:
1420···1960 Å continuum [8]
1420···1900 Å continuum [5] (Table I); [7] (Table 3)
1100···1940 Å selection of lines [15, 16, 3]
1000···1940 Å 1400 lines [7]
Ly continuum [24]
284···1400 Å most important lines [g] (Table XIII)
171···630 Å lines [12] (Table I), [36] (Table 3)
90···170 Å lines [13]
66···171 Å lines [21]

Soft X-ray spectrum is dominated by thermal bremsstrahlung; computed intensities for $\lambda < 20$ Å and $T = 5 \cdot 10^6$ K, $1.6 \cdot 10^7$ K and $5 \cdot 10^7$ K [41] and [g] Fig. 45.

Line emission is from H-like and He-like ions of most of the abundant solar elements (C to Ca) and from high ionization stages of Fe (XVII – XXV); tables of lines:

Lines	Ref.	Lines	Ref.
6 ···25 Å	26 (Table II)	1.62··· 8.42 Å	25
1.9··· 6.74 Å	28 (Tables II, III)	1.5 ···16 Å	13 (Figs. 1, 2)
1.82···3.24 Å	18a	1.46···22.09 Å	g (Table XII, refs.)

b) Hard X-ray bursts:

Table 3. Characteristics of hard X-ray bursts [c]; [f] chapt. 5; [g] p. 143.

normal energy range	10 keV ≦ E ≦ 100 keV
extreme limits	3 keV···2 MeV
peak flux at earth (E > 10 keV)	10^{-7}···10^{-5} erg cm^{-2} s^{-1}
differential spectrum at burst maximum	$dJ(E)/dE = C \cdot E^{-\gamma}$ [photons cm^{-2} s^{-1} $(keV)^{-1}$], with $2.5 \leqq \gamma \leqq 5$ for 10 keV ≦ E ≦ 60 keV, $\gamma \leqq 8$ for 60 keV < E ≦ 100 keV
mechanism	bremsstrahlung of non-thermal electrons
number of electrons (E > 10 keV)	10^{35}··· > 10^{36}
burst duration	a few seconds up to 1 min

c) Gamma-rays, line emission:

Table 4. Observed gamma-ray lines [6, 40, 20a].

Hα flare	Measured line energy; flux at 1 AU [photons $cm^{-2} s^{-1}$]			
1972, Aug. 4 max 06^h30 UT	(510.7±6.4) keV; (6.3±2.0)·10^{-2}	(2.24±0.02) MeV; (2.80±0.22)·10^{-1}	4.4 MeV; (3±1)·10^{-2}	6.1 MeV; (3±1)·10^{-2}
1972, Aug. 7 max 15^h30 UT	(508.1±5.8) keV; (3.0±1.5)·10^{-2}	(2.22±0.02) MeV; (6.9 ±1.1)·10^{-2}	4.4 MeV; <2·10^{-2}	6.1 MeV; <2·10^{-2}
1978, July 11 max 10^h53 UT		2.223 MeV; 1.0 ±0.29	4.43 MeV; 0.18±0.07	
mechanism	positron annihilation	de-excitation		
		^2H	^{12}C*	^{16}O*

Possible processes in solar flares resulting in gamma-ray lines [31···34]; gamma-rays above 8 MeV [9a].

d) Gamma-ray continuum (Aug. 4, 1972) [6] differential energy spectrum:
$dJ/dE = 0.4\,E^{-(3.42\pm0.3)}$ photons $cm^{-2} s^{-1} MeV^{-1}$ (0.36 MeV < E < 0.7 MeV),
$dJ/dE = k \exp(-E/E_0)$ photons $cm^{-2} s^{-1} MeV^{-1}$ (0.7 MeV < E < 7 MeV), $E_0 = (1.0\pm0.07)$ MeV.

3.1.2.7.3 Flare physics

Table 5. Physical characteristics [g]; [f] chapts. 5, 8.

Region	T [K]	N_e [cm^{-3}]
chromosphere	10^4 ($7\cdot10^3 \cdots 2\cdot10^4$)	$10^{12} \cdots 10^{13}$ (max. $4.4\cdot10^{13}$)
transition region (TR)	10^5	$5\cdot10^{11} \cdots 2\cdot10^{12}$
corona (XUV)	$10^6 \cdots 10^7$	$10^{10} \cdots 3\cdot10^{11}$
corona (X-rays)	$\cdots 4\cdot10^7$	$10^{10} \cdots 3\cdot10^{11}$

a) Densities, diagnostics:
- α) Chromosphere (visible emission) [g]:
  halfwidths of higher Balmer lines (≧H_{10}),
  intensity of Balmer continuum,
  Inglis-Teller formula.
- β) Transition region (EUV):
  line intensity ratios [16, 17],
  intensity ratio Si III 1301/1312 and Si III 1301/1296 [3, 4],
  density dependence of O VI 1401 Å line [3, 4]

Model of chromosphere and lower TR (T<10^5 K, N_e, N_H, ϱ) for O I – C IV lines and H-Ly emissions [23b].

- γ) Corona, XUV; N_e from intensity ratios:
  Fe XXIV lines [43, 44]
  Fe XIII, XIV, XV lines (12b]
  Ca XVII lines [14]
  Fe IX lines [18]
  Fe IX···Fe XV, Ni XI···Ni XVII lines and members of the Be, B, C, and N isoelectronic sequences [12c].
- δ) Corona, X-ray lines ; N_e from emission measure [27, 35, 44],
  from cooling time assuming conductive cooling [9, 11, 30].

b) Temperature, emission measure of soft X-ray flare:
- α) From line spectrum: ratio of line intensities [19, 22], line intensities [2, 27, 29, 45].
- β) From broad band measurements:
  Energy distribution in the X-ray continuum [25],
  intensity ratio of two bands (e.g. 0.5···3.0 Å and 1···8 Å) [g] p. 128ff.,
  intensity distribution in 7 channels in 2.6···10 Å [20] (two temperature model),
  intensity distribution in several bands in 0.5···60 Å [12a] (multithermal model).

3.1.2.7.4 Flare particle emission

Table 6. Particle emission.
Flux = number of particles per cm^2 s sr
γ = index of power law energy spectrum

Particles	Energy range	Flux at 1 AU cm^{-2} s^{-1} sr^{-1}	for	γ	for	Ref.
non-relativistic electrons						
in pure electron events (PE)	5 ⋯ > 100 keV	10⋯100 (< 1000)	$E > 40$ keV	2.5⋯4.6	$E < 100$ keV	23,23a
in proton-electron events (PR)		100⋯1000 (< 10^4)	$E > 40$ keV	≈ 3.0		
protons in						
PCA [1]) effects	5 ⋯100 MeV	10^{-3}⋯10^4	$E > 10$ MeV	1.3⋯5.0	$E = 20⋯80$ MeV	g
GLE [2]) effects	0.5⋯15 GeV					
relativistic electrons	> 0.3 MeV	10^{-2}⋯$5 \cdot 10^{-6}$ (median $3 \cdot 10^{-4}$) of proton flux at same energy	$E = 12⋯45$ MeV	{ 3.0 { > 3.0	$E =$ 3⋯12 MeV $E > 12$ MeV	39

[1]) List of proton events (PCA effects) [h], [42] Table I; distribution of size of events, see [g] Fig. 96, [42] Table II. PCA = Polar Cap Absorption (= absorption of galactic radio waves in the polar cap ionosphere due to enhanced ionization produced by incident energetic solar particles) [a] p. 173.
[2]) GLE = Ground Level Effect [= solar cosmic rays with E $\gtrsim$ 1 GeV measured at the ground (earth surface)].

Biological effective total radiation dosage of particles from very large flares: 700⋯1000 R behind 1 g cm^{-2} (1 R = $2.58 \cdot 10^{-4}$ C kg^{-1}) [37, 45].

Heavy nuclei from flares:
relative abundances of heavy nuclei in flare particle streams for $E > 15$ MeV/nucleon [16] Table II; [10] Table XV;
relative abundances for $E < 15$ MeV/nucleon increase with Z [1, 16].

3.1.2.7.5 References for 3.1.2.7

General references, current data series

a Bruzek, A., Durrant, C.J.: Illustrated Glossary for Solar and Solar-Terrestrial Physics, D. Reidel, Dordrecht 1977.
b Dodson, H.W., Hedeman, R.E.: WDC-A Report UAG – 14, NOAA Boulder, 1971; WDC-A Report UAG – 52, NOAA Boulder, 1975.
c Kane, S. (ed.): Solar Gamma-, X- and EUV Radiation (= Int. Astron. Union Symp. **68**), D. Reidel, Dordrecht 1975.
d Quart. Bull. Solar Activity, Zürich.
e Solar Geophysical Data, US Department of Commerce, NOAA Boulder, Col.
f Sturrock, P. (ed.): Solar Flares, Boulder, Col. 1979.
g Švestka, Z.: Solar Flares, D. Reidel, Dordrecht 1976.
h Švestka, Z., Simon, P. (eds.): Catalogue of Solar Particle Events (1955–1969), D. Reidel, Dordrecht 1975.
i Waldmeier, M.: Landolt-Börnstein, NS, Vol. VI/1, 1965.

Special references

1 Bertsch, D.L., Biswas, S., Reames, D.V.: Sol. Phys. **39** (1974) 479.
2 Bruzek, A.: in Xanthakis (ed.): Solar Physics, Interscience Publisher, London (1967) p. 399.
3 Cheng, C.C.: Sol. Phys. **56** (1978) 205.

4 Cheng, C.C.: Skylab/ATM Preprint 1979.
5 Cheng, C.C., Kjeldseth Moe, O.: Sol. Phys. **59** (1978) 361.
6 Chupp, E.L., Forrest, D.J., Suri, A.N.: Int. Astron. Union Symp. **68** (1975) 341.
7 Cohen, L., Feldman, U., Doschek, G.A.: Astrophys. J. Suppl. **37** (1978) 443.
8 Cook, J.W., Brueckner, G.E.: Astrophys. J. **227** (1979) 645.
9 Craig, I.J.D.: Sol. Phys. **31** (1973) 197.
9a Crannel, C.J., Crannel, H., Ramaty, R.: Astrophys. J. **229** (1979) 762.
10 Crawford, H.J., Price, P.B., Cartwright, B.G., Sullivan, J.D.: Astrophys. J. **195** (1975) 213.
11 Culhane, J.L., Vesecky, J.F., Phillips, K.J.H.: Sol. Phys. **15** (1970) 394.
12 Dere, K.P.: Astrophys. J. **221** (1978) 1062.
12a Dere, K.P., Horan, D.M., Kreplin, R.W.: Sol. Phys. **36** (1974) 459.
12b Dere, K.P., Horan, D.M., Kreplin, R.W.: Astrophys. J. **217** (1977) 976.
12c Dere, K.P., Mason, H.E., Widing, K.G., Bhatia, A.K.: Astrophys. J. **230** (1979) 288.
13 Doschek, G.A.: Int. Astron. Union Symp. **68** (1975) 165.
14 Doschek, G.A., Feldman, U., Dere, K.P.: Astron. Astrophys. **60** (1977) L11.
15 Doschek, G.A., Feldman, U., Rosenberg, F.D.: Astrophys. J. **215** (1977) 329.
16 Feldman, U., Doschek, G.A., Rosenberg, F.D.: Astrophys. J. **215** (1977) 652.
17 Feldman, U., Doschek, G.A.: Astron. Astrophys. **65** (1978) 215.
18 Feldman, U., Doschek, G.A., Widing, K.G.: Astrophys. J. **219** (1978) 304.
18a Feldman, U., Doschek, G.A., Kreplin, R.W.: Astrophys. J. **238** (1980) 365.
19 Grineva, Y.I., Karev, V.I., Korneev, V.V., Krutov, V.V., Mandelstam, S.L., Vainstein, L.A., Vasilyev, B.N., Zhitnik, I.A.: Sol. Phys. **29** (1973) 441.
20 Herring, J.R.H., Craig, I.J.D.: Sol. Phys. **28** (1973) 169.
20a Hudson, H.S., Bai, T., Gruber, D.E., Matteson, J.L., Nolan, P.L., Peterson, L.E.: Astrophys. J. **236** (1980) L91.
21 Kastner, S.O., Neupert, W.M., Swartz, M.: Astrophys. J. **191** (1974) 261.
22 Landini, M., Monsignori Fossi, B.C., Pallavacini, R.: Sol. Phys. **29** (1973) 93.
23 Lin, R.P.: Space Sci. Rev. **16** (1974) 189.
23a Lin, R.P.: Int. Astron. Union Symp. **68** (1975) 385.
23b Lites, B.W., Cook, J.W.: Astrophys. J. **228** (1979) 598.
24 Machado, M.E., Noyes, R.W.: Sol. Phys. **39** (1978) 129.
25 Meekins, J.F., Doschek, G.A., Friedman, H., Chubb, T.A., Kreplin, R.W.: Sol. Phys. **13** (1970) 198.
26 Neupert, W.M., Swartz, M., Kastner, S.O.: Sol. Phys. **31** (1973) 171.
27 Neupert, W.M., Thomas, R.J., Chapman, R.D.: Sol. Phys. **34** (1974) 349.
28 Parkinson, J.H., Wolff, R.S., Kestenbaum, H.L., Ku, W.H.-M., Lemen, J.R., Long, K.S., Novick, R., Suozzo, R.J., Weisskopf, M.C.: Sol. Phys. **60** (1978) 123.
29 Phillips, K.J.H., Neupert, W.M.: Sol. Phys. **32** (1973) 209.
30 Phillips, K.J.H., Neupert, W.M.,Thomas, R.J.: Sol. Phys. **36** (1974) 383.
31 Ramaty, R., Lingenfelter, R.E.: in Ramaty and Stone (eds.): High Energy Phenomena on the Sun, NASA SP **342** (1973) 301.
32 Ramaty, R., Kozlovsky, B., Lingenfelter, R.E.: Space Sci. Rev. **18** (1975) 341.
33 Ramaty, R., Lingenfelter, R.E.: Int. Astron. Union Symp. **68** (1975) 363.
34 Ramaty, R., Kozlovsky, B., Suri, A.N.: Astrophys. J. **214** (1977) 617.
35 Rust, D.M., Roy, J.R.: Sacramento Peak Obs. Contr. **221** (1974).
36 Sandlin, G.D., Brueckner, G.E., Scherrer, V.E., Tousey, R.: Astrophys. J. **205** (1976) L47.
37 Severny, A.B., Steshenko, N.V.: in C. de Jager (ed.): Solar Terrestrial Physics, Part I, D. Reidel, Dordrecht, (1972) p. 173.
38 Simnett, G.M.: Space Res. **13** (1972) 745.
39 Simnett, G.M.: Space Sci. Rev. **16** (1974) 257.
40 Talon, R., Vedrenne, G.: Int. Astron. Union Symp. **68** (1975) 315.
41 Tucker, W.H., Koren, M.: Astrophys. J. **168** (1971) 283.
42 Van Hollebecke, M.A.I., Ma Sung, L.S., McDonald, F.B.: Sol. Phys. **41** (1975) 189.
43 Widing, K.G.: Int. Astron. Union Symp. **68** (1975) 153.
44 Widing, K.G., Cheng, C.C.: Astrophys. J. **194** (1974) L111.
45 Winckler, J.R.: in LeGalley (ed.): Space Science (1963), chapt. 11.

3.1.2.8 Radio emission of the disturbed sun

See p. 286

3.2 The planets and their satellites

3.2.1 Mechanical data of the planets and satellites

3.2.1.1 The planets; orbital elements and related properties

Definitions

a, b	semi-major and semi-minor axis of the orbit
$e=\sqrt{a^2-b^2}/a$	eccentricity of the orbit
i	inclination of the orbit to the ecliptic
Ω	longitude of the ascending node of the orbit on the ecliptic, measured from the equinox
ω	argument of perihelion
$\varpi=\Omega+\omega$	longitude of perihelion, measured from the equinox along the ecliptic to the node, and then along the orbit from the node to perihelion
P	sidereal period = true period of the planet's revolution around the sun (with respect to the fixed star field)
S	synodic period = time of orbital revolution of a planet with respect to the sun-earth line (e.g. from conjunction to conjunction)
$n=\frac{2\pi}{P}$	mean daily angular motion of the planet
t_P	time in days since perihelion passage
$M=n\cdot t_P$	mean anomaly
f	true anomaly = angle between the perihelion point and the radius vector
$L=\varpi+M$	mean longitude of the planet in the orbit at a given epoch; L is reckoned in the same way as ϖ. L refers to the position of a fictive planet
$L'=\varpi+f$	true longitude in the orbit; L' refers to the actual position of the planet
$\bar{v}$	mean orbital velocity; $\bar{v}$ is the velocity defined by the equation of energy for radius vector $r=a$
E. T.	Ephemeris Time

The orbital elements of the nine major planets, given in Tables 1a···c, have been computed from the basic sources of the ephemerides, published annually in the "Astronomical Ephemeris" and in many other almanacs. These sources are

for the inner planets: Newcomb's theories [1] with Ross's corrections for Mars [2],

for the outer planets: numerical integrations of the equations of motion by Eckert, Brouwer, and Clemence[3].

For the inner planets, the data in Tables 1a and 1c are mean elements which contain the secular perturbations. Data for the outer planets (Tables 1b, 1c) are osculating elements. Epochs in Table 1c: the mean elements a and n (Mercury···Mars) are nearly constant, no statement of epoch is necessary; the osculating elements (Jupiter···Pluto) yield for 1981 Feb. 5. For further information, see [4].

Approximate orbital elements of the 9 major planets, representing the heliocentric motions over the period 1980–1984 and referred to the mean equinox and ecliptic 1950.0, can be found in [5].

Table 1. Orbital elements of the major planets.

a) Mean elements for epochs 1970, 1980, 1990 Jan. 0.5 E.T., referred to the mean equinox and ecliptic of the epoch

Jan. 0.5 E.T.		i	Ω	ϖ	L	e
Mercury	1970	7° 0′15″.0	47°58′32″.4	76°59′19″.2	47°58′57″.3	0.2056285
	1980	7 0 15.7	48 5 39.1	77 8 39.2	233 20 36.6	0.2056306
	1990	7 0 16.4	48 12 45.9	77 17 59.5	62 47 48.6	0.2056326
Venus	1970	3°23′39″.6	76°24′35″.0	131° 8′56″.3	265°24′52″.0	0.0067873
	1980	3 23 40.0	76 29 59.2	131 17 22.6	356 32 4.6	0.0067825
	1990	3 23 40.3	76 35 23.4	131 25 48.9	89 15 25.0	0.0067777
Earth	1970	–	–	102°25′28″.0	99°44′32″.1	0.0167217
	1980	–	–	102 35 47.1	99 19 34.9	0.0167176
	1990	–	–	102 46 6.4	99 53 46.1	0.0167134
Mars	1970	1°50′59″.5	49°19′34″.0	335°30′24″.4	12°40′30″.8	0.0933773
	1980	1 50 59.3	49 24 11.6	335 41 27.0	126 34 59.4	0.0933865
	1990	1 50 59.0	49 28 49.2	335 52 29.9	241 0 54.6	0.0933958

b) Osculating elements tabulated at intervals of 400 days, 0ʰ E.T.; ecliptic and mean equinox of the epoch [8]

0ʰ E.T.	i	Ω	ϖ	L	e	a		n
						AU	10^6 km	
Jupiter								
1978 Nov. 28	1°18′20″.0	100°13′16″.7	14°12′33″.5	114° 1′31″.8	0.0479233	5.202831	778.33	299″.13
1980 Jan. 2	1 18 20.3	100 14 7.8	14 27 52.2	147 16 4.4	0.0478482	5.203907	778.49	299.03
1981 Feb. 5	1 18 20.8	100 15 50.4	14 45 34.6	180 29 6.7	0.0477684	5.204829	778.63	298.95
1982 Mar. 12	1 18 20.5	100 17 57.1	15 0 52.2	213 41 46.7	0.0478508	5.204396	778.57	298.99
1983 Apr. 16	1 18 19.7	100 19 14.5	15 13 49.8	246 55 45.5	0.0479808	5.203394	778.42	299.08
1984 May 20	1 18 19.1	100 19 55.9	15 22 31.1	280 10 46.2	0.0480422	5.202763	778.32	299.13
1985 June 24	1 18 18.9	100 20 29.0	15 26 34.1	313 26 10.0	0.0480664	5.202562	778.29	299.15
1986 July 29	1 18 18.8	100 21 3.2	15 27 52.2	346 41 33.0	0.0480947	5.202646	778.31	299.14
1987 Sept. 2	1 18 18.6	100 21 39.6	15 28 48.4	19 56 44.9	0.0481365	5.202868	778.34	299.12
1988 Oct. 6	1 18 18.5	100 22 16.0	15 30 25.2	53 11 42.4	0.0481776	5.203096	778.37	299.10
1989 Nov. 10	1 18 18.3	100 22 50.5	15 32 11.4	86 26 29.0	0.0482071	5.203240	778.39	299.09
1990 Dec. 15	1 18 18.1	100 23 23.3	15 33 19.4	119 41 10.3	0.0482303	5.203260	778.40	299.09
Saturn								
1978 Nov. 28	2°29′11″.6	113°29′49″.9	94°57′37″.1	151°34′49″.8	0.0572728	9.580690	1433.3	119″.67
1980 Jan. 2	2 29 11.3	113 30 32.0	95 26 21.8	165 2 3.1	0.0561165	9.579658	1433.1	119.69
1981 Feb. 5	2 29 10.3	113 30 35.6	95 26 28.3	178 32 52.1	0.0543733	9.575616	1432.5	119.76
1982 Mar. 12	2 29 9.4	113 30 9.0	95 23 54.2	192 4 38.6	0.0525782	9.575006	1432.4	119.77
1983 Apr. 16	2 29 9.1	113 30 40.7	95 0 25.9	205 33 22.7	0.0513012	9.574363	1432.3	119.79
1984 May 20	2 29 8.5	113 31 51.6	94 0 36.7	218 59 34.1	0.0507887	9.569051	1431.5	119.88
1985 June 24	2 29 8.0	113 32 49.9	92 44 24.4	232 24 23.0	0.0511262	9.559049	1430.0	120.07
1986 July 29	2 29 8.5	113 33 14.4	91 44 25.8	245 48 12.6	0.0522040	9.546111	1428.1	120.32
1987 Sept. 2	2 29 10.6	113 33 15.5	91 25 8.0	259 11 4.6	0.0536135	9.532995	1426.1	120.57
1988 Oct. 6	2 29 13.7	113 33 20.9	91 47 20.0	272 33 13.3	0.0547872	9.522675	1424.6	120.76
1989 Nov. 10	2 29 16.3	113 33 47.5	92 31 2.6	285 55 7.0	0.0553313	9.517008	1423.7	120.87
1990 Dec. 15	2 29 17.2	113 34 30.0	93 11 42.4	299 17 12.1	0.0551927	9.516236	1423.6	120.88

continued

Table 1b, continued

0^h E.T.	i	Ω	ϖ	L	e	a AU	a 10^6 km	n
Uranus								
1978 Nov. 28	0°46′16″.8	74° 0′25″.6	169°55′ 9″.5	222°20′37″.0	0.0489528	19.20273	2872.7	42″.167
1980 Jan. 2	0 46 17.4	74 0 27.7	172 2 21.1	227 2 46.3	0.0500453	19.24464	2879.0	42.029
1981 Feb. 5	0 46 19.1	74 0 2.5	174 12 6.8	231 47 14.6	0.0505140	19.28093	2884.4	41.910
1982 Mar. 12	0 46 21.5	73 59 34.8	176 1 59.5	236 33 39.2	0.0502943	19.30589	2888.1	41.830
1983 Apr. 16	0 46 24.1	73 59 20.0	177 12 36.7	241 21 27.0	0.0494382	19.31577	2889.6	41.797
1984 May 20	0 46 26.2	73 59 24.4	177 23 28.7	246 9 43.9	0.0481381	19.30809	2888.5	41.822
1985 June 24	0 46 27.1	73 59 40.9	176 16 0.5	250 57 10.8	0.0467577	19.28202	2884.6	41.907
1986 July 29	0 46 26.3	73 59 52.4	173 54 27.7	255 42 22.7	0.0458147	19.24168	2878.5	42.039
1987 Sept. 2	0 46 24.0	73 59 39.8	171 7 17.4	260 24 16.9	0.0457303	19.19843	2872.0	42.181
1988 Oct. 6	0 46 21.0	73 59 1.3	169 6 31.0	265 3 6.5	0.0464286	19.16656	2867.3	42.286
1989 Nov. 10	0 46 18.8	73 58 19.9	168 33 17.6	269 40 18.8	0.0473717	19.15512	2865.6	42.324
1990 Dec. 15	0 46 17.8	73 58 5.2	169 20 48.8	274 17 37.7	0.0480289	19.16428	2866.9	42.294
Neptune *								
1978 Nov. 28	1°46′20″.9	131°32′ 4″.6	61°26′40″.9	258°17′ 6″.7	0.0100608	29.99401	4487.0	21″.601
1980 Jan. 2	1 46 22.5	131 31 23.2	60 51 58.7	260 37 13.1	0.0076571	30.06289	4497.3	21.527
1981 Feb. 5	1 46 22.9	131 31 36.8	51 2 39.1	263 1 6.2	0.0052576	30.14180	4509.2	21.442
1982 Mar. 12	1 46 21.8	131 32 43.4	24 17 30.5	265 28 37.6	0.0039709	20.21442	4520.0	21.365
1983 Apr. 16	1 46 19.5	131 34 31.4	358 15 33.8	267 59 13.2	0.0045910	30.26631	4527.8	21.310
1984 May 20	1 46 16.2	131 36 41.8	352 43 35.4	270 31 46.9	0.0060273	30.28472	4530.5	21.290
1985 June 24	1 46 12.9	131 38 47.4	0 30 45.4	273 4 28.9	0.0073793	30.26098	4527.0	21.315
1986 July 29	1 46 10.7	131 40 17.4	15 13 9.5	275 34 47.3	0.0085638	30.19726	4517.4	21.383
1987 Sept. 2	1 46 10.7	131 40 52.3	31 34 53.4	278 0 27.0	0.0096661	30.11309	4504.9	21.473
1988 Oct. 6	1 46 13.0	131 40 38.6	45 12 29.2	280 21 3.6	0.0104153	30.03998	4493.9	21.551
1989 Nov. 10	1 46 16.6	131 40 6.6	54 8 11.0	282 38 29.8	0.0103614	30.00308	4488.4	21.591
1990 Dec. 15	1 46 19.8	131 39 46.4	57 39 42.5	284 55 33.6	0.0093907	30.00911	4489.3	21.584
Pluto *								
1978 Nov. 28	17° 8′13″.6	109°53′49″.6	223° 0′52″.6	207°46′34″.0	0.2511010	39.64058	5930.2	14.217
1980 Jan. 2	17 8 14.1	109 56 28.0	222 58 17.8	209 26 18.2	0.2538654	39.78459	5951.7	14.139
1981 Feb. 5	17 8 12.7	110 0 28.4	223 7 58.1	211 8 37.3	0.2558651	39.88009	5966.0	14.089
1982 Mar. 12	17 8 8.3	110 5 1.3	223 26 30.8	212 52 10.2	0.2567071	39.91003	5970.5	14.073
1983 Apr. 16	17 8 1.7	110 9 14.0	223 49 28.2	214 35 33.0	0.2562024	39.86803	5964.2	14.095
1984 May 20	17 7 55.7	110 12 15.5	224 11 27.6	216 17 15.0	0.2543702	39.75881	5947.8	14.153
1985 June 24	17 7 54.9	110 13 19.9	224 26 18.2	217 55 42.6	0.2515564	39.60389	5924.7	14.237
1986 July 29	17 8 4.0	110 12 12.6	224 28 32.2	219 29 50.3	0.2485643	39.44665	5901.1	14.322
1987 Sept. 2	17 8 23.1	110 9 33.8	224 17 16.8	221 0 4.3	0.2465080	39.34267	5885.6	14.378
1988 Oct. 6	17 8 45.4	110 6 55.8	223 58 55.2	222 28 41.9	0.2462052	39.32997	5883.7	14.385
1989 Nov. 10	17 9 0.6	110 5 43.1	223 43 28.2	223 58 39.7	0.2476200	39.40467	5894.9	14.345
1990 Dec. 15	17 9 1.5	110 6 27.4	223 37 45.5	225 31 50.5	0.2500139	39.53002	5913.6	14.276

* For orbital relations between Neptune and Pluto, see 3.2.1.6.

c) Semi-major axes of the orbits, mean daily motions, periods

	a		n		P		S	$\bar{v}$
	AU	10^6 km			Ephemeris days	tropical years	Ephemeris days	km s^{-1}
Mercury	0.387099	57.9	4°092339	14732″42	87^{d}969	0^{a}24085	115^{d}88	47.9
Venus	0.723332	108.2	1.602131	5767.67	224.701	0.61521	583.92	35.0
Earth	1.000000	149.6	0.985609	3548.19	365.256	1.00004	—	29.8
Mars	1.523691	227.9	0.524033	1886.52	686.980	1.88089	779.94	24.1
Jupiter	5.204829	779	0.083043	298.95	—	11.869	398.9	13.1
Saturn	9.575616	1432	0.033267	119.76	—	29.628	378.0	9.6
Uranus	19.28093	2884	0.011642	41.91	—	84.665	369.6	6.8
Neptune	30.14180	4509	0.005956	21.44	—	165.49	367.5	5.4
Pluto	39.88009	5966	0.003914	14.09	—	251.86	366.7	4.7

Discoverer; date of discovery:

Uranus: W. Herschel; 1781 March 13,
Neptune: J. G. Galle; 1846 September 23,
Pluto: C. W. Tombough; 1930 January 21.

Invariable plane [7].

The invariable plane of the solar system is defined by the condition that the total angular momentum of the system about an axis perpendicular to this plane is a maximum, while it is zero about any axis lying in this plane. Since no actions within the system can alter the total angular momentum, this plane must be invariable [6].
i, Ω: inclination and longitude of node of the invariable plane to the ecliptic

Ecliptic and mean equinox	i	Ω
B 1950.0	1°35′23″	107° 7′.6
J 2000.0	1 35 14	107 36.5

The prefixes J and B are used to distinguish Julian and Besselian epochs
B 1950.0 = 1950 January 0^{d}923
J 2000.0 = 2000 January 1.5

For definition of the Besselian epoch, see 2.2.3.2; for Julian epoch, see 3.2.1.4.2.

The time changes of i and Ω are effects of the motion of the ecliptic plane in space and of the motion of the equinox on the ecliptic.

References for 3.2.1.1

1 Astron. Papers Wash. **6**, **7** (1895–1898).
2 Astron. Papers Wash. **9**, Part 2 (1917).
3 Astron. Papers Wash. **12** (1951).
4 Landolt-Börnstein, NS, Vol. VI/1 (1965). p. 151–152.
5 Planetary and Lunar Coordinates for the Years 1980–1984, London, Washington (1979).
6 Russell, H.N., Dugan, R.S., Stewart, J.Q.: Astronomy I. The Solar System, Ginn and Co., Boston (1926) 284.
7 Burkhardt, G.: Paper submitted to Astron. Astrophys. (1981).
8 Institute for Theoretical Astronomy, Leningrad.

3.2.1.2 Dimensions and mechanical properties, rotation of the planets

The diameters given in Table 2a (except the value for Pluto) are identical with the values recommended by IAU Comm. 4 to be used in the preparation of ephemerides, beginning with 1984 [8].

Table 2. Dimensions and mechanical properties of the planets.
a) Diameter, oblateness, gravity field

r adopted distance planet-observer
$d(r)$ apparent angular diameter at distance r
D_{equ} equatorial diameter
D_{pol} polar diameter
D diameter; for Earth to Saturn: $D=(2D_{equ}+D_{pol})/3$
$f=\frac{D_{equ}-D_{pol}}{D_{equ}}$ oblateness

$J_2=\frac{C-A}{\mathfrak{M}R_{equ}^2}$ ellipticity coefficient
$\frac{C}{\mathfrak{M}R_{equ}^2}$ coefficient of moment of inertia
$\mathfrak{M}$ mass
R_{equ} equatorial radius
A, C moments of inertia about equatorial and polar axis, respectively.

Planet	r [AU]	$d(r)$	Diameter				Oblateness		Gravity field			
			D_{equ} [km] D_{pol} [km]	D [km]	D [D_{Earth}]	Ref.	f	Ref.	J_2	Ref.	$\frac{C}{\mathfrak{M}\cdot R_{equ}^2}$	Ref.
Mercury	1.00	6″.73	–	4878	0.383	1⋯3	0	–	–	–	–	–
Venus	1.00	16.69	–	12104	0.950	4⋯6	0	–	–	–	–	–
Earth	1.00	equ. 17.588 pol. 17.529	12756.28 12713.51	12742.02	1.000	7	1:298.257	7	0.0010826	7	0.331	44
Mars	1.00	equ. 9.37 pol. 9.31	6794.4 6754.6	6781.1	0.532	8	opt. 1:171 dyn. 1:191	12	0.00196	36	0.38	44
Jupiter	5.20	equ. 37.84 pol. 35.46	142796 133800	139797	10.97	8, 9, 14	opt. 1:15.9 dyn. 1:15.5	13, 14	0.0147	9	<0.26	15
Saturn	9.54	equ. 17.33 pol. 15.47	120000 106900	115630	9.07	10	1:9.2	10	0.0167	37,38	<0.26	15
Uranus	19.2	3.65	50800 –	–	3.99	10	(1:50)	15, 48	uncertain $0.003\leqq J_2\leqq 0.012$	39⋯41	probably >0.3	15
Neptune	30.1	2.23	48600 –	–	3.81	10	(1:43)	15	0.005	42,43	probably >0.3	15
Pluto	39.4	(0.12)	–	(3500)	(0.27)	11, 46, 51	–	–	–	–	–	–

b) Mass, volume, density, surface gravity

V volume
$\bar{\varrho}$ mean density
g total acceleration, including centrifugal acceleration, at equator
g_z centrifugal acceleration at equator
v_e velocity of escape at equator

Planet	Reciprocal mass (including satellites)		Mass (excluding satellites)		V	$\bar{\varrho}$	g	g_z	g	v_e
	$\left[\frac{1}{\mathfrak{M}_\odot}\right]$	Ref.	$[\mathfrak{M}_{Earth}]$	g	$[V_{Earth}]$	$g\,cm^{-3}$	$cm\,s^{-2}$	$cm\,s^{-2}$	$[g_{Earth}]$	$km\,s^{-1}$
Mercury	6023600	16	0.0553	$3.302 \cdot 10^{26}$	0.056	5.43	370	0.00	0.38	4.25
Venus	408523.5	16	0.8150	$4.869 \cdot 10^{27}$	0.857	5.24	887	0.00	0.91	10.4
Earth	328900.5	16	1.0000	$5.974 \cdot 10^{27}$	1.000	5.515	978	− 3.39	1.00	11.2
Mars	3098710	16	0.1074	$6.419 \cdot 10^{26}$	0.151	3.93	371	− 1.71	0.38	5.02
Jupiter	1047.355	16	317.826	$1.8988 \cdot 10^{30}$	1320.6	1.33	2321	−225	2.37	57.6
Saturn	3498.5	16	95.145	$5.684 \cdot 10^{29}$	747.3	0.70	928	−175	0.95	33.4
Uranus	22869	16	14.559	$8.698 \cdot 10^{28}$	63.4	1.27	(838)	(− 37)	(0.86)	(20.6)
Neptune	19314	16	17.204	$1.028 \cdot 10^{29}$	55.5	1.71	(1154)	(− 22)	(1.18)	(23.7)
Pluto	(130000000)	17, 51	(0.0026)	$(1.5 \cdot 10^{25})$	(0.021)	(0.7)	—	—	—	—

Uranus, Neptune: The low accuracy in the values of oblateness and rotation-period causes large uncertainties in g, g_z, v_e.

Table 3. Rotation of the planets (see also 3.2.1.1.1).

Planet		Sidereal rotation period P					Inclination of equator to orbit	
		d	h	m	s	Ref.	i	Ref.
Mercury		58.65				18	≈ 2°	18
Venus		243.0	retrograde			19, 50	≈ 3°	6
Earth			23	56	4.099	20	23°27′	20
Mars			24	37	22.66	21, 22, 49	23°59′	20
Jupiter	System I		9	50	30.003	20	3°4′	20
	System II		9	55	40.632	20		
	IAU System III (1965)		9	55	29.7	23, 24		
Saturn	System I		10	14		25	26°44′	20
	System II		10	40		25		
Uranus		uncertain, probably near 15^h, see below				26···29, 45, 47	98°	20
Neptune		uncertain, see below				26, 30, 31, 45	29°	33
Pluto		$6\overset{d}{.}39$				32	probably greater than 50°	32

Uranus: New measurements [28, 29, 45, 47] yield periods between 13^h and 16^h; [26, 27] give values near 24^h. The older value $P = 10\overset{h}{.}8$ [34] is likely not correct.

Neptune: The older value $P = 15\overset{h}{.}8$ [35] is probably not correct. New results: $\approx 11^h$ [45], $\approx 18\overset{h}{.}5$ [30, 31], $\approx 22^h$ [26]. See also [15].

References for 3.2.1.2

1 Ash, M.E., Shapiro, I.I., Smith, W.B.: Science **174** (1971) 551.
2 Howard, H.T. et al.: Science **185** (1974) 179.
3 Fjeldbo, G. et al.: Icarus **29** (1976) 439.
4 Smith, W.B., Ingalls, R.P., Shapiro, I.I., Ash, M.E.: Radio Sci. **5** (1970) 411.
5 Campbell, D.B. et al.: Science **175** (1972) 514.
6 Marov, M.Ya.: Annu. Rev. Astron. Astrophys. **16** (1978) 141.
7 Bull. géodésique No. **118** (1975) 365, 403, 404.
8 Trans. Int. Astron. Union **16B** (1977) 60.
9 Null, G.W., Anderson, J.D., Wong, S.K.: Science **188** (1975) 476.
10 Dollfus, A.: Icarus **12** (1970) 101.
11 Cruikshank, D.P., Pilcher, C.B., Morrison, D.: Science **194** (1976) 835.
12 Köhler, H.W.: Der Mars, Vieweg, Braunschweig (1978) 75, 76.
13 Hubbard, W.B.: Icarus **30** (1977) 311.
14 Anderson, J.D., Null, G.W., Wong, S.K.: J. Geophys. Res. **79** (1974) 3661.
15 Cook, A.H.: Mon. Not. R. Astron. Soc. **187** (1979) 39p.
16 Trans. Int. Astron. Union **16B** (1977) 59, 63.
17 Harrington, R.S., Christy, J.W.: Astron. J. **85** (1980) 168.
18 Gault, D.E., Burns, J.A., Cassen, P., Strom, R.G.: Annu. Rev. Astron. Astrophys. **15** (1977) 99, 100.
19 Shapiro, I.I., Campbell, D.B., DeCampli, W.M.: Astrophys. J. **230** (1979) L 123.
20 Explanatory Supplement to The Astronomical Ephemeris and The American Ephemeris, London (1961, third impression 1974) 490, 491.
21 Mariner 9: de Vaucouleurs, G., Davies, M.E., Sturms, F.M.: J. Geophys. Res. **78** (1973) 4395.
22 Viking 1, 2: Michael, W.H. et al.: Science **194** (1976) 1337.
23 Riddle, A.C., Warwick, J.W.: Icarus **27** (1976) 457.
24 Seidelmann, P.K., Divine, N.: Geophys. Res. Lett. **4** (1977) 65.
25 Cragg, T.A.: Publ. Astron. Soc. Pac. **73** (1961) 318.
26 Hayes, S.H., Belton, M.J.S.: Icarus **32** (1977) 383.
27 Trafton, L.: Icarus **32** (1977) 402.

28 Brown, R.A., Goody, R.M.: Astrophys. J. **217** (1977) 680.
29 Trauger, J.T., Roesler, F.L., Münch, G.: Astrophys. J. **219** (1978) 1079.
30 Cruikshank, D.P.: Astrophys. J. **220** (1978) L57.
31 Slavsky, D., Smith, H.J.: Astrophys. J. **226** (1978) L49.
32 Andersson, L.E., Fix, J.D.: Icarus **20** (1973) 279.
33 Eichelberger, W.S., Newton, A.: Astron. Papers Wash. **9**, Part 3 (1926).
34 Moore, J.H., Menzel, D.H.: Publ. Astron. Soc. Pac. **42** (1930) 330.
35 Moore, J.H., Menzel, D.H.: Publ. Astron. Soc. Pac. **40** (1928) 234.
36 Anderson, J.D.: EOS Trans. American Geophys. Union **55** (1974) 515.
37 Jeffreys, H.: Mon. Not. R. Astron. Soc. **114** (1954) 433.
38 Seidelmann, P.K.: Celestial Mech. **16** (1977) 165.
39 Dunham, D.W.: PhD thesis, Yale University (1971).

continued

3.2.1.3 Satellites and ring systems of the planets

3.2.1.3.1 Orbital elements, diameters, masses of the satellites

Table 4. a) Orbital elements of the satellites. (For definitions, see also 3.2.1.1.)

R_{pl} equatorial radius of the planet
P sidereal period around the planet
i_E inclination of the satellite's orbit to planet's equator
i_O inclination of orbit to planet's orbital plane

Satellite	a			P		e	Ref.	i_E	i_O	Ref.
	10^3 km	[R_{pl}]	Ref.	d	Ref.					
Earth										
Moon [1])	384.40	60.268		27.321661		0.0549		18°3···28°6	5°1	
Mars										
M 1 Phobos	9.38	2.761	1	0.3189	8	0.015	1	1°1		12
2 Deimos	23.46	6.906	1	1.262	8	0.00052	1	0°9···2°7		12
Jupiter										
1979 J 1	127.8	1.79	39	0.294	39	0	34	0°		34
J 5 Amalthea	181.3	2.539	2	0.498	2	0.0028	9	0.4		13
1979 J 2	221.7	3.105	39, 44	0.675	39, 44	—	—	≈1.25		44
1 Io	421.6	5.905	2	1.769	2	0.0000	9	0.00		13
2 Europa	670.9	9.397	2	3.551	2	0.0003	9	0.02		13
3 Ganymede	1070	14.99	2	7.155	2	0.0015	9	0.09		13
4 Callisto	1880	26.33	2	16.689	2	0.0075	9	0.43		13
13 Leda	11094	155.4	3	239	3	0.148	3		27°	14
6 Himalia	11470	160.6	2	250.6	2	0.158	10		28	14
10 Lysithea	11710	164.0	2	260	2	0.130	10		29	14
7 Elara	11740	164.4	2	260.1	2	0.207	10		28	14
12 Ananke	20700	290	2	617	2	0.17	10		147	14
11 Carme	22350	313	2	692	2	0.21	10		163	14
8 Pasiphae	23300	326	2	735	2	0.38	10		148	14
9 Sinope	23700	332	2	758	2	0.28	10		153	14
Saturn										
1980 S 28	138	2.29	49	0.60	49	—	—	0°		49
1980 S 27	140	2.33	50	0.61	50	—	—	0		50
1980 S 26	142	2.37	50	0.63	50	—	—	0		50
1966 S 2 =1980 S 1	151	2.52	5, 41	0.694	5, 41, 42	0	5	0		5, 41
1980 S 3	151	2.52	41	0.694	41, 42	—	—	0		41
S 1 Mimas	186	3.100	2	0.942	2	0.0201	10	1.5		14
S 2 Enceladus	238	3.967	2	1.370	2	0.0044	10	0.0		14
S 3 Tethys	295	4.917	2	1.888	2	0.0000	10	1.1		14

continued

40 Whitaker, E.A., Greenberg, R.J.: Mon. Not. R. Astron. Soc. **165** (1973) 15p.
41 Nicholson, P.D., Persson, S.E., Matthews, K., Goldreich, P., Neugebauer, G.: Astron. J. **83** (1978) 1240.
42 Brouwer, D., Clemence, G.M.: in Planets and Satellites, (Kuiper-Middlehurst, ed.) Chicago (1961) 76.
43 Gill, J.R., Gault, B.L.: Astron. J. **73** (1968) S95 (Abstract).
44 Cole, G.H.A.: The Structure of Planets (1978) 130, 178.
45 Münch, G., Hippelein, H.: Astron. Astrophys. **81** (1980) 189.
46 Arnold, S.J., Boksenberg, A., Sargent, W.L.W.: Astrophys. J. **234** (1979) L 159.
47 Brown, R.A., Goody, R.M.: Astrophys. J. **235** (1980) 1066.
48 Franklin, F.A., Avis, Ch.C., Colombo, G., Shapiro, I.I.: Astrophys. J. **236** (1980) 1031.
49 De Vaucouleurs, G.: Astron. J. **85** (1980) 945.
50 Zohar, S., Goldstein, R.M., Rumsey, H.C.: Astron. J. **85** (1980) 1103.
51 Bonneau, D., Foy, R.: Astron. Astrophys. **92** (1980) L 1.

b) Diameters, masses, densities of the satellites.

D diameter
$\mathfrak{M}$ mass
$\bar{\varrho}$ mean density
$\mathfrak{M}_{pl}$ mass of the respective planet
m_{opp} mean opposition visual magnitude

D		$\mathfrak{M}$			$\bar{\varrho}$	m_{opp}	Discovery	Satellite
km	Ref.	$[\mathfrak{M}_{pl}]$	g	Ref.	$g\,cm^{-3}$			
								Earth
3476		0.01230	$7.35\cdot10^{25}$		3.34	$-12^{m}.7$	—	Moon
								Mars
$27\times22\times19$	15, 16	—	—		—	11.6	1877 Hall	M 1 Phobos
$15\times12\times11$	15, 16	—	—		—	12.7	1877 Hall	2 Deimos
								Jupiter
<40	34	—	—		—	—	1979 [34]	1979 J 1
240	17	—	—		—	13.0	1892 Barnard	J 5 Amalthea
≈80	39, 44	—	—		—	—	1980 [39, 44]	1979 J 2
3650	18	$4.70\cdot10^{-5}$	$8.92\cdot10^{25}$	22	3.5	5.0	1610 Galilei	1 Io
3120	18	$2.56\cdot10^{-5}$	$4.86\cdot10^{25}$	22	3.1	5.3	1610 Galilei	2 Europa
5280	18	$7.84\cdot10^{-5}$	$14.89\cdot10^{25}$	22	1.9	4.6	1610 Galilei	3 Ganymede
4840	18	$5.60\cdot10^{-5}$	$10.63\cdot10^{25}$	22	1.8	5.6	1610 Galilei	4 Callisto
≈10	19	—	—		—	20	1974 Kowal	13 Leda
170	19	—	—		—	14.8	1904 Perrine	6 Himalia
≈24	19	—	—		—	18.4	1938 Nicholson	10 Lysithea
80	19	—	—		—	16.4	1905 Perrine	7 Elara
≈20	19	—	—		—	18.9	1951 Nicholson	12 Ananke
≈30	19	—	—		—	18.0	1938 Nicholson	11 Carme
≈36	19	—	—		—	17.7	1908 Melotte	8 Pasiphae
≈28	19	—	—		—	18.3	1914 Nicholson	9 Sinope
								Saturn
(100)	49	—	—	—	—	—	1980 Voyager 1	1980 S 28
(200)	46	—	—	—	—	(15)	1980 Voyager 1	1980 S 27
(250)	46	—	—	—	—	(15)	1980 Voyager 1	1980 S 26
140×70	46	—	—	—	—	14	1976 [1])	1966 S 2
								=1980 S 1
(200)	46	—	—	—	—	15.5	1980 Cruikshank	1980 S 3
(360)	20	$6.59\cdot10^{-8}$	$3.7\cdot10^{22}$	23	(1.5)	12.9	1789 W. Herschel	S 1 Mimas
(600)	20	$1.48\cdot10^{-7}$	$8.4\cdot10^{22}$	23	(0.7)	11.8	1789 W. Herschel	S 2 Enceladus
1040	20	$1.10\cdot10^{-6}$	$6.3\cdot10^{23}$	23	1.1	10.3	1684 Cassini	S 3 Tethys

continued

Table 4a continued

Satellite	a			P		e	Ref.	i_E	i_O	Ref.
	10^3 km	[R_{pl}]	Ref.	d	Ref.					
Saturn										
1980 S 13	306	5.10	48	2.00	48	—	—	—		—
1980 S 6	377	6.28	43, 47	2.736	43, 47	—	—	0°		41
S 4 Dione	377	6.283	2	2.737	2	0.0022	10	0.0		14
S 5 Rhea	527	8.783	2	4.518	2	0.0010	10	0.4		14
S 6 Titan	1222	20.37	2	15.95	2	0.0289	10	0.3		14
S 7 Hyperion	1481	24.68	2	21.28	2	0.1042	10	0.4		14
S 8 Japetus	3560	59.33	2	79.33	2	0.0283	10		18°.4	14
S 9 Phoebe	12933	215.56	33	549.2	33	0.1591	33		174.8	33
Uranus										
U 5 Miranda	130	5.118	2	1.413	2	0.017	11	3°.4		11
1 Ariel	192	7.559	2	2.520	2	0.0028	10	0		10
2 Umbriel	267	10.51	2	4.144	2	0.0035	10	0		10
3 Titania	438	17.24	2	8.706	2	0.0024	10	0		10
4 Oberon	586	23.07	2	13.46	2	0.0007	10	0		10
Neptune										
N 1 Triton	354	14.57	2	5.877	2	0.00	10	160°		14
2 Nereid	5510	226.7	6	360.1	6	0.75	6		28°	14
Pluto										
P 1 Charon	20	11.4	37, 45	6.387	37, 45	0	37, 45	?		—

[1]) For more details, see 3.2.1.5.

Temporary nomenclature for satellites. – After a proposal of Aksnes and Franklin [4], a provisional satellite nomenclature system has been established: the observed objects are referred to by the year of the observation, a letter for the planet, and a number.

When the perturbations of the satellite's orbit are caused mainly by the attraction of the sun (this holds for orbits with relatively large semimajor axes), then the inclination i_E of satellite orbit to the equator of the planet is not fixed, as it is nearly the case at satellites with smaller orbital axes. At these outer satellites, however, the inclination i_O of the orbit with respect to the planet's orbital plane remains essentially constant.

3.2.1.3.2 Ring systems of the planets Jupiter, Saturn, Uranus (see also 3.2.1.6.3)

Jupiter

A faint ring in the planet's equatorial plane was discovered by Voyager 1 (March 1979) and verified by Voyager 2 (July 1979). Distance of the outer edge from the center of the planet ≈ 1.8 Jupiter radii $(R_J) = 128.3 \cdot 10^3$ km, well within the Roche limit (2.4 R_J). Approximate width; 6000 km, thickness: $\leqq 30$ km [24, 35].

Saturn

Table 5a. Boundaries of Saturn's rings [25].

	Apparent radii at 9.5388 AU (mean opposition distance)	True radii	
		[R_{Sat}]	10^3 km
Outer edge of the moderately bright ring A	19″.82	2.29	137
Inner edge of ring A	17.57	2.03	122
Outer edge of the very bright ring B	16.87	1.95	117
Inner edge of ring B	13.21	1.53	91.6
Inner edge of the faint ring C	10.5	1.21	72.8
Equatorial radius of Saturn (R_{Sat})	8.65	1.00	60.0

Table 4b continued

D		$\mathfrak{M}$			$\bar{\varrho}$	m_{opp}	Discovery	Satellite
km	Ref.	[$\mathfrak{M}_{pl}$]	g	Ref.	g cm^{-3}			
								Saturn
—	—	—	—	—	—	—	1980 ²)	1980 S 13
(80)	46	—	—	—	—	17	1980 ³)	1980 S 6
1000	20	$2.04\cdot10^{-6}$	$1.2\cdot10^{24}$	23	2.2	10.4	1684 Cassini	S 4 Dione
1600	20	$3.8\cdot10^{-6}$	$2.2\cdot10^{24}$	36	1.0	9.7	1672 Cassini	S 5 Rhea
5800	20	$2.41\cdot10^{-4}$	$1.4\cdot10^{26}$	22	1.3	8.4	1655 Huyghens	S 6 Titan
224	20	—	—	—	—	14.2	1848 Bond	S 7 Hyperion
1450	20	$5.0\cdot10^{-6}$	$2.8\cdot10^{24}$	36	1.8	10.2···11.9	1671 Cassini	S 8 Japetus
(240)	20	—	—	—	—	16.5	1898 Pickering	S 9 Phoebe
								Uranus
(300)	19	—	—		—	16.5	1948 Kuiper	U 5 Miranda
(800)	19	—	—		—	14.4	1851 Lassell	1 Ariel
(550)	19	—	—		—	15.3	1851 Lassell	2 Umbriel
(1000)	19	—	—		—	14.0	1787 Herschel	3 Titania
(900)	19	—	—		—	14.2	1787 Herschel	4 Oberon
								Neptune
4400	21	$(2\cdot10^{-3})$	$(2\cdot10^{26})$	22	(4.6)	13.6	1846 Lassell	N 1 Triton
(300)	19	—	—		—	18.7	1949 Kuiper	2 Nereid
								Pluto
2000 > D	7, 38	—	—		—	17	1978 Christy	P 1 Charon
> 1200	45							

¹) Fountain, Larson, on plates taken in 1966. ²) Reitsema, Smith, Larson. ³) Laques, Lecacheux.

A. Dollfus has realized that Saturn's satellite Janus = S 10 = 1966 S 1 does not exist [41].

The two satellites 1980 S 1 and 1980 S 3 are moving in essentially the same orbit.

The orbit of 1980 S 6 is almost identical with Dione's orbit; difference in orbital longitude about 70°. This is near the 60° difference between Dione and its L 4 Lagrangian point.

Radio occultation measurements of Voyager 1 yield a value of 5140 km for the diameter of Titan without atmosphere, and a corresponding mean density of 1.9 g cm^{-3} [51].

For orbital relations, see 3.2.1.6.2 .

Pictures taken by Voyager 1 [46] show that the rings, especially B and C, consist of hundreds of individual concentric components. Cassini's division, separating the A and B rings, is not empty; there are at least 20 ringlets within its boundaries.

Thickness of rings: probably between 1 and 3 km [26, 27].
Mass of rings: between 10^{-6} and 10^{-5} $\mathfrak{M}_{Sat}$ [28, 36].

Position of the ring-plane [29]:
i inclination of the plane of the rings to the ecliptic
Ω longitude of the ascending node of the plane of the rings on the ecliptic

The orientation of the ring-plane in space is completely determined by *i* and Ω.

Epoch	Ecliptic and mean equinox of epoch	
	i	Ω
1975.0	28°065	169°163
1980.0	28.064	169.233
1985.0	28.063	169.302
1990.0	28.063	169.372

Three extremely faint rings have the IAU designations D, E, and F. Ring D lies inside, rings F and E outside the bright rings A, B, C.

Table 5b. Saturn's rings D, E, F.

Ring	Distance from the center of Saturn $R_{Sat}=1$	Ref.	Discovery
D	inside the bright rings	30, 46	Guérin, 1969
F	2.34, narrow ≈3000 km beyond the visible edge of the A ring	36, 46	Pioneer 11, 1979
E	from 3.3 to about 6	30, 40, 43, 46	Feibelman, 1966

Uranus

Uranus is encircled by at least 9 narrow rings, all near the plane of the Uranian equator. The inner 8 rings are roughly circular, the outermost ring ε seems to be elliptical.

The table gives provisional values of the mean ring radii, $\bar{R}$, after occultation observations in March 1977 (discovery) and April 1978 [31, 32].

Ring	6	5	4	α	β	η	γ	δ	ε
$\bar{R}$ [10^3 km]	42.02	42.34	42.66	44.83	45.80	47.31	47.75	48.42	elliptical[1])

[1]) Semi-axes still uncertain between 50.9 and $51.7 \cdot 10^3$ km.

3.2.1.3.3 References for 3.2.1.3

1 Born, G.H., Duxbury, T.C.: Celestial Mech. **12** (1975) 77.
2 Brouwer, D., Clemence, G.M. in: Planets and Satellites (Kuiper-Middlehurst, eds.), Chicago (1961) 31.
3 Aksnes, K.: Astron. J. **83** (1978) 1249.
4 Aksnes, K., Franklin, F.A.: Icarus **36** (1978) 107.
5 Fountain, J.W., Larson, S.M.: Icarus **36** (1978) 92.
6 Rose, L.E.: Astron. J. **79** (1974) 489.
7 Christy, J.W., Harrington, R.S.: Astron. J. **83** (1978) 1005.
8 Sinclair, A.T.: Vistas Astron. **22** (1978) 133.
9 Landolt-Börnstein, 6. Edition, Vol. III (1952) 81.
10 Landolt-Börnstein, NS, Vol. VI/1 (1965) 158.
11 Whitaker, E., Greenberg, R.: Commun. Lunar Planet. Lab. **10** (1973) 70.
12 Porter, J.G.: J. Brit. Astron. Assoc. **70** (1960) 33.
13 Annuaire du Bureau des Longitudes, éphémérides 1979; Paris 1978, 61.
14 Kovalevsky, J., Sagnier, J.-L. in: Planetary Satellites (Burns, J.A., ed.), Tucson, Arizona (1977) 57.
15 Pollack, J.B. et al.: J. Geophys. Res. **78** (1973) 4313.
16 Duxbury, T.C. in: Planetary Satellites (Burns, J.A., ed.), Tucson, Arizona (1977) 354, Table 15.2.
17 Rieke, G.H.: Icarus **25** (1975) 333.
18 Pioneer 10/11 and Voyager. Science **204** (1979) 964.
19 Morrison, D., Cruikshank, D.P., Burns, J.A. in: Planetary Satellites (Burns, J.A., ed.), Tucson, Arizona (1977) 12, Table 1.4.
20 Hunter, D.M., Morrison, D. (eds.): NASA Conference Publication 2068 (1978) 238.
21 Cruikshank, D.P., Stockton, A., Dyck, H.M., Becklin, E.E., Macy, W.: Icarus **40** (1979) 104.
22 Trans. Int. Astron. Union. **16B** (1977) 60, 66.
23 Kozai, Y.: Ann. Tokyo Astron. Obs. **5** (1957) 73.
24 Pioneer 10/11 and Voyager, Science **204** (1979) 955.
25 Cook, A.F., Franklin, F.A., Palluconi, F.D.: Icarus **18** (1973) 317.
26 Lumme, K., Irvine, W.M.: Astron. Astrophys. **71** (1979) 123.
27 Dollfus, A.: Astron. Astrophys. **75** (1979) 204.

Gondolatsch

28 McLaughlin, W.I., Talbot, T.D.: Mon. Not. R. Astron. Soc. **179** (1977) 619.
29 Danjon, A.: Astronomie Générale, Seconde Edition, Paris (1979) 372.
30 Smith, B.A. in: NASA Conference Publication (Hunter, D.M., Morrison, D., eds.) 2068 (1978) 105.
31 Elliot, J.L., Dunham, E., Wasserman, L.H., Millis, R.L., Churms, J.: Astron. J. **83** (1978) 980.
32 Nicholson, P.D., Persson, S.E., Matthews, K., Goldreich, P., Neugebauer, G.: Astron. J. **83** (1978) 1240.
33 Rose, L.E.: Astron. J. **84** (1979) 1067.
34 Jewitt, D.C., Danielson, G.E., Synnott, S.P.: Science **206** (1979) 951.
35 Owen, T., Danielson, G.E., Cook, A.T., Hansen, C., Hall, V.L., Duxbury, T.C.: Nature **281** (1979) 442.
36 Science **207** (1980) 400, 401, 415, 434.
37 Harrington, R.S., Christy, J.W.: Astron. J. **85** (1980) 168.
38 Walker, A.R.: Mon. Not. R. Astron. Soc. **192** (1980) 47 p.
39 IAU Circ. No. 3470 (1980).
40 Feibelman, W.A., Klinglesmith, D.A.: Science **209** (1980) 277.
41 Larson, S., Fountain, J.W.: Sky Telesc. **60** (1980) 356.
42 IAU Circ. No. 3534 (1980).
43 Lecacheux, J., Laques, P., Vapillon, L., Auge, A., Despiau, R.: Icarus **43** (1980) 111.
44 Synnott, S.P.: Science **210** (1980) 786.
45 Bonneau, D., Foy, R.: Astron. Astrophys. **92** (1980) L 1.
46 Beatty, J.K.: Sky Telesc. **61** (1981) 7.
47 Reitsema, H.J., Smith, B.A., Larson, S.M.: Icarus **43** (1980) 116.
48 IAU Circ. No. 3549 (1980).
49 IAU Circ. No. 3539 (1980).
50 IAU Circ. No. 3532 (1980).
51 Tyler, G.L. et al.: Science **212** (1981) 201.

3.2.1.4 Earth data

3.2.1.4.1 Figure, mass, gravity

Table 6. Earth: semi-diameters a, b, oblateness f, gravitation constant, dynamical form-factor J_2.

a equatorial radius
b polar radius
$f=\frac{a-b}{a}$ flattening
GM geocentric gravitational constant
J_2 the dynamical form-factor for the earth is the coefficient of the second harmonic in the expression for the earth's gravitational potential

System	a m	b m	f	GM $m^3 s^{-2}$	J_2
International Ellipsoid of Reference (Madrid 1924 ≈ Hayford 1909) [1]	6378388	6356912	1:297.0	($398625 \cdot 10^9$)	(0.001091)
Geodetic Reference System 1967 (= IAU System Astron. Const. 1964) [2]	6378160	6356775	1:298.247	$398603 \cdot 10^9$	0.0010827
System of best values, IAG 1975 (= IAU System Astron. Const. 1976) [3]	6378140	6356755	1:298.257	$398600.5 \cdot 10^9$	0.00108263
Geodetic Reference System 1980 [7]	6378137	6356752	1:298.257	$398600.5 \cdot 10^9$	0.00108263

Mass and density of the earth [6]:

Ratio of mass of sun to that of earth	332946.0	mean density of the earth	$5.515\ g\ cm^{-3}$
Mass of the sun	$1.9891 \cdot 10^{33}$ g	ratio of mass of sun to that of earth + moon	328900.5
Mass of the earth	$5.9742 \cdot 10^{27}$ g		

Standard gravity:

International gravity formula

$$g_\varphi = g_{Eq}(1 + B_2 \sin^2 \varphi + B_4 \sin^2 2\varphi)\ [cm\ s^{-2}]$$

Table 7 gives the constants in three systems which have been used to give standard gravity at sea-level g_{φ}, in different geographic latitudes φ. g_{Eq} is the normal acceleration of gravity at the equator.

Table 7.

System	g_{Eq}	B_2	B_4
IUGG Stockholm 1930	978.049	+0.0052884	−0.0000059
Geodetic Reference System 1967	978.032	+0.0053023	−0.0000059
Geodetic Reference System 1980	978.0327	+0.0053024	−0.0000058

3.2.1.4.2 Rotation of the earth, precession

The speed of rotation of the earth is not constant; the small variations in the period of rotation are secular, irregular, and periodic (see 2.2.6).

The ratio of the mean solar day to the period of rotation is constant; numerical value	1.002737811911
Accordingly, period of rotation of the earth in mean solar time	$0\overset{d}{.}997269663237$ $=23^h56^m4\overset{s}{.}098904$ $=86164\overset{s}{.}098904$
Rate of rotation per 1 s mean solar time	$7.292115\cdot10^{-5}$ rad $=15\overset{''}{.}041067$
Owing to the precession of the equinox, the sidereal day is slightly shorter than the true period of the earth's rotation. Length of the mean sidereal day (i.e. time between two successive transits of the vernal equinox) in mean solar time	$0\overset{d}{.}9972695664$ $=23^h56^m4\overset{s}{.}09054$ $=86164\overset{s}{.}09054$
Speed of rotation at equator	465.12 m s^{-1}
Centrifugal acceleration at equator	−3.39 cm s^{-2}
Obliquity of the ecliptic (angle between celestial equator and earth's orbit) [4]	$\varepsilon=23°26'21\overset{''}{.}448-0\overset{''}{.}46815\,t$
Annual rates of precession [4]	
general precession in longitude	$p=50\overset{''}{.}2910+0\overset{''}{.}000222\,t$
lunisolar precession in longitude	$\psi=50\overset{''}{.}3878+0\overset{''}{.}000049\,t$
planetary precession in right ascension	$\chi=0\overset{''}{.}1055-0\overset{''}{.}000189\,t$
general precession in right ascension	$m=46\overset{''}{.}1244+0\overset{''}{.}000279\,t$ $=3\overset{s}{.}07496+0\overset{s}{.}0000186\,t$
precession in declination	$n=20\overset{''}{.}0431-0\overset{''}{.}000085\,t$
angle between fixed and moving ecliptic	$\pi=0\overset{''}{.}4700-0\overset{''}{.}000007\,t$
longitude of node of moving on fixed ecliptic	$\Pi=174°52\overset{'}{.}58+0\overset{'}{.}5482\,t$

t = number of Julian years of 365.25 days from J 2000.0.

The epoch designated J 2000.0 shall be 2000 January $1\overset{d}{.}5$; the new standard equinox (position of the first point of Aries) shall correspond to this instant. The new standard epoch is one Julian century after 1900 January $0\overset{d}{.}5$, which corresponds to the fundamental epoch of Newcomb's planetary theories.

Period of precession (Platonic Year) ≈25700 a

The rate of lunisolar precession and the mean obliquity of the ecliptic have secular variations from dynamical causes. So the lunisolar motion of the celestial pole and the consequent rotation of the mean equator do not have a fixed period. The expressions in powers of the time give accurate results for only a few centuries on either side of the epoch.

3.2.1.4.3 Orbital motion of the earth

Lengths of the years in ephemeris days (see also 2.2.3.2).
T in Julian centuries of 36525 days, from 1900 January 0.5.

Tropical year (equinox to equinox) [5]	$365\overset{d}{.}24219878 - 0\overset{d}{.}00000614\,T$
Sidereal year (fixed star to fixed star) [4, 8]	$365\overset{d}{.}25636590 + 0\overset{d}{.}00000011\,T$
Anomalistic year (perihelion to perihelion) [4, 8]	$365\overset{d}{.}25962642 + 0\overset{d}{.}00000316\,T$
Julian year	$365\overset{d}{.}25 = 31557600$ SI seconds

The length of the tropical year results from Newcomb's expression for the mean longitude of the sun. – The length of the sidereal year is derived from the length of the tropical year by eliminating the effect of the precessional motion of the equinox. – The length of the anomalistic year differs from the length of the sidereal year on account of the motion of earth's perihelion in the orbital plane, caused by the perturbative forces of planets and moon and by the effect of relativity.

Sidereal period of perihelion	≈112000 a
Tropical period of perihelion	≈ 21000 a
Mean orbital speed of the earth	29.8 km s^{-1}
Mean centripetal acceleration	0.594 cm s^{-2}
Distance of earth from sun	
at perihelion	$147.1 \cdot 10^6$ km
mean distance	$149.6 \cdot 10^6$ km
at aphelion	$152.1 \cdot 10^6$ km

3.2.1.4.4 References for 3.2.1.4

1 Jordan-Eggert-Kneissl, Handbuch der Vermessungskunde, Band V (1969) 857.
2 Bull. Géodésique No. **103** (1972) 85.
3 Bull. Géodésique No. **118** (1975) 365.
4 Lieske, J.H., Lederle, T., Fricke, W., Morando, B: Astron. Astrophys. **58** (1977) 1.
5 Newcomb, S.: Astron. Papers Wash. **6**, Part 1 (1895).
6 Trans. Int. Astron. Union **16** B (1977) 59, 63.
7 Moritz, H.: Bull. Géodésique **54** (1980) 395.
8 Lederle, T.: Private communication (1980).

3.2.1.5 The moon

3.2.1.5.1 Distance, size, gravity, librations

Distance and parallax

Mean distance from earth = semi-major axis of moon's orbit [1] $a_☾ = 384400$ km

By definition, $a_☾$ is not the semi-major axis of an unperturbed Keplerian ellipse, but the longer axis of Hill's variational curve. This curve is a closed oval with the longer axis perpendicular to the direction of the sun. Unlike a Kepler-orbit, the variational orbit takes into account a major part of the solar perturbations (the variation). Hill's variational curve is the basic orbit in Brown's Moon-theory.

$a_☾$ has been determined trigonometrically (via $\pi_☾$, see below) [2]; now, the mean distance from the earth to the moon (in [km]) belongs to the results of lunar distance-measurements by radar and laser beam techniques [3, 4].

Extreme values for the moon's distance $r_☾$ from the earth
at perigee 356410 km
at apogee 406740 km

Mean equatorial horizontal parallax [1]

$$\sin \pi_☾ = \frac{R_E}{a_☾}, \quad R_E = \text{equatorial radius of the earth}$$

$$(\sin \pi_☾)'' = \sin \pi_☾ \cdot 206264''.8 = 57'\ 2''.45$$

$$\pi_☾ = (\sin \pi_☾)'' + 0''.157 = 57'\ 2''.61$$

Radius, figure, mass, density

Geocentric apparent semi-diameter of the lunar disk at mean distance 15′ 32″.6

Mean radius [5, 6] $R_{☾} = 1738.0\,\text{km} = 0.2725\, R_E$

The center of mass is displaced toward the earth by about 2.5 km relative to the center of figure [6].

Mean elevations with respect to a 1738 km sphere [6]

farside terrae	+1.8 km	ringed maria	−4.0
nearside terrae	−1.4	other maria	−2.3

surface $F_{☾} = 3.788 \cdot 10^7\ \text{km}^2 = 0.0743\, F_E$

volume $\text{Vol}_{☾} = 2.199 \cdot 10^{10}\ \text{km}^3 = 0.0202\, \text{Vol}_E$

mass [7, 8] $\mathfrak{M}_{☾} = 0.012\,300\, \mathfrak{M}_E = 1{:}81.301\, \mathfrak{M}_E = 7.3483 \cdot 10^{25}\ \text{g}$

mean density $\bar{\varrho}_{☾} = 3.34\ \text{g cm}^{-3}$

surface gravity $g_{☾} = 162.0\ \text{cm s}^{-2} = 0.166\, (g_{equ})_E$

surface escape velocity $v_e = 2.37\ \text{km s}^{-1}$

Gravity field of the moon [5, 8, 9, 22]

Notations

Axes and moments of inertia

The semi-major axes of the moon are denoted by a, b and c, where a is along the direction of the line joining the moon and the earth, c is along the axis of rotation of the moon, and b lies in the equatorial plane of the moon tangential to the orbit and therefore perpendicular to both a and c. The moments of inertia about the axes a, b and c are respectively A, B, and C [10].

Physical libration parameters

$$\beta = \frac{C-A}{B}, \qquad \gamma = \frac{B-A}{C}$$

Lower spherical harmonic coefficients for the lunar gravitational potential

$$J_2 = -C_{20} = \frac{1}{\mathfrak{M}_{☾} R_{☾}^2}\left[C - \frac{B+A}{2}\right] \qquad C_{22} = \frac{1}{4\,\mathfrak{M}_{☾} R_{☾}^2}[B-A]$$

$\beta = 631.3 \cdot 10^{-6}$ $\qquad J_2 = +202.7 \cdot 10^{-6}$

$\gamma = 227.8 \cdot 10^{-6}$ $\qquad C_{22} = +\ 22.3 \cdot 10^{-6}$

Coefficient of moment of inertia about the rotation axis: $\dfrac{C}{\mathfrak{M}_{☾} R_{☾}^2} = 0.392$

The coefficient of moment of inertia is critical for understanding the density distribution in the lunar interior. The value for a homogeneous sphere is 0.40; the value 0.392 indicates a moon which is somewhat more dense toward the center.

Inclination of lunar equator, librations

Inclination of equator to ecliptic [5,8]: $I = 1°32′32″.7$

Libration	In longitude	In latitude
optical		
selenocentric displacement, maximum	±7°53′	±6°51′
period	1 sidereal month	
physical		
displacement, principal term	±66″	±105″
period	1 year	1 month

3.2.1.5.2 Orbital motion

Orbital elements

The ephemeris of the moon, published annually in the astronomical almanacs, is based on Brown's theory of the moon as given in the Improved Lunar Ephemeris [11···13]. Brown's orbital elements, given here, are mean elements as defined in 3.2.1.1. T is measured in Julian centuries of 36525 ephemeris days, beginning on 1900 January 0.5 E.T. For definitions of the symbols, see 3.2.1.1.

Lunar ephemerides for space and ranging purposes have been constructed by direct numerical integration of the equations of motion for the moon and planets. These ephemerides are being continuously improved to fit the observations, especially the lunar laser ranging measures [14, 15, 9].

Activities for producing ephemerides of the moon from analytical theories are in progress, especially an improvement of Brown's theory [16] and a new analytical theory [17].

$$L = 270^\circ\,26'\,3''.69 + 1732564379''.31\,T$$
$$\varpi = 334^\circ\,19'\,46''.75 + 14648522''.52\,T$$
$$\Omega = 259^\circ\,10'\,59''.79 - 6962911''.23\,T$$

$e = 0.0549005$; $0.044 \leqq e \leqq 0.067$ (due to perturbations by the sun)
$i = 5^\circ\,8'\,43''.4$; $4^\circ\,59' \leqq i \leqq 5^\circ\,19'$ (period 173^d) (perturbations by the sun)

$\Delta L/\mathrm{d} = 13^\circ\,10'\,35''.03/\mathrm{d}$ = mean daily motion, corresponding to the T coefficient of L
$\Delta L/\mathrm{h} \approx 0^\circ.549/\mathrm{h} \approx 33'/\mathrm{h}$ = hourly variation

Inclination of the moon's equator to its mean orbital plane

$I + i = 6^\circ\,41'$

Variation of the inclination of moon's mean orbital plane to the earth's equator:

between $23^\circ\,27' - 5^\circ\,9' = 18^\circ\,18'$
and $23^\circ\,27' + 5^\circ\,9' = 28^\circ\,36'$

Period of rotation of moon's perigee (direct)	$3232^d \approx$ 8.85 tropical years
Period of moon's node (retrograde)	$6798^d \approx$ 18.61 tropical years
Moon's mean orbital speed	1.023 km s^{-1}
Mean centripetal acceleration	0.272 cm s^{-2}

Lengths of the mean months in ephemeris days; Saros (see also 2.2.3.3)

Sidereal month (fixed star to fixed star) = moon's rotation period	$27^d.321661$
Tropical month (equinox to equinox)	27.321582
Anomalistic month (perigee to perigee)	27.554551
Draconic month (node to node)	27.212220
Synodic month (new moon to new moon)	29.530589

These mean values vary by only a few hundredths of a second per century, but in a given lunation appreciable changes may occur. For example:

Length of	may vary by	on account of
sidereal month	7^h	perturbation of the lunar motion due to the sun
synodic month	13^h	eccentricity of the lunar orbit

Saros-period [18, 19] = 223 synodic months (or lunations) = $6585^d.32$
$\approx$ 239 anomalistic months = $6585^d.54$
$\approx$ 19 ecliptic years = $6585^d.78$
$\approx$ 18 years + 11 days

1 ecliptic year = time between two passages of the sun through the ascending (or descending) node of the moon's orbit = $346^d.62$

Main periodic terms in the moon's motion [20, 11, 21]

		with:	
Principal elliptic term in longitude	$+22\,639'' \sin M_☾$	$M_☾$	moon's mean anomaly
Principal elliptic term in latitude	$+18\,461'' \sin u$	$M_\odot$	sun's mean anomaly
Evection	$+\ 4\,586'' \sin (2A - M_☾)$	A	moon's mean age
Variation	$+\ 2\,370'' \sin 2A$	u	distance of mean moon from ascending node
Annual inequality	$-\ 668'' \sin M_\odot$		
Parallactic inequality	$-\ 125'' \sin A$		

3.2.1.5.3 References for 3.2.1.5

1 Trans. Int. Astron. Union **12B** (1966) 595.
2 Fischer, I.: Bull. Géodésique Nr. **71** (1964) 37.
3 Yaplee, B.S. et al.: Symposium Int. Astron. Union **21** (1965) 81.
4 Bender, P.L.: Rev. Geophys. Space Phys. **13** (1975) 271, 290.
5 Trans. Int. Astron. Union **16B** (1977) 60.
6 Kaula, W.M., Schubert, G., Lingenfelter, R.E.: 5. Lunar Sci. Conf. Vol. **3** (1974) 3049.
7 Trans. Int. Astron. Union **16B** (1977) 58, 62, 63.
8 Seidelmann, P.K.: Celestial Mech. **16** (1977) 165.
9 Williams, J.G. in: Scientific applications of lunar laser ranging (Mulholland, J.D., ed.) (1977) 45.
10 Cook, A.H.: Mon. Not. R. Astron. Soc. **150** (1970) 187.
11 Improved lunar ephemeris 1952–1959, Washington (1954).
12 Eckert, W.J., Van Flandern, T.C., Wilkins, G.A.: Mon. Not. R. Astron. Soc. **146** (1969) 473.
13 The Astronomical Ephemeris for the year 1979, 542.
14 Mulholland, J.D., Shelus, P.J.: Moon **8** (1973) 532.
15 Bender, P.L. et al.: Science **182** (1973) 229.
16 Gutzwiller, M.C.: Astron. J. **84** (1979) 889.
17 Henrard, J.: Ciel Terre **89** (1973) 1.
18 Russell, H.N., Dugan, R.S., Stewart, J.Q.: Astronomy I, The Solar System, Ginn and Co., Boston (1926) 226–228.
19 Blanco, V.M., McCusky, S.W.: Basic Physics of the Solar System, Reading MA (1961) 97.
20 Brown, E.W.: Tables of the motion of the Moon, New Haven (1919).
21 Danjon, A.: Astronomie générale, Seconde édition, Paris (1959) 296.
22 Mulholland, J.D.: Rev. Geophys. Space Phys. **18** (1980) 549.

3.2.1.6 The orbital relations

3.2.1.6.1 The planets

The orbits (see 3.2.1.1) and rotation states (see 3.2.1.2) of several planets are characterized by resonant relations. First, we have the orbit-spin 3:2 resonance of Mercury with its rotation period (58.644 ± 0.009) d [26] which is very nearly 2/3 of the orbital period (88 days). This resonance is established by the solar tidal effect and the eccentric orbit of Mercury [15]. The period of the retrograde rotation of Venus of about 250 days as estimated by early radar observations [4] is close to the value of 243.16 days required for Venus to be locked into a spin-orbit resonance with the earth. In such a resonance the same face of Venus is oriented towards the earth at each conjunction of these two planets [15]. However, recent analysis of radar observations of Venus [30] has yielded a spin period of (243.01 ± 0.03) d which is significantly different from the resonant value. In the outer solar system we encounter the 3:2 orbit-orbit resonance of Neptune and Pluto. In all orbital resonances a librating argument ($\phi = 3\lambda_{\text{Pluto}} - 2\lambda_{\text{Neptune}} - \varpi_{\text{Pluto}}$ in this case; symbols see Table 1) can be defined such that ϕ is a slowly oscillating parameter with period much larger than the orbital periods of the bodies locked in resonances. Because of such libration, Neptune and Pluto can never come closer together than 18 AU even though the perihelion of Pluto is within the aphelion of Neptune [6].

3.2.1.6.2 The satellites

There are several stable orbit-orbit resonances in the satellite systems of the outer planets (Tab. 1) (For orbital elements, see 3.2.1.3.1). Roy and Ovenden [29] have pointed out that the occurrence of nearly-commensurate satellite pairs is more frequent than one would expect from pure chance. Such a behaviour of the satellite orbits may be the result of tidal evolution [14, 18] or some other dynamical processes in the early solar system. Titan and Hyperion are locked into a 3:4 resonance, Enceladus-Dione, Mimas-Tethys, Io-Europa, and Europa-Ganymede are all locked in 2:1 resonances. The orbit-orbit resonances of the three Galilean satellites are particularly interesting. Sinclair [33] has found that the three resonance parameters $\lambda_I - \lambda_E + \varpi_I$, $\lambda_I - \lambda_E + \varpi_E$, and $\lambda_E - 2\lambda_G + \varpi_E$ are also librating about constant values. Based on these relations the eccentricities of Io and Europa are found to be 0.0043 and 0.011, respectively. With these non-zero eccentricities the tidal energy dissipation rates for Io and Europa (which are in synchronous rotation) can be estimated [28]; and it is found that the heating rate of Io required by the Io-Europa-Ganymede resonant relation may be high enough to cause substantial melting of its interior. The occurrence of active volcanism on Io as predicted by Peale et al. [28] has been confirmed by the Voyager 1 close-up observations of Io [34]. The 4:3 orbit-orbit resonance of the Hyperion-Titan pair is also the result of tidal interaction [18]. In addition to resonances, tidal actions have been found to cause secular changes in the orbits of Triton [23] and Phobos [31, 32]. Both satellites appear to spiral towards the central planets with lifetimes of the order of 10^8 a. The retrograde orbit of Triton is indicative of its capture origin. Similarly, the two groups of outer Jovian satellites (one prograde and one retrograde) have been suggested to be of capture origin. Greenberg [21] has reviewed this problem at some length. Christy and Harrington [5] have reported the detection of a satellite possibly in synchronous rotation around Pluto (orbital period $\approx$6.4 d and mean distance $\approx$15000...20000 km). They have deduced that the combined mass of Pluto and its satellite should be about 0.0017 earth masses, and if the diameter is approximately 3000 km [8] the mean density of Pluto is about 0.7 g cm^{-3}. The diameter of the satellite is estimated to be around 0.4 that of Pluto and the mass 0.05...0.10 that of Pluto. These values are of course still uncertain.

Table 1. A summary of the resonant relations in the planetary and satellite systems.

Librating argument:
- Ω = longitude of the ascending node
- ϖ = longitude of the perihelion $= \omega + \Omega$, with ω = longitude of the perihelion in the orbit
- λ = mean longitude $= M + \varpi$, with M = mean anomaly

The indices indicate the corresponding satellite or planet.

Mean value = mean value of the librating argument

Ampl = amplitude P = period

Planet or satellite	Librating argument	Mean value	Libration		Ref.
			Ampl.	P [a]	
Titan Rhea	$\varpi_T - \varpi_R$	0	9°.5	38	35
Hyperion Titan	$4\lambda_H - 3\lambda_T - \varpi_H$	180°	36°	1.75	36
Enceladus Dione	$2\lambda_D - \lambda_E - \varpi_E$	0	< 1°	12	Jefferys, W.H. in 21
Mimas Tethys	$4\lambda_T - 2\lambda_M - \Omega_T - \Omega_M$	0	97°	71	1
Neptune Pluto	$3\lambda_P - 2\lambda_N - \varpi_P$	180°	76°	$2 \cdot 10^4$	6
Io Europa Ganymede	$\lambda_I - 3\lambda_E + 2\lambda_G$	180°	$\lesssim$ 0°.03	$\approx$ 6	10
Miranda Ariel Umbriel	$\lambda_M - 3\lambda_A + 2\lambda_U$ (circulates)	—	—	12	19

3.2.1.6.3 The planetary rings (For data, see 3.2.1.3.2)

Rings within the Roche limit (defined as $R_{Roche}=2.44(\varrho_p/\varrho_s)^{1/3}R_p$, where ϱ_p and ϱ_s are respectively the densities of the planet and the small particles and R_p the planetary radius) are found around Jupiter, Saturn, and Uranus. While the rings of Saturn have been well studied, the discoveries of the rings of Jupiter [34] and Uranus [3, 11, 24] occurred only recently. The structures of the ring systems appear to be marked by resonant effects of the satellites. First, the Cassini division separating the A and B rings of Saturn is very near the position corresponding to 2:1 resonance with Mimas. Dynamical models have been constructed to explain such a resonant perturbation effect [13, 16]. Second, the radial spacing and narrow structures of the Uranian rings have been suggested to be related to a series of three-body resonances involving Miranda and Ariel and possibly other combinations [9], and the perturbation effects by a number of small satellites within the ring system have also been proposed [17]. Despite various interestings proposals, a consistent theory for the formation of the Uranian rings remains to be worked out.

The ring of Jupiter as discovered by the Voyager 1 imaging experiment [34] is located at a radial distance of about 1.8 planetary radii. Its thickness is estimated to be less than 30 km and the width more than 9000 km. No detailed information about its radial structure is yet available.

3.2.1.6.4 References for 3.2.1.6

1 Allan, R.R.: Astron. J. **74** (1969) 497.
2 Allen, C.W.: Astrophysical Quantities. The Athlone Press, 3. Edition (1976) p. 140.
3 Bhattacharyya, I.C., Kuppuswamy, K.: Nature **267** (1977) 332.
4 Carpenter, R.L.: Astron. J. **69** (1964) 2.
5 Christy, J.W., Harrington, R.S.: Astron. J. **83** (1971) 1005.
6 Cohen, C.J., Hubbard, E.C.: Astron. J. **70** (1964) 10.
7 Cook, A.F., Franklin, F.A., Palluconi, F.D.: Icarus **18** (1973) 317.
8 Cruikshank, D.P., Pilcher, C.B., Morrison, D.: Science **194** (1976) 835.
9 Dermott, S.F., Gold, T.: Nature **267** (1977) 590.
10 De Sitter, W.: Mon. Not. R. Astron. Soc. **91** (1931) 706.
11 Elliot, J.L., Dunham, E., Mink, D.: Nature **267** (1977) 330.
12 Elliot, J.L., Dunham, E., Wasserman, L.H., Millis, R.L., Churms, J.: Astron. J. **83** (1978) 980.
13 Franklin, F.A., Colombo, G., Cook, A.F.: Icarus **15** (1971) 80.
14 Goldreich, P.: Mon. Not. R. Astron. Soc. **130** (1965) 159.
15 Goldreich, P., Peale, S.J.: Annu. Rev. Astron. Astrophys. **8** (1968) 287.
16 Goldreich, P., Tremaine, S.: Icarus **34** (1978) 240.
17 Goldreich, P., Tremaine, S.: Nature **277** (1979) 97.
18 Greenberg, R.: Astron. J. **78** (1973) 338.
19 Greenberg, R.: Icarus **29** (1976) 427.
20 Greenberg, R. in: Jupiter, The Univ. of Arizona Press (Gehrels, T., ed.) (1976) p. 122.
21 Greenberg, R.: Vistas in Astronomy **21** (1977) 209.
22 Klassen, K.P.: J. Geophys. Res. **80** (1978) 2415.
23 McCord, T.B.: Astron. J. **71** (1966) 585.
24 Millis, R.L., Wasserman, L.H., Birch, P.V.: Nature **267** (1977) 331.
25 Morrison, D., Cruikshank, D.P., Burns, J.A. in: Planetary Satellites (Burns, J.A., ed.), Univ. of Arizona Press (1977) p. 6.
26 Murray, J.B., Dollfus, A., Smith, B.: Icarus **17** (1972) 576.
27 Nicholson, P.D., Persson, S.E., Matthews, K., Goldreich, P., Neugebauer, G.: Astron. J. **83** (1978) 1240.
28 Peale, S.J., Cassen, P., Reynolds, R.T.: Science **203** (1979) 892.
29 Roy, A.E., Ovenden, M.W.: Mon. Not. R. Astron. Soc. **114** (1954) 232.
30 Shapiro, I.I., Campbell, D.B., De Campli, W.M.: Astrophys. J. **230** (1979) L 123.
31 Shor, V.A.: Celestial Mech. **12** (1975) 61.
32 Sinclair, A.T.: Mon. Not. R. Astron. Soc. **155** (1972) 249.
33 Sinclair, A.T.: Mon. Not. R. Astron. Soc. **171** (1975) 59.
34 Smith, B.A. et al.: Science **204** (1979) 13.
35 Struve, G.: Astron. J. **38** (1928) 193.
36 Woltjer, J., jr.: Ann. Sternwacht Leiden **XVI** (1928) Pt. 3.

3.2.2 Physics of the planets and satellites

3.2.2.1 Introduction

Planetary science has advanced significantly during the last decade. In part this is due to recent efforts in space research. Not only have close-up spacecraft observations been made for all the terrestrial planets, Jupiter and the Galilean satellite system, in-situ measurements on the lunar surface have also been carried out by astronauts. Ground-based observations involving various new techniques have also increased our knowledge about the atmospheres and surfaces of planets and satellites tremendously.

Because of such rapid progress the nature of planetary astronomy has now been transformed into a field of interdisciplinary interest. In this chapter we want to review the basic data together with some new information on the atmospheric, geological and surface properties, and internal structures of the various planetary bodies. The satellite data are emphasized here since we believe that they are as interesting as the planetary.

General tables for orbital elements, dimensions, and mechanical data: see 3.2.1.

3.2.2.2 Internal compositions and structures

3.2.2.2.1 The terrestrial planets

The terrestrial planets are mainly made up of the oxides of Al, Ca, Fe, Si, Mg, and also free Fe and Ni. The best studied planet, of course, is the earth, for which the lower and upper mantles can be described as a combination of SiO_2, MgO, and FeO, whereas the chemical composition of the whole planet may be approximated by a mixture of carbonaceous chondrite, ordinary chondrite, and iron meteorite with FeS, Fe, Ni, and Si contained mostly in the core [2]. Such estimates are given in Table 1. Models of the density distribution and internal structure are constrained by the moment of inertia of the planet, its free oscillations, and the dispersion of seismic body waves (see [60] for general reference).

Table 1. Composition of meteorites, the earth and moon (in percent by weight) from Anderson et al. [2]. See also 3.3.2 and 3.4.2

Component	Carbonaceous chondrite I	Ordinary chondrite	Iron meteorite	Mix	Reduced carbonaceous chondrite I	Lower mantle	Lunar mantle	Upper mantle
	1	2	3	4	5	6	7	8
SiO_2	32.5	38.8	–	49.6 (35.5)	47.7 (32.9)	51	49.1	49.6
MgO	21.9	24.3	–	32.0 (21.6)	42.0 (28.9)	28	33.8	41.1
FeO	14.5	12.1	–	18.4 (12.4)	10.3 (7.1)	21	17.1	9.3
total	68.9	75.2	–	100.0 (67.5)	100.0 (68.9)	100	100.0	100.0
FeS	23.8	6.0	3.7	39.4 (12.8)	(–)			
Fe	0.2	11.8	89.1	55.4 (18.0)	83.3 (25.9)			
Ni	0.0	1.4	7.2	5.2 (1.7)	5.5 (1.7)			
Si	–	–	–	– –	11.3 (3.5)			
total	24.0	19.2	100.0	100.0 (32.5)	100.0 (31.1)			

On the basis of seismic data the interior of the earth can be divided into eight sections as depicted in Fig. 1. The gradual increase of the density from the upper mantle to the lower mantle is caused by the increasing compression together with phase transitions. But the interface (Mohorovicic discontinuity) between the crust and the upper mantle is determined by the chemical composition difference between magnesia and iron oxide-rich mantle basalt and silicate-rich crustal granite. The whole crustal layer (thickness 30 km in the continental region and 5 km in oceanic region) is derived from volcanism as a result of chemical differentiation of the upper mantle. The sharp transition between the lower mantle and the outer core at a depth of 2800 km is ascribed to the formation of a liquid metal outer core. While Fe, Ni, and Si, and also S are the major constituents of the earth's core, their exact compositional distribution is not well determined; however, Anderson [1] has pointed out that the properties of the core are consistent with Fe–Ni (totaling 88% of the core content) in cosmic proportion together with 11% of S. For more detailed information, see Vol. VI/1.

Models for other terrestrial planets have also been constructed [50]. As a reflection of the density variations among these planets, Mercury with an average density of 5.4 g cm^{-3} has a large metallic iron core containing about 60% of the total mass, the core of Venus has about 25% of the total mass, and that of Mars about 15%.

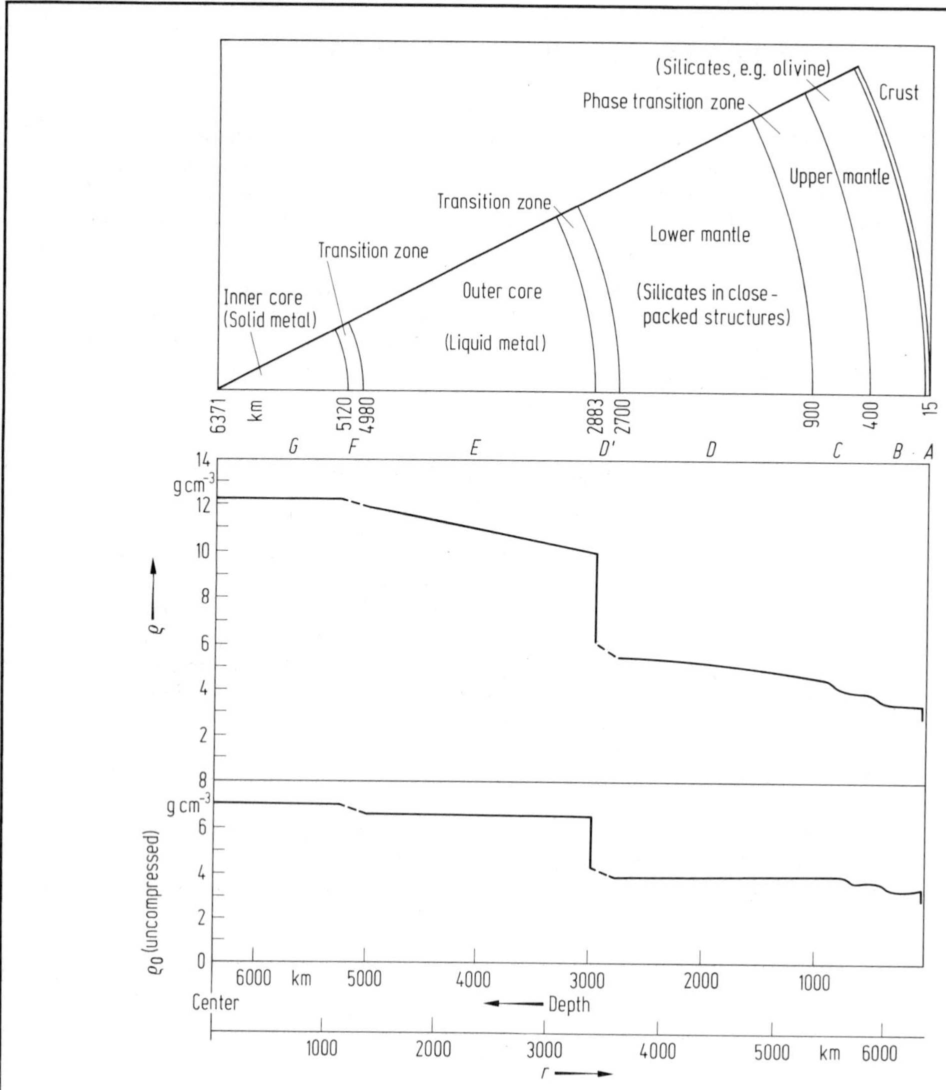

Fig. 1. Major subdivisions of the earth's interior with estimated densities [60]. r=distance from the center.

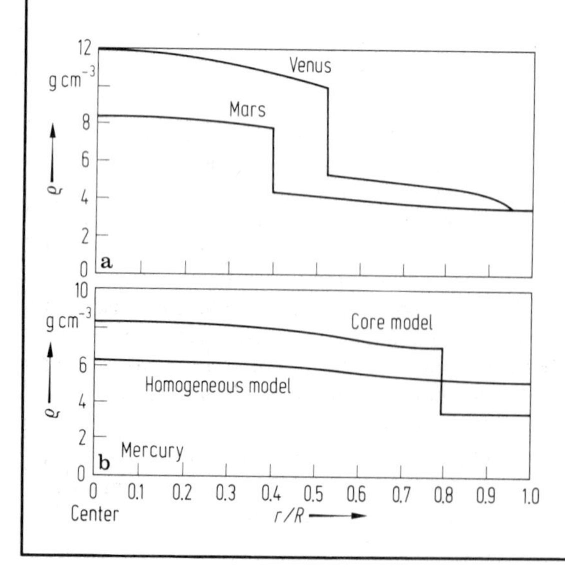

Fig. 2. Density distributions (a) for core models of Venus and Mars, and (b) for core and homogeneous models of Mercury. From Reynolds and Summers [50]. r=distance from the center, R=radius of the respective planet.

More recent model calculations [62] suggested the core of Mercury to be of pure iron composition, of Mars to be Fe–FeS, and of Venus to be Fe–Ni. The presence of a magnetic dipole field at Mercury [41] may require dynamo action and hence a partially molten core [20]; however, thermal model calculations [62] suggest that such a molten core may not exist at the present time. Furthermore, it was found that all major melting in the mantle of Mars should have disappeared about 10^9 years ago, whereas extensive melting may still be present in the case of Venus. See Fig. 2.

3.2.2.2.2 The outer planets

Of the outer planets, Jupiter and Saturn are hydrogen-helium-rich, whereas Uranus and Neptune are C, N, O-rich. However, their exact chemical compositions remain unclear. For example, estimates of the composition of Jupiter vary from purely hydrogen and helium with 22% He by mass [12], 30···35% He by mass [25], to a mixture of H, He, rocky material (SiO_2, MgO, FeS, FeO), and also ice (CH_4, NH_3, H_2S, H_2O) [45, 46, 66]. Figure 3 illustrates two possible interior structures of Jupiter and Saturn based on the numerical data of Zharkov and Trubitsyn. In their models, the He/H ratio is assumed to be of solar abundance and a rock-ice core is assumed to form in the centre. The phase transition from the H_2 liquid state to H-liquid metal takes place at 0.765 planetary radii for Jupiter and 0.465 planetary radii for Saturn. The introduction of the rock-ice core, besides being required by fitting of the model with observations, may be also a result of the planetary accretion of Jupiter and Saturn. Podolak and Cameron [46] and Cameron and Pollack [8] have discussed this problem in detail.

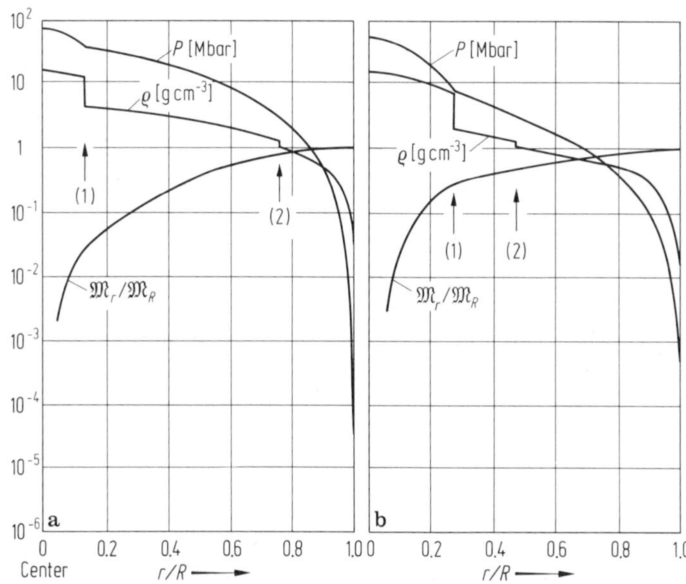

Fig. 3. Interior models of (a) Jupiter and (b) Saturn. Adapted from Zharkov and Trubitsyn [66].
(1): Limit of rock-ice core
(2): Phase transition from H_2 liquid state to H-liquid metal
r = distance from the center, R = radius of the respective planet,
$\mathfrak{M}_r/\mathfrak{M}_R$ = mass ratio, $\mathfrak{M}_R$ = total mass of the planet, $\mathfrak{M}_r$ = mass of part of the planet up to the distance r.

As indicated in early works [11, 50] Uranus and Neptune are both rich in the C, N, O elements and contain no more than 23% of hydrogen by mass for Uranus and 14% for Neptune. More recent calculations have involved various combinations of H_2O, CH_4, NH_3, H_2, and He [44, 46], but the estimates for the relative abundance of H and He and the C, N, O elements remain essentially the same. An interesting development in these calculations is that Podolak [44] has found that a fit of the model with the values of the dynamical oblateness ε ($=0.03\pm0.01$) given by Dollfus [13] and the gravity harmonic coefficient J_2($=0.005$) derived by Whitaker and Greenberg [65] cannot be achieved if the rotation period of Uranus is fixed to be (10.8 ± 0.5) hours. However, better agreement can be obtained if the CH_4/H_2 ratio is enriched 100 times more compared to the solar abundance and the rotation period is increased to the order of 18 hours. Redeterminations of the rotation period of Uranus have given new values ranging between (15.57 ± 0.8) hours and (24 ± 3) hours [7, 24, 64]. An interior model of Uranus is given in Fig. 4.

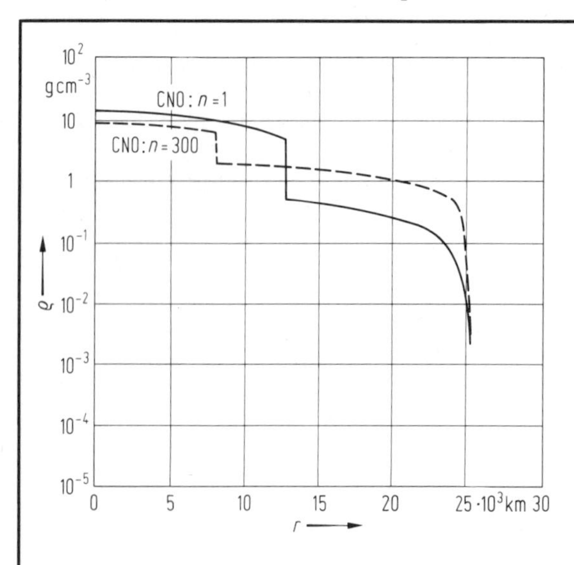

Fig. 4. Interior model of Uranus for CNO enhancement parameter $n=1$ and $n=300$ (Enhancement parameter n means that the ratio $CH_4:H_2$ in the envelope is taken to be n times the solar value). r=distance from the center. From Podolak [44].

Infrared observations [3, 5, 27] have shown that Jupiter and Saturn are emitting 2···3 times the amount of energy they absorb from the sun. These internal energy sources might derive from gravitational contractions of the planets by about 1 mm a^{-1} [23, 26, 58]. See also 3.2.2.4, Table 5.

Neptune has been observed to radiate about 2.4 times as much energy as it absorbs from the sun, but Uranus is probably in equilibrium with the incident solar flux [34, 35, 38, 40]. These thermal budgets should impose additional constraints on models of the outer planets.

3.2.2.2.3 The satellites

Physical data of satellites: see Table 4b in 3.2.1.3.1. Photometric data of satellites: see Table 2 in 3.2.2.3.2. The satellites can be broadly divided into two classes: bodies of rocky composition (with average density $\gtrsim 3$ g cm^{-3}) and those of icy composition or a mixture of ice and non-volatiles such as hydrous silicates and iron oxides (with average density $\lesssim 2$ g cm^{-3}). Based on Apollo magnetometer and seismic data, lunar models have been constructed by various investigators [14, 59, 63]. Though differing in some detailed aspects, the general conclusion seems to be that the moon has a basaltic crust of 20···60 km thickness (covering the crust is a surface layer of breccias and broken rocks, 1···2 km thick), a solid mantle, 65 km thick, of magnesium-rich pyroxanite and/or dunite, and then a core of 700 km radius [53, 63]. From the attenuation of seismic S-waves it is further deduced that the outer core is partially molten (S=secondary, or transversal).

Even though the two inner Galilean satellites Io and Europa have similar densities and sizes, their internal structure could be very different from that of the moon because of intense tidal heating [42]. The active volcanism on Io as observed by Voyager 1 [57] has confirmed this view. According to [42] Io should have a large molten core and a solid outer shell with thickness estimated to be ≈18 km.

Lewis [33] has considered the internal structures and temperature profiles of icy satellites assuming a mixture of H_2O, NH_3, CH_4, FeO, SiO_2, and MgO. The internal heat source is supplied by the decay of long-lived radionuclides (K, U, Th) contained in the rocky material. If the heat budget (with the omission of radioactive and convective heat transfer in the satellite interiors) is balanced by the heat flux conducted to the surface and surface radiation, Lewis found that objects of radii larger than 1000 km (i.e. Ganymede, Callisto, Rhea, and Titan) would have thin solid icy crusts, extensive and nearly isothermal liquid mantles and solid rocky cores whereas objects with radii less than 500 km (i.e. Enceladus, Tethys, Dione, and the Uranian satellites) would be unmelted. Consolmagno and Lewis [9] have considered the icy Galilean satellites in more detail. Taking into account the various proportions of H_2O and silicates (60% of H_2O by weight for Callisto, 20% for Ganymede and 10% for Europa), their thermal history calculations indicate that Callisto should have a thick icy crust containing undifferentiated silicate material, while Europa would only have a thin crust with most of the rocky material differentiated into the central core. In the model of Consolmagno and Lewis [9], in which solid state convection in the icy crust is neglected, liquid mantles of water are found to exist in Europa, Ganymede, and Callisto. But Reynolds and Cassen [49] have pointed out that, if such a solid state convection effect is taken into account, the liquid mantles of these icy satellites would be solidified. The calculations by Thurber et al. [61] have lent support to this view. See Figs. 5 and 6.

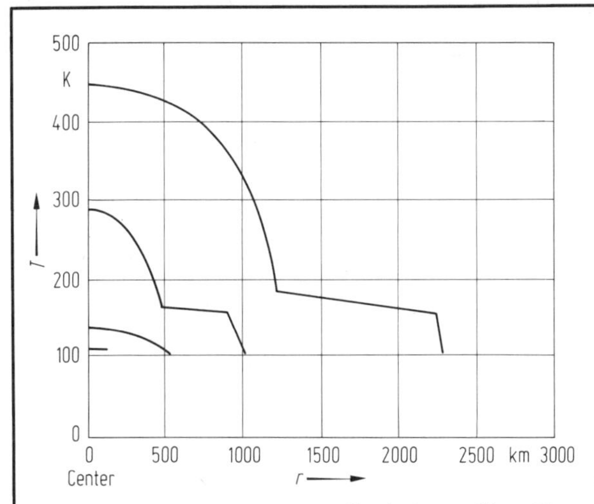

Fig. 5. Schematic temperature profiles in icy satellites. The surface temperatures for four objects of radii $r=100$, 500, 1000, and 2300 km are taken as $T=112$ K corresponding to typical surface temperatures for satellites of Jupiter. Note that the 100 and 500 km objects are unmelted, but the 1000 and 2300 km objects have thin crusts, extensive nearly isothermal mantles, and dense solid cores. Temperature gradients within the core are highly uncertain. From Lewis [33]. r = distance from the center.

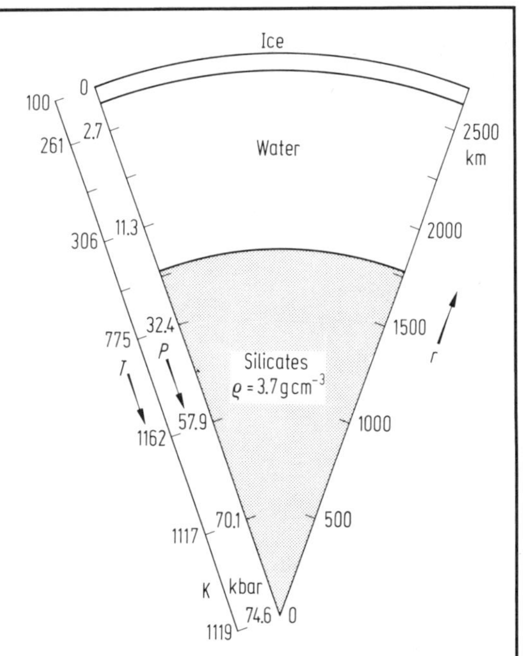

Fig. 6. Model for Ganymede's present interior structure. From Consolmagno and Lewis [9] and Johnson [28]. r = distance from the center.

3.2.2.2.4 The planetary rings (see also 3.2.1.3.2)

From infrared observations [30, 32, 43] and the high value of the geometrical albedo obtained from photometric observations [19] the surface composition of the Saturnian ring particles has been identified to be H_2O ice. In contrast, optical observations [54, 55] have established that the ring particles of Uranus must be very dark. The fact that their optical albedo is only of the order of a few percent suggests that the Uranian ring particles may be of carbonaceous chondritic composition. The albedo of the ring particles of Jupiter has not yet been determined and hence their surface composition remains unknown at the present time.

Table 2. Physical properties of the Saturnian ring system.

Physical parameter	Estimated value or property	Observation or calculation	Ref.
Surface composition of the ring particle	H_2O ice (plus meteoritic material?)	1⋯3 μm reflectivity spectrum	32, 43, 30
Size of the surface grain	$\langle r' \rangle \approx 25 \cdots 125$ μm	Mie scattering computation of the 1⋯3 μm reflectivity spectrum	48
Average particle size	$\langle r \rangle \lesssim 500$ μm or $\gtrsim 2$ cm	Eclipse cooling and heating curves as deduced from 10⋯20 μm observations	4, 37
	$\langle r \rangle_{ice} \approx 2 \cdots 15$ cm	Radar and radio observations	48 refs.
	$\langle r \rangle_{metal} \approx 1$ m	Dual-frequency radar observations	21, 22
Albedo	$A_V \approx 0.63$ (visual)	Photometric observations	19, 10
	$A_B \approx 0.49$ (blue)		
Volume density	$D \approx 10^{-2} \cdots 10^{-3}$	Opposition effect if interpreted as due to interparticle shadowing	52, 6, 29, 47

$$D = \frac{N \cdot V_p}{V_{ring}}$$

with N = total number of particles in the ring
V_p = average volume of one particle $= \frac{4}{3}\pi r^3$
V_{ring} = total volume of the ring.

continued

Table 2, continued

Physical parameter	Estimated value or property	Observation or calculation	Ref.
Ring mass	$\mathfrak{M}_{total}=(6.2\pm2.4)\cdot10^{-6}\,\mathfrak{M}_{Saturn}$	Gravitational perturbation calculation	36
	$\mathfrak{M}_{B\text{-}ring}=6\cdot10^{-6}\,\mathfrak{M}_{Saturn}$	Matching of the position of the Cassini division with the 2:1 resonant effect of Mimas	18
Ring thickness	$Z_R=$ 1.56±0.54 km (blue)	Observations during passage of the earth through the ring plane	31
	0.80±1.14 km (red)		
	2.8 ±1.4 km (yellow)		16
	1.3 ±0.3 km		17
Optical thickness	$\tau_{D'}\approx10^{-7}\ldots10^{-2}$	Stellar occultation by the ring system and edge-on observation of the ring plane	56, 15
	$\tau_A\approx0.5$		6
	$\tau_{CASS}\approx0.06$		17
	$\tau_B\approx1.0$		6
	$\tau_D\approx0.03$		15

For more recent results, see appendix 3.2.2.7.

3.2.2.2.5 References for 3.2.2.2

1 Anderson, D.L.: Annu. Rev. Earth Planet. Sci. **5** (1977) 179.
2 Anderson, D.L., Sammis, C., Jordon, T.: Science **171** (1971) 1103.
3 Armstrong, K.R., Harper, D.A., jr., Low, F.J.: Astrophys. J. **179** (1972) L 89.
4 Aumann, H.H., Kieffer, H.M.: Astrophys. J. **186** (1973) 305.
5 Aumann, H.H., Gillespie, C.M., jr., Low, F.J.: Astrophys. J. **157** (1969) L 69.
6 Bobrov, M.S. in: Surfaces and Interiors of Planets and Satellites (Dollfus, A., ed.), Academic Press, New York (1970) p. 377.
7 Brown, R.A., Goody, R.M.: Astrophys. J. **217** (1977) 680.
8 Cameron, A.G.W., Pollack, J.B. in: Jupiter (Gehrels, T., ed.), Univ. of Arizona Press (1976) p. 61.
9 Consolmagno, G.J., Lewis, J.S. in: Jupiter (Gehrels, T., ed.), Univ. of Arizona Press (1976) p. 1035.
10 Cook, A.F., Franklin, F.A., Palluconi, F.D.: Icarus **18** (1973) 317.
11 De Marcus, W.C., Reynolds, R.T.: Mem. Soc. R. Sci. Liège, Ser. (5) **7** (1963) 196.
12 De Marcus, W.C.: Astron. J. **63** (1958) 2.
13 Dollfus, A.: Icarus **12** (1970) 101.
14 Dyal, P., Parkin, C.W.: Scientific American **225**, **2** (1971) 62.
15 Ferrin, I.: Icarus **22** (1974) 159.
16 Focas, J.H., Dollfus, A.: Astron. Astrophys. **2** (1969) 251.
17 Fountain, J.W., Larson, S.M.: Icarus **36** (1978) 92.
18 Franklin, F.A., Colombo, G., Cook, A.F.: Icarus **15** (1971) 80.
19 Franklin, F.A., Cook, A.F.: Astron. J. **70** (1965) 704.
20 Fricker, P.E., Reynolds, R.T., Summers, A.L., Cassen, P.: Nature **259** (1976) 293.
21 Goldstein, R.M., Morris, G.A.: Icarus **20** (1973) 260.
22 Goldstein, R.M., Green, R.R., Pettengill, G.H., Campbell, D.B.: Icarus **30** (1977) 104.
23 Graboske, H.C., Pollack, J.B., Grossman, A.S., Olness, R.J.: Astrophys. J. **199** (1975) 265.
24 Hayes, S.H., Belton, M.J.S.: Icarus **32** (1977) 383.
25 Hubbard, W.B.: Astrophys. J. **152** (1968) 745.
26 Hubbard, W.B.: Astrophys. J. **155** (1969) 333.
27 Ingersoll, A.P., Münch, G., Neugebauer, G., Diner, D.J., Orton, G.S., Schupler, B., Schroeder, M., Chase, S.C., Ruiz, R.D., Trafton, L.M.: Science **188** (1975) 472.
28 Johnson, T.V.: Annu. Rev. Earth Planet. Sci. **6** (1978) 93.
29 Kawata, Y., Irvine, W.M. in: Exploration of the Solar System (Woszezyk, A., Ivaniszewska, C., eds.), IAU Symp. **65** (1974) p. 441.
30 Kieffer, H.H. in: The Rings of Saturn (Palluconi, D., Pettengill, C.H., eds.), NASA-SP 343, Washington (1974).
31 Kiladze, R.I.: Byull. Abastuman. Astrofiz. Obs **37** (1969) 151.
32 Kuiper, G.P., Cruikshank, D.P., Fink, U.: Sky and Telescope **39** (1970) 14.
33 Lewis, J.S.: Icarus **15** (1971) 174.
34 Loewenstein, R.F., Harper, D.A., Moseley, H.: Astrophys. J. **218** (1977) L 145.

35 Low, F.J., Rieke, G.M., Armstrong, K.R.: Astrophys. J. **183** (1973) L 105.
36 McLaughlin, W.I., Talbot, T.D.: Mon. Not. R. Astron. Soc. **179** (1977) 619.
37 Morrison, D.: Icarus **22** (1974) 57.
38 Morrison, D., Cruikshank, D.P.: Astrophys. J. **179** (1973) 329.
39 Morrison, D., Cruikshank, D.P., Burns, J. in: Planetary Satellites (Burns, J.A., ed.), Univ. of Arizona Press (1977) p. 12.
40 Murthy, R.E., Trafton, L.M.: Astrophys. J. **193** (1974) 253.
41 Ness, N.F., Behannon, K.W., Lepping, R.P., Whang, Y.C.: J. Geophys. Res. **80** (1975) 2708.
42 Peale, S.J., Cassen, P., Reynolds, R.T.: Science **203** (1979) 892.
43 Pilcher, C.B., Chapman, C.R., Lebofsky, L.A., Kieffer, H.H.: Science **167** (1970) 1372.
44 Podolak, M.: Icarus **27** (1976) 473.
45 Podolak, M.: Icarus **30** (1977) 155.
46 Podolak, M., Cameron, A.G.W.: Icarus **22** (1974) 123.
47 Pollack, J.B.: Space Sci. Rev. **18** (1975) 3.
48 Pollack, J.B., Summers, A., Baldwin, B.: Icarus **25** (1975) 263.
49 Reynolds, R.T., Cassen, P.: Geophys. Res. Lett. **6** (1979) 121.
50 Reynolds, R.T., Summers, A.L.: J. Geophys. Res. **70** (1965) 199.
51 Reynolds, R.T., Summers, A.L.: J. Geophys. Res. **74** (1969) 2494.
52 Seeliger, H.N.: Abh. Bayer. Akad. Wiss. 2K l, **16** (1887) 467.
53 Short, N.M.: Planetary Geology, Prentice-Hall, Inc., New Jersey (1975).
54 Sinton, W.M.: Science **198** (1977) 503.
55 Smith, B.A.: Nature **268** (1977) 32.
56 Smith, B.A., Cook, A.F., Feibelman, R.F., Beebe, R.F.: Icarus **25** (1975) 466.
57 Smith, B.A., et al.: Science **204** (1979) 13.
58 Smoluchowski, R.: Nature **215** (1967) 691.
59 Sonett, C.P., Colburn, D.S., Dyal, P., Parkin, C.W., Smith, B.F., Schubert, G., Schwartz, K.: Nature **230** (1971) 359.
60 Stacey, F.D.: Physics of the Earth, John Wiley and Sons, Inc., New York (1969) pp. 65–124.
61 Thurber, C.M., Hsui, A.T., Toksöz, M.N.: EOS, Trans. Amer. Geophys. Union **60** (1979) 307.
62 Toksöz, M.N., Hsui, A.T., Johnson, D.H.: The Moon and Planets **18** (1978) 281.
63 Toksöz, M.N., Press, F., Anderson, K., Dainty, A., Latham, G., Ewing, M., Dorman, J., Lammlein, D., Nakamura, Y., Sutton, G., Duennebier, F.: The Moon **4** (1972) 490.
64 Trafton, L.: Icarus **32** (1977) 402.
65 Whitaker, E.A., Greenberg, R.J.: Mon. Not. R. Astron. Soc. **155** (1973) 15.
66 Zharkov, V.N., Trubitsyn, V.P. in: Jupiter (Gehrels, T., ed.), Univ. of Arizona Press (1976) p. 133.

3.2.2.3 Surface properties

Table 3. Photometric data for the planets [1].

V_{opp} = visual magnitude at opposition
$V(1,0)$ = visual magnitude at unit distance (= reduced to 1 AU distance from both earth and sun)
P_v = geometrical visual albedo
q = phase integral
A = Bond albedo

	V_{opp}	$V(1,0)$	Colour index		P_v	q	A
			$B-V$	$U-B$			
Mercury	$-0^m.17$	$-0^m.36$	$0^m.91$	$0^m.4$	0.096	0.58	0.056
Venus	−3.81	−4.34	0.79	0.5	0.6	1.2	0.72
Earth	−3.87	−3.9	0.2		0.37	1.05	0.39
Mars	−2.01	−1.51	1.37	0.6	0.154	1.02	0.16
Jupiter	−2.55	−9.25	0.83	0.4	0.44	1.6	0.70
Saturn	+0.67	−9.0	1.04	0.6	0.47	1.6	0.75
Uranus	+5.52	−7.15	0.56	0.3	0.57	1.6	0.90
Neptune	+7.84	−6.90	0.41	0.2	0.51	1.6	0.82
Pluto	+14.90	−1.0	0.80	0.3	0.12	1.2	0.145

3.2.2.3.1 The terrestrial planets

The crater is the fundamental landform on the terrestrial planets. The sizes of craters range from a few hundred km to under one km in radius. From stratigraphy, the crater frequency distribution, and crater morphology, relative ages of the planetary surfaces may be derived. This method has been employed to study the surfaces of the moon, Mercury, and Mars. With the return of samples of lunar rocks from the Apollo missions, an absolute chronology has been established for the geological history of the moon. Of particular interest is that a sharp cutoff in the crystallization ages of lunar samples is found at about $3.9 \cdot 10^9$ a [66]. The Rb/Sr dating method shows that the ages of mare basalt are clustered around 3.3 and $3.7 \cdot 10^9$ a whereas those of the highland rocks are clustered around $3.9 \cdot 10^9$ a. This might mean that there was a terminal episode (cataclysm) of intense impact cratering at $\approx 3.95 \cdot 10^9$ a having a duration of about 10^8 a [47, 76], even though other time variations of the flux and size distribution of the stray bodies might produce similar effects [2, 23]. After the Mariner 10 flyby of Mercury, the suggestion was made that the striking similarity in the surfaces of the moon and Mercury might be the result of similar history of early bombardment [40]. It is still to be determined [8, 77] whether an absolute chronology can be established for the moon, Mercury, and other planets such that they followed more or less the same surface history of (1) accretion and differentiation; (2) terminal heavy bombardment; (3) formation of the Caloris basin on Mercury and Imbrium and other basins on the moon; (4) flooding of the basins and other areas by volcanism; and (5) light cratering and development of a lunar-like regolith [40].

1. Mercury

Trask and Guest [69] have constructed a geological terrain map of Mercury and have found that, besides a similarity in general morphology, three major surface units have close analogues on the moon. They are the inter-crater plains, the heavily cratered plains and the smooth plains which are comparable to the lunar pre-Imbrium plains, highlands and light plains, respectively. A new landform has been identified in the region antipodal to the Caloris region. The hilly and lineated terrain contains crater rims broken into hills and depressions. Schultz and Gault [57] have suggested that this particular landform may be related to the seismic effects of the Caloris impact. Another surface feature unique to Mercury is the linear lobate scarps transecting both craters and inter-crater areas. It seems that the global distribution of such scarps with heights ranging between a few hundred meters to about 3 km and lengths varying between 20 and over 500 km may be caused by planet-wide compressive stresses [63], i.e. a decrease in the radius of Mercury by $1 \cdots 2$ km. Thermal history calculation by Solomon [62] indicates that cooling of the lithosphere could indeed lead to reduction of Mercury's radius by 2 km. (See Fig. 7.)

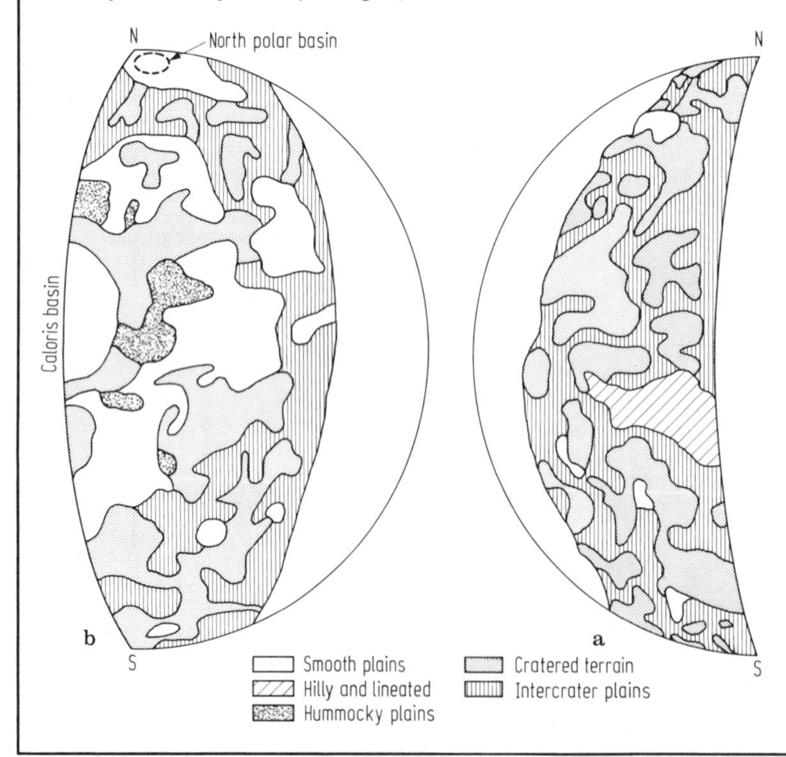

Fig. 7. Generalized geological terrain map of Mercury based on Mariner 10 photographs. (a) Incoming hemisphere; (b) outgoing hemisphere. From Gault et al. [19].

2. Venus

On Venus, two roundish and rough spots (alpha and beta) about 1000 km across, perhaps of volcanic origin, together with a number of large topographic features have been identified by ground-based radar observations [6, 20, 21, 32, 48].

Besides these seemingly tectonic structures, craters and impact basins are also apparent in the radar maps [22]. While the Pioneer Venus spacecraft radar mapping experiment has revealed further details of the surface topography [49] the Venera 9 and 10 gamma ray spectrometers have indicated the surface rocks to have radioactive elements (K, U, Th) similar to those of terrestrial basaltic rocks [33, 64].

3. Mars

A more complete photographic survey of Mars made by the Mariner 9 spacecraft has shown that its surface can be divided into two geological provinces: the northern hemisphere which is mainly smooth plains similar to the lunar maria, and the southern hemisphere which is densely cratered like the lunar highlands [43, 61]. Based on crater counts in different regions, Soderblom et al. [61] have argued that the Martian and lunar impact flux histories are nearly the same. While this may be true as far as cratering processes are concerned, the obliteration history of Mars (at least for the small craters with diameters less than 20 km) is very different from that of the moon. As indicated by the presence of dust storms, sand dunes, and channels, aeolian and aqueous processes must be important in modification of the Martian surface. That there is a break in the size distribution of Martian craters at 20···50 km diameter [23, 45] and that the craters of this size appear to be fresher than the larger ones [42] could be the result of such erosion processes which may appear as relative short episodes of intense obliteration [9]. The aqueous process may be caused by the seasonal warm-up of the Martian atmosphere leading to release of volatiles at present frozen in the polar caps into the atmosphere. A number of physical models have been postulated for the periodic warming of the Martian atmosphere [41, 56, 75]. (See Fig. 8.)

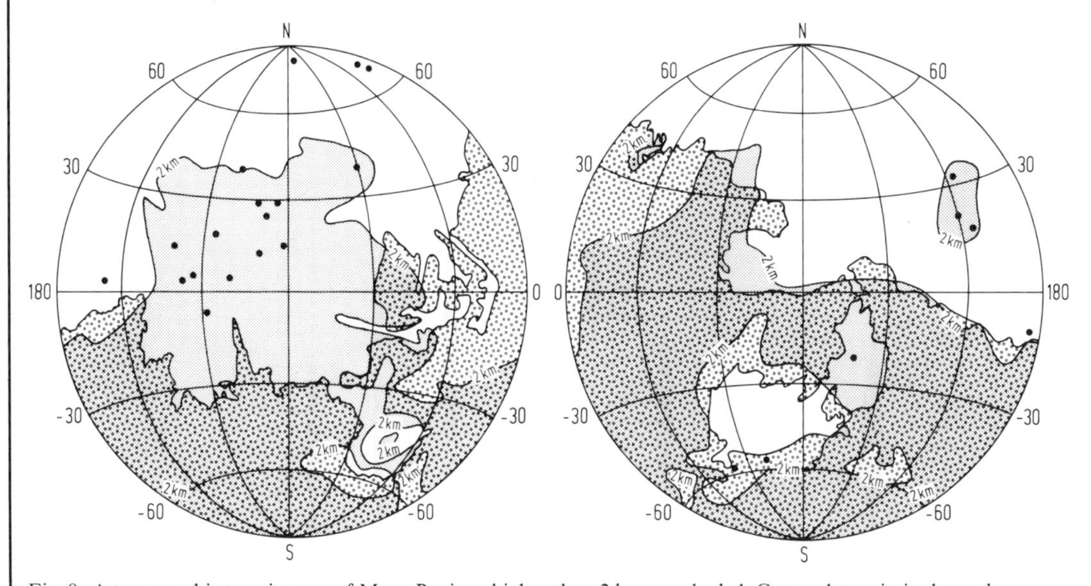

Fig. 8. A topographic terrain map of Mars. Regions higher than 2 km are shaded. Cratered terrain is shown by a circle pattern. The more prominent volcanos are shown by solid circles. From Mutch and Head [43].

Shield volcanos (the largest, Olympus Mons, is 22 km high and 500 km across) are another interesting feature of Mars [61, 71]. According to Soderblom et al. [61] the Martian volcanos show a range of ages with less than $0.75 \cdot 10^9$ a for the Tharsis volcanics and $0.75 \cdots 1.5 \cdot 10^9$ a for the Elysium volcanics. Neukum and Wise [44], however, favour older ages for the volcanos with an end of volcanic and tectonic activities set at $2.5 \cdot 10^9$ a ago. The Martian cratering chronology has been studied in more detail by Neukum and Hiller [44a] attempting to narrow down the discrepancy in the various studies. Even so, the uncertainty in the age determination is still large (see Figs. 9 and 10).

Fig. 9: see p. 159.

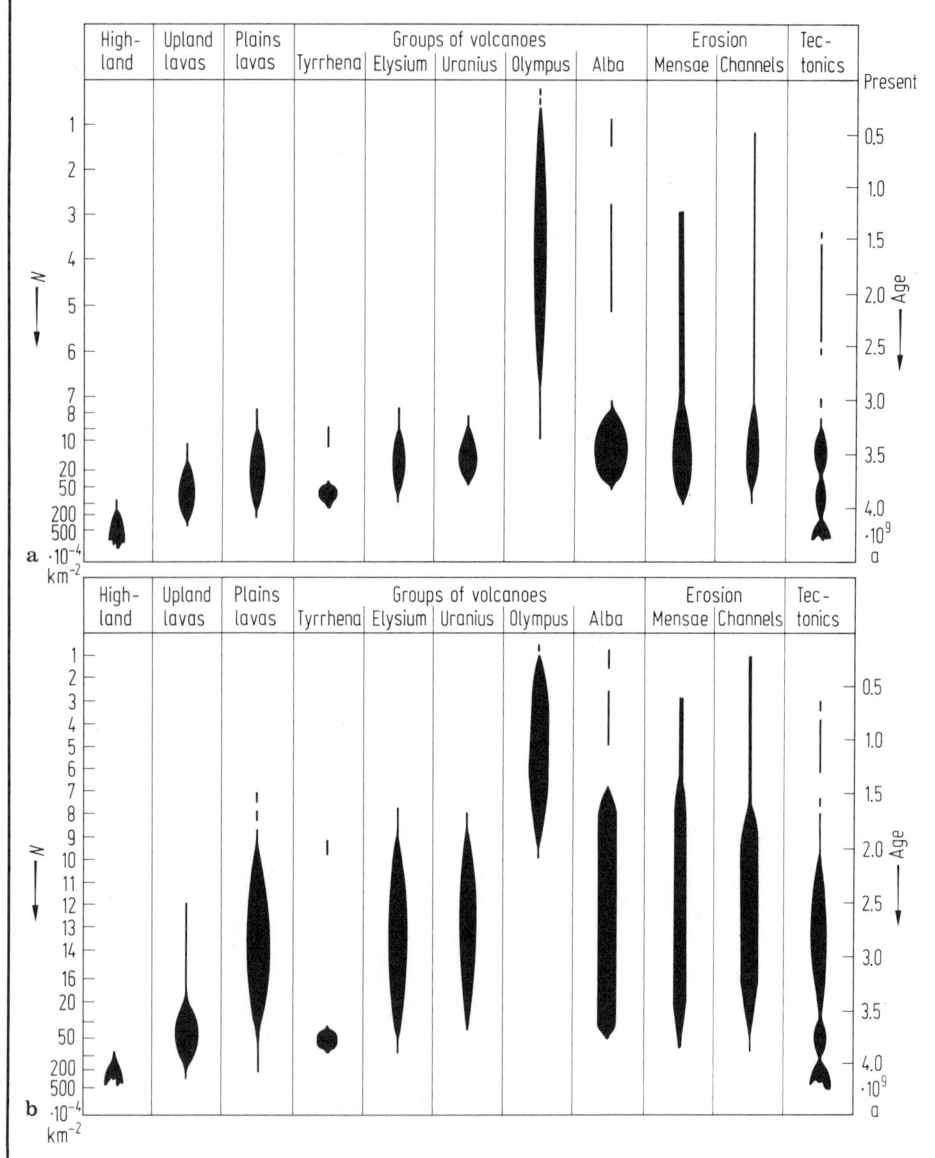

Fig. 10. Martian geologic activity: the diagrams show in a qualitative way absolute ages and periods of geologic activity.
(a) Absolute ages based on the model I of cratering chronology of Neukum and Hiller [44a].
Notice that most major events took place before $3.0 \cdot 10^9$ a ago. Exception is the Olympus volcanoes group and Alba Patera, active through recent history. Channel ages of approx. 10^9 a mark young floor filling but not original excavation.
(b) Absolute ages based on the model II of Neukum and Hiller [44a]. This chronology gives longer periods of activity of the individual geological event. As in (a), separation into 2 epochs of activity is still detectable (division here at an age of about $1.5 \cdot 10^9$ a).
N = cumulative crater number density at diameter $D = 1$ km.

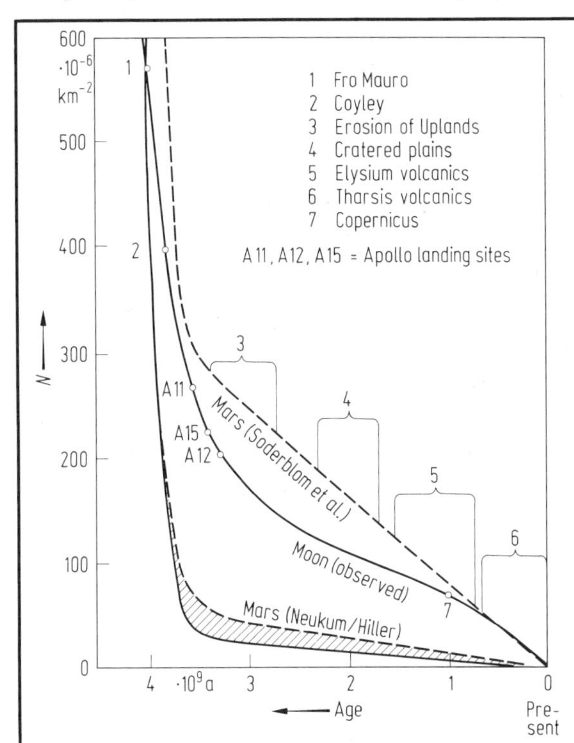

Fig. 9. Comparison of impact flux history derived for Mars with that observed for the moon. The present view (Neukum and Hiller [44a]) is that the Martian ages are much older than the previous estimates (Soderblom et al. [61]). N = crater number density at diameter D = 4...10 km.

3.2.2.3.2 The satellites

Photometric data for the satellites: Table 4, next page.

1. The moon

The moon is covered by a porous layer (typically a few meters deep) of fine dust particles with diameters mostly less than 1 mm. Underlying this regolith blanket is the so-called "megaregolith" of coarser material extending to a depth of kilometers. Due to the effect of meteoritic impacts, there is a continuous size distribution of particles from submicron to meter-size. For the lunar rocks with diameter larger than a few cm, the dominant types are the breccias, which are heterogeneous assemblages of the regolith formed by impact remelting [30]. From the measurements of the transition depth between bowl-shaped and flat-floored craters Quaide and Oberbeck [54] have derived a value ranging between 5 m and 15 m for the regolith layer in the mare region with 8 m as the average depth. However, the deep drill core tubes from sites of Apollo 15, 16 and 17 give a lower value of 2.4···3.2 m. All together this means a regolith growth rate of about 2 m/10^9 a. Various techniques have been applied to determine the exposure age of individual rocks on the lunar surface and typical ages of 50···500·10^6 a have been estimated for different samples. This large spread in exposure ages is consistent with the "gardening" of the lunar surface by meteoritic impacts [31]. The geological provinces of the moon are shown in Fig. 11.

From absolute dating of lunar samples plus study of the surface morphology the geological history of the moon has been reconstructed. In brief, after formation at 4.6·10^9 a ago the moon went through a period of planetary differentiation process between 4.6 and 4.3·10^9 a. Records of bombardment by large planetesimals were kept after the formation of the solid anorthositic crust. The last such large impact event (the Imbrian system) has been dated at about 4.0···3.9·10^9 a ago. Volcanism occurring between 3.7 and 3.2·10^9 a ago then filled the impact basin and other low land areas with basaltic lava [15, 16, 47]. Flooding of the maria concentrated mostly on the near side of the lunar surface so that the corresponding topography is distinctly smoother than that on the far side. For large circular basins (Smythii, Crisium, Nectaris, Serenitatis, Imbrium, Humorum, and Orientale) where the basaltic inner fill may be tens of kilometers thick, large positive gravity anomalies have been found [39] indicating mass concentrations (mascons) in these areas. Laser altimetry [28] and radar sounding experiments [51] have provided information on the latitude profiles of the lunar surface. Topographic relief changes in the mare surfaces average about 150 m over areas of 200···600 km on a side; but the relief changes in the highland areas vary from 0.6 km to 1.5 km over areas of 1600 km^2 [58]. Furthermore, the far side is observed to be more than 4 km higher than the near side indicating a displacement of the mass centre of the moon of 2 km away from the earth. More information concerning the chemical composition, geological properties, and the interaction between the solar wind and cosmic ray particles and lunar rocks can be found in the works of El-Baz [15, 16]; Short [58]; Taylor [65]; Levinson and Taylor [31]; Langevin and Arnold [30]; Walker [74]; and Toksöz [67], among others. See also Fig. 12, p. 161, and Fig. 13, p. 162.

Table 4. Photometric data for the satellites [12, 70]. For explanation of symbols, see next page.

Planet	Satellite [1])	V_0	$V(1,0)$	ΔV	P_v [2])	Θ_{max}	Θ_{min}	$U-B$	$B-V$	I	$I-H$	$I-K$	$I-L$
Earth	Moon	$-12^{m}.74$	$+0^{m}.21$	–	0.12	–	–	$0^{m}.46$	$0^{m}.92$				
Mars	M1 Phobos	+11.6	+12.1	–	0.06	–	–		0.6				
	M2 Deimos	12.8	+13.3	–	0.07	–	–		0.6				
Jupiter	J5 Amalthea	13.0 [2])	+ 6.3	?	0.10	?	?						
	J1 Io	5.0	− 1.68	0.21	0.63	100°±20°	300°±20°	1.3	1.15	$3^{m}.7\pm0.1$	$+0^{m}.35\pm0.01$	$+0^{m}.43\pm0.02$	$+0^{m}.43\pm0.03$
	J2 Europa(L)	5.3	− 1.41	0.34	0.64	80°±10°	280°±10°	0.5	0.89	4.1±0.1	−0.31±0.02	−0.66±0.03	−2.90±0.04
	J2 Europa(T)										−0.37±0.02	−0.91±0.04	−3.25±0.05
	J3 Ganymede(L)	4.6	− 2.09	0.16	0.43	60°±10°	270°±20°	0.53	0.81	3.6±0.1	−0.10±0.03	−0.18±0.04	−2.08±0.14
	J3 Ganymede(T)										−0.07±0.02	−0.14±0.03	−1.58±0.04
	J4 Callisto	5.6	− 1.05	0.16	0.17	220°±10°	110°±30°	0.55	0.88	4.4±0.1	+0.27±0.01	+0.34±0.02	−0.67
	J6 Himalia	14.8	+ 8.0		0.03								
	J7 Elara	16.4	+ 9.3		0.03								
	J8 Pasiphae	17.7	+11.0										
	J9 Simope	18.3	+11.6										
	J10 Lysithea	18.4	+11.7										
	J11 Carme	18.0	+11.3										
	J12 Ananke	18.9	+12.2										
	J13 Leda	20 :	+13.3										
Saturn	S1 Mimas	12.9	+ 3.3	?		?	?	?	0.65±0.1	?	?	?	?
	S2 Enceladus	11.8	+ 2.2	0.4		270° (?)	90° (?)	?	0.62	10.6±0.1	−0.16±0.05	−0.33±0.07	> −0.66
	S3 Tethys	10.3	+ 0.7	≈0.15		90°	270°	0.34	0.74	9.3±0.1	−0.20±0.05	−0.36±0.07	> −0.46
	S4 Dione	10.4	+ 0.89	≈0.4	0.60	120°±30°	290°±30°	0.30	0.71	9.6±0.2	−0.20±0.05	−0.30±0.07	> −0.42
	S5 Rhea	9.7	+ 0.21	0.2	0.60	90°± 5°	270°± 5°	0.35	0.76	8.6±0.1	−0.05±0.05	−0.29±0.07	−1.89±0.18
	S6 Titan	8.4	− 1.56	<0.02	0.21	–	–	0.75	1.30	–	–	–	–
	S7 Hyperion	14.2	+ 4.61	0.05...0.10		120°, 300°	30°, 240°	0.33	0.78	13.0±0.1	+0.15±0.05	−0.03±0.07	> 0.55
	S8 Iapetus(L)	10.2	+ 1.48	1.9	0.12	90°± 5°	270°± 5°	0.38	0.78	10.3±0.1	+0.20±0.03	+0.28±0.04	−0.82±0.25
	S8 Iapetus(T)	11.9	+ 1.6 [2])	–		–	–	0.29	0.69	9.4±0.1	−0.11±0.02	−0.24±0.03	> −1.35
	S9 Phoebe	16.5	+ 6.9 [2])	≈0.3		90°, 180°	270°, 0°	0.30	0.63	–	–	–	–
Uranus	U5 Miranda	16.5	+ 3.8	–		?	?	?	?	?	?	?	?
	U1 Ariel	14.4	+ 1.7	–		?	?	?	?	?	?	?	?
	U2 Umbriel	15.3	+ 2.6	–		?	?	?	?	?	?	?	?
	U3 Titania	14.01	+ 1.30	–		?	?	0.25	0.62	12.6±0.1	+0.23±0.10	−0.07±0.10	–
	U4 Oberon	14.20	+ 1.49	–		?	?	0.24	0.65	12.9±0.1	+0.27±0.10	+0.15±0.10	+ 1.25±0.10
Neptune	N1 Triton	13.6	− 1.16	–		?	?	0.40	0.77	12.2±0.1	+0.31±0.08	+0.07±0.10	+ 1.17±0.10
	N2 Nereid	18.7	+ 4.0	–		?	?						

[1]) (L) = leading side; (T) = trailing side.

[2]) From Morrison et al. [37a].

Symbols in Table 4:

V_0 = visual magnitude, for solar phase angle $\alpha = 0°$
$V(1,0)$ = visual magnitude at unit distance (= reduced to 1 AU distance from both earth and sun)
ΔV = visual light curve amplitude
P_v = geometrical visual albedo
Θ = longitude of satellite (= 0° at superior geocentric conjunction; = 90° at eastern elongation)
$\Theta_{max} = \Theta$ with maximum brightness; $\Theta_{min} = \Theta$ with minimum brightness
U, B, V, I, H, K, L = magnitudes and colours in the U, B, V... system (see 4.2.5.12)

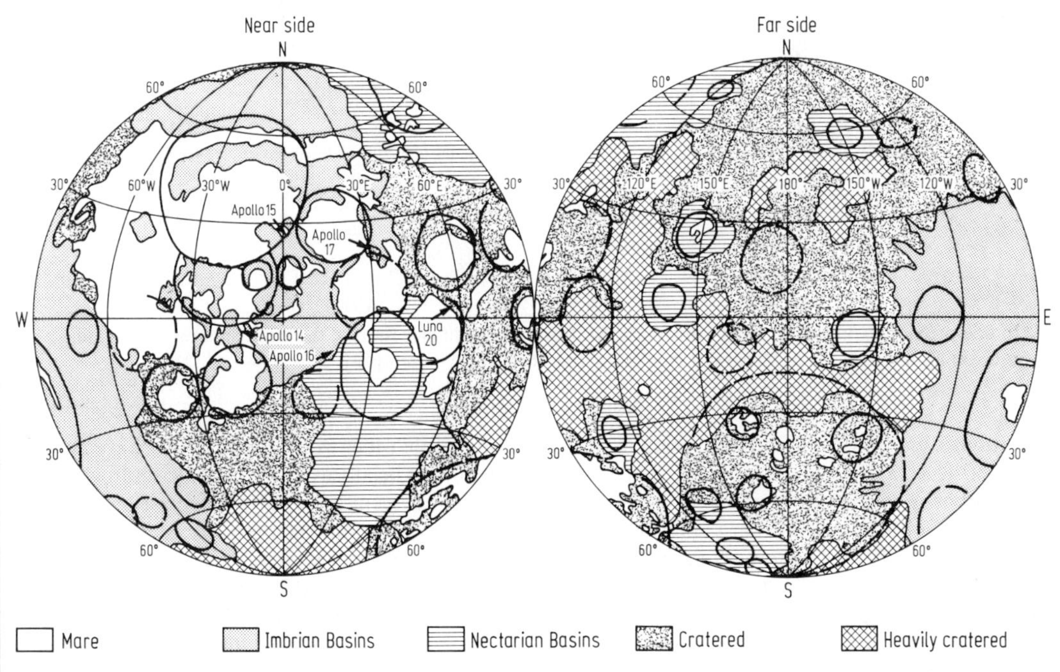

Fig. 11. Map of the major geologic provinces of the moon showing that the lunar physiography is controlled by the large circular basins and their overlapping ejecta blankets.
Note asymmetry in the distributions of maria (more on the near side than on the far side). From Howard et al. [25].

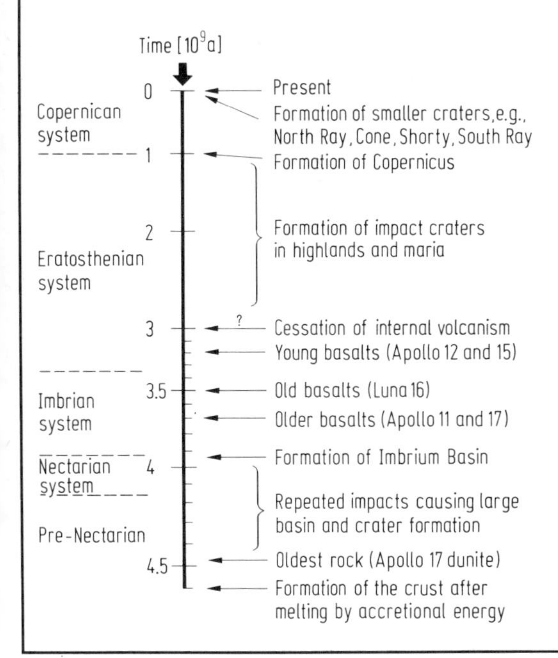

Fig. 12. Schematic illustration of the evolution of the lunar surface. The absolute time scale in [10^9 a] as deduced from age and exposure data techniques is shown along the middle line. On the right, major events and episodes of material emplacement are shown, on the left is the lunar wide relative age scheme based mainly on superposition relationships of geologic units. From El-Baz [16].

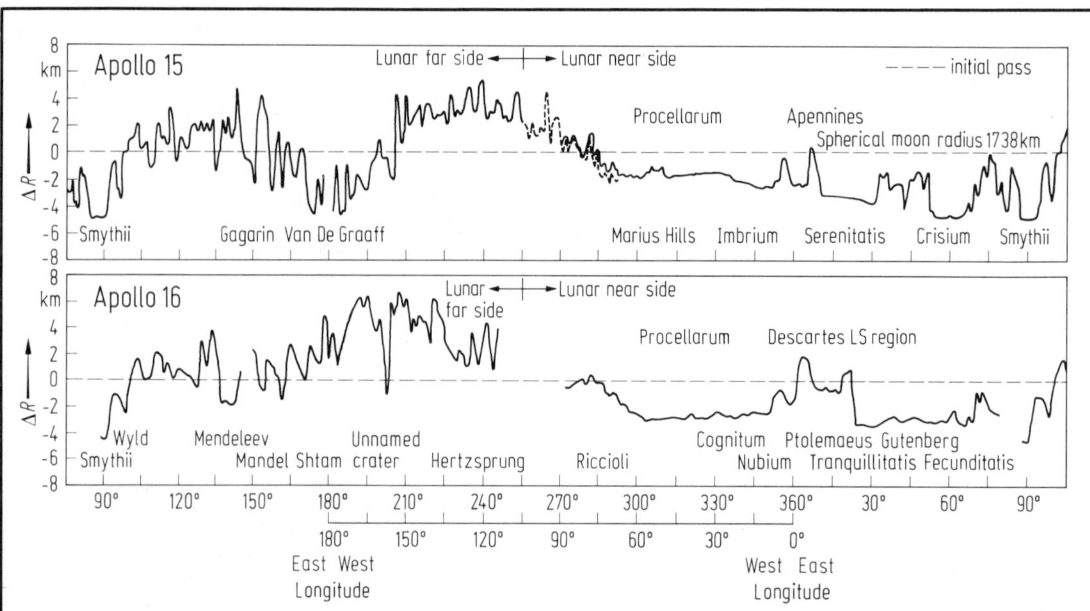

Fig. 13. Altitude profiles (expressed as height difference ΔR from the mean elevation of the lunar spheroid) determined by the laser beam altimeter in the SIM bays of Apollo 15 and 16. From Sjogren [59].

2. Phobos and Deimos

Observations from Mariner 9 and the Viking orbiters have shown that the two Martian satellites are heavily cratered objects [53, 73]. Crater counts indicate that these bodies should be at least $3 \cdot 10^9$ a old [51]. In addition to craters, Phobos is covered by a system of trough-like grooves which seem to be surface expressions of deep fractures within Phobos resulting from the formation of the Stickney crater; no such grooves are observed on the surface of Deimos [72]. The low geometric albedo (≈ 0.065), small density (2 g cm^{-3}) and UV reflectance spectrum of Phobos all indicate that Phobos is made of carbonaceous chondritic material [46, 51, 68]. Photometric and polarimetric measurements indicate that both Phobos and Deimos are covered by a layer of regolith [70, 71] as in the case of the moon. The Viking orbiter observations have further shown that the regolith of Phobos may be several hundred meters deep and that of Deimos at least 5 m deep [72]. (See also Figs. 14 and 15.)

Fig. 14: see p. 163.

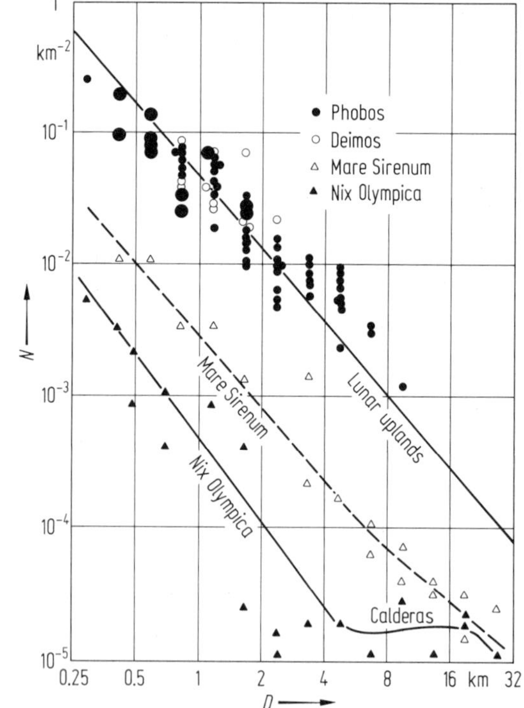

Fig. 15. Crater number densities, N, for selected areas of Phobos and Deimos compared with those in the lunar highlands and in two regions on Mars. From Pollack et al. [53]. The size of the full circles for Phobos reflects the quality of the determination. D = diameter.

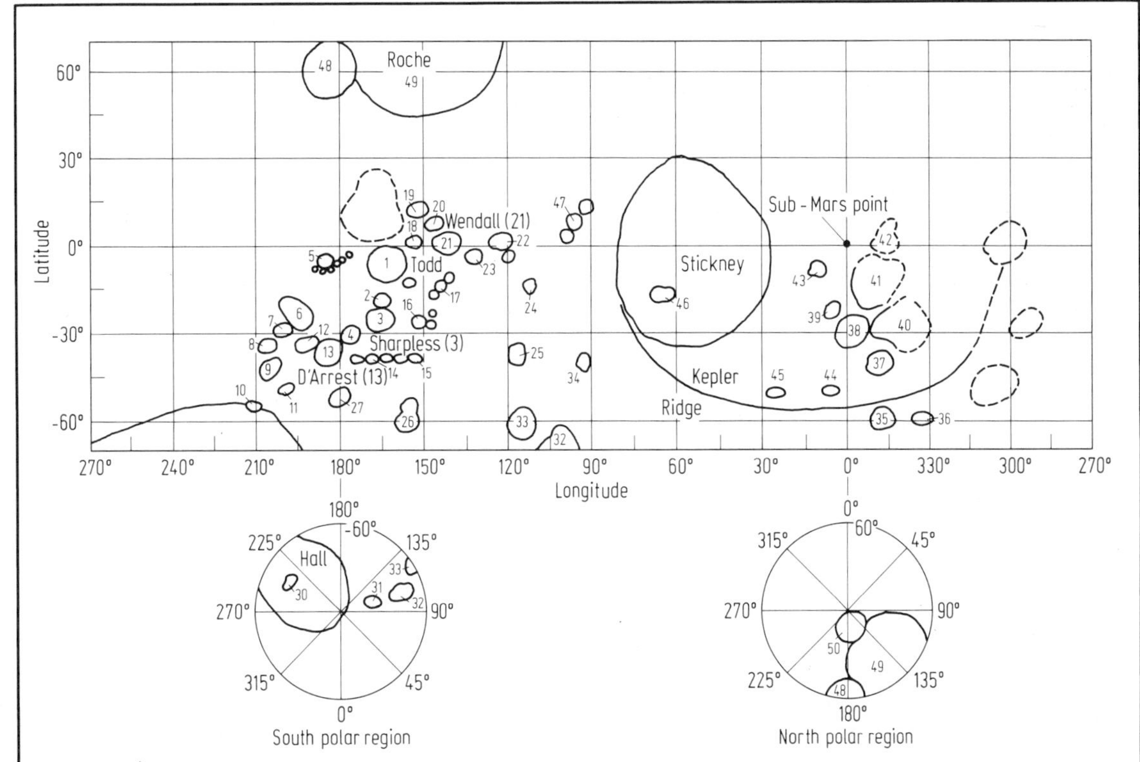

Fig. 14. Map of Phobos from the Mariner 9 observations. From Duxbury [14]. Features are numbered for identification.

3. Satellites of the outer planets

For the synchronously rotating satellites of Jupiter and Saturn, albedo variations are found between the leading sides (in the orbital direction) and the trailing sides (opposite to the orbital direction). Generally, the trailing sides are darker except for Callisto and Iapetus which show the opposite effect. Such brightness variations may be caused by sputtering of the surface by the corotating magnetospheric plasma, bombardment by interplanetary meteoroids, accretion of the impact fragments from the outer satellites or a combination of these processes [11, 34, 52]. (See Figs. 16 and 17.)

The Voyager mission has revealed many new features of the Galilean satellites. The most important, of course, is the discovery of active volcanos on Io [60]. This explains the red, sulfur-rich surface [17] and the production of a sulfur ion nebula in the vicinity of Io's orbit [3, 4, 29]. No classic forms of impact craters can be found over Io's surface; and based on the absence of craters of 1···2 km diameter it has been deduced that Io's surface must be no more than 10^6 years old [60]. Only a few features suggestive of large impacts are found on Europa's surface but a system of intersecting linear features 50···200 km wide and several thousand km long is to be seen.

Ganymede has a more moon-like surface which is covered with diffuse bright spots with rayed impact crater structures. Besides cratered terrain, grooved terrain with systems of shallow grooves which may be of tectonic origin has also been identified. Callisto is the most heavily cratered Galilean satellite. Its surface, like the cratered terrain of Ganymede, may be dated back to the heavy bombardment event of $4 \cdot 10^9$ a ago. There is also a large multi-ring structure reminiscent of the Caloris basin on Mercury and Mare Orientale of the moon. Large impact structures appear to be common but no large topographic relief exists. Presumably this is due to the nature of the icy crust of the satellite.

At small phase angle and opposition, the brightness of planetary bodies may increase to various degrees according to the physical nature of the surface material brightness (the opposition effect). For Ganymede and Europa, the surges at opposition are comparable to what would be expected from a frost-covered satellite. But in the case of Callisto, the dark side (leading side) shows a strong opposition effect which is consistent with the presence of a microscopically porous surface layer on this hemisphere [77]. Infrared observations have established that water ice covers 50···100% of Europa's surface, 20···65% of Ganymede's surface, and 5···25% of Callisto's surface [50, 52]. This, at first sight, is inconsistent with the increasing abundance of water ice from Europa and Callisto implied by the density variation. Consolmagno and Lewis [10] have suggested that this effect may simply be due to the fact that the crusts of the icier satellites should be less differentiated, so that Callisto's surface may still retain a large portion of primordial non-volatile materials: hence the rocky appearance. On the other hand, Pollack et al. [52] have argued that preferential accretion of the impact ejecta from the outer satellites by Callisto plus the degree of water recoating as a function of thickness of the satellite icy crusts may provide the answer.

The Saturnian satellites Tethys, Dione, Rhea, and Iapetus show absorption features attributed to water frost or ice [18]. Iapetus, which displays the largest brightness variation between the leading and trailing sides, is also peculiar in its surface composition in the sense that the leading side may consist mostly of rocky material [27]. For further details, see Figs. 18···20. Though at larger solar distances, the Uranian satellites Titania, Oberon, and the Neptunian satellite Triton have a non-volatile surface composition of a quite different kind according to the 1.2···2.2 μm observations of Cruikshank et al. [13]. For more recent results, see appendix 3.2.2.7.

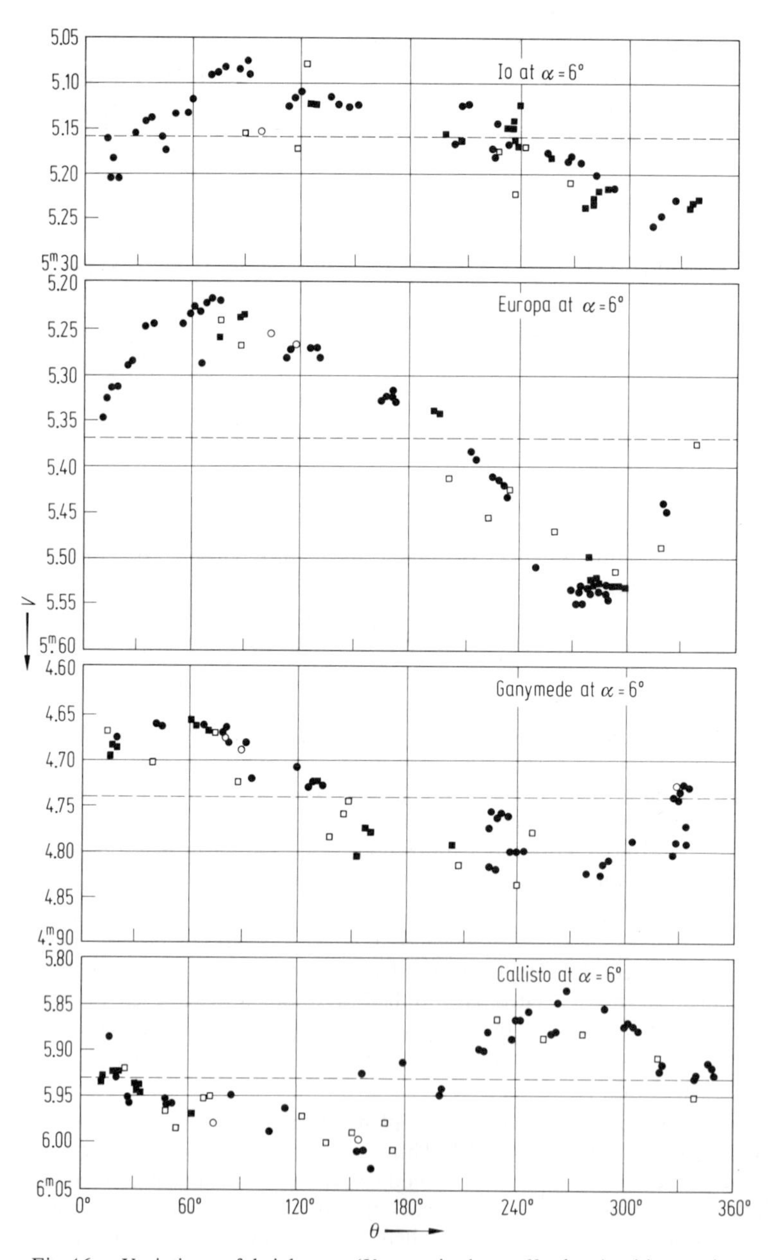

Fig. 16. Variations of brightness (V magnitude at 6° phase) with rotation phase for the Galilean satellites. θ = rotational phase angle. From Morrison and Morrison [37] and Johnson [26].

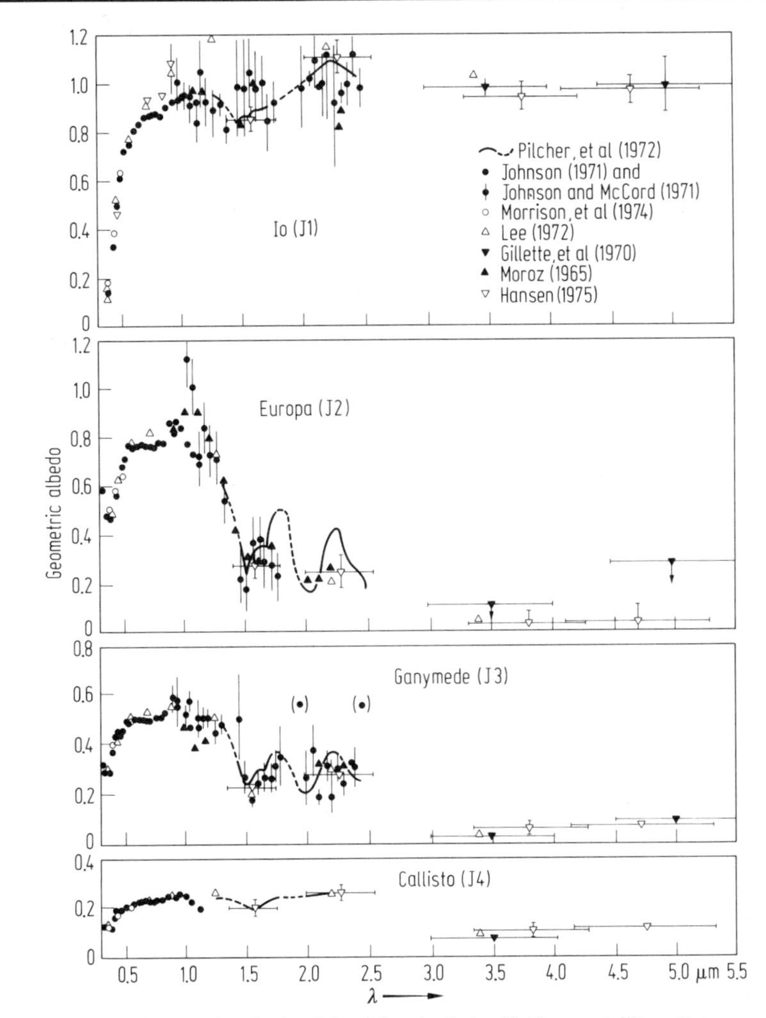

Fig. 17. Geometric albedo (0.3...5.5 μm) of the Galilean satellites. Data are scaled at 0.56 μm except for those of Pilcher et al. [50] (scaled near 1.25 μm) and Moroz [35], scaled to match near 1.0 μm. From Johnson and Pilcher [27].

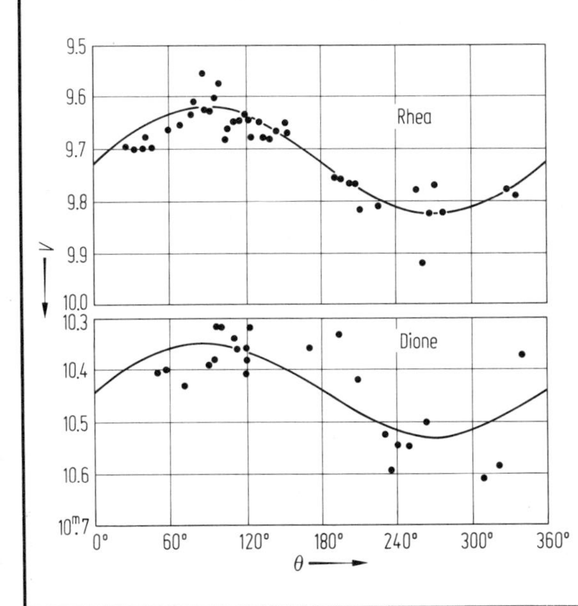

Fig. 18. Orbital photometric variations of Rhea and Dione in the Y band. From Morrison and Cruikshank [36]. V = magnitude of mean opposition, θ = orbital phase.

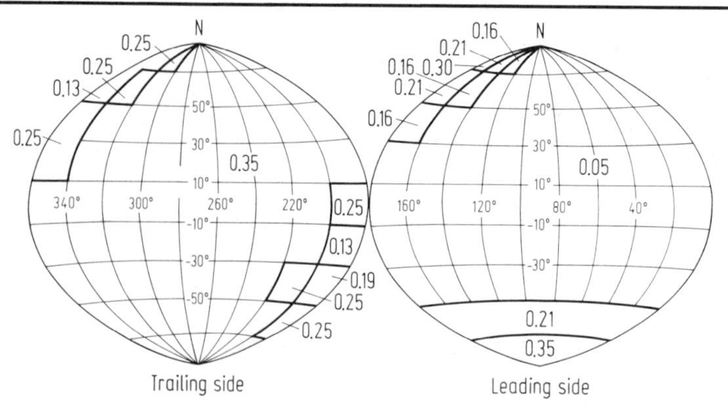

Fig. 19. Model albedo distribution for Iapetus, shown on an Aitoff equal-areas projection. Each area is labelled by its visual geometric albedo. The coordinate system is defined such that the longitude on the satellite is equal to the orbital longitude, measured prograde from superior geocentric conjunction, and the axis of rotation is assumed normal to the satellite's orbital plane. Longitude increases to the east as seen from the earth, and north is at the top. From Morrison et al. [38].

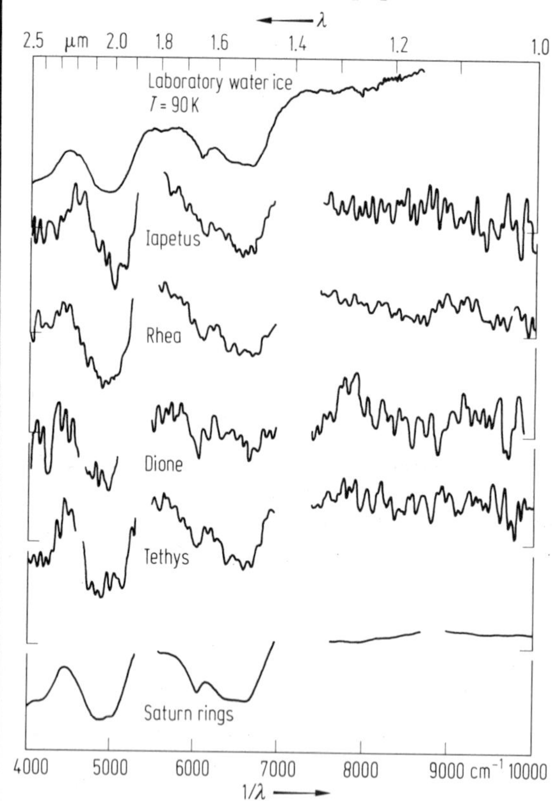

Fig. 20 Spectra of the satellites and rings of Saturn showing absorptions attributed to water frost or ice. From Fink et al. [18].

3.2.2.3.3 References for 3.2.2.3

General reference

a Planetary Satellites (Burns, J.A., ed.), Univ. of Arizona Press (1977).

Special references

1 Allen, C.W.: Astrophysical Quantities, third edition, Univ. of London (1976) p. 144.
2 Baldwin, R.B.: Icarus **23** (1974) 157.
3 Bridge, M.S., Belcher, J.W., Lazarus, A.J., Sullivan, J.D., McNutt, R.L., Begenal, F., Scudder, J.D., Sittler, E.C., Siscoe, G.L., Vasyliunas, V.M., Goertz, C.K., Yeates, C.M.: Science **204** (1979) 987.

4 Broadfoot, A.L., Belton, M.J.S., Takacs, P.Z., Sandel, B.R., Shemansky, D.E., Holberg, J.B., Ajello, J.M., Atreya, S.K., Donahue, T.M., Moos, H.W., Bertaux, J.L., Blamont, J.E., Strobel, D.F., McConnell, J.C., Dalgarno, A., Goody, R., McElroy, M.B.: Science **204** (1979) 979.

5 Brown, W.E., jr., Adams, G.F., Eggleton, R.E., Jackson, P., Jordan, R., Kobrick, M., Peeples, W.J., Phillips, R.J., Porcello, L.J., Schaber, G., Sill, W.R., Thompson, T.W., Ward, S.H., Zelenka, J.S.: Proc. 5th Lunar Sci. Conf. **3** (1974) 3037.

6 Campbell, D.B., Dyce, R.B., Ingalls, R.P., Pettengill, G.H., Shapiro, I.I.: Science **175** (1972) 514.

7 Carr, M.H.: J. Geophys. Res. **78** (1973) 4049.

8 Chapman, C.R.: Icarus **28** (1976) 523.

9 Chapman, C.R., Jones, K.L.: Annu. Rev. Earth Planet. Sci. **5** (1977) 515.

10 Consolmagno, G.J., Lewis, J.S. in: Jupiter (Gehrels, T., ed.), Univ. of Arizona Press (1976) p. 1035.

11 Cook, A.F., Franklin, F.A.: Icarus **13** (1970) 282.

12 Cruikshank, D.P.: Rev. Geophys. Space Physics **17** (1979) 165.

13 Cruikshank, D.P., Pilcher, C.B., Morrison, D.: Astrophys. J. **217** (1977) 1006.

14 Duxbury, T.C. in: [a] p. 346.

15 El-Baz, F.: Annu. Rev. Astron. Astrophys. **12** (1974) 135.

16 El-Baz, F.: Icarus **25** (1975) 495.

17 Fanale, F.P., Johnson, T.V., Matson, D.L. in: [a] p. 379.

18 Fink, U., Larson, H.P., Gautier, T.N., Treffers, R.R.: Astrophvs. J. **207** (1976) L 63.

19 Gault, D.E., Burns, J.A., Cassen, P., Strom, R.G.: Annu. Rev. Astron. Astrophys. **15** (1977) 97.

20 Goldstein, R.M.: Radio Sci. **69** D (1965) 1623.

21 Goldstein, R.M., Rumsey, H.C.: Icarus **17** (1972) 699.

22 Goldstein, R.M., Green, R.R., Rumsey, H.C.: J. Geophys. Res. **81** (1976) 4807.

23 Hartmann, W.K.: Icarus **5** (1966) 565.

24 Hartmann, W.K.: Icarus **24** (1975) 181.

25 Howard, K.A., Wilhelms, D.E., Scott, D.H.: Rev. Geophys. Space Phys. **12** (1974) 309.

26 Johnson, T.V.: Annu. Rev. Earth. Planet. Sci. **6** (1978) 93.

27 Johnson, T.V., Pilcher, C.B. in: [a] p. 264.

28 Kaula, W.M., Schubert, G., Lingenfelter, R.E., Sjogren, W.L., Wollenhaupt, W.T.: Proc. 4th Lunar Sci. Conf. **3** (1973) 2811.

29 Kupo, I., Mekler, Y., Eviatar, A.: Astrophys. J. **205** (1976) L 51.

30 Langevin, Y., Arnold, J.R.: Annu. Rev. Earth. Planet. Sci. **5** (1977) 449.

31 Levinson, A., Taylor, R.: Moon Rocks and Minerals, Pergamon Press, New York (1970) p. 189.

32 Malin, M.C., Saunders, R.S.: Science **196** (1977) 987.

33 Marov, M.Ya.: Annu. Rev. Astron. Astrophys. **16** (1978) 141.

34 Mendis, D.A., Axford, W.I.: Annu. Rev. Earth Planet. Sci. **2** (1974) 419.

35 Moroz, V.I.: Soviet Astron. – Astron. J. **9** (1965) 999.

36 Morrison, D., Cruikshank, D.P.: Space Sci. Rev. **15** (1974) 641.

37 Morrison, D., Morrison, N.D. in: [a] p. 363.

37a Morrison, D., Cruikshank, D.P., Burns, J. in: [a] p. 12.

38 Morrison, D.T., Jones, J., Cruikshank, D.P., Murphy, R.E.: Icarus **24** (1975) 157.

39 Muller, P.M., Sjogren, W.L.: Science **161** (1968) 680.

40 Murray, B.C., Strom, R.G., Trask, U.J., Gault, D.E.: J. Geophys. Res. **80** (1975) 2508.

41 Murray, B.C., Ward, W.R., Yeung, S.C.: Science **180** (1973) 638.

42 Murray, B.C., Soderblom, L.A., Sharp, R.P., Cutts, J.A.: J. Geophys. Res. **76** (1971) 313.

43 Mutch, T.A., Head, J.W.: Rev. Geophys. Space Phys. **13** (1975) 411.

44 Neukum, G., Wise, D.U.: Science **194** (1976) 1381.

44a Neukum, G., Hiller, K.: J. Geophys. Res. **86** (1981) 3097.

45 Öpik, E.J.: Science **153** (1966) 255.

46 Pang, K.D., Pollack, J.B., Veverka, J., Lane, A.J., Ajello, J.M.: Science **199** (1978) 64.

47 Papanastassiou, D.A., Wasserburg, G.J.: Earth Planet. Sci. Lett. **11** (1971) 37.

48 Pettengill, G.H.: Annu. Rev. Astron. Astrophys. **16** (1978) 265.

49 Pettengill, G.H., Ford, P.G., Brown, W.E., Kaula, W.M., Keller, C.H., Masursky, H., McGill, G.E.: Science **203** (1979) 806.

50 Pilcher, C.B., Ridway, S.T., McCord, T.B.: Science **178** (1972) 1087.

51 Pollack, J.B. in: [a] p. 319.

52 Pollack, J.B., Witteborn, F.C., Erickson, E.F., Strecker, D.W., Baldwin, B.J., Bunch, T.E.: Icarus **36** (1978) 271.

53 Pollack, J.B., Veverka, J., Noland, M., Sagan, C., Hartmann, W.K., Duxbury, T.C., Born, G.H., Milton, D.J., Smith, B.A.: Icarus **17** (1972) 394.

54 Quaide, W.L., Oberbeck, V.R.: J. Geophys. Res. **73** (1968) 5247.
55 Rumsey, H.C., Morris, G.A., Green, R.R., Goldstein, R.M.: Icarus **23** (1974) 81.
56 Sagan, C., Toon, O.B., Gierasch, P.J.: Science **181** (1973) 1045.
57 Schultz, P.H., Gault, D.E.: The Moon **12** (1975) 159.
58 Short, N.M.: Planetary Geology, Prentice-Hall, Inc., New Jersey (1975) p. 188.
59 Sjogren, W.L. in: Short, N.M.: Plantary Geolgy Prentice-Hall, Inc., New Jersey (1975) p. 188.
60 Smith, B.A., Soderblom, L.A., Johnson, T.V., Ingersoll, A.P., Collins, S.A., Shoemaker, E.M., Hunt, G.E., Masursky, H., Carr, M.H., Davies, M.E., Cook II, A.F., Boyce, J., Danielson, G.E., Owen, T., Sagan, C., Beebe, R.F., Veverka, J., Strom, R.G., McCauley, J.F., Morrison, D., Briggs, G.A., Suomi, V.E.: Science **204** (1979) 951.
61 Soderblom, L.A., Condit, C.D., West, R.A., Herman, B.M., Kreidler, T.J.: Icarus **22** (1974) 239.
62 Solomon, S.C.: Icarus, **28** (1976) 509.
63 Strom, R., Trask, N.J., Guest, J.E.: J. Geophys. Res. **80** (1975) 2478.
64 Surkov, Yu.A.: Proc. 8th Lunar Sci. Conf. (1977) 2665.
65 Taylor, S.R.: Lunar Science: A post-Apollo View, Pergamon Press, New York (1975).
66 Tera, F., Papanastassiou, D.A., Wasserburg, G.J.: Lunar Science IV, Lunar Science Institute, Houston, Texas (1973) p. 723.
67 Toksöz, M.N.: Annu. Rev. Earth Planet. Sci. **2** (1974) 151.
68 Tolson, R.H., Duxbury, T.C., Born, G.H., Christensen, E.J., Diehl, R.E., Farless, D., Hildebrand, C.E., Mitchell, R.T., Molko, P.M., Morabito, L.A., Palluconi, F.D., Reichert, R.J., Taraji, H., Veverka, J., Neugebauer, G., Findlay, J.T.: Science **199** (1978) 62.
69 Trask, N.J., Guest, J.E.: J. Geophys. Res. **80** (1975) 2461.
70 Veverka, J. in: [a] p. 171.
71 Veverka, J. in: [a] p. 210.
72 Veverka, J.: Vistas in Astronomy **22** (1978) 163.
73 Veverka, J., Duxbury, T.: J. Geophys. Res. **82** (1977) 4213.
74 Walker, R.M.: Annu. Rev. Earth Planet. Sci. **3** (1975) 99.
75 Ward, W.R.: Science **181** (1973) 620.
76 Wetherill, G.W.: Proc. Soviet-Amer. Conf. Cosmochem. of the Moon and Planets, June 4–8, 1974, Nauka, Moscow (1975) p. 411–424 (also NASA SP-370, 1977, p. 553).
77 Wetherill, G.W.: Icarus **28** (1976) 537.

3.2.2.4 Temperatures of the planets and satellites

For details, see also 3.2.2.5.

Table 5. Temperatures of the planets.
T_{eff} = effective temperature
$E_{int}/E_{\odot}$ = ratio of internal heat source to absorbed solar radiation

Planet	T_{eff} [K]		$E_{int}/E_{\odot}$	Ref.
Mercury	600 (subsolar point)		–	1
	100 (dark side)			
Venus	240 (cloud top)		–	
Earth	295 (subsolar point)		–	
	280 (dark side)			
Mars	250		–	
Jupiter	125 ±3	(105) *)	1.9±0.2	9
Saturn	90 ±5			18
	97 ±4	(71)	≈3.5	2
Uranus	58 ±2			17
	58 ±3			4
	60 ±4	(57)	≈0	15
Neptune	55.5±2.3	(45)	$2.4^{+1.3}_{-0.9}$	12
	57 ±4			15
Pluto	–	(42)	–	–

*) Bracketed values are predicted temperatures for the outer planets from Owen [16].

Table 6. Estimated temperatures of satellites [10].

A = visual albedo (Bond albedo)

T_{max} = maximum temperature for the subsolar point of a slowly rotating satellite in [K] $= 394.0\,(1-A)^{1/4} r^{-1/2}$ with r in [AU]

T_{av} = average temperature of a rapidly rotating sphere $= T_{max}/\sqrt{2}$

Satellite		A	T_{max} [K]	T_{av} [K]
	Moon	0.067	387	274
M1	Phobos	(0.3)	(292)	(207)
M2	Deimos	(0.3)	(292)	(207)
J5	Amalthea	(0.5)	(145)	(103)
J1	Io	0.54	143	101
J2	Europa	0.73	125	88
J3	Ganymede	0.34	156	110
J4	Callisto	0.15	166	117
S1	Mimas	(0.7)	(94)	(67)
S2	Enceladus	(0.7)	(94)	(67)
S3	Tethys	0.77	88	62
S4	Dione	0.66	98	69
S5	Rhea	0.30	117	83
S6	Titan	0.24	119	84
S7	Hyperion	–	–	–
S8	Iapetus	0.15	122	87
U1	Ariel	(0.7)	(67)	(47)
U2	Umbriel	(0.7)	(67)	(47)
U3	Titania	(0.7)	(67)	(47)
U4	Oberon	(0.7)	(67)	(47)

Table 7. Measured brightness temperatures T_B of satellites.

λ = wavelength; $\Delta\lambda$ = bandwidth

Satellites		λ [μm]	$\Delta\lambda$ [μm]	T_B [K]	Ref.
M1	Phobos (10.9 km)	10.2	4.3	296±15	5
J1	Io (1820 km)	8.4	0.8	149± 3	7
		11.0	5.0	138± 4	13
		21.0	10.0	128± 5	14
J2	Europa (1500 km)	8.4	0.8	134± 3	7
		11.0	5.0	129± 4	13
		21.0	10.0	121± 5	14
J3	Ganymede (2635 km)	8.4	0.8	145± 3	7
		11.0	5.0	145± 4	13
		21.0	10.0	138± 5	14
		28 mm		55±14	17
J4	Callisto (2500 km)	8.4	0.8	160± 3	7
		11.0	5.0	153± 5	13
		21.0	10.0	151± 7	14
		28 mm		88±18	17
S6	Titan (2900 km)	8.4	0.8	143± 5	6
		11.0	2.0	131± 2	6
		21.0	10.0	91± 2	14
		3 mm		200	19
		37 mm		103	3

References for 3.2.2.4

1 Allen, C.W.: Astrophysical Quantities (third edition), The Athlone Press (1976) p. 149.
2 Aumann, H.H., Gillespie, C.M., jr., Low, F.J.: Astrophys. J. **157** (1969) L69.
3 Briggs, F.H.: Icarus **22** (1973) 48.
4 Fazio, G.G., Traub, W.A., Wright, E.L., Low, F.J., Trafton, L.: Astrophys. J. **209** (1976) 633.
5 Gatley, I., Kieffer, H., Miner, E., Neugebauer, G.: Astrophys. J. **190** (1974) 497.
6 Gillett, F.C., Forrest, W.J., Merrill, K.M.: Astrophys. J. **184** (1973) L 93.
7 Gillett, F.C., Merrill, K.M., Stein, W.A.: Astrophys. Lett. **6** (1970) 247.
9 Ingersoll, A.P., Münch, G., Neugebauer, G., Orton, G.S. in: Jupiter (Gehrels, T., ed.), Univ. of Arizona Press (1976) p. 197.
10 Kuiper, G.P.: Landolt-Börnstein, NS, Vol. VI/1 (1965) p. 171.
11 Loewenstein, R.F., Harper, D.A., Moseley, S.H., Telesco, C.M., Thronson, H.A., jr., Hildebrand, R.H., Whitcomb, S.E., Winston, R., Stiening, R.F.: Icarus **31** (1977) 315.
12 Loewenstein, R.F., Harper, D.A., Moseley, S.H.: Astrophys. J. **218** (1977) L 145.
13 Morrison, D. in: Planetary Satellites (Burns, J.A., ed.), Univ. of Arizona Press. (1976) p. 272.
14 Morrison, D., Cruikshank, D.P., Murphy, R.E.: Astrophys. J. **173** (1972) L 143.
15 Murphy, R.E., Trafton, L.M.: Astrophys. J. **193** (1974) 253.
16 Owen, T.: Rev. Geophys. Space Sci. **13** (1975) 416.
17 Pauliny-Toth, I.K., Witzel, A., Gorgolewski, S.: Astron. Astrophys. **34** (1974) 129.
18 Tokunaga, A.T. in: The Saturn System (Hunter, D.M., Morrison, D., eds.) NASA Conf. Publ. 2068 (1978) p. 53.
19 Ulich, B.L., Conklin, E.K., Dickel, J.R. quoted in: The Saturn System (Hunten, D.M., Morrison, D., eds.), NASA Conf. Publ. 2068 (1978) p. 128.

3.2.2.5 Atmospheres

3.2.2.5.1 Mercury

The atmosphere of Mercury is very thin. An upper limit of 10^6 molecules per cm^3 for the dayside surface density has been inferred by the occultation measurement of Mariner 10 [32]. For a surface temperature of 500 K this would mean the corresponding surface pressure is not more than 10^{-10} mbar. Such an exosphere, if it exists, may be sustained by accretion of solar wind ions or radiogenic outgassing with ^4He, ^{40}Ar, and ^{20}Ne as the possible major constituents [2, 38, 54]. Due to lateral transport of the atmospheric constituents between the dayside and the nightside, the surface density would have a $T^{-5/2}$ dependence on the surface temperature [41], and the corresponding dayside surface density of He has been estimated to be $1.3 \cdot 10^3\ cm^{-3}$ as compared with $2.6 \cdot 10^5\ cm^{-3}$ for the nightside by Hartle et al [38].

3.2.2.5.2 Venus

In contrast to Mercury, Venus has a thick CO_2-rich atmosphere displaying cloud patterns of long duration in the UV [25]. These planetary-scale markings have the shape of a dark horizontal Y or V with brightness contrast estimated to be about 20%. The equatorially symmetric features move with a 4 day retrograde rotation period and a propagation speed of approximately 100 m s^{-1} in the equatorial region [8, 25, 72, 113]. Ground-based differential Doppler shift measurements [111] and the Venera 8 Doppler drift experiment [64] have established that this rapid retrograde zonal wind involves actual mass motion of the Venus atmosphere and not just some sort of planetary-scale wave phenomenon. Even though Venus itself has a very slow rotation period of 243 days, there is no indication from infrared observations of a sharp temperature transition across the day-night line [26, 73]. This suggests that the temperature may be equalized by large-scale atmospheric circulation such as the 4-day rotation at altitudes below the cloud tops. The pressure level of the visible cloud tops is estimated to be $\approx$100 mbar. The temperature at this altitude ($\approx$65 km) is close to 240 K [122] but it increases rapidly to about 750 K at the surface as indicated by both infrared observations [3] and various Venus probe measurements [31, 63]. A greenhouse effect due to blocking of re-emission of the absorbed solar radiation by the thick layer of CO_2 and H_2O has been proposed to explain this elevated surface temperature [84, 92]. Because of the high surface pressure and temperature, a chemical equilibrium between the atmosphere and the polar lithosphere may be established modifying the surface mineralogy and yielding traces of hydrochloric acid (HCl) and hydrofluoric acid (HF) in the lower atmosphere [22, 59, 115]. One important product of such an interaction would be sulfuric acid H_2SO_4. Based on polarimetric and spectroscopic measurements the cloud particles have been identified to be micron-sized liquid droplets of H_2SO_4 [86, 95, 122]. This identification has been essentially confirmed by the Venera and Pioneer Venus data [53, 63], but solid sulfur particles of 10-micron size may also exist between 45 and 60 km altitude. Since H_2SO_4 and its dissociation product SO_2 are good absorbers at infrared wavelengths, this sulfur cloud may contribute significantly to the greenhouse effect. The infrared instrument on board the Pioneer Venus orbiter spacecraft has discovered extremely localized and warm features at the north pole with a peak temperature of 260 K [102, 103]. Surrounding these hot spots is a collar of high and cool atmosphere indicating vortex-type circulation at the poles [71, 101]. Also, the meridional circulation pattern appears to be composed of three cells (stratospheric cell, driving cell, and surface layer cell) with the middle cell in the clouds of solar radiation absorbing gases and particulates driving the circulation [94].

3.2.2.5.3 Mars

As with Venus, the atmosphere of Mars is rich in CO_2 but the corresponding surface pressure is a factor of 10^4 smaller and there is no dense cloud layer covering the planetary surface except those due to occasional atmospheric activity (i.e. dust storms). The average temperature at the surface is so low that most of the H_2O would be trapped in the polar region or under the surface forming a permafrost [27]. Only the polar caps are cold enough to freeze carbon dioxide (CO_2) and suggestions have been made that pressure of the Martian CO_2 atmosphere is maintained by vapour equilibrium with the polar ice caps [58]. This model is supported by the facts that the minimum temperature of the south pole has been measured to be 150 K [74] which is very near the equilibrium temperature for an average CO_2 partial pressure of 6.5 mbar [52] and that the southern polar cap shows absorption features of solid CO_2 in infrared [39]. Murray and Malin [71] have further identified the northern ice cap to be the permanent CO_2 reservoir. The exact nature of the polar caps and the surface-atmosphere exchange is, however, still a matter of debate [47, 68]. In addition to CO_2, the water vapour is also controlled by conditions at the polar caps as indicated by the seasonal variation of its partial pressure. For example, the Mariner 9 interferometer spectrometer experiment has shown that during summer in the southern hemisphere there is between 10 and 20 μm of precipitable water over the summer polar region and the subsolar region but that water vapour was below 10 μm over the northern winter polar region. During warm periods in Mars large quantities of water may be released resulting in surface flooding. At the present time, Mars may be in the middle of an ice age with most of the water trapped underground.

The isotope ratios of carbon, oxygen, nitrogen, argon, and xenon in the atmospheres of the terrestrial planets are listed in Table 9. While the values of $^{12}C/^{13}C$ and $^{16}O/^{18}O$ are quite similar to each other, there seems to be a gradual enrichment of ^{40}Ar relative to ^{36}Ar from Venus to Mars. Such a difference, if established by further measurements, may be the result of different processes involved in the formation of these planets or evolution of their atmospheres.

Table 8. Surface atmospheric composition of the terrestrial planets from in-situ measurements.

Composition	Venus *)	Earth	Mars **)
CO_2	96.4%	0.03%	95.32%
N_2	3.4%	78.08%	2.7%
CO	(20±3) ppm	1 ppm	0.07%
O_2	(69.3±1.27) ppm	20.95%	0.13%
H_2O	≈ 0.1%	0.1···2.8%	0.03% +)
Ar	67 ppm	0.93%	1.6%
Ne	$4.3\pm^{6}_{4}$ ppm	18.2 ppm	2.5 ppm
Kr	–	1.14 ppm	0.3 ppm
Xe	–	0.087 ppm	0.08 ppm
Surface pressure	90 bar	1 bar	6.5 mbar

*) Measurements made at 24 km altitude by the Pioneer Venus sounder probe [82].
**) Measurements by the molecular analysis experiments on the Viking landers [81].
+) Variable

Table 9. Isotope ratios in the atmospheres of the terrestrial planets.

Ratio	Venus [75, 116]	Earth [81]	Mars [81]
$^{12}C/^{13}C$	≈ 90	89	90 ± 9
$^{16}O/^{18}O$	≈ 500	499	500 ± 50
$^{14}N/^{15}N$	–	277	165 ± 16
$^{20}Ne/^{22}Ne$	≈ 14	9.8	10 ± 3
$^{40}Ar/^{36}Ar$	≈ 1	292	3000 ±600
$^{129}Xe/^{132}Xe$	–	0.97	2.5 ± 0.5

3.2.2.5.4 The outer planets

The upper atmosphere of Jupiter is dominated by the zonal structures of various shades of colour: red, brown, yellow, blue, etc. The chemistry leading to such coloration is not yet clear at this moment [96]. Because the Jovian atmosphere may be contaminated by the sulfur from Io, the free radicals CH_3S (yellow), $(CH_3)_3CS$ (red), and S_2 (purple) may be important in acting as chromophores. Weidenschilling and Lewis [118] have constructed a thermodynamic equilibrium model for the Jovian atmosphere with solar composition. They predicted the formation of three cloud layers of NH_3, NH_4SH, and H_2O at $T \approx 150$ K, 200 K, and 270 K, respectively. The tri-modal distribution of 5 μm emission temperatures and the variations of the colour of cloud patterns with emission temperatures [104] provide evidence for such a multi-layered cloud structure. The latitudinal structures of dark belts and light zones with rapid motion (rotational speed ≈ 100 m s^{-1} in the case of the westerly equatorial jet near the equator) have been investigated by the Pioneer polarimetric experiment [20] and the cloud tops corresponding to zones are generally a few kilometers higher than those of the belts. The Great Red Spot is found to have a cloud top a few kilometers above the surrounding top of the South Tropical Zone [20, 34]. The Great Red Spot is now believed to be an anticyclonic (high pressure) meteorological system, although the details of its formation remain to be investigated [48, 65, 97]. Besides the Great Red Spot, other zones also appear to have red spots of smaller size. It may be inferred that the red spots and the white zones are regions of active convection and rising motion, and the brown-blue belts are regions of cloud dispersal and sinking motion [48].

In the Jovian atmosphere the thermal inversion at 160 km altitude has been suggested to be caused by absorption of solar radiation by a layer of particulate absorbers (Axel [1], Tomasko [106]). This "Axel" dust is supposedly the photolysis products of CH_4. In other outer planets the presence of a layer of haze or aerosols may be also important for the thermal structure of the upper atmospheres. For example, in the Saturnian atmosphere model which is constrained by data from infrared observations [13] an optically thick NH_3 cloud is capped by an ammonia haze with a mixture of hydrogen. The properties of such dust layers have been considered by Podolak and Danielson [83]. Structures of Jupiter's and Saturn's atmospheres are given in Fig. 21 and 22. For more recent results, see appendix 3.2.2.7 .

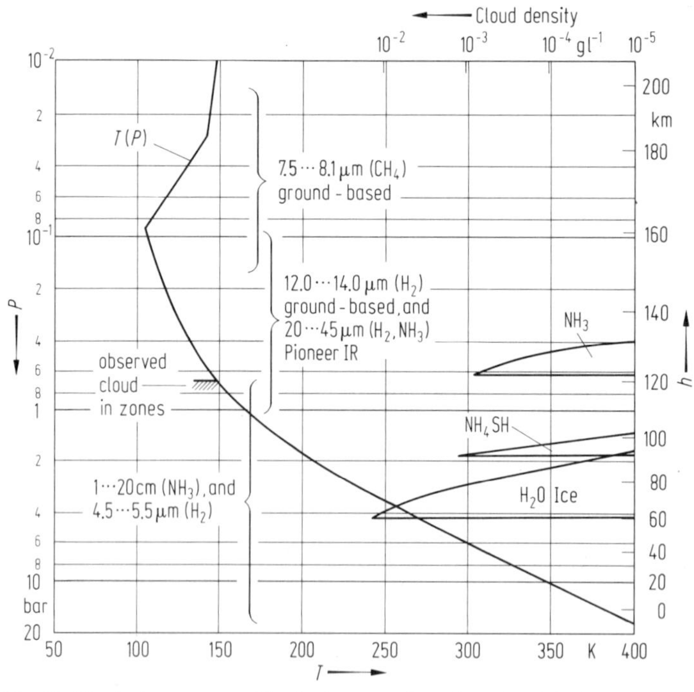

Fig. 21. Vertical structure of Jupiter's atmosphere from $P=0.01$ bar to $P=20$ bar. The temperature profile for $P \leqq 1.0$ bar is from Orton [77, 78]. The temperature profile at $P > 0.7$ bar is assumed to be adiabatic, consistent with radio emissions at 1...20 cm wavelength (Gulkis and Poynter [36]). Cloud densities are from the chemical equilibrium model of Weidenschilling and Lewis [118] for a solar composition atmosphere with the temperature profile shown. From Ingersoll [47]. h=altitude.

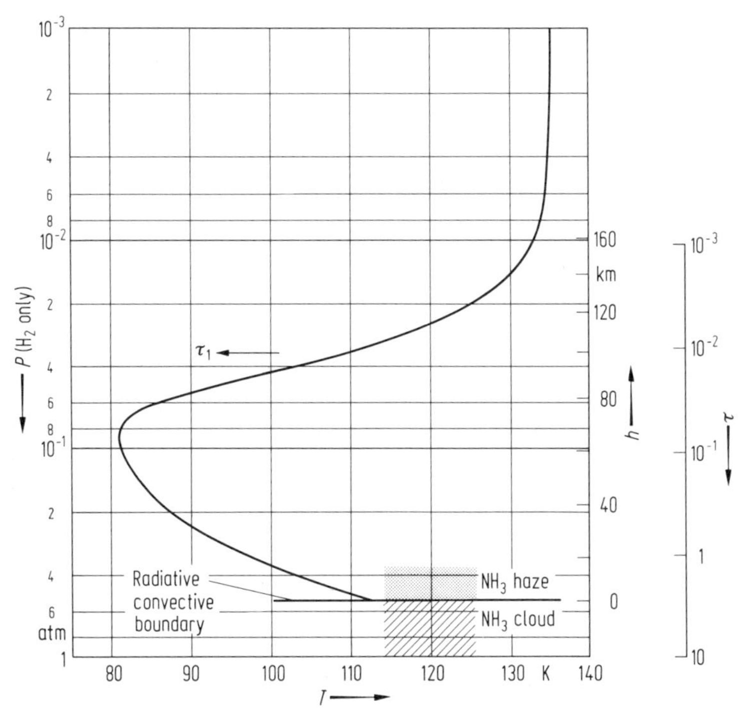

Fig. 22. The theoretical variations of temperature, T, with pressure, P, altitude, h, and Rosseland mean optical depth, τ, in the radiative part of the atmosphere of Saturn. Calculations stop at the radiative-convective boundary. τ_1 is the characteristic optical depth for energy deposition caused by absorption of solar ultraviolet light by Axel dust. The vertical extent of the regions of the ammonia haze, which produces the absorption features at 9.5 μm, and the region of thick clouds in the convective region is indicated. Note that the left ordinate refers to the partial pressure of H_2 only. From Caldwell [12].

The atmospheres of Uranus and Neptune are rich in CH_4 and there are efforts continually being made to obtain a more accurate value for the CH_4/H_2 ratio. The current value varies from 1 to 3 times the solar composition value [7] to one order of magnitude higher [19]. Prinn and Lewis [87] and Danielson [23] have considered the structure of the Uranian atmosphere and they have found the formation of a cloud top of CH_4 droplets to be an important feature in such atmospheres. For a solar composition atmosphere the CH_4 cloud is formed at 50 K and 0.4 bar level [87], whereas an CH_4-enrichment model gives 90 K and 4 bar level at the cloud top [23]. Observations of Uranus and Neptune in the 5···22 μm spectral region have indicated a strong thermal inversion in the atmosphere of Neptune but not of Uranus [35]. This difference may be caused by a number of reasons (internal heat sources, chemical composition, seasonal changes, etc.); the exact interpretation remains unclear, however [35, 44].

Table 10: see next page.

Table 11. Abundances in the outer solar system (from Owen [80]).

Planet	Column abundance			Relative abundance	
	H_2 km atm	NH_3 m atm	CH_4 m atm	H/C	D/H *)
Jupiter	75±15	13±3	50±15	3000±300	$(5.1\pm0.7)\cdot10^{-5}$
Saturn	75±20	2±1	60±12	2500±400	$(5.5\pm2.9)\cdot10^{-5}$
Uranus	225±75	< 2.5	$\gtrsim 6\cdot10^3$	< 100	$(3.0\pm1.2)\cdot10^{-5}$
Neptune	225±75	–	$\gtrsim 6\cdot10^3$	< 100	–
Pluto	–	<10	< 2?	–	–

*) From Macy and Smith [61].

Table 10. Composition of the Jovian atmosphere (from Ridgway et al. [91]).
λ = region where detected.

Gas	Abundance relative to H_2	λ µm	Reference to first detection
1. Identification firm			
H_2	1	0.8	51
HD	$2 \cdot 10^{-5}$	0.7	110
He	0.05···0.15	–	79, 43, 17
CH_4	$7 \cdot 10^{-4}$	0.7	120
CH_3D	$3 \cdot 10^{-7}$	5.0	5, 6
$^{13}CH_4$	$^{12}C/^{13}C \approx 110 \pm 35$	1.1	33
NH_3	$2 \cdot 10^{-4}$	0.7	120
H_2O	$1 \cdot 10^{-6}$	5.0	57
C_2H_6	$4 \cdot 10^{-4}$	10.0	89
2. Recent reports *)			
CO	$2 \cdot 10^{-9}$	5.0	4
GeH_4	$7 \cdot 10^{-10}$	5.0	56
HCN	$1 \cdot 10^{-7}$	5.0	56
C_2H_2 **)	$8 \cdot 10^{-5}$	10.0	89
PH_3	$4 \cdot 10^{-7}$	10.0	90

*) The Voyager 1 infrared spectroscopy and radiometry observations show clear evidence of H_2, CH_4, C_2H_2, C_2H_6, CH_3D, NH_3, PH_3, H_2O, and GeH_4. The He/H_2 ratio is determined to be 0.11 ± 0.03 (Hanel et al. [37]).

**) Recent work by Tokunaga et al. [105] indicates lower abundances. Orton and Aumann [78] give a mixing ratio of $5 \cdot 10^{-8} \cdots 6 \cdot 10^{-9}$.

Table 12. Composition of the Saturnian atmosphere.

Molecule	Abundance relative to H_2	Ref.
HD	$(6.6 \pm 3.1) \cdot 10^{-5}$	98
	$(5.1 \pm 0.7) \cdot 10^{-5}$	112
He	0.05···0.15	(solar abundance and values derived for Jupiter)
CH_4	$(3.5 \cdots 3.9) \cdot 10^{-3}$	83
	$2.1 \cdot 10^{-3}$	60, 30
$^{13}CH_4$	$^{12}C/^{13}C = 89^{+25}_{-18}$	21
NH_3	$2.6 \cdot 10^{-5}$	121
C_2H_6	$1.8 \cdot 10^{-6}$	12
H_2S	$1.4 \cdot 10^{-8}$	12
PH_3	–	30
CH_3D	–	30
C_2H_2	–	69

3.2.2.5.5 The moon

Like in the case of Mercury, solar wind hydrogen, helium, and argon are important sources of the lunar atmosphere. Theoretical models of the helium concentration [40] are consistent with the measured values of $4 \cdot 10^4$ cm^{-3} on the nightside and a factor of 20 lower on the dayside as observed by the Apollo 17 lunar surface mass spectrometer [42] if the solar wind influx is the main supplier of the helium. In the case of ^{40}Ar the density is $1.6 \cdot 10^3$ cm^{-3} on the dayside and $4 \cdot 10^4$ cm^{-3} on the nightside (and $1.3 \cdot 10^2$, $4 \cdot 10^4$, respectively, in the case of ^{36}Ar; see Kumar [54]). At the same time, atomic hydrogen, which may be directly converted from the solar wind protons via thermal accomodation on the lunar surface, was found to have surface density much lower than the predicted value of 10 cm^{-3}. Hodges [40] has argued that the hydrogen may have recombined to form H_2, forming an H_2 atmosphere satisfying the lower limits for the dayside minimum of $6 \cdot 10^3$ cm^{-3} [29] and nightside maximum of $3.5 \cdot 10^4$ cm^{-3} [42].

3.2.2.5.6 The Galilean satellites

Upper limits of the surface pressures have been derived for Io and Ganymede from stellar occultations. The lower limit of 10^{-3} mbar as given by Carlson et al. [18] for Ganymede has prompted Yung and McElroy [123] to construct a model atmosphere with O_2 (from photolysis of H_2O) as the main constituent. The recent Voyager 1 UV measurement [9], however, has indicated a surface pressure much lower than that estimated by Carlson et al. The surface pressure of Io has been evaluated to be no more than 10^{-4} mbar [100] and a number of atmospheric models including NH_3, CH_4, and N_2 as the main atmospheric constituents [67, 117, 119] have been proposed. Ground based observations have discovered that Io emits extensive clouds of sodium [10, 11], potassium [109], and sulfur ions [55] in the vicinity of its orbit. Fanale et al. [28] have suggested that Io's surface may be covered by a layer of evaporated salts rich in sodium and sulfur such that sputtering of the surface material by energetic magnetospheric particles would generate a large escaping flux of Na and S atoms [62]. After the Voyager 1 flyby of Jupiter it seems clear that the atmosphere of Io is supplied by the volcanic outgassing and SO_2 may be the most abundant gas [100]. In these circumstances the surface pressure of Io may be determined by vapour equilibrium of SO_2 and its atmosphere could be thick near the subsolar point and the sites of volcanos (surface pressure $\gtrsim 10^{-3}$ mbar). However, in the dark side where the temperature drops to about 90 K the atmosphere should be exospheric in nature [49].

The sulfur escape flux from the surface of Io has been estimated to be about 10^{10} cm^{-2} s^{-1} [9] and that of sodium about $10^6 \cdots 10^7$ cm^{-2} s^{-1} [28]. During the Pioneer 10 spacecraft flyby of Jupiter the UV photometer data indicated the presence of a torus of neutral hydrogen extending 120° around Io's orbit [16, 50]. If the observed emission was indeed Lyman α, the total population of the cloud was about $2 \cdot 10^{33}$ atoms and the total production rate $2 \cdot 10^{28}$ atoms s^{-1}. The Voyager 1 UV instrument has not detected such a hydrogen torus during its encounter with Jupiter.

3.2.2.5.7 Titan

Titan is the only satellite with a substantial atmosphere, and visible observations have indicated the presence of an optically thick layer of aerosols (or blue absorbing Axel dust) [24, 83, 88]. The CH_4 column abundance is estimated to be 2 km atm [108], and on the basis of absorption features at the (3–0) S 1 quadrupole line region Trafton [107] has derived a column abundance of 5 km atm for H_2. But Münch et al. [70] have recently derived an upper limit of 1 km atm. Hence there may be still some question as to whether Titan indeed has an extensive H_2 atmosphere. In addition to this problem, the structure of the Titan atmosphere as a whole remains somewhat uncertain. Low surface pressure models assuming that the surface of Titan is CH_4 ice and that the atmosphere is in vapour equilibrium with the surface have been proposed by Danielson et al. [24] and Caldwell [14, 15]. The column abundance of 2 km atm hence fixes the surface temperature to be about 80 K and the surface pressure to be 20 mbar. On the other hand, models with thicker atmospheres and higher surface temperatures have been proposed [45, 46, 85]. A recent model by Hunten [46] which assumes the surface temperature to be 200 K [114] and the surface pressure to be 20 bar describes the Titan atmosphere as mainly N_2 (from photolysis of NH_3), 0.25% CH_4, and 0.5% H_2. In this case CH_4 clouds will form at about 100 km altitude and temperature of about 80 K, and the H_2 escape flux is estimated to be about $9 \cdot 10^9$ cm^{-2} s^{-1} [45]. It is thus possible that a torus of neutral hydrogen may form in the vicinity of the orbit of Titan as postulated by McDonough and Brice [66]. For more recent results, see appendix 3.2.2.7.2.

3.2.2.5.8 References for 3.2.2.5

General references

a Planetary Satellites (Burns, J.A., ed.), Univ. of Arizona Press (1977).
b Jupiter (Gehrels, T., ed.), Univ. of Arizona Press (1976).
c The Saturn System (Hunten, D.M., Morrison, D., eds.), NASA Conf. Publ. 2068 (1978).

Special references

1 Axel, L.: Astrophys. J. **173** (1972) 451.
2 Banks, P.M., Johnson, H.E., Axford, W.I.: Comm. Astrophys. Space Phys. **2** (1970) 214.
3 Bareth, F.T., Barrett, A.H., Copeland, I., Jones, D.C., Lilley, A.E.: Science **139** (1963) 908.
4 Beer, R.: Astrophys. J. **200** (1975) L 167.
5 Beer, R., Taylor, F.W.: Astrophys. J. **179** (1973) 309.
6 Beer, R., Taylor, F.W.: Astrophys. J. **182** (1973) L131.
7 Belton, M.J.S., Hayes, S.H.: Icarus **24** (1975) 348.
8 Belton, M.J.S., Smith, G.R.: J. Atmosph. Sci. **33** (1976) 1394.
9 Broadfoot, A.L., Belton, M.J.S., Takacs, P.Z., Sandel, B.R., Shemansky, D.E., Holberg, J.B., Ajello, J.M., Atreya, S.K., Donahue, T.M., Moos, H.W., Bertaux, J.L., Blamont, J.E., Strobel, D.F., McConnell, J.C., Dalgarno, A., Goody, R., McElroy, M.B.: Science **204** (1979) 979.
10 Brown, R.A. in: Exploration of the Planetary System (Woszczyk, A., Iwaniszewska, C., eds.), D. Reidel Publ. Co., Dordrecht, Holland (1974) p. 527.
11 Brown, R.A., Yung, Y.L. in: [b] p. 1102.
12 Caldwell, J.: Icarus **30** (1977) 493.
13 Caldwell, J.: Icarus **32** (1977) 190.
14 Caldwell, J. in: [a] p. 438.
15 Caldwell, J. in: [c] p. 113.
16 Carlson, R.W., Judge, D.L.: J. Geophys. Res. **79** (1974) 3623.
17 Carlson, R.W., Judge, D.L. in: [b] p. 418.
18 Carlson, R.W., Bhattacharyya, J.C., Smith, B.A., Johnson, T.V., Hidayat, B., Smith, S.A., Taylor, G.E., O'Leary, B., Brinkmann, R.T.: Science **182** (1973) 53.
19 Cess, R.D., Owen, T.: Astrophys. J. **197** (1975) 37.

20 Coffeen, D.L.: J. Geophys. Res. **79** (1974) 3645.
21 Combes, M., Maillard, J., De Bergh, C.: Astron. Astrophys. **61** (1977) 531.
22 Connes, P., Connes, J., Benedict, W.S., Kaplan, L.D.: Astrophys. J. **147** (1967) 1230.
23 Danielson, R.E.: Icarus **30** (1977) 462.
24 Danielson, R.E., Caldwell, J., Larach, D.R.: Icarus **20** (1973) 437.
25 Dollfus, A.: J. Atmosph. Sci. **32** (1975) 1060.
26 Drake, F.D.: Publ. Natl. Radio Astron. Obs. **1** (1962) 165.
27 Fanale, F.P., Cannon, W.A.: J. Geophys. Res. **79** (1974) 3397.
28 Fanale, F.P., Johnson, T.V., Matson, D.L. in: [a] p. 379.
29 Fastie, W.G., Feldman, P.D., Henry, R.C., Moos, H.W., Barth, C.A., Thomas, G.E., Donahue, T.M.: Science **182** (1973) 710.
30 Fink, U., Larson, H.: Bull. American Astron. Soc. **9** (1977) 535.
31 Fjeldbo, G., Kliore, A.J., Eshleman, R.: Astron. J. **76** (1971) 123.
32 Fjeldbo, G., Kliore, A.J., Sweetnam, D., Esposito, P., Seidel, B.: Icarus, **29** (1976) 439.
33 Fox, K., Owen, T., Mantz, A.W., Rao, K.R.: Astrophys. J. **176** (1972) L81.
34 Gehrels, T. in: [b] p. 531.
35 Gillett, F.C., Rieke, G.H.: Astrophys. J. **218** (1977) L141.
36 Gulkis, S., Poynter, R.: Phys. Earth Planetary Interiors **6** (1972) 36.
37 Hanel, R., Conrath, B., Flasar, M., Kunde, V., Lowman, P., Maguire, W., Pearl, J., Pirraglia, J., Samuelson, R., Gautier, D., Gierasch, P., Kumar, S., Ponnamperuma, C.: Science **204** (1979) 972.
38 Hartle, R.E., Curtis, S.A., Thomas, G.E.: J. Geophys. Res. **80** (1975) 3689.
39 Herr, K.C., Pimentel, G.C.: Science **166** (1969) 496.
40 Hodges, R.R.: J. Geophys. Res. **78** (1973) 8055.
41 Hodges, R.R., Johnson, F.S.: J. Geophys. Res. **73** (1968) 7307.
42 Hoffman, J.H., Hodges, R.R., Evans, D.E.: Proc. 4th Lunar Sci. Conf. **3** (1973) 2865.
43 Houck, J.R., Pollack, J., Shaak, D., Reed, R.A., Summers, A.: Science **189** (1975) 720.
44 Hunt, G.E.: Nature **272** (1978) 403.
45 Hunten, D.M. in: [a] p. 420.
46 Hunten, D.M. in: [c] p. 127.
47 Ingersoll, A.P.: J. Geophys. Res. **79** (1974) 3403.
48 Ingersoll, A.P.: Space Sci. Rev. **18** (1976) 603.
49 Ip, W.-H., Axford, W.I.: Nature **283** (1980) 180.
50 Judge, D.L., Carlson, R.W.: Science **183** (1974) 317.
51 Kiess, C.C., Corliss, C.H., Kiess, H.K.: Astrophys. J. **132** (1960) 221.
52 Kliore, A., Cain, D.L., Levy, G.S., Eshleman, V.R., Fjeldbo, G., Drake, F.D.: Science **149** (1965) 1243.
53 Knollenberg, R.G., Hunten, D.M.: Science **203** (1979) 792.
54 Kumar, S.: Icarus **28** (1976) 579.
55 Kupo, I., Mekler, Y., Eviatar, A.: Astrophys. J. **205** (1976) L51.
56 Larson, H.P., Fink, U., Treffers, R.R.: Astrophys. J. **211** (1976) 972.
57 Larson, H.P., Fink, U., Treffers, R.R., Gautier, T.N.: Astrophys. J. **197** (1975) L137.
58 Leighton, R.B., Murray, B.C.: Science **153** (1966) 136.
59 Lewis, J.S.: Icarus **11** (1969) 367.
60 Macy, W.: Icarus **29** (1976) 49.
61 Macy, W., Smith, W.H.: Astrophys. J. **222** (1978) L73.
62 Matson, D.L., Johnson, T.V., Fanale, F.P.: Astrophys. J. **192** (1974) L43.
63 Marov, M.Ya.: Annu. Rev. Astron. Astrophys. **16** (1978) 141.
64 Marov, M.Ya., Avduevsky, V.S., Borodin, N.F., Ekonomov, A.P., Kerzhanovich, V.V., Lysov, V.P., Moshkin, B.Ye., Rozhdestvensky, M.K., Ryabov, O.L.: Icarus **20** (1973) 407.
65 Maxworthy, T.: Planet. Space Sci. **21** (1973) 623.
66 McDonough, T.R., Brice, N.M.: Icarus **20** (1973) 136.
67 McElroy, M.B., Yung, Y.L.: Astrophys. J. **196** (1975) 227.
68 Miller, S.L., Smythe, W.D.: Science **170** (1970) 531.
69 Moos, H.W., Clarke, J.T.: Astrophys. J. **229** (1979) L107.
70 Münch, G., Trauger, J.T., Roesler, F.L.: Astrophys. J. **216** (1977) 963.
71 Murray, B.C., Malin, M.C.: Science **182** (1973) 437.
72 Murray, B.C., Wildey, R.L., Westphal, J.A.: J. Geophys. Res. **68** (1963) 4813.
73 Murray, B.C., Belton, M.J.S., Danielson, G.E., Davies, M.E., Gault, D., Hapke, B., O'Leary, B., Strom, R.G., Suomi, V., Trask, N.: Science **183** (1974) 1307.

74 Neugebauer, G., Münch, G., Chase, S.C., jr., Hatzenbeler, H., Miner, E., Schofield, D.: Science **166** (1969) 98.
75 Niemann, H.B., Hartle, R.E., Kasprzak, W.T., Spencer, N.W., Hunten, D.M., Carignan, G.R.: Science **203** (1979) 770.
76 Orton, G.S.: Icarus **26** (1975) 125.
77 Orton, G.S.: Icarus **26** (1975) 142.
78 Orton, G.S., Aumann, H.H.: Icarus **32** (1977) 431.
79 Orton, G.S., Ingersoll, A.P. in: [b] p. 206.
80 Owen, T.: Rev. Geophys. Space Phys. **13** (1975) 416.
81 Owen, T., Biemann, K., Rushneck, D.R., Biller, J.E., Howarth, D.W., Lafleur, A.L.: J. Geophys. Res. **82** (1977) 4635.
82 Oyama, V.I., Carle, G.C., Woeller, F., Pollack, J.B.: Science **203** (1979) 802.
83 Podolak, M., Danielson, R.: Icarus **30** (1977) 479.
84 Pollack, J.B.: Icarus **10** (1969) 314.
85 Pollack, J.B.: Icarus **19** (1973) 43.
86 Pollack, J.B., Erickson, E.F., Witteborn, F.C., Chackerian, Ch., jr., Summers, A.L., van Camp, W., Baldwin, B.J., Augason, G.C., Caroff, L.J.: Icarus **23** (1974) 8.
87 Prinn, R.G., Lewis, J.S.: Astrophys. J. **179** (1973) 333.
88 Rages, K., Pollack, J.B. in: [c] p. 149.
89 Ridgway, S.T.: Astrophys. J. **187** (1974) L41.
90 Ridgway, S.T., Wallace, L., Smith, G.: Astrophys. J. **207** (1976) 1002.
91 Ridgway, S.T., Larson, H.P., Fink, U. in: [b] p. 406.
92 Sagan, C.: JPL Tech. Rep. 32–34, Calif. Inst. Tech. (1960).
93 Sagan, C.: Icarus **1** (1962) 151.
94 Seiff, A.: Science News **115** (1979) 372.
95 Sill, G.T.: Comm. Lunar Planet. Lab. **9** (1972) 191.
96 Sill, G.T.: [b] p. 372.
97 Smith, B.A., Hunt, G.E. in: [b] p. 564.
98 Smith, W., Macy, W.: Bull. American Astron. Soc. **9** (1977) 516.
99 Smith, B.A., Smith, S.A.: Icarus **17** (1972) 218.
100 Smith, B.A., Soderblom, L.A., Johnson, T.V., Ingersoll, A.P., Collins, S.A., Shoemaker, E.M., Hunt, G.E., Masursky, H., Carr, M.H., Davies, M.E., Cook II, A.F., Boyce, J., Danielson, G.E., Owen, T., Sagan, C., Beebe, R.F., Veverka, J., Strom, R.G., McCauley, J.F., Morrison, D., Briggs, G.A., Suomi, V.E.: Science **204** (1979) 951.
101 Suomi, V.E., Limaye, S.S.: Science **201** (1978) 1009.
102 Taylor, F.W., McCleese, D.J., Diner, D.J.: Nature **279** (1979) 613.
103 Taylor, F.W., Diner, D.J., Elson, L.S., Hanner, M.S., McCleese, D.J., Martonchik, J.V., Peichley, P.E., Houghton, J.T., Delderfield, J., Schofield, J.T., Bradley, S.E., Ingersoll, A.P.: Science **203** (1979) 779.
104 Terrile, R.J., Westphal, J.A.: Icarus **30** (1977) 274.
105 Tokunaga, A.T., Knacke, R.F., Owen, T.: Astrophys. J. **209** (1976) 294.
106 Tomasko, M.G. in: [b] p. 486.
107 Trafton, L.M.: Astrophys. J. **175** (1972) 285.
108 Trafton, L.M.: Astrophys. J. **175** (1972) 295.
109 Trafton, L.M.: Nature **258** (1975) 690.
110 Trauger, J., Roesler, F., Mickelson, M.: Bull. American Astron. Soc. **9** (1977) 516.
111 Traub, W.A., Carleton, N.P.: Bull. American Astron. Soc. **4** (1972) 371.
112 Trauger, J.T., Roesler, F., Carleton, N.P., Traub, W.A.: Astrophys. J. **184** (1973) L137.
113 Travis, L.D., Coffeen, D.L., Hansen, J.E., Kawabata, K., Lucis, A.A., Lane, W.A., Limaye, S.S.: Science **203** (1979) 781.
114 Ulich, B.L., Conklin, E.K., Dickel, J.R.: (1978), quoted in Hunten [46].
115 Urey, H.C.: The Planets, Yale Univ. Press (1952).
116 Von Zahn, U., Fricke, K.H., Hoffmann, H.-J., Pelka, K.: Geophys. Res. Lett. **6** (1979) 337.
117 Webster, D.L., Alskne, A.Y., Whitten, R.C.: Astrophys. J. **174** (1972) 685.
118 Weidenschilling, S.J., Lewis, J.S.: Icarus **20** (1973) 465.
119 Whitten, R.C., Reynolds, R.T., Michelson, P.F.: Geophys. Res. Lett. **2** (1975) 49.
120 Wildt, R.: Nachr. Ges. Wiss. Göttingen **1** (1932) 87.
121 Woodman, J., Trafton, L., Owen, T.: Icarus **32** (1977) 314.
122 Young, A.T.: Icarus **18** (1973) 564.
123 Yung, Y.L., McElroy, M.B.: Icarus **30** (1977) 97.

3.2.2.6 Magnetic fields

3.2.2.6.1 The planets

In addition to the earth, the planets Mercury and Jupiter have been found, as a result of in situ measurements, to have magnetic fields of internal origin In-situ measurements also suggest that Venus and Mars may have magnetic fields. However, it is difficult to distinguish between fields of internal and external origin on the basis of measurements made to date.

There is indirect evidence for a Saturnian magnetic field based on the detection of kilometric and decametric radio emissions [3, 9] similar to those found in the case of the earth and Jupiter, and which are apparently the result of magnetospheric wave-particle interactions. There is some evidence for similar emissions from Uranus [4]. First current estimates for the magnetic fields of these bodies are given in Table 13, based on a review by Russell [16].

The magnetic fields of Mercury, the earth, Jupiter, and Saturn are all probably the result of dynamo action produced by fluid motions in the electrically conducting inner cores of these planets [14, 10]. Permanent magnetism of crustal regions may make a significant contribution to the relatively weak field of Mercury [17, 20] but it is of negligible significance in the case of the earth and entirely absent in the case of Jupiter. There are, of course, additional external contributions to the magnetic field in all three cases which are associated with stresses exerted by the solar wind, trapped hot plasma, and (especially in the case of Jupiter) the rotation of the planet. For more recent results, see appendix 3.2.2.7.2.

The earth's field has been investigated in great detail and has been shown to have a secular variation with occasional reversals of the dipole field [19]. The medium- and large-scale features of the geomagnetic field tend to drift westwards at a rate of about $0\overset{\circ}{.}05$ per year [11, 19].

It has been very difficult on this basis of only two encounters of Mariner 10 to get more than a rough estimate of the dipole moment of Mercury, but in the case of Jupiter it has so far been possible, on the basis of four spacecraft encounters, to determine the dipole moment quite accurately and to show that the ratio of dipole: quadrupole: octopole moments are 1.00:0.25:0.20 (in comparison with 1.00:0.14:0.10 for the earth) [1, 18].

3.2.2.6.2 The moon

The moon has no detectable dipole magnetic field; however, orbital magnetic measurements have shown the presence of an irregular surface magnetization which may have been the result of a magnetizing field of the order of 1 Gauss [8]. There is no evidence for any dynamo action in the moon at present time, since the upper limit on the dipole moment of 10^{19} Gauss cm^3 [16] is very low indeed. However, there may well have been dynamo action in the past which could have accounted for the remnant surface magnetism [15].

3.2.2.6.3 The satellites of Jupiter

To date, there are no measurements of the magnetic fields of the Galilean satellites. Because of their large sizes and possible existence of liquid inner cores there may be dynamo action in the satellites' interiors generating magnetic fields. Neubauer [13] has used a similarity law derived by Busse [5] to make some estimates for the dynamo magnetic moments (M) and surface magnetic fields (B_s) for various Jovian satellites and Titan; see Table 14.

Table 13. Planetary magnetic fields.
M = magnetic dipole moment
B_s = equatorial surface field

Planet	M [Gauss cm^3]	B_s [Gauss]	Ref.
Mercury	$(4.9 \pm 0.2) \cdot 10^{22}$	$3.4 \cdot 10^{-3}$	12
Venus	$23 \cdot 10^{22}$	$\approx 10^{-5}$	7
Earth	$8.0 \cdot 10^{25}$	0.31	2
Mars	$2.5 \cdot 10^{22}$	$6.4 \cdot 10^{-4}$	6
Jupiter	$1.5 \cdot 10^{30}$	4	18
Saturn	$4 \cdot 10^{28}$	≈ 0.2	21,22
Uranus	?	?	4

Table 14. Estimated magnetic fields of Jovian satellites [13].
M = magnetic dipole moment
B_s = equatorial surface field

Satellite	M [Gauss cm^3]	B_s [γ] = [10^{-5} Gauss]
Io	$1.35 \cdot 10^{23}$	2200
Europa	$2.88 \cdot 10^{22}$	810
Ganymede	$4.0 \cdot 10^{22}$	220
Callisto	$9.33 \cdot 10^{21}$	63
Titan	$1.63 \cdot 10^{22}$	100

These numbers should be compared with those derived for Mercury and Venus, for example (Table 13).

3.2.2.6.4 References for 3.2.2.6

1 Acuna, N.H., Ness, N.F.: J. Geophys. Res. **81** (1976) 2917.
2 Allen, C.W.: Astrophysical Quantities (third edition), The Athlone Press. (1976) p. 136.
3 Brown, L.W.: Astrophys. J. **198** (1975) L89.
4 Brown, L.W.: Astrophys. J. **207** (1976) L209.
5 Busse, F.H.: Phys. of Earth Planet. Interior **12** (1976) 350.
6 Dolginov, Sh. Sh., Yeroshenko, Ye.G., Zhuzgov, L.N.: J. Geophys. Res. **81** (1976) 3353.
7 Dolginov, Sh.Sh., Zhuzgov, L.N., Sharova, V.A., Buzin, V.B., Yeroshenko, Ye.G.: Lunar and Planetary Sci. **IX** (1978) 256.
8 Fuller, M.: Rev. Geophys. Space Phys. **12** (1974) 23.
9 Kaiser, M.L., Stone, R.G.: Science **189** (1975) 285.
10 Levy, E.H.: Annu. Rev. Earth Planet. Sci. **4** (1976) 159.
11 Nagata, T.: J. Geomag. Geoelect. **17** (1965) 263.
12 Ness, N.F.: Space Sci. Rev. **21** (1978) 527.
13 Neubauer, F.M.: Geophys. Res. Lett. **5** (1978) 905.
14 Parker, E.N. in: Physics of Solar Planetary Environments (Williams, D.J. ed.) American Geophysical Union (1976) p. 812.
15 Runcorn, S.K.: Science **199** (1978) 771.
16 Russell, C.T.: Rev. Geophys. Space Phys. **17** (1979) 295.
17 Sharpe, H.N., Strangway, D.W.: Geophys. Res. Lett. **3** (1976) 285.
18 Smith, E.J., Davis, L., Jones, D.E. in: Jupiter (Gehrels, T., ed.), Univ. of Arizona Press (1976) p. 783.
19 Stacey, F.D.: Physics of the Earth, John Wiley & Sons Inc., New York (1969) chap. 5.
20 Stephenson, A.: Earth Planet. Sci. Lett. **28** (1976) 454.
21 Smith, E.J. et al.: Science **207** (1980) 407.
22 Acuña, M.H., Ness, N.F.: Science **207** (1980) 444.

3.2.2.7 Appendix to 3.2.2

3.2.2.7.1 Jupiter

The Jupiter encounter of the Voyager 2 spacecraft in September 1979 has provided further information on Jupiter and the Galilean satellites. Together with Voyager 1, these observations thus constitute an almost continuous close-up monitoring – over a 6 month period – of the dynamics of the Jovian atmosphere. For example, some variations in the velocity profiles of the differential zonal atmospheric motion have been discerned. Preliminary discussions of the Voyager results concerning the Jovian atmosphere can be found in the imaging science team preliminary reports [15, 16]. These will be our major source of references for this section if not otherwise stated.

1. The satellites

Voyager 2 observations confirmed the ground-based measurements [10, 12] and Voyager 1 result that Amalthea, orbiting synchronously at 2.54 R_J (R_J=Jovian radius), has a very low value (0.04···0.06) of average albedo and that it is very red. The current best estimates for Amalthea's dimensions are (270±15) km (long axis), (170±15) km (intermediate axis), and (155±10) km (polar axis). The long axis points towards Jupiter as expected.

Seven out of the eight volcanic plumes on Io as observed by Voyager 1 were found to be still active during the Voyager 2 encounter. That no active plume smaller than 70 km was seen on the limb in either encounter suggests that the large plumes might be far more common. From the time-lapse sequence of images of volcanic eruptions, some plumes were observed to show brightening as the phase angle approached 160°. This effect, which appears stronger in blue and violet, is probably caused by strong forward light scattering of small micron-sized SO_2 condensates.

Only three probable craters approximately 20 km in diameter have been identified on the surface of Europa. This implies that the surface of Europa is much younger than that of Ganymede and Callisto, but is probably at least several hundred million years old. The dark markings which crisscross Europa's smooth surface may be the result of its evolutionary history. One possible scenario according to the Voyager imaging team report can be described in the following: During the final stage of accretion, which was accompanied by heavy surface bombardment, the satellite was still covered by a liquid water ocean. Consequently most of the ancient craters were obliterated. Subsequent freezing of this liquid layer would then result in large-scale expansion of the thin crust (50···100 km thick) producing fracture pattern all over the surface. To account for the presence of the dark markings – presumably representing the record of such crustal expansion – the amount of surface expansion should be 5···15% of the surface area.

While Ganymede was found to possess a large range of crater densities over its surface (the dark cratered terrains have crater density comparable to that of Callisto and that of the smooth terrain is comparable to those of the lunar maria and old Martian plains, see 3.2.2.3), the whole surface of Callisto is nearly as heavily cratered as the densely cratered highlands of the moon, Mars and Mercury. This implies that the surface of Callisto might, in fact, retain the record of the early heavy bombardment. Like Europa, but to a varying degree, the presence of icy crusts could play a role in modifying the structure of the craters on Callisto as well as Ganymede. Note that besides bombardment by stray bodies the satellites surfaces are also subject to sputtering by magnetospheric particles [3] and also poleward migration of volatiles as perhaps evidenced by Ganymede's bluish-white polar caps [16].

Finally, two new small satellites were discovered near the outer edge of the Jovian ring at 1.79 R_J. 1979 J 1 with a 25 km diameter has an orbital period of 7^h09^m, and 1979 J 3 with 40 km diameter has an orbital period of $7^h04^m30^s$. Voyager observations also discovered another new satellite, 1979 J 2, whith an orbital period of $16^h11^m21^s$. See also 3.2.1.3.1.

2. The ring system

After the first discovery of the Jupiter's ring by Voyager 1 more spectacular photographs were obtained with large phase angles when the Voyager 2 spacecraft was in Jupiter's shadow. Examination of these pictures [11] has revealed that the ring has a sharp outer edge at 1.8 R_J and the ring system can be divided into four distinct components:

a) A narrow bright segment about (800 ± 100) km wide, inner edge at 1.7 R_J;
b) A fainter segment about 6000 km wide, inner edge at 1.68 R_J;
c) A very faint inner "ring" that appears to be a continuous sheet of dispersed material, probably extending all the way to the planet;
d) an even fainter outer edge of small particles that extends to 1.8 R_J and slightly beyond.

From the nature of the brightness of the scattering at high phase angles, the mean radius of the particles contributing to the above intensity variation is derived to be about 4 µm. The small satellites orbiting just near the outer edge are the most likely source of these dust particles.

The normal optical thickness of the bright ring has been estimated to be $\tau \approx 3 \cdot 10^{-5}$ if the geometrical albedo is assumed to be 0.04 [8].

3.2.2.7.2 Saturn

In November 1980 the Voyager 1 spacecraft made detailed observations of the Saturnian system. These results together with the data obtained from the Saturn encounter of Pioneer 11 spacecraft in 1979, have yielded a wealth of new information on the structures of the planet, its satellites and the ring system. In the following we have summarized some of the highlights of those observations, obtained from the preliminary Voyager reports and other sources. For further information on the Pioneer 11 Saturn flyby observations, the readers are referred to the special issue of Science **207** (1980) and J. Geophys. Res. **85** (1980). When not otherwise specified, the description here is derived from the Voyager Bulletin, Mission Status Report No. 55···60 (1980). [Preliminary results from the Voyager 1 observations can be found in: Science **212** (1981).]

1. The planetary magnetic field

Somewhat surprisingly the dipole field of Saturn was found to be smaller than originally expected. The Pioneer 11 magnetometer experiments [1, 17] reported a value of 0.2 R_S^3 G km³ for the dipole moment (R_S=Saturnian radius in [km]). The dipole tilt angle is consistent with 0° and the ratio of the quadrupole to dipole moment is 0.07, a value that is smaller than those for earth and Jupiter. The quadrupole moment can be eliminated if it is assumed that the magnetic center is offset from the center of Saturn. This offset has a magnitude of 0.04 R_S and is principally in the polar direction. While the equatorial field strength is 0.2 G, the field strength is 0.63 G at the north pole and 0.48 G at the south pole as a consequence of the oblateness of Saturn (1/10.4) and the dipole offset. It has been suggested that the weaker quadrupole moment of Saturn as compared with that of Jupiter may be related to the difference in the size of their central metallic hydrogen cores ($R \approx 0.2 \cdots 0.5\, R_S$ for Saturn and $0.2 \cdots 0.7\, R_J$ for Jupiter) within which the dynamo is supposed to operate. The small dimension of Saturn's core is also consistent with its small dipole field.

2. Satellites

For the first time the surface morphology of several satellites has been surveyed. Preliminary reports from the Voyager 1 imaging experiment indicate that while the basic landforms are impact craters there are large differences between the satellites, perhaps as a result of different evolutionary histories. For example, an impact crater was discovered on the leading face of Mimas with a diameter (=130 km) about 1/4 that of the satellite. The corresponding ratio of crater diameter to satellite diameter may be the largest in the solar system. In contrast, Enceladus appears to have a relatively smooth surface. A large valley about 750 km long and 60 km wide is located on the planet-facing side of Tethys, which is also densely cratered. There is a distribution of bright wispy structures on the surface of Dione which may be caused by frost deposits. Rhea appears to be the most heavily cratered satellite in the Saturnian system.

The theory that Titan's atmosphere should be composed of mainly N_2 and that the surface pressure should be high (a few bars or more) has been confirmed by the Voyager observations but further data analysis is required to gain the exact values of the surface temperature and pressure. Acetylene (C_2H_2), ethylene (C_2H_4), ethane (C_2H_6) and cyanide (HCN) have also been identified in the atmosphere of Titan. In addition, the discovery of a rather extensive hydrogen cloud in association with Titan has been reported (see also [9]).

Titan apparently is not the only satellite that emits neutral gas into the Saturnian magnetosphere, as a large torus of oxygen ions is observed between 4 and 8 R_S [5, 21, 22]. Photodissociation [2] and energetic particle impact sputtering [3] of the ice on the surface of the ring particles, and on Tethys and Dione could all contribute to the production of these oxygen ions with a source strength estimated to be $\approx 10^{24} \cdots 10^{26}$ ions s^{-1} [5, 22]. The long-wave channel of the ultraviolet photometer onboard Pioneer 11 also detected emission from a ring cloud, presumed to be atomic hydrogen, with enhancement in the vicinity of the B ring [9]. UV instrumentations carried on rockets have also detected hydrogen Lyman α emission in the vicinity of the ring system [20].

As a result of recent ground-based observations as well as in-situ measurements, a number of new satellites have been found. The well established ones are included in Table 4 in 3.2.1.3.1.

3. The ring system

A better view of the narrow F ring at 2.32 R_S first discovered by Pioneer 11 – as provided by the Voyager 1 imaging experiment – has shown that this ring actually composes three components at the time of the Voyager observation. Each is 30 to 50 km wide and the outer two appear to be interwoven. This peculiar behaviour may be caused by electromagnetic interaction between the charged dust particles and the planetary magnetic field, and the gravitational effect of the two "herding" satellites could also play a role. The broad ring system has now been resolved into a series of 500⋯1000 (if not more) narrow rings. The material in the Cassini division (with a width of $4 \cdot 10^3$ km and normal optical opacity of 0.1) is also found to be collected into several rings. Unless some other dynamical effect is involved, the resonance theory for the formation of Cassini division and other gross brightness variations of the Saturn rings might have to be extended in order to account for the presence of these many ringlets. Spoke-like features were observed in the B ring and a comparison of the inbound and outbound observations has indicated that these are caused by the injection of small dust particles. This phenomenon might result from electrostatic charging of the large ring particles which in turn causes the micron-sized particles to escape from their surfaces. Ground-based observations have revealed the presence of a very tenuous dust ring between 3.5 and 5 R_S [14] which may be in part responsible for the absorption of trapped radiation particles in this region [4, 13, 18, 19].

3.2.2.7.3 References for 3.2.2.7

1 Acuña, M.H., Ness, N.F.: Science **207** (1980) 444.
2 Carlson, R.W.: Nature **283** (1980) 461.
3 Cheng, A.F., Lanzerotti, L.J.: J. Geophys. Res. **83** (1978) 2597.
4 Fillius, W., Ip, W.H., McIlwain, C.E.: Science **207** (1980) 425.
5 Frank, L.A., Burek, B.G., Ackerson, K.L., Wolfe, J.H., Mihalov, J.D.: J. Geophys. Res. **85** (1980) 5695.
6 Gehrels, T. et al.: Science **207** (1980) 434.
7 Goldreich, P., Tremaine, S.: Nature **277** (1979) 97.
8 Jewitt, D.C., Danielson, G.E.: Contribution No. 3447, Division of Geological and Planetary Sciences, Calif. Inst. Tech. (1980).
9 Judge, D.L., Wu, F.-M., Carlson, R.W.: Science **207** (1980) 431.
10 Millis, R.L.: Icarus **33** (1978) 319.
11 Owen, T., Danielson, G.E., Cook, A.F., Hansen, C., Hall, V.L., Duxbury, T.C.: Nature **281** (1979) 442.
12 Rieke, G.H.: Icarus **25** (1975) 233.
13 Simpson, J.A., Bastian, T.S., Chenette, D.L., Lentz, G.A., McKibben, R.B., Pyle, K.R., Tuzzolino, A.J.: Science **207** (1980) 411.
14 Sky and Telescope **59** (1980) 296.
15 Smith, B.A. et al.: Science **204** (1979) 951.
16 Smith, B.A. et al.: Science **206** (1980) 927.
17 Smith, E.J., Davis, L., jr., Jones, D.E., Coleman, P.J., jr., Colburn, D.S., Dyal, P., Sonett, C.P.: Science **207** (1980) 407.
18 Trainor, J.H., McDonald, F.B., Schardt, A.W.: Science **207** (1980) 421.
19 Van Allen, J.A., Thomsen, M.F., Randall, B.A., Rairden, R.L., Grosskreutz, C.L.: Science **207** (1980) 415.
20 Weiser, H., Vitz, R.C., Moos, H.W.: Science **197** (1977) 755.
21 Wolfe, J.H., Mihalov, J.D., Collard, H.R., McKibbin, D.D., Frank, L.A., Intriligator, D.S.: Science **207** (1980) 403.
22 Wu, F.-M., Judge, D.L., Carlson, R.W.: J. Geophys. Res. **85** (1980) 5853.

3.3 Small bodies in the solar system

3.3.1 The asteroids (minor planets)

3.3.1.1 Representative orbits

In March 1979 the list of permanently numbered asteroids contained 2125 objects. Orbits of high accuracy are known for almost all of these objects. A list of osculating orbital elements of the numbered asteroids appears annually, together with ephemerides [1]; for recently numbered orbits see [2]. Since the Keplerian orbits vary according to planetary attractions, each set of osculating elements refers to an epoch, which is usually the beginning of a day in ephemeris time (ET).

Table 1. Osculating orbital elements of selected asteroids. Ecliptic and equinox 1950.0; values from [1, 2].

M mean anomaly
ω argument of perihelion
Ω longitude of ascending node
i inclination to ecliptic
ϕ angle of eccentricity
n mean daily motion [″/day]
a semi-major axis [AU]
$e=\sin\phi$ numerical eccentricity
$q=a(1-e)$ minimum of heliocentric distance
$\varpi=\omega+\Omega$ longitude of perihelion
$L=M+\varpi$ mean longitude of asteroid

No.	Name	Epoch 0^hET	M	ω	Ω	i	ϕ	n	a AU	Remarks
1	Ceres	1962 Dec. 2	348°58	71°34	80°49	10°61	4°50	770″719	2.767	1
2	Pallas	1962 Dec. 2	338.94	310.14	172.84	34.83	13.61	769.962	2.769	1, 2
4	Vesta	1962 Dec. 2	263.31	148.68	103.68	7.13	5.10	977.543	2.362	1
221	Eos	1947 Jan. 15	319.62	192.13	142.34	10.85	5.50	678.284	3.013	3
434	Hungaria	1968 May 24	302.56	123.68	174.93	22.51	4.23	1308.945	1.944	4
1221	Amor	1964 June 14	27.02	25.89	171.09	11.91	25.85	1333.049	1.921	5
1862	Apollo	1973 Aug. 26	9.46	285.25	35.57	6.36	34.05	1991.441	1.470	6
2062	Aten	1978 Oct. 19	194.89	147.82	108.07	18.94	10.51	3735.582	0.966	7
2102	Tantalus	1979 Nov. 23	206.24	61.66	93.74	64.01	17.36	2421.534	1.290	6, 2
1566	Icarus	1969 Jan. 19	230.85	31.04	87.63	22.95	55.75	3170.636	1.078	6, 8
1362	Griqua	1941 Jan. 6	294.84	263.37	121.46	24.06	19.99	597.123	3.281	9, 10
1373	Cincinnati	1941 Jan. 6	293.61	99.05	298.07	38.90	18.76	563.201	3.411	11, 2
153	Hilda	1947 Jan. 15	121.75	49.17	228.42	7.84	8.83	447.653	3.975	12, 9, 11
279	Thule	1973 Nov. 14	336.38	192.70	74.59	2.34	1.87	403.870	4.258	9, 10, 11
588	Achilles	1973 Oct. 5	290.01	130.59	315.91	10.34	8.57	301.096	5.178	13, 9, 11
617	Patroclus	1974 June 2	291.38	305.24	43.77	22.05	8.09	296.867	5.228	14, 9, 11
944	Hidalgo	1976 Dec. 8	354.99	57.39	20.96	42.40	41.04	249.901	5.864	15, 2
2060	Chiron	1977 Sep. 14	229.07	339.10	208.71	6.92	22.25	70.007	13.695	16

Remarks to Table 1

1 Ceres, Pallas, and Vesta are the three largest bodies in the main belt of asteroid motion.
2 High-inclination orbit.
3 Name-giving member of a typical Hirayama family [a], see 3.3.1.2.
4 A large inclination is typical for many objects with $a\approx1.9$ AU.
5 Earth-approaching asteroid with $q>1$ AU, Mars Crosser [3, a].
6 Apollo-type asteroid [a], $q<1$ AU, $a>1$ AU.
7 Aten-type asteroid, $a<1$ AU, cf. [d] p. 253.
8 $q=0.19$ AU is especially small.
9 n corresponds to a low-order resonance ratio with respect to Jupiter, see "commensurabilities" in [4] p. 164.
10 Value of n is rare according to the frequency distribution, see 3.3.1.2.
11 The object is able to avoid close approaches to Jupiter, cf. [3, 5, a].
12 Hilda-type asteroid, 3/2 resonance case [a].
13 Trojan asteroid, preceding Jupiter [a], 1/1 resonance case.
14 Trojan asteroid, following Jupiter [a], 1/1 resonance case.
15 Unusual orbit reaching to Saturn's orbit.
16 Remote orbit, mainly between the orbits of Saturn and Uranus, $q=8.5$ AU, cf. [d] p. 436.

3.3.1.2 Statistics of orbits

Two tables in [4] show the frequency distribution of the elements n, Ω, ϖ, i, and ϕ, derived from 1637 reliable orbits of asteroids. A survey of faint asteroids [6] extends the basis of statistics for n. Fig. 1 represents numbered asteroids [1] and the quality classes 1 and 2 in [6]. In case of most asteroids, the theory of secular variations [7] allows the removal of long-period effects caused by planetary attraction. If this is done [8], proper elements A and B replace $e=\sin\phi$ and $\sin i$, respectively. Families of asteroids (Table 2) appear as concentrated clusters of points in the a-A-B space [8], cf. [c] p.177, and [6, 9, 10, a, d].

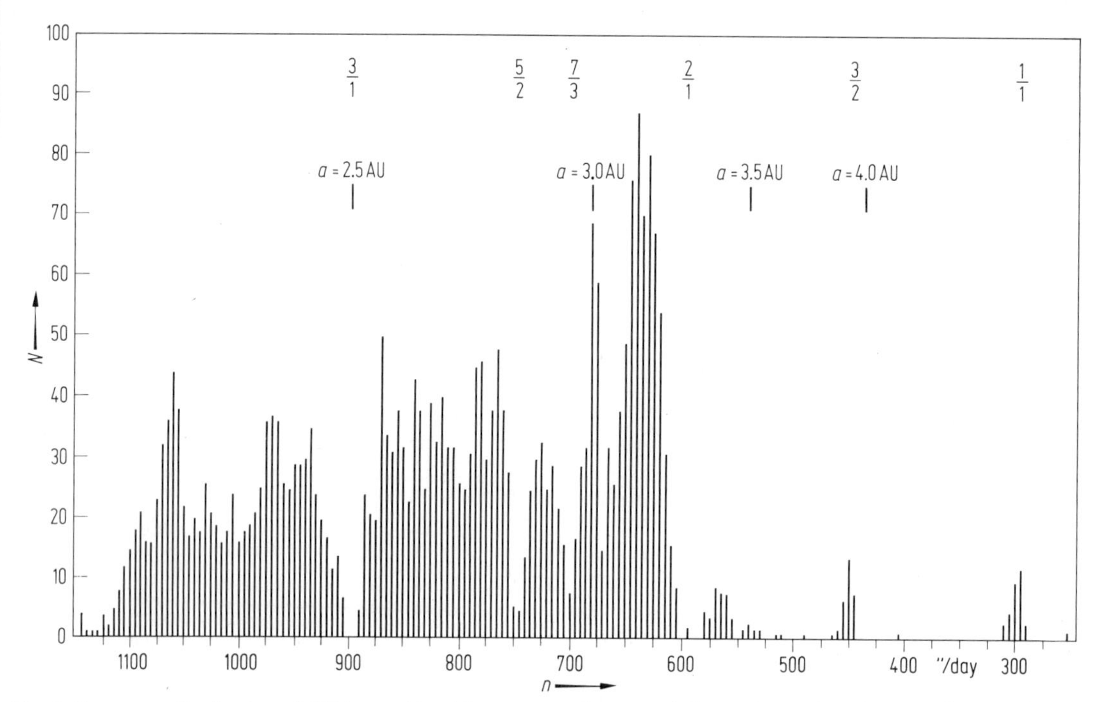

Fig. 1. Frequency distribution of 3077 asteroids with respect to the mean daily motion, n. Numbers N refer to 5" intervals of n. Simple ratios of commensurability to Jupiter [4] and selected values of the semi-major axis, a, are marked in the upper part of the figure. This corresponds to Fig. 1 in [5], but was updated according to [1, 6] by I. Seckel and H. Scholl.

Table 2. Some populous families of asteroids.
Mean values of a, A, B from [6, 8, 9]; some of the families listed here may actually consist of several families.
a = semi-major axis [AU]
$A=e_1$ = proper eccentricity
$B=\sin i_1$ = proper inclination

Name of family	Mean value of			Number of members			Number of the representative asteroid
	a AU	A	B	[8]	[9]	[6]	
Themis	3.14	0.16	0.03	53	67	63	(24)
Eos	3.02	0.07	0.18	58	66	23	(221)
Koronis	2.88	0.05	0.04	33	37	32	(158)
Maria	2.55	0.10	0.26	17	20	3	(170)
Flora	2.23	0.14	0.08	125	156	36	(8)
Nysa-Hertha	2.40	0.17	0.05	9	10	77	(44)
Medea	3.16	0.13	0.09			32	(212)
Vesta	2.33	0.09	0.12			28	(4)

3.3.1.3 Sizes and physical characteristics

Most asteroids are too small for a direct telescopic measurement of size; the four values measured by Barnard [4] are subject to systematic errors of unknown amount [3].

Table 3 gives results from three other methods; the diameters D [km] represent mean values if the bodies have an irregular shape. D_{occ} results from observation of a stellar occultation at several places. D_{rad} is derived indirectly from the reflected and re-emitted energy of solar radiation. D_{pol} follows from observed brightness and reflectivity of surface, found from measurements of polarization at different angles of aspect, according to an empirical law [11].

Very few asteroids have caused observable gravitational effects that allow a mass determination, see Table 4 and [a, d] for masses and densities.

Table 3. Diameters of asteroids.
D = diameter (mean value, see text)
D_{occ} = results from occultation [16]
D_{rad} = results from radiometry [17]
D_{pol} = results from polarimetry [11], implicit data

Object		D_{occ} km	D_{rad} km	D_{pol} km
(1)	Ceres		1017	957
(2)	Pallas	538	585	635
(3)	Juno		247	247
(4)	Vesta		531	558
(5)	Astraea		122	112
(7)	Iris		205	211
(8)	Flora		161	148
(10)	Hygiea		450	
(511)	Davida		340	
(704)	Interamnia		338	

Table 4. Masses and densities.

Object	Mass 10^{24} g	Adopted D km	Density g cm^{-3}	Ref.
(1) Ceres	1.17 ± 0.06	987 ± 150	2.3 ± 1.1	d
(2) Pallas	0.21 ± 0.04	538 ± 50	2.6 ± 0.9	d
(4) Vesta	0.27 ± 0.02	544 ± 80	3.3 ± 1.5	d
Total [1])	3.0			18

[1]) Total mass of asteroids, estimate.
For the estimation of error, see Schubart and Matson in [d] p. 84.

Table 5. Properties of asteroid surfaces.
The blue magnitude B (1,0) is reduced to unit distance from earth and sun, and to phase angle 0°, using an adopted phase factor [19]. All the other quantities and the types are taken from the TRIAD file [20]. The respective contributors for the UBV colour indices, for the visual geometric albedo P_V, determined from radiometric measurements, and for the spectrophotometric parameter "depth", are listed in [21]. See also in the text.

Object	B(1,0)	$U-B$	$B-V$	P_V	Depth	Type
(1) Ceres	4^{m}48	0^{m}42	0^{m}72	0.059	1.00	C
(2) Pallas	5.02	0.29	0.65	0.093	1.00	U
(3) Juno	6.51	0.42	0.82	0.162	0.88	S
(4) Vesta	4.31	0.49	0.78	0.255	0.71	U
(5) Astraea	8.13	0.42	0.83	0.125	0.87	S
(6) Hebe	6.98	0.39	0.83	0.162	0.88	S
(7) Iris	6.84	0.47	0.84	0.196	0.88	S
(8) Flora	7.73	0.46	0.88	0.133	0.84	S
(9) Metis	7.78	0.50	0.87	0.118	0.98	S
(10) Hygiea	6.50	0.31	0.69	0.050	1.00	C
(24) Themis	8.27	0.34	0.69		1.00	C
(44) Nysa	7.85	0.24	0.70	0.480	1.00	E
(135) Hertha	9.24	0.27	0.69	0.124		M
(158) Koronis	10.78	0.38	0.84	0.159		S
(221) Eos	8.94	0.38	0.81	0.123	1.00	U
(349) Dembowska	7.24	0.54	0.95	0.278	0.73	R
(433) Eros	12.40	0.50	0.88	0.180	0.90	S
(511) Davida	7.36	0.35	0.71	0.040	1.00	C
(704) Interamnia	7.24	0.25	0.64	0.043	0.89	U

Table 5 lists further physical parameters. The reduced brightness $B(1,0)$ appears together with *UBV* colours. The values can be considered as mean values with respect to rotational variation. The geometric albedo is a measure of the mean reflectivity of the surface back in the direction of incoming radiation [12]. "Depth", by the deviation from 1.00, gives a measure of the strength of the "olivine-pyroxene absorption feature" near 0.95 μm in the spectrum of reflectivity [13···15]. Some of these and other parameters allow the automatic derivation of a "taxonomy of asteroids" [13] with types C, S, M, E, and R. U means unclassifiable; C is the most frequent type; E and R are rare, cf. [d]. For data on distant asteroids, see [d] p. 417.

Table 6 gives the maximum amplitude of the observed rotational variation in brightness and the respective period. Irregular shape of body, differences in reflectivity over the surface, or both can cause these variations. For possible relations between meteorites, comets, and certain types of asteroids, for mechanisms of heating bodies of asteroidal size, and for other questions about origin and evolution of asteroids, see [b, d].

Table 6. Evidence on axial rotation.
Data from TRIAD file [20]; the contributor is E. Tedesco [21].

Object	Maximum of brightness change	Period of rotation
(1) Ceres	$0^m.04$	$9^h.078$
(2) Pallas	0.15	7.811 [6]) [7])
(3) Juno	0.15	7.213
(4) Vesta [1])	0.14	5.342
(5) Astraea	0.27	16.812 [6])
(6) Hebe	0.19	7.274 [6])
(7) Iris	0.29	7.135
(8) Flora	0.04	13.6
(15) Eunomia	0.53	6.081
(39) Laetitia	0.54	5.138
(43) Ariadne	0.66	5.751
(44) Nysa	0.50	6.44
(433) Eros [2])	1.50	5.270 [6])
(624) Hektor [3])	1.09	6.923
(1566) Icarus [4])	0.22	2.273 [5])
(1620) Geographos [4])	2.03	5.223 [6])
(1685) Toro [4])	0.80	10.196 [6])
(1864) Daedalus [4])	0.80	8.57

[1]) The period may be twice the value given here [22]. However, the brightness changes of Vesta are best explained by differences in reflectivity [23], and then the shorter period is the more simple hypothesis.
[2]) Earth-approaching asteroid, cf. [4,24].
[3]) Trojan asteroid, cf. [4].
[4]) Apollo-type asteroid, cf. [a,4].
[5]) Very short period of rotation.
[6]) Well-determined sidereal period.
[7]) Corrected value, cf. [25].

3.3.1.4 References for 3.3.1

General references

a Chapman, C.R., Williams, J.G., Hartmann, W.K.: The Asteroids, Annu. Rev. Astron. Astrophys. **16** (1978) 33.
b Delsemme, A.H., ed.: Comets, Asteroids, Meteorites – interrelations, evolution and origins, University of Toledo, Ohio (1977).
c Gehrels, T., ed.: Physical Studies of Minor Planets, NASA special publication SP-267, Washington, D.C. (1971).
d Gehrels, T., ed.: Asteroids, The University of Arizona Press, Tucson, Ariz. (1979).

Special references

1 Ephemerides of Minor Planets for 1979, Inst. of Theoretical Astron., Acad. of Sci. USSR, Leningrad (1978).
2 Minor Planet Circulars, (MPC), Minor Planet Center, Cincinnati Observatory, Ohio (1947···1978) and Smithsonian Astrophys. Obs., Cambridge, Mass. (since Aug. 1978).
3 Schubart, J.: Naturwiss. **62** (1975) 565.
4 Gondolatsch, F.: Landolt-Börnstein, NS, Vol. VI/1 (1965) 163···166.
5 Froeschlé, C., Scholl, H.: Astron. Astrophys. **72** (1979) 246.
6 van Houten, C.J., van Houten-Groeneveld, I., Herget, P., Gehrels, T.: Astron. Astrophys. Suppl. **2** (1970) 339.
7 Brouwer, D., van Woerkom, A.J.J.: Astron. Papers Wash. **13** (1950) part 2.
8 Brouwer, D.: Astron. J. **56** (1951) 9.

9 Arnold, J.R.: Astron. J. **74** (1969) 1235.
10 Gradie, J., Zellner, B.: Science **197** (1977) 254.
11 Zellner, B. et al.: Proc. Lunar Sci. Conf. **8**th (1977) 1091.
12 Allen, C.W.: Astrophysical Quantities, 3rd ed., Athlone Press, University of London (1973) 142.
13 Bowell, E., Chapman, C.R., Gradie, J.C., Morrison, D., Zellner, B.: Icarus **35** (1978) 313.
14 Chapman, C.R., Morrison, D., Zellner, B.: Icarus **25** (1975) 104.
15 Pieters, C., Gaffey, M.J., Chapman, C.R., McCord, T.B.: Icarus **28** (1976) 105.
16 Wasserman, L.H. et al.: Astron. J. **84** (1979) 259.
17 Morrison, D.: Icarus **31** (1977) 185.
18 Kresák, L.: Bull. Astron. Inst. Czech. **28** (1977) 65.
19 Gehrels, T., Gehrels, N.: Astron. J. **83** (1978) 1660.
20 Tucson Revised Index of Asteroid Data, (TRIAD), status of June 1979 as published in [d] p. 1011.
21 Bender, D., Bowell, E., Chapman, C.R., Gaffey, M.J., Gehrels, T., Zellner, B., Morrison, D., Tedesco, E.: Icarus **33** (1978) 630.
22 Taylor, R.C.: Astron. J. **78** (1973) 1131.
23 Degewij, J., Tedesco, E.F., Zellner, B.: Icarus **40** (1979) 364.
24 Dunlap, J.L.: Icarus **28** (1976) 69.
25 Schroll, A., Haupt, H.F., Maitzen, H.M.: Icarus **27** (1976) 147.

3.3.2 Meteors and meteorites

3.3.2.0 Definitions

The following definitions are agreed upon [1]:

A. Meteor: in particular, the light phenomenon which results from the entry of a solid particle into the earth's atmosphere from space; more generally, as a noun or an adjective, any physical object or phenomenon associated with such an event.
B. Meteoroid: a solid object. moving in interplanetary space, of a size considerably smaller than an asteroid and considerably larger than an atom or molecule.
C. Meteorite: any object defined under B which has reached the surface of the earth without being completely vaporized.
D. Meteoric: the adjectival form pertaining to definitions A and B.
E. Meteoritic: the adjectival form pertaining to definition C.
F. Fireball: a bright meteor with luminosity which equals or exceeds that of the brightest planets.
G. Micrometeorite: a very small meteorite or meteoritic particle with a diameter in general less than a millimeter.
H. Dust: when used with D or E – finely divided solid matter, with particle sizes in general smaller than micrometeorites.
I. Absolute magnitude: the stellar magnitude any meteor would have if placed in the observer's zenith at a height of 100 km.
K. Trajectory: the line of motion of the meteor relative to the earth, considered in three dimensions.
L. Path: the projection of the trajectory on the celestial sphere, as seen by the observer.
M. Train: anything (such as light or ionization) left along the trajectory of the meteor after the head of the meteor has passed.
N. Persistent: an adjectival form for use with M indicating durations of some appreciable length.
O. Wake: train phenomena of very short duration, in general much less than a second.
P. Radiant: the point where the backward projection of the meteor trajectory intersects the celestial sphere.
Q. Earth-point: the point where the forward, straight-line projection of the meteor trajectory intersects the surface of the earth.
R. Zenith attraction: the effect of the earth's gravity on a meteoric body increasing the velocity and moving the radiant towards the zenith.
S. Orbit: the line of motion of a meteoric body when plotted with reference to the sun as origin of coordinates.
T. Shower: for use with A or D – a number of meteors with approximately parallel trajectories.
U. Stream: for use with A or D – a group of meteoric bodies with nearly identical orbits.

3.3.2.1 Meteors

3.3.2.1.1 Significance of meteor study

Meteor study is the most efficient technique to obtain information on distribution, orbits and to some degree chemical composition of meteoroids in the mass range of $\approx 10^{-5}$ g$\cdots 10^{8}$ g. It also gives insight into the mechanics of comets producing meteor streams. Meteor studies also are relevant for atmospheric physics and chemistry since aerosols in the stratosphere and upper atmosphere originate from meteoroids.

3.3.2.1.2 Orbits

Fig. 1 shows distributions of orbital elements of meteors observed by different techniques. Orbital elements of observed and recovered meteorites are given in Table 1. Extensive tabulations of orbits in [10⋯12] and rates in [13⋯15], cf. [16].

Table 1. Orbital elements of fallen and recovered meteorites with observed meteor phenomena.
a = semi-major axis q = aphelion distance = $a(1+e)$
e = eccentricity Q = perihelion distance = $a(1-e)$
i = inclination

	Date of fall		True radiant		Orbital elements					Photo	Ref.
	y/month/d	local time	α	δ	a AU	e	Q AU	q AU	i		
Sikhote Alin	1947/2/12	10^h38^m	28°	80°	2.16	0.54	0.99	3.34	9°.4	–	4
Pribram	1959/4/7	20^h30^m	191.5	17.7	2.42	0.67	0.79	4.05	10.4	+	5
Lost City	1970/1/3	20^h14^m	315.0	39.1	1.66	0.42	0.96	2.36	12.0	+	6
Innisfree	1977/2/6	2^h17^m	6.7	66.2	1.87	0.47	0.99	2.76	12.3	+	7
Dhajala	1976/1/28	20^h40^m	156.7	59.4	1.80	0.59	0.73	2.84	27.6	–	8

Fig. 1: see next page.

3.3.2.1.3 Classification: p. 190

3.3.2.1.4 References for 3.3.2.1

1 Millman, P.M.: Meteoritics **2** (1963) 7.
2 Ceplecha, Z. in: Comets, Asteroids, Meteorites (Delsemme, A.H., ed.), University of Toledo Press (1977) 143.
3 Hughes, D.W. in: Cosmic Dust (McDonnell, J.A.M., ed.), Wiley Sons, New York (1978) 123.
4 Fesenkov, V.G.: Meteoritika **9** (1951) 27.
5 Ceplecha, Z.: Bull. Astron. Inst. Czech. **12** (1961) 21.
6 McCrosky, R.E., Posen, A., Schwartz, G., Shao, C.J.: J. Geophys. Res. **76** (1971) 4090.
7 Halliday, I., Blackswell, A.T., Griffin, A.A.: J. R. Astron. Soc. Canada **75** (1978) 15.
8 Ballath, G.M., Bhatnagar, F., Bhandari, N.: Icarus **33** (1978) 361.
9 Verniani, F.: Space Sci. Rev. **10** (1969) 230.
10 Ceplecha, Z., McCrosky, R.E.: J. Geophys. Res. **81** (1976) 6257.
11 McCrosky, R.E.: Smithsonian Astrophys. Obs., Spec. Rep. **252** (1967).
12 Hawkins, G.S., Wouthworth, R.B.: Smithsonian Contrib. Astrophys. **4** (1961) 85.
13 Millman, P.M., McIntosh, B.A.: Canadian J. Phys. **42** (1964) 1730.
14 McCrosky, R.E., Posen, A.: Smithsonian Astrophys. Obs., Spec. Rep. **273** (1968).
15 Keay, C.S.L., Ellyett, C.D.: Mem. R. Astron. Soc. **72** (1969) 185.
16 Hoffmeister, C. in: Landolt-Börnstein, NS, Vol. VI/1 (1965) p. 195.

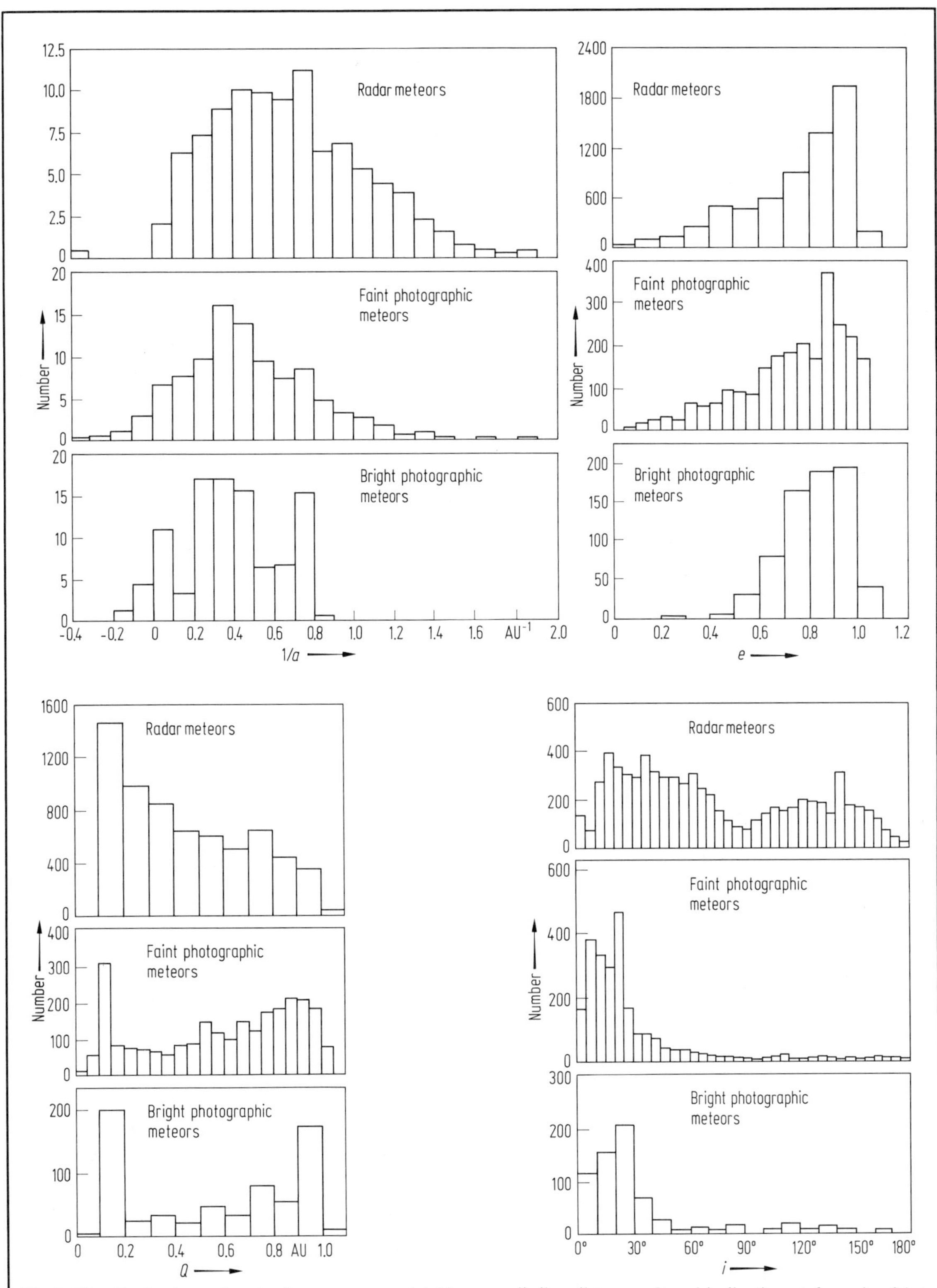

Fig. 1. The distributions of semimajor axes, *a*, eccentricities, *e*, perihelion distances, *Q*, and inclinations, *i*, for radar, faint photographic and bright photographic meteors [3].

3.3.2.1.3 Classification

Ceplecha [2] proposed a classification of meteoroid populations based on observation of meteor phenomena covering the entire mass range from $10^{-4} \cdots 10^{8}$ g which also attempts to identify properties and compositions of meteoroids producing certain groups of meteors (Table 2). A different classification and density identification has been proposed by Verniani [9].

Table 2. Meteoroid populations among photographic meteors and the relationship of meteor groups found in different photographic data [2].
The data are based on photographic observations of ≈2600 sporadic meteors.
The percentage number does not contain major classical showers. The second line gives typical mass ranges.
a, *e*, *i*: characteristic orbital elements. The values of the maxima of statistical distributions of *a*, *e*, and *i* are given,if not stated otherwise.
ϱ: bulk density of meteoroid; *σ*: ablation coefficient.

	Super-Schmidt camera meteors				Small-camera meteors				Prairie-Network fireballs					Properties of the meteoroid material		
Mass range	10^{-3} to 10 g				10^{-1} to 10^{3} g				[d]) 10^{2} to 10^{6} g							
Group	% obs.	*a* AU	*e*	*i*	% obs.	*a* AU	*e*	*i*	Group	% obs.[a])	*a* AU	*e*	*i*	*ϱ* g cm^{-3}	*σ* s^{2} cm^{-2}	Assumed composition
"asteroidal" meteors	< 1 [c])	2.4	0.64	15°	5 [b])	2.5	0.64	10°	I	32	2.4	0.68	6°	3.7	$10^{-11.5}$	ordinary chondrites
A	54	2.3	0.61	1°	37	2.5	0.64	4°	II	37	2.3	0.61	5°	2.1	$10^{-11.3}$	carbonaceous chondrites
B	6	2.4	0.92	5°	7	2.5	0.90	6°						1.0	$10^{-11.1}$	dense cometary material
C_1	9	2.2	0.80	6°	16	2.5	0.80	5°	IIIA	9	2.4	0.82	4°	0.6	$10^{-10.9}$	regular cometary material
C_2	31	≈∞	0.99	[g])	30	≈∞	0.99	[g])	IIIAi[f])	9	≈∞	0.99	[g])	0.6	$10^{-10.9}$	regular cometary material
D (above C)	< 1 [e])	3.3	0.70	25°	5 [b])	3.1	0.77	10°	IIIB	13	3.0	0.70	13°	0.2	$10^{-10.7}$	soft cometary material of Draconid type

[a]) Rough estimates from the numbers of computed trajectories (42, 37, 6, 6, 9 % respectively), assuming twofold reduction preference of group I over group III.
[b]) Median from 12 cases.
[c]) Only two meteors recognized as asteroidal (No. 7946 and 19816); their average elements are given.
[d]) The largest bodies photographed within fireball networks so far have the mass of the order of 10^{8} g.
[e]) Draconid shower orbit. (2 Super-Schmidt members of the stream)
[f]) All high-inclined near-parabolic orbits.
[g]) Random.

3.3.2.2 Meteorites

3.3.2.2.1 Definition

A meteorite is an extraterrestrial solid object which survived passage through the earth's atmosphere and reached the earth's surface as a recoverable object [2].

3.3.2.2.2 Significance of meteorite study

Before the Apollo and Luna missions, meteorites were the only material from outside the earth accessible for laboratory investigation. In contrast to terrestrial rocks, most meteorites have preserved records of early solar system processes, some meteorites possibly even of pre- or extra-solar system phenomena. Their detailed study continues to provide basic information about the history and origin of the solar system. They also serve as interplanetary space probes to record cosmic rays.

3.3.2.2.3 Orbits

Orbital elements of the three recovered meteorites for which the glowing passage through the atmosphere has been photographed by more than one camera, Lost City, Pribram and Innisfree, are included in 3.3.2.1.2, Table 1.

3.3.2.2.4 Classification and chemical composition

Meteorites are classified into two major categories (Table 3): a) undifferentiated meteorites containing chondrules, and b) differentiated meteorites lacking chondrules. Chondrules are round or elliptical objects with diameters of about 1 mm and mostly consist of olivine and orthopyroxene.

Table 3. Classification of meteorites and characteristic minerals (based on refs. [1···3]).

a) Undifferentiated meteorites: chondrites		
Enstatite chondrites:	E (enstatite)	enstatite, nickel-iron, troilite
Ordinary chondrites:	H (bronzite)	olivine, bronzite, nickel-iron
	L (hypersthene)	olivine, hypersthene, nickel-iron
	LL (amphoterite)	olivine, hypersthene, plagioclase
Carbonaceous chondrites:	CV, CO (C3)	olivine, pyroxene, organics
	CM (C2)	magnetite, phyllosilicates, organics
	CI (C1)	magnetite, phyllosilicates, organics
b) Differentiated meteorites:		
Silica-rich: achondrites		
Calcium-poor:	aubrites	enstatite, forsterite
	diogenites	bronzite, clinobronzite
	chassignites	olivine
	ureilites	olivine, nickel-iron, diamond
Calcium-rich:	angrites	titanaugite, olivine
	nakhlites	diopside, olivine
	eucrites	pigeonite, bytownite [2]
	howardites	hypersthene, bytownite [2]
Metal-rich		
Stony-irons:	mesosiderites	pyroxene, nickel-iron, bytownite [2]
	pallasites	olivine, nickel-iron
	siderophyre	bronzite, tridymite
	londranite	pyroxene, nickel-iron, olivine
Irons [1]:	hexahedrites	kamacite
	octahedrites	kamacite, taenite
	ataxites	kamacite, taenite

[1]) More detailed subdivision, using chemical data, see Table 4. [2]) Calcium-rich plagioclase.

a) Undifferentiated meteorites are further subdivided using chemical criteria into 5 groups; the members of each group appear to be genetically related. Figure 2 shows that the groups each populate a distinct field in a compositional diagram [4]. The carbonaceous chondrites are further subdivided into four groups on the basis of carbon and water contents [5, 6] and also of other criteria (e.g. chondrule/matrix ratio) [2,7]; they are named CV, CO, CM, and CI, the second letter representing the name of a prominent group member. CI-chondrites do not contain chondrules, but based on chemical and mineralogical arguments are classified as chondrites. Introduced by Van Schmus and Wood [4], a second dimension, the petrologic type, is used to discern highly metamorphosed meteorites (type 6) from very weakly or non-metamorphosed meteorites (type 1), with intermediate types 2···5. Distinguishing criteria are given in [4].

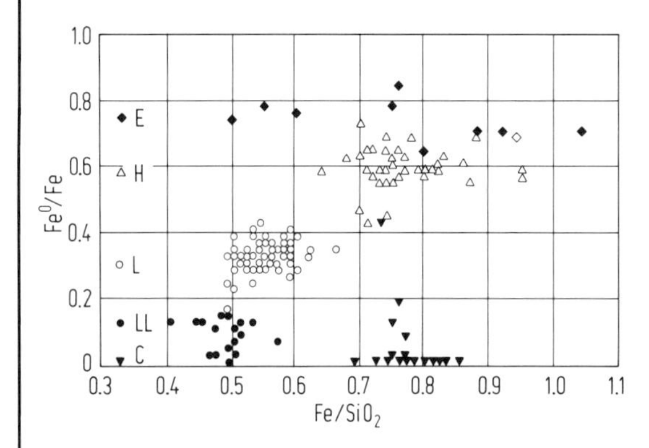

Fig. 2. Plot of metallic-iron/total-iron ratios (Fe^0/Fe) versus Fe/SiO_2 ratios for the chemical groups of chondrites [4] (Falls only). Symbols: E = Enstatite, H = High-, L = Low-iron, LL = Low-metal Low-iron and C = Carbonaceous chondrites.

b) Differentiated meteorites are subdivided into silica-rich and metal-rich (>6% metallic Fe) populations.

Silica-rich meteorites are further subdivided mainly on the basis of the Ca/Mg and Fe/(Fe+Mg) ratios (Fig. 3) reflecting genetic relationships: extremes in Ca/Mg point towards magmatic fractionation. The Fe/(Fe+Mg) ratio reflects the degree of oxidation and tends to be lower in the solid than in the liquid phase of a cooling magma. Wasson [2] gives a comprehensive list of distinguishing properties of silica-rich meteorites.

Metal-rich meteorites are further subdivided into iron and stony-iron meteorites. The classification of iron-meteorites is based on structure (width of kamacite lamellae, which depends on Ni-content, cooling rate and nucleation temperature) and Ni, Ga, Ge, and Ir concentrations [2].

Table 4a. Classification and properties of iron-meteorites [2].
Freq.: observed frequency corr.: correlation (positive, negative or absent)
Bandwidth: width of kamacite lamellae

Group	Freq. %	Bandwidth mm	Structure (see Table 4b)	Ni wt%	Ga ppm	Ge ppm	Ir ppm	Ge–Ni corr.
IA	17.1	1.0 ···3.1	Om–Ogg	6.4··· 8.7	55 ···100	190 ···520	0.6 ···5.5	neg.
IB	1.7	0.01 ···1.0	D–Om	8.7···25	11 ···55	25 ···190	0.3 ···2.0	neg.
IIA	8.3	50	H	5.3··· 5.7	57 ···62	170 ···185	2 ···60	pos.?
IIB	2.6	5 ···15	Ogg	5.7··· 6.4	46 ···59	107 ···183	0.01···0.5	neg.
IIC	1.5	0.06 ···0.07	Opl	9.3···11.5	37 ···39	88 ···114	4 ···11	pos.
IID	2.6	0.4 ···0.9	Of–Om	9.8···11.3	70 ···83	82 ···98	3.5 ···18	pos.
IIE	2.3	0.1 ···2	Anom[a])	7.5··· 9.7	21 ···28	60 ···75	1 ···8	abs.
IIIA	24.9	0.9 ···1.3	Om	7.1··· 9.3	17 ···23	32 ···47	0.17···19	pos.
IIIB	7.0	0.6 ···1.3	Om	8.4···10.5	16 ···21	27 ···46	0.01···0.17	neg.
IIIC	1.5	0.2 ···0.4	Off–Of	10 ···13	11 ···27	8 ···70	0.07···0.55	abs.
IIID	1.1	0.01 ···0.05	D–Off	16 ···23	1.5 ···5.2	1.4 ···4.0	0.02···0.07	neg.
IIIE	1.7	1.3 ···1.6	Og	8.2··· 9.0	17 ···19	34 ···37	0.05···6	abs.
IIIF	1.1	0.5 ···1.5	Om–Og[b])	6.8··· 7.8	6.3 ···7.2	0.7 ···1.1	1.3 ···7.9	abs.
IVA	8.3	0.25 ···0.45	Of	7.4··· 9.4	1.6 ···2.4	0.09···0.14	0.4 ···4	pos.
IVB	2.3	0.006···0.03	D	16 ···26	0.17···0.27	0.03···0.07	13 ···38	pos.

[a]) Also Om and Og. [b]) Also Anom.

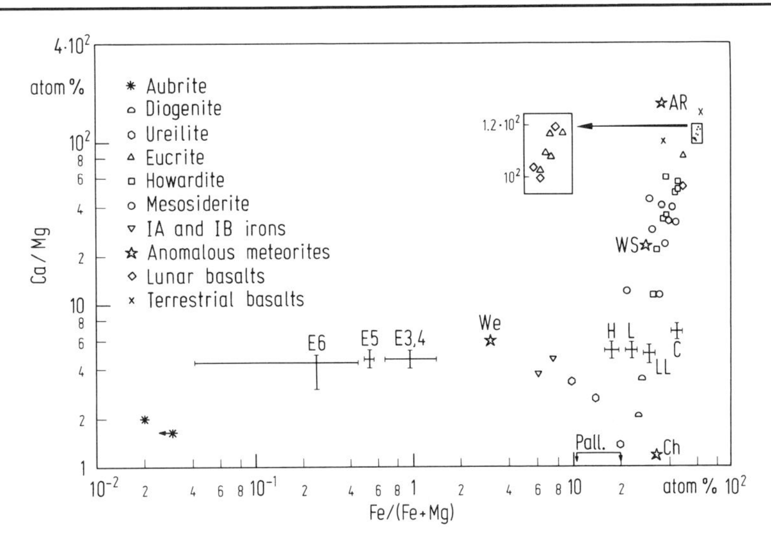

Fig. 3. Plot of Ca/Mg vs. Fe/(Fe+Mg) ratio of silicate-rich meteorites and three lunar basaltic rocks. Chondritic groups are shown as ranges, and differentiated meteorites are plotted individually. In most meteorites the Fe/(Fe+Mg) ratio is a measure of the degree of oxidation of the mineral assemblages. The variation in Ca/Mg ratio is probably mainly a result of igneous differentiation. It rises from very low values in ultramafic, high-melting silicates (such as those in the pallasites) to very high values in low-melting basaltic silicates as found in the eucrites. The anomalous meteorites are abbreviated: Angra dos Reis (stone) = AR; Chassigny = Ch; Weatherford = We; Weekeroo Station (now included in a new iron meteorite group IIE) = WS [2]. Letters E, H, ...: see Fig. 2 and Table 5.

Table 4b. Structural nomenclature [3].

Symbol	Name	Remarks
H	hexahedrites	no octahedral orientation even in large sections
Ogg	coarsest octahedrites	taenite may or may not be present
Og	coarse octahedrites	
Om	medium octahedrites	
Of	fine octahedrites	
Off	finest octahedrites	distinct bands of kamacite
Opl	plessitic octahedrites	kamacite sparks and spindles
D	ataxites	well-developed, slowly annealed plessite, kamacite spindles very rare
Anom	anomalous	includes all irons which demand individual descriptions

c) Chemical compositions of meteorite groups are given in Table 5. The chemical composition of CI carbonaceous chondrites resembles most closely the abundance of the non-volatile elements in the sun and is a major source for the reference chemical composition of the solar system as a whole [8···10], cf. 3.4.

The distribution of trace elements in mineral phases has been shown to be indicative for meteorite forming and alteration processes [11···15]. Among the minor elements, the rare earths are highly diagnostic. Their abundance patterns (usually normalized to that of CI chondrites) in meteorites and their minerals reflect condensational as well as magnetic and metamorphic processes and may serve to estimate the physical parameters involved, pressure, temperature etc. [11, 16···26]. Comprehensive discussions and references in [1, 2, 27].

Table 5. Average chemical composition of chondrites, achondrites, silicates of stony-irons and irons. Only groups with more than 4 members are listed.

Group	C	Na	Mg	Al	Si	P	S	Ca	Fe	Ni	K	Ti	Cr	Mn	Co	Ref.
	wt%										ppm					
Chondrites																
E types 3, 4	0.39	0.82	10.7	0.79	17.0	0.21	5.9	0.84	33.0	1.81	860	570	3210	3200	850	27
E type 5	0.42	0.52	11.1	0.82	16.4	0.13	5.7	0.85	31.4	1.71	780	690	3300	1800	850	27
E type 6	0.33	0.52	13.3	0.85	19.5	0.12	3.3	0.90	25.5	1.53	680	620	3300	1800	850	27
H	0.11	0.58	14.2	1.01	17.1	0.11	2.0	1.19	27.6	1.68	800[d])	620	3400	2250	830	27
L	0.10	0.66	15.2	1.10	18.7	0.11	2.2	1.28	21.8	1.15	830	660	3800	2450	550	27
LL	0.12	0.66	15.3	1.12	18.8	0.11	2.3	1.25	20.0	0.91	810	840	3700	2560	480	27
CV	0.134	0.33	14.6	1.64	15.5	0.097	2.2	1.80	22:8	0.097	300	960	3550	1530	960	6, 78
CO	0.42	0.43	14.7	1.46	15.8	0.14	2.2	1.60	25.8	1.40	840	820	3360	1660	740	6, 132, 78
CM	2.33	0.38	11.8	1.08	13.1	0.11	3.4	1.34	21.9	1.24	405	540	3100	1630	560	27
CI	3.15	0.51	14.5	0.85	10.3	0.10	5.9	1.06	18.4	1.02	520	430	2400	1900	500	27
Achondrites																
Aubrites	0.07	0.35	23.2	0.50	27.7	0.10	0.4	0.80	1.0	0.016	250[e])	340	500	1400	4	27
Diogenites	0.04	0.003	15.9	0.80	24.7	0.18	0.4	1.02	13.5	0.04	23	1200	6800	2480	19	27, 139
Ureilites	3.0	0.03	22.6	0.30	19.1	0.03	0.5	0.62	14.5	0.13	150	880	4900	2890	114	27
Eucrites	0.06	0.28	4.3	6.50	23.0	0.03	0.2	7.65	14.4	0.0013	360	4600	2300	3900	2	27
Howardites	0.11	0.19	9.4	4.20	22.9	0.03	0.3	4.47	13.9	0.0013	310[f])	2200	4600	3900	19	27
Stony-irons																
Pallasites[a])	0.07	0.05	28.0	0.01	8.0	0.08	0.2	0.005	52.2	0.0053	37	25	4650	2040		27, 139
Mesosiderites[b])	0.08	0.13	8.9	4.63	9.1	0.36	1.1	4.05	48.2	1.7	160	960	2460	3250	2800	27, 139
Irons	0.04		1[c])	1[c])	10[c])	0.2	0.1		90.6	7.9		5	50	1	5000	3

[a]) Olivine.
[b]) Silicates.
[c]) ppm.
[d]) Beardsley: 1160 ppm.
[e]) Bishopville: 1140 ppm.
[f]) Zmenj: 1100

3.3.2.2.5 Mineralogy and petrology

The knowledge of the chemical compositions and compositional trends of minerals, their relationships and descriptions of the structures of a meteorite as a whole and of its lithologic units, is the key to understanding its origin and history and the genetic relationship to other meteorites. Certain mineral systems may be used to indicate pressure, temperature, oxygen fugacity, cooling rate etc. [1, 2, 28···30, 156, 157]. The studies of mineral systems and of other features in meteorites (chondrules, fractures, inclusions, flow structures etc.) bear implication on the accretional and shock and temperature history (cf. [2]). Table 6 lists the common meteoritic minerals [138].

Table 6. The more common meteoritic minerals and their formulas (modified from [2]).

Mineral	Formula	Mineral	Formula
Chlorapatite	$Ca_5(PO_4)_3Cl$	Orthopyroxene	$(Mg, Fe)SiO_3$
Chromite	$FeCr_2O_4$	Enstatite	$MgSiO_3$
Clinopyroxene	$(Mg, Fe, Ca)SiO_3$	Bronzite	$Mg_{0.8\cdots0.9}\,Fe_{0.2\cdots0.1}\,SiO_3$
Augite	$(Ca, Mg, Fe^{2+}, Al)_2\,(Si, Al)_2O_6$	Hypersthene	$Mg_{0.1\cdots0.8}\,Fe_{0.9\cdots0.2}\,SiO_3$
Diopside	$Ca_{0.5}Mg_{0.5}SiO_3$	Ferrosilite	$FeSiO_3$
Hedenbergite	$Ca_{0.5}Fe_{0.5}SiO_3$	Pentlandite	$(Fe, Ni)_9S_8$
Pigeonite	$Ca_x(Mg,Fe)_{1-x}SiO_3 (x \approx 0.1)$	Plagioclase	$NaAlSi_3O_8\cdots CaAl_2Si_2O_8$
Cohenite	Fe_3C	Albite	$NaAlSi_3O_8$
Cristobalite	SiO_2	Anorthite	$CaAl_2Si_2O_8$
Daubreelite	$FeCr_2S_4$	Quarz	SiO_2
Diamond	C	Rutile	TiO_2
Farringtonite	$Mg_3(PO_4)_2$	Schreibersite	$(Fe, Ni)_3P$
Graphite	C	Serpentine	$Mg_6Si_4O_{10}(OH)_8$
Ilmenite	$FeTiO_3$	Taenite	$Fe_{<0.8}Ni_{>0.2}$
Kamacite	$Fe_{0.93\cdots0.96}Ni_{0.07\cdots0.04}$	Tridymite	SiO_2
Magnetite	Fe_3O_4	Troilite	FeS
Melilite	$Ca_2\,(Mg, Al)\,(Si, Al)_2O_7$	Whitlockite	$Ca_9MgNa\,(PO_4)_7$
Nickel-iron	(Fe, Ni)	Wollastonite	$CaSiO_3$ (only in Ca-Al-rich inclusions)
Oldhamite	CaS		
Olivine	$(Mg, Fe)_2SiO_4$		
Fayalite	Fe_2SiO_4		
Forsterite	Mg_2SiO_4		

3.3.2.2.6 Organic matter

The study of organic matter in meteorites (mostly carbonaceous chondrites) is relevant to a) the search for extraterrestrial life and b) the investigation of abiotic processes which form organic matter. Nagy [31] gives an evaluation of the present status of the field. Two principal processes may abiotically form organic compounds in the observed compositions and quantities: a) Fischer-Tropsch-type synthesis [32, 33] or b) Miller-Urey-type synthesis [34, 35]. Biological processes have been excluded [36], although some authors discuss the possibility of extraterrestrial microfossils [31].

3.3.2.2.7 Rare gases

Rare gases are useful for measuring time (cf. 3.5) and nuclear effects since their initial concentrations are low allowing the detection of effects which involve only $\gtrsim 10000$ atoms/g. To define the isotopic and elemental composition of the various noble gas components, step- and linear-heating extraction techniques and the analyses of mineral separates (in cases involving chemical treatments) and of grain size fractions and combinations thereof are applied. A general discussion of noble gases in meteorites is found in [37⋯41, 161] and a full compilation of light noble gas concentrations in stony meteorites in [42].

Rare gas components which are generated within the meteorite by radioactive decay (radiogenic rare gases) or by interaction of the meteoroid with high energy cosmic rays (spallogenic rare gases) are discussed in 3.5. Other components can be distinguished by means of their particular isotopic composition and/or their local distribution. The main types are a) solar and b) planetary rare gases, in special cases accompanied by "exotic" components (cf. 3.3.2.2.8). Fig. 4 gives the absolute and relative abundances of He, Ne, Ar, Kr, and Xe of solar- and planetary-type rare gases (see next page).

Surface-correlated solar-type gas has been incorporated into the surfaces of the constituents of meteorites by irradiation with solar particles and subsequently has been at most moderately redistributed and fractionated [43⋯49]. The irradiation happened either in free space [51] or within a lunar-type regolith developed on a meteoritic parent body. At the same time solar flare particle tracks were induced [50⋯53].

Planetary-type rare gas is dissolved and probably derived by fractional processes from solar-type gas [40, 54⋯61]; probably it is concentrated in minute acid-resistent C-rich grains [94, 164, 165]. Its elemental composition is similar to that found in the terrestrial and Martian atmospheres [64].

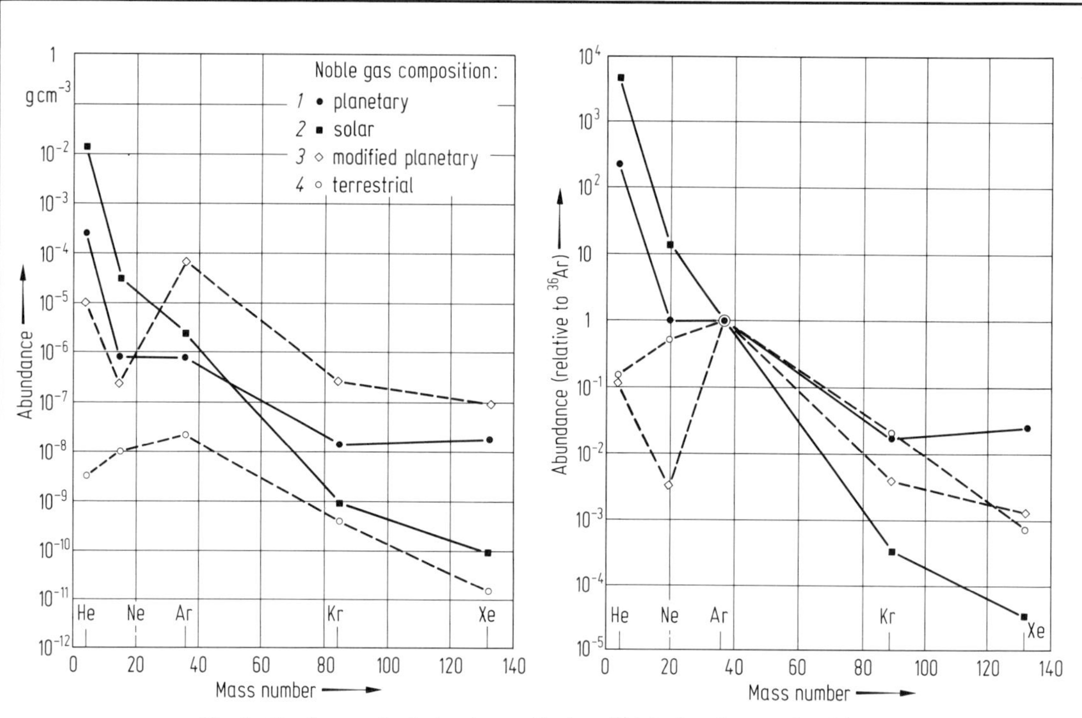

Fig. 4. Absolute and relative (normalized to ^{36}Ar) abundances of noble gases:
1 Planetary type (CM = chondrite Murray, average from [40, 54, 62]),
2 solar type (aubrite Pesyanoe [48]),
3 modified planetary type (diamonds from the ureilite Haverö [163]),
4 terrestrial atmosphere [63, 64].

3.3.2.2.8 Isotopic anomalies

Deviations from the cosmic isotopic abundance pattern [10], which cannot be explained as resulting from radioactive decay, mass fractionation or in-situ energetic particle interactions, are termed isotopic anomalies. Their presence indicates that the source region of certain meteorites, or of parts of them, has chemically and isotopically not been in equilibrium with the material which formed the sun. The magnitudes of the anomalies range from less than 1 part per 10^4 to several parts per 100.

The isotopic anomaly which is most elaborated and probably involves the largest quantity of anomalous material is that of oxygen [65···70]. Fig. 5 [71] unambiguously demonstrates large-scale isotopic inhomogeneities within the planetary system recorded in undifferentiated meteorites.

Other isotopic anomalies have been reported for the following elements (found in many cases in calcium-aluminium-rich inclusions in the Allende C3V chondrite [72···77]): neon [57, 58, 79, 80, 80a], magnesium [81···84], solicon [85], sulfur [86], calcium [87, 88], krypton [89], strontium [97], tellurium [90], barium [91, 92], xenon [93···95, 165], neodymium [91], samarium [92, 96], mercury [98], uranium [99, 99a].

Although there appears to be a general agreement that isotopic anomalies reflect nucleosynthesis and condensation in supernovae, no widely accepted scenario has yet been developed for the relationship in time and space between measured isotopic anomalies and the possible production processes [89, 95, 100···105, 166].

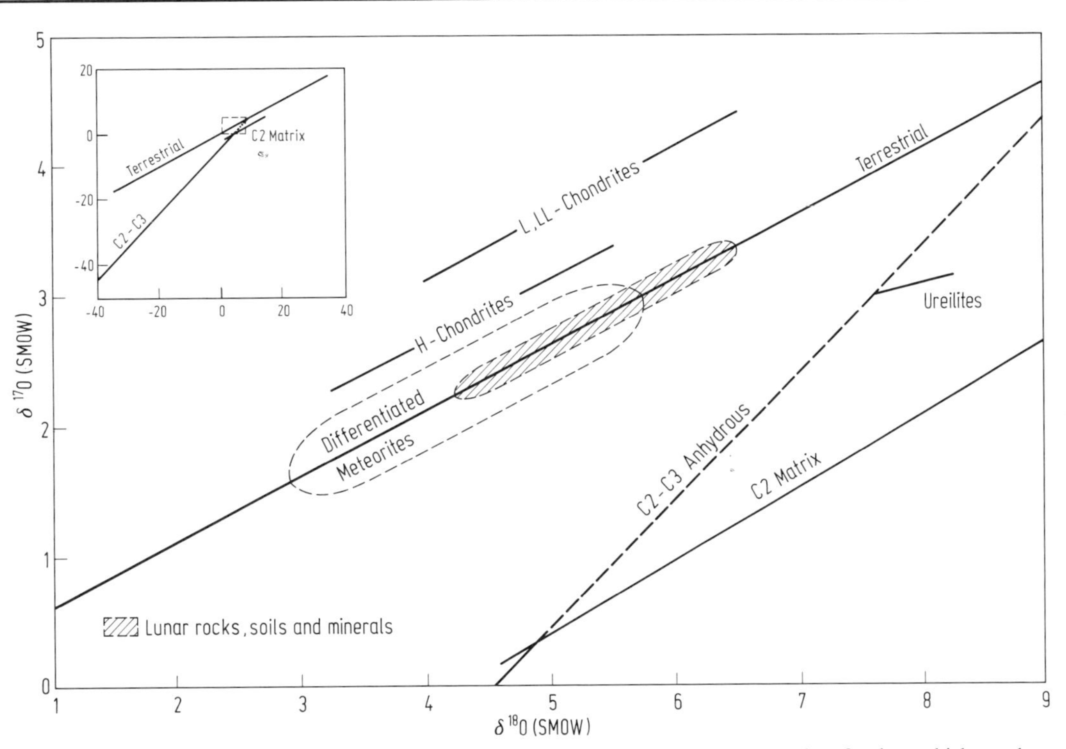

Fig. 5. Oxygen isotopic compositions for various classes of solar system materials expressed as δ-values which are the per mil deviations from "Standard Mean Ocean Water" (SMOW) of the ratios $^{17}O/^{16}O$ ($=\delta^{17}O$) and $^{18}O/^{16}O$ ($=\delta^{18}O$). Two sources of compositional variations are evident. Fractionations produce variations that trend along the solid lines (slope 1/2). The offset of the different fractionation lines indicates that these groupings cannot be related only by fractionation but must also reflect different admixtures of two independent components. Refractory (anhydrous) inclusions in C2 and C3 meteorites populate the mixing line (slope approximately unity) shown in the small-scale inset (the dashed line shown in the main figure is an extension of this line but is not populated by these inclusions). The largest admixture, up to almost 5%, of ^{16}O has been found in spinels from calcium-aluminum-rich inclusions of the C3V chondrite Allende given by the lowest values in the inset.

3.3.2.2.9 Origin of meteorites

Exposure ages of the order of 10^7 a (cf. 3.5) require a source region for meteorites away from the earth's orbit and provide evidence for shielding against galactic cosmic ray irradiation for the largest portion of the lifetime of meteorites, that is, evidence that meteorites are fragments of larger "parent bodies". (The plural is suggested e.g. by the oxygen isotopic inhomogeneity.) Model calculations [106···109] and chemical considerations (e.g. [110, 111]) suggest that meteorites are derived from the asteroid belt. Chondritic meteorites may have a cometary origin [112···115], but at present no unambiguous identification of individual meteorites or groups with specific source regions appears to be possible [116, 117]. Estimates for the maximum sizes of the parent bodies vary from some 10 km to $\gtrsim$1000 km [116, 118···120]. To produce differentiated meteorites in such relatively small bodies, effective internal heat sources (e.g. ^{26}Al [121]), electrical induction heating [122, 124] or gas friction heating events [116] have been proposed [123]. The cooling history of the parent body can be inferred from metallurgical [125, 126] or radiometric [127] studies. Various models have been put forward for the formation of parent bodies [110, 116, 119, 128···137, 158, 159].

3.3.2.2.10 Meteorites on earth

A. Distinguishing meteoritic features

a) Fusion crust: During meteor phase the skin of the luminous body is ablated. Then the meteorite has flight markings and an up to 1 mm thick dark crust [140].

b) Density: Metal-rich meteorites have densities in the range $4 \cdots 8$ g cm^{-3}.

c) Chondrules: They are present in ≈80% of falls.

d) Ni/Co and Fe/Ni ratios: Both are ≈17 [1] with rare exceptions.

e) Noble gas composition: All meteorites contain spallogenic isotopes, some solar and other non-terrestrial types of rare gas.

B. Falls and finds

According to Wasson's list [2], out of the total of ≈1650 known meteorites, 42.5% are observed and witnessed falls, the remainder are finds. 71% of falls are H- and L-group chondrites. For metal-rich meteorites the find/fall ratio is ≈14, for silica-rich meteorites ≈1 because the silica-rich meteorites a) have a higher fall frequency, b) are less resistant against terrestrial weathering and c) are more difficult to recognize among terrestrial rocks. Recently Antarctica has been found to contain a wealth of meteorite finds [141].

C. Meteorite nomenclature

Meteorites are named after geographic landmarks, the nearest city or village to their place of fall or find [2, 160].

D. Terrestrial ages

The present-day amount of a radioactive isotope of proper half-life, which has been produced by spallation reactions in orbit, depends on the time the meteorite has spent on earth shielded from galactic cosmic rays. Useful chronometers are $3 \cdot 10^5$ a ^{36}Cl, 5700 a ^{14}C, and 270 a ^{39}Ar. The production rates for the chronometer isotopes depend on the shielding depth of the sample in orbit and on its chemistry and in cases are difficult to estimate. Terrestrial ages of meteorites may be used to assess the rate of fall of various meteorite types. Compilations of terrestrial ages are given in [142···147, 167].

E. Meteorite craters

Meteoroids hit the earth's atmosphere at $11 \cdots 30$ km s^{-1} and are slowed down by the dense air below ≈100 km altitude. If the meteoroid's mass is less than ≈10^4 kg, it hits the earth's surface with a terminal velocity of $0.1 \cdots 0.3$ km s^{-1} and no explosion crater will result from the impact. Larger meteoroids can create impact craters [148, 149, 172].

By analogy to the moon, during the first 500 million years of the earth's history it has been heavily bombarded by many and large meteorites [150]; the craters, however, have been eroded and eliminated by geologic processes since then.

Grieve and Robertson [151, 152] give an update of the terrestrial cratering record. 13 craters are associated with meteorites, for 5 more craters traces of meteoritic matter have been identified [153···155, 162]. 78 structures probably have an impact origin, cf. [168···172].

3.3.2.2.11 References for 3.3.2.2

1 Mason, B.: Meteorites, J. Wiley and Sons Inc. New York (1962) 274.
2 Wasson, J.: Meteorites, Springer-Verlag, Berlin (1974).
3 Buchwald, V.F.: Handbook of iron meteorites, Center for Meteorite Studies, Arizona, University of California Press, Berkeley (1975).
4 Van Schmus, W.R., Wood, J.A.: Geochim. Cosmochim. Acta **31** (1967) 747.
5 Wiik, H.B.: Geochim. Cosmochim. Acta **9** (1956) 279.
6 Mason, B.: Space Sci. Rev. **1** (1963) 621.
7 Van Schmus, W.R. in: Meteorite Research (Millman, P.M., ed.) D. Reidel, Dordrecht (1969) 480.
8 Urey, H.C.: Rev. Geophys. **2** (1964) 1.
9 Anders, E.: Geochim. Cosmochim. Acta **35** (1971) 516.
10 Cameron, A.G.W.: Space Sci. Rev. **15** (1973) 121.
11 Gast, P.W. in: The Nature of the Solid Earth (Robertson, E.C., ed.), McGraw-Hill, New York, N.Y. (1971) 19.

12 Grossman, L.: Geochim. Cosmochim. Acta **36** (1972) 597.
13 Grossman, L.: Geochim. Cosmochim. Acta **37** (1973) 1119.
14 Grossman, L., Clark, S.P.: Geochim. Cosmochim. Acta **37** (1973) 635.
15 Wänke, H., Baddenhausen, H., Palme, H., Spettel, B.: Earth Planet. Sci. Lett. **23** (1974) 1.
16 Masuda, A., Nakamura, N., Tanaka, T.: Geochim. Cosmochim. Acta **37** (1973) 239.
17 Grossman, L., Ganapathy, R.: Geochim. Cosmochim. Acta **40** (1976) 331.
18 Nakamura, N.: Geochim. Cosmochim. Acta **38** (1974) 757.
19 Boynton, W.V.: Geochim. Cosmochim. Acta **39** (1975) 569.
20 Chou, C.L., Baedecker, P.A., Wasson, J.T.: Geochim. Cosmochim. Acta **40** (1976) 85.
21 Ikramuddin, M., Matza, S.D., Lipschutz, M.E.: Geochim. Cosmochim. Acta **41** (1977) 1247.
22 Nagasawa, H., Blanchard, D.P., Jacobs, J.W., Brannon, J.C., Philpotts, J.A., Onuma, N.: Geochim. Cosmochim. Acta **41** (1977) 1587.
23 McSween, H.Y., jr.: Geochim. Cosmochim. Acta **41** (1977) 477.
24 Matza, S.D., Lipschutz, M.E.: Geochim. Cosmochim. Acta **42** (1978) 1655.
25 Takahashi, H., Janssen, M.J., Morgan, J.W., Anders, E.: Geochim. Cosmochim. Acta **42** (1978) 97.
26 Martin, P.M., Mason, B.: Nature **249** (1974) 333.
27 Mason, B.: Handbook of Elemental Abundances in Meteorites, Gordon and Breach Science Publ., N.Y. (1971).
28 Ramdohr, P.: The Opaque Minerals in Stony Meteorites, Akademie-Verlag, Berlin (1973).
29 Keil, K. in: Handbook of Geochemistry (Wedepohl, K.H., ed.) Vol. I, Springer-Verlag Berlin, Heidelberg, New York (1969) 78.
30 Kurat, G.: Geochim. Cosmochim. Acta **31** (1967) 491.
31 Nagy, B.: Carbonaceous Meteorites, Elsevier Sci. Publ. Comp. Amsterdam (1975).
32 Gelpi, E., Han, J., Nooner, D., Oro, J.: Geochim. Cosmochim. Acta **34** (1970) 965.
33 Studier, M.H., Hayatsu, R., Anders, E.: Geochim. Cosmochim. Acta **36** (1972) 189.
34 Miller, S.L., Urey, H.C.: Science **130** (1959) 245.
35 Urey, H.C., Lewis, J.S.: Science **152** (1966) 102.
36 Kvenvolden, K.A., Lawless, J., Pering, K., Peterson, E., Flores, J., Ponnamperuma, G., Kaplan, I.R., Moore, C.B.: Nature **228** (1976) 923.
37 Kirsten, T., Krankowsky, D., Zähringer, J.: Geochim. Cosmochim. Acta **27** (1963) 13.
38 Reynolds, J.H.: Annu. Rev. Nucl. Sci. **17** (1967) 253.
39 Zähringer, J.: Geochim. Cosmochim. Acta **32** (1968) 209.
40 Mazor, E., Heymann, D., Anders, E.: Geochim. Cosmochim. Acta **34** (1970) 781.
41 Nord, R., Zähringer, J.: Ocherki Sovvemennoy Geokhimii i Analyticheshoy Khimic, Nauka, Moscow (1972) 59.
42 Schultz, L., Kruse, H.: Nuclear Track Detection **2** (1978) 65.
43 Signer, P., Suess, H.E. in: Earth Science and Meteorites (Geiss, J., Goldberg, E.D., eds.), North Holland Publ. Amsterdam (1963) 241.
44 Suess, H.E., Wänke, H., Wlotzka, F.: Geochim. Cosmochim. Acta **28** (1964) 595.
45 Wänke, H.: Z. Naturforsch. **20a** (1965) 946.
46 Zähringer, J.: Earth Planet Sci. Lett. **1** (1966) 20.
47 Müller, H.W., Zähringer, J. in: Meteorite Research (Millman, P.M., ed.), Reidel Publ., Dordrecht (1969) 845.
48 Marti, K.: Science **166** (1969) 1263.
49 Arrhenius, G., Alfvén, H.: Earth Planet. Sci. Lett. **10** (1971) 253.
50 Pellas, P., Poupeau, G., Lorin, J.C., Reeves, H., Audouze, J.: Nature **223** (1969) 272.
51 Lal, D., Rajan, R.S.: Nature **223** (1969) 269.
52 Wilkening, L., Lal, D., Reid, A.M.: Earth Planet. Sci. Lett. **10** (1971) 334.
53 Poupeau, G., Kirsten, T., Steinbrunn, F., Storzer, D.: Earth Planet Sci. Lett. **24** (1974) 229.
54 Zähringer, J.: Z. Naturforsch. **17a** (1962) 460.
55 Kuroda, P.K., Manuel, O.K.: Nature **227** (1970) 1113.
56 Fanale, F.P., Cannon, W.A.: Geochim. Cosmochim. Acta **36** (1972) 319.
57 Black, D.C.: Geochim. Cosmochim. Acta **36** (1972) 347.
58 Black, D.C.: Geochim. Cosmochim. Acta **36** (1972) 377.
59 Manuel, O.K., Henneke, E.W., Sabu, D.D.: Nature **240** (1972) 99.
60 Lancet, M.S., Anders, E.: Geochim. Cosmochim. Acta **37** (1973) 1371.
61 Begemann, F., Weber, H.W., Hintenberger, H.: Astrophys. J. **203** (1976) L 155.
62 Reynolds, J.H.: Phys. Rev. Lett. **4** (1960) 351.

63 Nier, O.A.: Phys. Rev. **79** (1950) 450.
64 Anders, E., Owen, T.: Science **198** (1977) 453.
65 Clayton, R.N., Grossman, L., Mayeda, T.K.: Science **182** (1973) 485.
66 Clayton, R.N., Mayeda, T.K.: Geophys. Res. Lett. **4** (1977) 295.
67 Clayton, R.N., Grossman, L., Mayeda, T.K., Onuma, N. in: Proc. Soviet. Am. Conf. Cosmochem. Moon Planets, NASA SP-370 (1977) 781.
68 Clayton, R.N., Onuma, N., Grossman, L., Mayeda, T.K.: Earth Planet. Sci. Lett. **34** (1977) 209.
69 Clayton, R.N., Mayeda, T.K.: Proc. Lunar Sci. Conf. 6th (1975) 1761.
70 Clayton, R.N., Onuma, N., Mayeda, T.K.: Earth Planet. Sci. Lett. **30** (1976) 10.
71 Podosek, F.A.: Annu. Rev. Astron. Astrophys. **16** (1978) 293.
72 Clarke, R.S., jr., Jarosewich, E., Mason, B., Nelen, J., Gomez, M., Hyde, J.R. : Smithsonian Contrib. Earth No. 5 (1970).
73 Marvin, U.B., Wood, J.A., Dickey, J.S., jr.: Earth Planet. Sci. Lett. **7** (1970) 346.
74 El Goresy, A., Nagel, K., Ramdohr, P.: The Allende meteorite: Fremdlinge and their noble relatives, Lunar and Planet. Sci. IX, The Lunar and Planetary Institute, Houston (1978) 282.
75 Grossman, L.: Geochim. Cosmochim. Acta **39** (1975) 433.
76 McSween, H.Y., jr.: Geochim. Cosmochim. Acta **41** (1977) 1777.
77 Wark, D.A., Lovering, J.F.: Proc. Lunar Sci. Conf. 8th (1977) 95.
78 McCarthy, T.S., Ahrens, L.H.: Earth Planet. Sci. Lett. **14** (1972) 97.
79 Eberhardt, P.: Earth Planet. Sci. Lett. **24** (1974) 182.
80 Niederer, F., Eberhardt, P.: Meteoritics **12** (1977) 327.
80a Eberhardt, P., Jungck, M.H.A., Meier, F.O., Niederer, F.: Astrophys. J. **234** (1979) L169.
81 Gray, C.M., Compston, W.: Nature **251** (1974) 495.
82 Lee, T., Papanastassiou, D.A., Wasserburg, G.J.: Astrophys. J. Lett. **211** (1977) L 107.
83 Wasserburg, G.J., Lee, T., Papanastassiou, D.A.: Geophys. Res. Lett. **4** (1977) 299.
84 Lorin, J.C., Michel-Levy, M.: Lunar Planet. Sci. IX, The Lunar and Planetary Institute, Houston (1978) 660.
85 Clayton, R.N., Mayeda, T.K., Epstein, S.: Lunar Planet. Sci. IX, The Lunar and Planetary Institute, Houston (1978) 186.
86 Rees, C.E., Thode, H.G.: Geochim. Cosmochim. Acta **41** (1977) 1679.
87 Lee, T., Papanastassiou, D.A., Wasserburg, G.J.: Astrophys. J. Lett. **220** (1978) L21.
88 Lee, T., Russel, W.A., Wasserburg, G.J.: Astrophys. J. Lett. **228** (1979) L93.
89 Lewis, R.S., Gros, J., Anders, E.: J. Geophys. Res. **82** (1977) 779.
90 Ballad, R.V., Oliver, L.L., Downing, R.G., Manuel, O.K.: Nature **277** (1979) 615.
91 McCulloch, M.T., Wasserburg, G.J.: Geophys. Res. Lett. **5** (1978) 599.
92 Lugmair, G.W., Marti, K., Scheinin, N.B.: Lunar Planet. Sci. IX, The Lunar and Planetary Institute, Houston (1978) 672.
93 Manuel, O.K., Wright, R.J., Miller, D.K., Kuroda, P.K.: Geochim. Cosmochim. Acta **36** (1972) 961.
94 Lewis, R.S., Srinivasan, B., Anders, E.: Science **190** (1975) 1251.
95 Srinivasan, B., Anders, E.: Science **201** (1978) 51.
96 McCulloch, M.T., Wasserburg, G.J.: Astrophys. J. Lett. **220** (1978) L 15.
97 Papanastassiou, D.A., Huneke, J.C., Esat, T.M., Wasserburg, G.J.: Lunar Planet. Sci. IX, The Lunar and Planetary Institute, Houston (1978) 859.
98 Jovanovic, S., Reed, G.W.: Science **193** (1976) 888.
99 Arden, J.W.: Nature **269** (1977) 788.
99a Tatsumoto, M., Shimamura, T.: Meteoritics **14** (1979) 543.
100 Manuel, O.K., Sabu, D.D.: Science **195** (1977) 208.
101 Clayton, D.D.: Icarus **32** (1977) 255.
102 Clayton, D.D.: Space Sci. Rev. **24** (1979) 147.
103 Cameron, A.G.W., Truran, J.W.: Icarus **30** (1977) 447.
104 Heymann, D., Dziczkaniec, M., Walker, A., Huss, G., Morgan, J.A.: Astrophys. J. **225** (1978) 1030.
105 Lattimer, J.M., Schramm, D.N., Grossmann, L.: Astrophys. J. **219** (1978) 230.
106 Arnold, J.R. in: Isotopic and Cosmic Chemistry (Craig, H., Miller, S.L., Wasserburg, G.J., eds.), North Holland, Amsterdam (1964) 347.
107 Wetherill, G.W. in: Origin and Distribution of the Elements (Ahrens, L.H., ed.), Pergamon Press, London (1968) 423.
108 Zimmerman, P.D., Wetherill, G.W.: Science **182** (1973) 51.
109 Chapman, C.R. in: Comets, Asteroids, Meteorites (Delsemme, A.H., ed.), University of Toledo (1977) 265.

Jessberger

110 Anders, E.: Space Sci. Rev. **3** (1964) 583.
111 Larimer, J.W.: Geochim. Cosmochim. Acta **31** (1967) 1215.
112 Yokoyama, Y., Mabuchi, H., Labeyrie, J. in: Origin and Distribution of the Elements (Ahrens, L.H., ed.), Pergamon Press, London (1968) 445.
113 Millman, P.M. in: From Plasma to Planet (Elvius, A., ed.), Wiley, New York (1972) 157.
114 Wetherill, G.W. in: Comets, Asteroids, Meteorites (Delsemme, A.H., ed.), University of Toledo (1977) 283.
115 Scholl, H., Froeschle, G. in: Comets, Asteroids, Meteorites (Delsemme, A.H., ed.), University of Toledo (1977) 293.
116 Alfvén, H., Arrhenius, G.: Evolution of the solar system, NASA SP-345, Washington, D.C. (1976).
117 Dohnanyi, J.S. in: Cosmic Dust (McDonnell, J.A.M., ed.), J. Wiley Sons, New York (1978) 527.
118 Scott, E.R.D. in: Comets, Asteroids, Meteorites (Delsemme, A.H., ed.), University of Toledo (1977) 439.
119 Urey, H.C.: Proc. Conf. on the Origin of the Solar System, Nice/France (1972).
120 Hartmann, W.K. in: Comets, Asteroids, Meteorites (Delsemme, A.H., ed.), University of Toledo (1977) 277.
121 Papanastassiou, D.A., Lee, T., Wasserburg, G.J. in: Comets, Asteroids, Meteorites (Delsemme, A.H., ed.), University of Toledo (1977) 343.
122 Sonett, C.P., Smith, B.F., Colburn, D.S., Schubert, G., Schwartz, K.: Proc. Lunar Sci. Conf. 3rd (1972) 2309.
123 Sonett, C.P., Herbert, F.L. in: Comets, Asteroids, Meteorites (Delsemme, A.H., ed.), University of Toledo (1977) 429.
124 Sonett, C.P., Colburn, D.S., Schwartz, K., Keil, K.: Astrophys. Space Sci. **7** (1970) 446.
125 Buseck, P.R., Goldstein, J.I.: Bull. Geol. Soc. Am. **80** (1969) 2141.
126 Powell, B.N.: Geochim. Cosmochim. Acta **33** (1969) 789.
127 Pellas, P., Storzer, D. in: Comets, Asteroids, Meteorites (Delsemme, A.H., ed.), University of Toledo (1977) 335.
128 Urey, H.C., Craig, H.: Geochim. Cosmochim. Acta **4** (1953) 36.
129 Wood, J.A.: Icarus **2** (1963) 152.
130 Larimer, J.W., Anders, E.: Geochim. Cosmochim. Acta **34** (1970) 367.
131 Anders, E.: Annu. Rev. Astron. Astrophys. **9** (1971) 1.
132 Mason, B.: Meteoritics **6** (1971) 59.
133 De, R.B.: Proc. Lunar Sci. Conf. 8th (1977) 87.
134 Larimer, J.W.: Space Sci. Rev. **15** (1973) 103.
135 Grossman, L., Larimer, J.W.: Rev. Geophys. Space Phys. **12** (1974) 71.
136 Arrhenius, G.: Chemical aspects of the formation of the solar system, NATO Advanced Study Institute on the Origin of the Solar System (Dermott, S.F., ed.), Wiley, New York (1977) 221.
137 Kurat, G., Hoinkes, G., Fredrikson, K.: Earth Planet. Sci. Lett. **26** (1975) 140.
138 Mason, B.: Meteoritics **7** (1972) 309.
139 Wood, J.A. in: The Moon, Meteorites and Comets, The solar system III (Middlehurst, B., Kuiper, B.P., eds.), Vol. IV, University Chicago Press (1963) 337.
140 Krinov, E.L.: Principles of Meteoritics, Pergamon, London, New York (1960).
141 Yanai, K., Cassidy, W.A., Funaki, M., Glass, B.P.: Proc. Lunar Planet. Sci. Conf. 9th (1978) 977.
142 Fireman, E.L., DeFelice, J.: J. Geophys. Res. **65** (1960) 3035.
143 Suess, H.E., Wänke, H.: Geochim. Cosmochim. Acta **26** (1962) 475.
144 Goel, P.S., Kohman, T.P.: Science **136** (1962) 875.
145 Vilcsek, E., Wänke, H. in: Radioactive Dating, IAEA, Vienna (1963) 381.
146 Chang, C., Wänke, H. in: Meteorite Research (Millman, P. M., ed.) (1969) 397.
147 Boeckl, R.: Nature **236** (1972) 25.
148 Heide, F.: Kleine Meteoritenkunde, Springer-Verlag, Heidelberg (1957).
149 Dence, M.R.: Proc. 24th Intern. Geol. Cong. **15** (1972) 77.
150 Neukum, G., Wise, D.U.: Science **194** (1976) 1381.
151 Grieve, R.A.F., Robertson, P.B.: Icarus **38** (1979) 212.
152 Grieve, R.A.F., Robertson, P.B.: Icarus **38** (1979) 230.
153 Grieve, R.A.F.: Proc. Lunar Sci. Conf. 9th (1978) 2579.
154 Palme, H., Janssens, M.J., Takahashi, H., Anders, E., Hertogen, J.: Geochim. Cosmochim. Acta **42** (1978) 313.
155 Lambert, P. in: Impact and Explosion Cratering (Roddy, D.J., Pepin, R.O., Berill, R.B., eds.), Pergamon, Elmsford (1977) 44.
156 Scott, E.R.D.: Geochim. Cosmochim. Acta **41** (1977) 349.
157 Smith, B.A., Goldstein, J.I.: Geochim. Cosmochim. Acta **41** (1977) 1061.

158 Herndon, J.M., Suess, H.E.: Geochim. Cosmochim. Acta **41** (1977) 233.
159 McSween, H.Y., jr., Richardson, S.M.: Geochim. Cosmochim. Acta **41** (1977) 1145.
160 Buchwald, V.F., Wasson, J.T.: Meteoritics **7** (1972) 17.
161 Heymann, D. in: Handbook of Elemental Abundances in Meteorites, (Mason, B., ed.), Gordon and Breach Sci. Publ., New York (1971) 22.
162 El Goresy, A., Chao, E.C.T.: Earth Planet. Sci. Lett. **31** (1976) 330.
163 Weber, H.W., Begemann, F., Hintenberger, H.: Earth Planet. Sci. Lett. **29** (1976) 81.
164 Otting, W., Zähringer, J.: Geochim. Cosmochim. Acta **31** (1967) 1949.
165 Reynolds, J.H., Frick, U., Neil, J.M., Phinney, D.L.: Geochim. Cosmochim Acta **42** (1978) 1775.
166 Heymann, D., Dziczkaniec, M.: Science **191** (1976) 79.
167 Fireman, E.L., Rancitelli, L.A., Kirsten, T.: Science **203** (1979) 453.
168 Nininger, H.H. in: The Moon, Meteorites, and Comets (Middlehurst, B.M., Kuiper, B.P., eds.) (1963) 162.
169 Krinov, E.L. in: The Moon, Meteorites, and Comets (Middlehurst, B.M., Kuiper, B.P., eds.) (1963) 183.
170 Dietz, R.S. in: The Moon, Meteorites, and Comets (Middlehurst, B.M., Kuiper, B.P., eds.) (1963) 285.
171 Shoemaker, E.M. in: The Moon, Meteorites, and Comets (Middlehurst, B.M., Kuiper, B.P., eds.) (1963) 301.
172 Chao, E.C.T.: Geologica Bavarica **75** (1977) 421.

3.3.3 Comets

3.3.3.1 Mechanical data

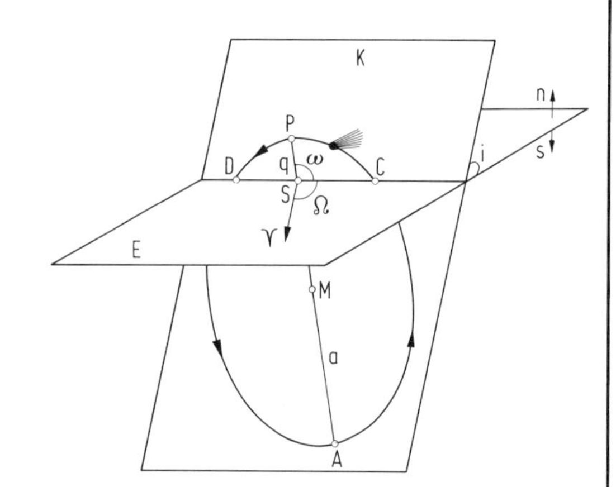

Fig. 1. The orbit of a comet and the plane of ecliptic.

K	plane of orbit of comet
E	plane of ecliptic
n, s	north, south
S	sun
P	perihelion
A	aphelion
C	ascending node
D	descending node
M	center of orbit
a = AM	semi-major axis
q = PS	perihelion distance
Q = AS	aphelion distance
ϒ	first point of Aries

Definitions

Orbital Elements:	T	time of perihelion passage
	$\omega \sphericalangle$ CSP	argument of perihelion measured in the plane of the comet's orbit
	$\Omega \sphericalangle$ ϒSC	longitude of the ascending node measured in the plane of the ecliptic
	i	inclination of the orbit to the ecliptic, an angle between 0° and 180° (the motion is direct if $i < 90°$, and retrograde if $i > 90°$)
	q = PS $= a(1-e)$	perihelion distance [AU]
	e = SM/PM $= \sin\varphi$	eccentricity (φ = angle of eccentricity)
	P	period [a]

Additional Elements: $a^3 = P^2$

$Q = a(1+e) = 2a - q$	aphelion distance [AU]
$\pi = \omega + \Omega$	longitude of perihelion
$n = 0.985\,608/P$	mean daily motion in degrees
R	earth—sun = radius vector of the earth
r	comet—sun = radius vector of the comet
Δ	earth—comet = geocentric distance of the comet
α	phase angle = angle between earth, comet, and sun
$\cos\alpha = \frac{\Delta^2 + r^2 - R^2}{2r\Delta}$;	$\sin\frac{1}{2}\alpha = \frac{1}{2}\sqrt{\frac{(R+r-\Delta)(R-r+\Delta)}{r\Delta}}$
$L = \Omega + \text{arc tan}(\tan\omega \cos i)$	ecliptical longitude of perihelion
$B = \text{arc sin}(\sin\omega \sin i)$	ecliptical latitude of perihelion

Discovery
Until about 1970, mostly visually by amateurs; presently mostly on photographic plates as by-products of other research projects.

Number of comets
Chinese records since 11th century; more than 500 comets are recorded before 1600 AD ([z] Table 1). A total of 10^{11} comets is estimated to move around the sun with a total mass comparable to that of the earth.
Halley's comet: See Appendix p. 375 in Subvol. c

Table 1. Number of appearances of comets with known orbits [G, z].
Interval: compiled according to year of perihelion passage.
N total number of appearances
N_0 total number of discoveries
N_1 total number of newly discovered short-period comets ($P < 200$ a; $e < 0.97$)
$N(e=1.0)$ number of comets with parabolic orbits
$N(e>1.0)$ number of comets with hyperbolic orbits

Interval	N	N_0	N_1	$N(e=1.0)$	$N(e>1.0)$
0 ··· 500	6				
500 ··· 1000	12				
1000 ··· 1500	25				
1500 ··· 1600	14				
1600 ··· 1700	21				
1700 ··· 1800	63				
1800 ··· 1820.0	24	22	5	13	0
1820 ··· 1840	31	22	1	17	0
1840 ··· 1860	74	61	9	27	5
1860 ··· 1880	73	52	5	29	2
1880 ··· 1900	106	78	15	29	14
1900 ··· 1910.0	41	29	6	8	10
1910 ··· 1920	47	28	6	5	7
1920 ··· 1930	50	29	9	6	6
1930 ··· 1940	54	30	4	5	5
1940 ··· 1950	75	47	11	12	12
1950 ··· 1960	83	41	7	7	15
1960 ··· 1970	94	42	8	12	4
1970 ··· 1979	132	63	18	10	16

Nomenclature

Preliminary: name or names (up to three) of discoverer(s), year of discovery and lowercase letter in order of discovery within that year. Final: year of perihelion passage followed by roman numeral in order of perihelion passage: comet Arend-Roland 1956 h = 1957 III. Short-period comets ($P<200$ years) are marked by the letter P followed by the discoverer's (or computer's) name: comet 1910 II = P/Halley.

Comets are often called "new" when they have extremely large semi-major axes (possibly coming for the first time from the Öpik-Oort cloud into the inner solar system), and "old" when they have made many approaches to the sun (short-period comets).

Catalogues

Most recent catalogue [G] lists orbital elements for 1027 cometary apparitions of 658 individual comets observed between 87 BC and the end of 1978. Discussion: [b, l, w, r]. Some data are given in Tables 2···4, and Fig. 2.

Table 2. Distribution of orbital forms for 1027 apparitions of 658 individual comets observed between 86 BC and the end of 1978 [G].

Orbital form	Eccentricity	N	%
Elliptical orbits	$e<1.0$	275	42
short-period comets ($P<200$ a)	$e<0.97$	72*) 41**) } 113	17
long-period comets ($P<200$ a)	$0.97<e<1.0$	162	25
Parabolic orbits	$e=1.0$	285	43
Hyperbolic orbits	$e>1.0$	98	15
Strongly hyperbolic orbits	$e>1.006$	0	0

*) Observed at two or more apparitions.
**) Observed at only one apparition.

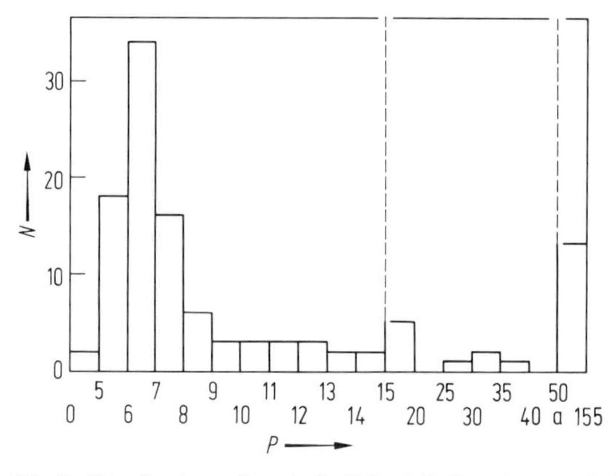

Fig. 2. Distribution of period, P in [a], for short-period comets.

Table 3. Periodic comets of more than one appearance, observed before 1979 *).
Symbols: see 3.3.3.1 (definitions and orbital elements)

No.	Name	Appearance			T	P	q	e	ω	Ω	i	Q
		first	number	last		a	AU					AU
1	Encke	1786 I	51	1977 XI	1977.63	3.31	0.341	0.846	186°0	334°2	11°9	4.10
2	Grigg–Skjellerup	1902 II	13	1977 VI	1977.28	5.10	0.993	0.665	359.3	212.6	21.1	4.93
3	Tempel 2	1873 II	16	1977 d	1978.14	5.27	1.369	0.548	190.9	119.2	12.5	4.69
4	Honda–Mrkos–Pajdusakova	1948 XII	5	1974 XVI	1974.99	5.28	0.579	0.809	184.6	233.0	13.1	5.49
5	Neujmin 2	1916 II	2	1927 I	1927.04	5.43	1.338	0.567	193.7	328.0	10.6	4.84
6	Brorsen	1846 III	5	1879 I	1879.24	5.46	0.590	0.810	14.9	102.3	29.4	5.61
7	Tempel 1	1867 II	6	1977 i	1978.03	5.50	1.497	0.519	179.1	68.3	10.5	4.73
8	Clark	1973 V	2	1978 g	1978.90	5.51	1.557	0.501	209.0	59.1	9.5	4.68
9	Tuttle–Giacobini–Kresák	1858 III	6	1978 r	1978.98	5.58	1.124	0.643	49.4	153.3	9.9	5.17
10	Tempel–Swift	1869 III	4	1908 II	1908.76	5.68	1.153	0.638	113.5	291.1	5.4	5.22
11	Wirtanen	1947 XIII	5	1974 XI	1974.51	5.87	1.256	0.614	351.8	83.5	12.3	5.26
12	D'Arrest	1851 II	13	1976 XI	1976.62	6.23	1.164	0.656	178.9	141.4	16.7	5.61
13	Du Toit–Neujmin–Delporte	1941 VII	2	1970 XIII	1970.77	6.31	1.677	0.509	115.7	187.9	2.9	5.15
14	De Vico–Swift	1844 I	3	1965 VII	1965.31	6.31	1.624	0.524	325.4	24.4	3.6	5.21
15	Pons–Winnecke	1819 III	18	1976 XIV	1976.91	6.36	1.254	0.635	172.4	92.7	22.3	5.61
16	Forbes	1929 II	5	1974 IX	1974.38	6.40	1.533	0.555	259.9	25.2	4.6	5.36
17	Kopff	1906 IV	11	1977 V	1977.18	6.43	1.572	0.545	162.9	120.3	4.7	5.34
18	Schwassmann–Wachmann 2	1929 I	8	1974 XIII	1974.70	6.51	2.142	0.386	357.3	126.0	3.7	4.83
19	Giacobini–Zinner	1900 III	10	1978 h	1979.12	6.52	0.996	0.715	172.0	195.1	31.7	5.99
20	Wolf–Harrington	1924 IV	6	1977 j	1978.20	6.53	1.615	0.538	187.0	254.2	18.5	5.38
21	Churyumov–Gerasimenko	1969 IV	2	1976 VII	1976.27	6.59	1.298	0.631	11.3	50.4	7.1	5.73
22	Biela	1772	6	1852 III	1852.73	6.62	0.861	0.756	223.2	247.3	12.6	6.19
23	Tsuchinshan 1	1965 I	3	1978 e	1978.35	6.65	1.499	0.576	22.8	96.2	10.5	5.58
24	Perrine–Mrkos	1896 VII	5	1968 VIII	1968.84	6.72	1.272	0.643	166.1	240.2	17.8	5.85

continued

*) Cometary statistics will change as new comets are found or as old ones are re-observed. Comet Schwassmann-Wachmann 3 e.g., (No. 5 in Table 4), was last seen in 1930, and re-discovered in 1979 (orbital elements: T=1979 Sept. 2.814; $\omega=198°8$, $\Omega=69°4$; $i=11°4$; $q=0.9392$ AU; $e=0.692$; $P=5.32$ a).

Table 3 (continued).

No.	Name	Appearance first	number	last	T	P a	q AU	e	ω	Ω	i	Q AU
25	Reinmuth 2	1947 VII	5	1974 VI	1974.35	6.74	1.941	0.456	45.4	296.1	7.0	5.19
26	Johnson	1949 II	5	1977 I	1977.02	6.76	2.196	0.386	206.2	117.8	13.9	4.96
27	Borrelly	1905 II	9	1974 VII	1974.36	6.76	1.316	0.632	352.7	75.1	30.2	5.84
28	Harrington	1953 VI	2	1960 VII	1960.49	6.80	1.582	0.559	232.8	119.2	8.7	5.60
29	Gunn	1969 II	2	1976 III	1976.11	6.80	2.445	0.319	197.5	68.0	10.4	4.74
30	Tsuchinshan 2	1965 II	3	1978 p	1978.72	6.83	1.785	0.504	203.2	287.6	6.7	5.41
31	Arend–Rigaux	1950 VII	5	1977 k	1978.09	6.83	1.442	0.600	329.0	121.5	17.9	5.76
32	Brooks 2	1889 V	11	1974 I	1974.01	6.88	1.840	0.491	198.2	176.3	5.6	5.39
33	Finlay	1886 VII	9	1974 X	1974.50	6.95	1.096	0.699	322.1	41.8	3.6	6.19
34	Taylor	1916 I	2	1977 II	1977.03	6.98	1.951	0.466	355.6	108.2	20.5	5.35
35	Holmes	1892 III	5	1972 I	1972.08	7.05	2.155	0.414	23.4	327.5	19.2	5.20
36	Daniel	1909 IV	5	1964 II	1964.30	7.09	1.662	0.550	10.8	68.5	20.1	5.72
37	Shajn–Schaldach	1949 VI	3	1978 I	1979.02	7.25	2.223	0.407	215.3	167.2	6.2	5.27
38	Faye	1843 III	17	1977 IV	1977.16	7.39	1.610	0.576	203.7	199.1	9.1	5.98
39	Ashbrook–Jackson	1948 IX	5	1977 g	1978.63	7.43	2.284	0.400	349.0	2.1	12.5	5.33
40	Whipple	1933 IV	7	1977 h	1978.23	7.44	2.469	0.352	190.0	188.3	10.2	5.15
41	Harrington–Abell	1954 XIII	4	1976 VIII	1976.31	7.59	1.776	0.540	138.5	336.8	10.2	5.95
42	Reinmuth 1	1928 I	6	1973 IV	1973.22	7.63	1.995	0.485	9.5	121.1	8.3	5.76
43	Kojima	1970 XII	2	1977 r	1978.39	7.85	2.399	0.393	348.6	154.1	0.9	5.50
44	Oterma	1942 VII	3	1958 IV	1958.44	7.88	3.388	0.144	354.9	155.1	4.0	4.53
45	Arend	1951 X	4	1975 VI	1975.39	7.98	1.847	0.538	46.9	355.7	20.0	6.14
46	Schaumasse	1911 VII	6	1960 III	1960.29	8.18	1.196	0.705	51.9	86.2	12.0	6.92
47	Jackson–Neujmin	1936 IV	3	1978 q	1978.98	8.37	1.425	0.654	196.3	163.1	14.1	6.82
48	Wolf	1884 III	12	1976 II	1976.07	8.42	2.501	0.396	161.1	203.8	27.3	5.78

continued

Table 3 (continued).

No.	Name	Appearance			T	P	q	e	ω	Ω	i	Q
		first	number	last		a	AU					AU
49	Comas Sola	1927 III	7	1977 n	1978.73	8.94	1.870	0.566	42.8	62.4	13.0	6.74
50	Kearns–Kwee	1963 VII	2	1972 XI	1972.91	9.01	2.229	0.485	131.2	315.4	9.0	6.43
51	Denning–Fujikawa	1881 V	2	1978 n	1978.75	9.01	0.779	0.820	334.0	41.0	8.7	7.88
52	Swift–Gehrels	1889 VI	2	1972 VII	1972.66	9.23	1.354	0.692	84.4	314.2	9.3	7.44
53	Neujmin 3	1929 III	3	1972 IV	1972.37	10.57	1.976	0.590	146.9	150.2	3.9	7.66
54	Klemola	1965 VI	2	1976 X	1976.61	10.94	1.766	0.642	148.9	181.6	10.6	8.09
55	Gale	1927 VI	2	1938	1938.46	10.99	1.183	0.761	209.1	67.3	11.7	8.70
56	Vaisala 1	1939 IV	4	1971 VII	1971.70	11.28	1.866	0.629	49.7	134.7	11.5	8.19
57	Slaughter–Burnham	1958 VI	2	1970 V	1970.28	11.62	2.543	0.504	44.3	346.1	8.2	7.72
58	Van Biesbroeck	1954 IV	3	1977 s	1978.92	12.39	2.395	0.553	134.3	148.6	6.6	8.31
59	Wild 1	1960 I	2	1973 VIII	1973.50	13.29	1.981	0.647	167.9	358.2	19.9	9.24
60	Tuttle	1790 II	9	1967 V	1967.24	13.77	1.023	0.822	206.9	269.8	54.4	10.46
61	Du Toit 1	1944 III	2	1974 IV	1974.25	14.97	1.294	0.787	257.2	22.1	18.7	10.86
62	Schwassmann–Wachmann 1	1925 II	4	1974 II	1974.12	15.03	5.448	0.105	14.5	319.6	9.7	6.73
63	Neujmin 1	1913 III	4	1966 VI	1966.94	17.93	1.543	0.775	346.8	347.2	15.0	12.16
64	Crommelin	1818 I	4	1956 VI	1956.82	27.89	0.743	0.919	196.0	250.4	28.9	17.65
65	Tempel–Tuttle	1366	4	1965 IV	1965.33	32.91	0.982	0.904	172.6	234.4	162.7	19.56
66	Stephan–Oterma	1867 I	2	1942 IX	1942.96	38.84	1.595	0.861	358.3	78.6	17.9	21.34
67	Westphal	1852 IV	2	1913 VI	1913.90	61.86	1.254	0.920	57.1	347.3	40.9	30.03
68	Olbers	1815	3	1956 IV	1956.46	69.57	1.178	0.930	64.6	85.4	44.6	32.65
69	Pons–Brooks	1812	3	1954 VII	1954.39	70.92	0.774	0.955	199.0	255.2	74.2	33.49
70	Brorsen–Metcalf	1847 V	2	1919 III	1919.79	71.95	0.484	0.972	129.6	311.2	19.2	34.11
71	Halley	−86	27	1910 II	1910.30	76.08	0.587	0.967	111.7	57.8	162.2	35.32
72	Herschel–Rigollet	1788 II	2	1939 VI	1939.60	154.90	0.748	0.974	29.3	355.3	64.2	56.94

Table 4. Periodic comets of only one appearance, observed before 1979.
Symbols: see 3.3.3.1 (definitions and orbital elements).

No.	Name		T	P a	q AU	e	ω	Ω	i	Q AU
1	1766 II	Helfenzrieder	1766.32	4.51	0.403	0.852	178°.1	76°.1	7°.9	5.05
2	1819 IV	Blanpain	1819.89	5.10	0.892	0.699	350.2	79.2	9.1	5.03
3	1945 II	Du Toit 2	1945.30	5.28	1.250	0.588	201.5	358.9	6.9	4.81
4	1884 II	Barnard 1	1884.63	5.38	1.279	0.583	301.0	6.1	5.5	4.86
5	1930 VI	Schwassmann–Wachmann 3 *)	1930.45	5.43	1.011	0.673	192.3	77.1	17.4	5.17
6	1886 IV	Brooks 1	1886.43	5.44	1.325	0.571	176.8	54.5	12.7	4.86
7	1770 I	Lexell	1770.62	5.60	0.674	0.786	224.9	133.9	1.6	5.63
8	1783	Pigott	1783.89	5.89	1.459	0.552	354.6	58.0	45.1	5.06
9	1978 i	Haneda–Campos	1978.77	5.97	1.101	0.665	240.4	131.6	6.0	5.48
10	1975 IV	West–Kohoutek–Ikemura	1975.15	6.12	1.398	0.582	358.0	84.7	30.1	5.29
11	1978 b	Wild 2	1978.45	6.17	1.491	0.557	39.9	136.1	3.3	5.23
12	1975 III	Kohoutek	1975.05	6.23	1.568	0.537	169.8	273.2	5.4	5.21
13	1977 XIII	Tritton	1977.82	6.34	1.438	0.580	147.7	300.0	7.0	5.41
14	1951 IX	Harrington-Wilson	1951.83	6.36	1.664	0.515	343.0	127.8	16.4	5.20
15	1890 VII	Spitaler	1890.82	6.37	1.818	0.471	13.4	45.9	12.8	5.06
16	1882 III	Barnard 3	1892.95	6.52	1.432	0.590	170.0	207.3	31.3	5.55
17	1896 V	Giacobini	1896.83	6.65	1.455	0.588	140.5	194.2	11.4	5.62
18	1918 III	Schorr	1918.74	6.66	1.882	0.468	279.1	118.3	5.6	5.20
19	1978 k	Giclas	1978.89	6.68	1.732	0.512	247.2	141.5	8.5	5.36
20	1974 IX	Longmore	1974.84	6.98	2.402	0.342	196.3	15.0	24.4	4.90
21	1895 II	Swift	1895.64	7.20	1.298	0.652	167.8	171.1	3.0	6.16
22	1894 I	Denning	1894.11	7.42	1.147	0.698	46.3	85.1	5.5	6.46
23	1977 o	Schuster	1978.02	7.47	1.628	0.574	353.9	50.8	20.4	6.01
24	1906 VI	Metcalf	1906.77	7.78	1.631	0.584	199.8	195.2	14.6	6.22
25	1973 XI	Gehrels 2	1973.92	7.94	2.348	0.410	183.3	215.6	6.7	5.61
26	1977 VII	Gehrels 3	1977.31	8.11	3.424	0.152	231.5	242.6	1.1	4.65
27	1975 VII	Smirnova–Chernykh	1975.60	8.53	3.567	0.145	90.2	77.1	6.6	4.78
28	1975 I	Boethin	1975.01	11.05	1.094	0.780	11.1	27.0	5.9	8.83
29	1977 XII	Sanguin	1977.71	12.50	1.811	0.664	162.1	182.3	18.6	8.96
30	1846 VI	Peters	1846.42	12.71	1.527	0.720	339.6	261.9	30.5	9.37
31	1973 I	Gehrels 1	1973.07	14.52	2.935	0.507	28.9	14.6	9.6	8.97
32	1977 III	Kowal	1977.15	15.11	4.664	0.237	178.0	28.4	4.4	7.56
33	1961 X	Van Houten	1961.32	15.62	3.957	0.367	14.4	22.9	6.7	8.54
34	1977 I	Chernykh	1978.12	15.93	2.568	0.594	266.7	134.1	5.7	10.09
35	1827 II	Pons–Gambart	1827.43	57.46	0.807	0.946	19.2	319.3	136.5	28.97
36	1921 I	Dubiago	1921.34	62.35	1.115	0.929	97.4	66.5	22.3	30.33
37	1846 IV	De Vico	1846.18	76.30	0.664	0.963	12.9	79.0	85.1	35.31
38	1942 II	Vaisala 2	1942.13	85.42	1.287	0.934	335.2	171.6	38.0	37.50
39	1862 III	Swift–Tuttle	1862.64	119.98	0.963	0.960	152.8	138.7	113.6	47.69
40	1889 III	Barnard 2	1889.47	145.35	1.105	0.960	60.2	271.9	31.2	54.18
41	1917 I	Mellish	1917.27	145.36	0.190	0.993	121.3	88.0	32.7	55.10
42	1937 II	Wilk [26a]	1937.14	187.38	0.619	0.981	31.5	57.6	26.0	64.87

*) Re-discovered in 1979, see footnote of Table 3.

Perihelion distance (q)
Long-period comets: $\bar{q}=1.08$ AU; short-period comets: $\bar{q}=1.61$ AU (Fig. 3). Concentration near 1 AU is due to observational selection effects. Record largest $q=6.88$ AU (comet 1975 II); of the 9 "sun-grazers" with $q<0.01$ AU, 8 are clearly related (Table 6).

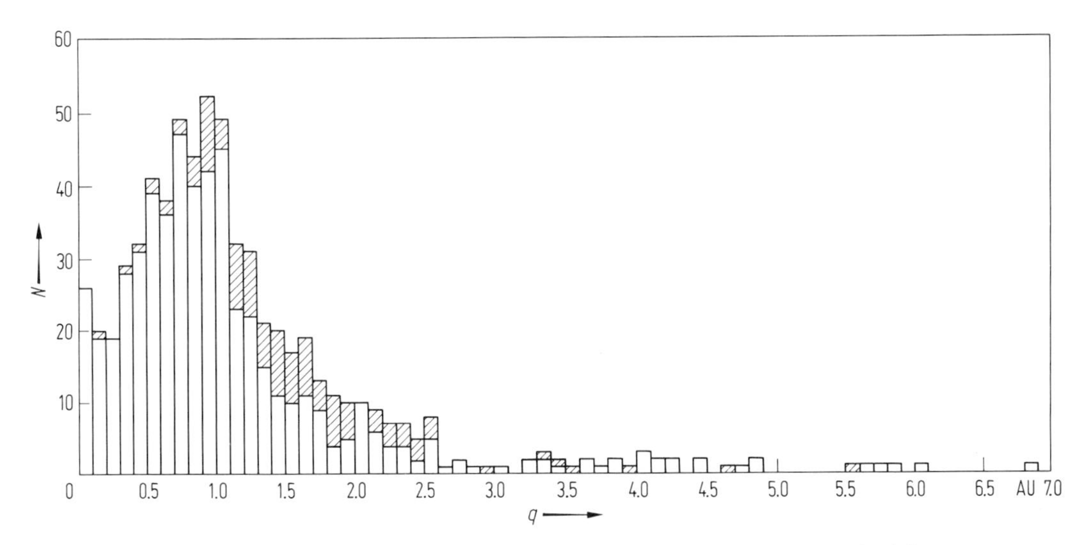

Fig. 3. Distribution of perihelion distance, q in [AU]; short-period comets shaded.

Table 5. Greatest values of perihelion distance [G].
Symbols: see 3.3.3.1 (orbital elements).

Comet		q AU	e	P a
1936 I	van Biesbroeck	4.043	1.0020	
1729		4.051	1.0	
1956 I	Haro–Chavira	4.076	1.0046	
1942 VIII	Oterma	4.113	1.0	
1925 VI	Shajn–Comas Sola	4.180	1.0024	
1959 X	Humason	4.266	1.0008	
1972 IX	Sandage	4.275	1.0062	
1957 VI	Wirtanen	4.447	1.0027	
1954 V	Abell	4.495	1.0027	
1977 III	P/Kowal	4.664	0.2370	15.1
1948 III	Johnson	4.708	1.0	
1973 X	Sandage	4.812	1.0001	
1972 XII	Araya	4.860	0.9999	
1974 II	P/Schwassmann–Wachmann 1	5.447	0.1054	15.0
1977 IX	West	5.606	1.0026	
1976 XII	Lovas	5.715	1.0031	
1976 IX	Lovas	5.875	1.0039	
1974 XII	van den Berg	6.018	1.0039	
1975 II	Schuster	6.880	1.0021	

Table 6. Least values of perihelion distance ("sun-grazers") [G].
Symbols: see 3.3.3.1 (orbital elements).

Comet		q AU	e	P a	ω	Ω	i
1887 I	Great Southern Comet (Thome)	0.00483	1.0		83°51	3°88	144°3
1963 V	Pereyra	0.00506	0.999946	903	86.15	7.23	144.5
1880 I	Great Southern Comet (Gould)	0.00549	1.0		86.24	7.07	144.6
1843 I	Great March Comet	0.00552	0.99991	513	82.63	2.82	144.3
1680		0.00622	0.99998		350.62	275.93	60.6
1945 VII	Du Toit	0.00751	1.0		72.06	350.50	141.8
1882 II	Great September Comet	0.00775	0.99990	759	69.58	346.95	142.0
1965 VIII	Ikeya–Seki	0.00778	0.99991	880	69.05	346.29	141.8
1970 VI	White–Ortiz–Bolelli	0.00887	1.0		61.29	336.31	139.0

Table 7. Distribution of nonperiodic comets with respect to their perihelion distance q [h].
q = perihelion distance
M_r = "reduced brightness" (at $\Delta = r = 1$ AU, see 3.3.3.2) in [mag]
$= m - 5 \log \Delta_0 - 10 \log r_0$,
with m = apparent magnitude at heliocentric distance r_0 and geocentric distance Δ_0.
It is assumed that the brightness follows an inverse 4th power law with heliocentric distance.

	Number of comets							
q [AU]	With type I tail (certain cases)		With type II tail only (including dubious type I cases)		Without tail		Total	
	$M_r \leqq 7.5$	$M_r > 7.5$	$M_r \leqq 7.5$	$M_r > 7.5$	$M_r \leqq 7.5$	$M_r > 7.5$	$M_r \leqq 7.5$	$M_r > 7.5$
0···1.2	28	9	15	31	2	27	45	67
1.2···2	4	0	7	9	6	6	17	15
2 ···3	0	0	10	2	5	1	15	3
3 ···4	0	0	5	0	1	0	6	0
4 ···5	0	0	5	0	1	0	6	0
> 5	0	0	0	0	0	0	0	0
Total	32	9	42	42	15	34	89	85

Inclination (i)
Long-period comets: $\bar{i} = 92°4$ (Fig. 4); nearly randomly distributed, i.e., the number varies with $\sin i$. Short-period comets: $\bar{i} = 20°3$; only four orbits are retrograde. For variation of i with period P, see Fig. 5. For the 98 comets with $P < 30$ a, $\bar{i} = 12°7$.

Aphelion distance (Q)
Of the 113 short-period comets, 50 (44%) have Q between 5···6 AU (Fig. 6), corresponding to Jupiter's distance from the sun (5.2 AU). The 78 (69%) with $Q < 7$ AU can be classified as belonging to a "Jupiter family"; their ω is strongly concentrated near 0° and 180° (Fig. 7), so that near perihelion and aphelion they are also near their nodes, and comets almost always come closer to Jupiter than to any other giant planet. No other planetary "family" is generally recognized [r].

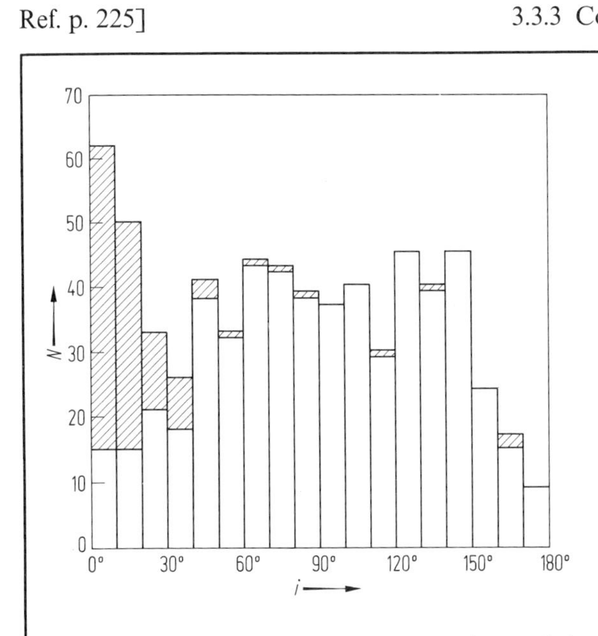

Fig. 4. Distribution of inclination, i in [°]; short-period comets shaded.

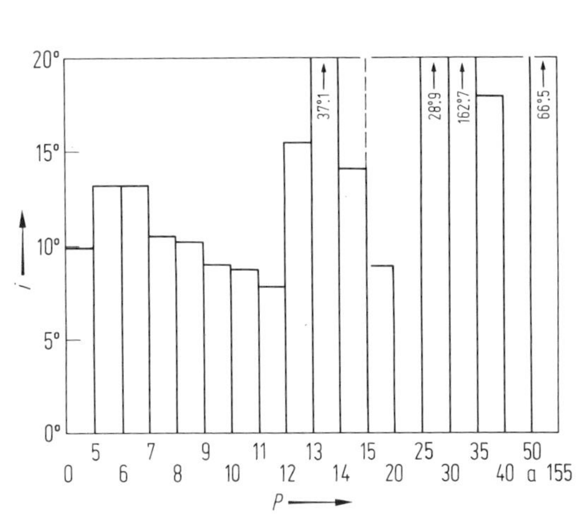

Fig. 5. Variation of inclination, i in [°], with period, P, for short-period comets.

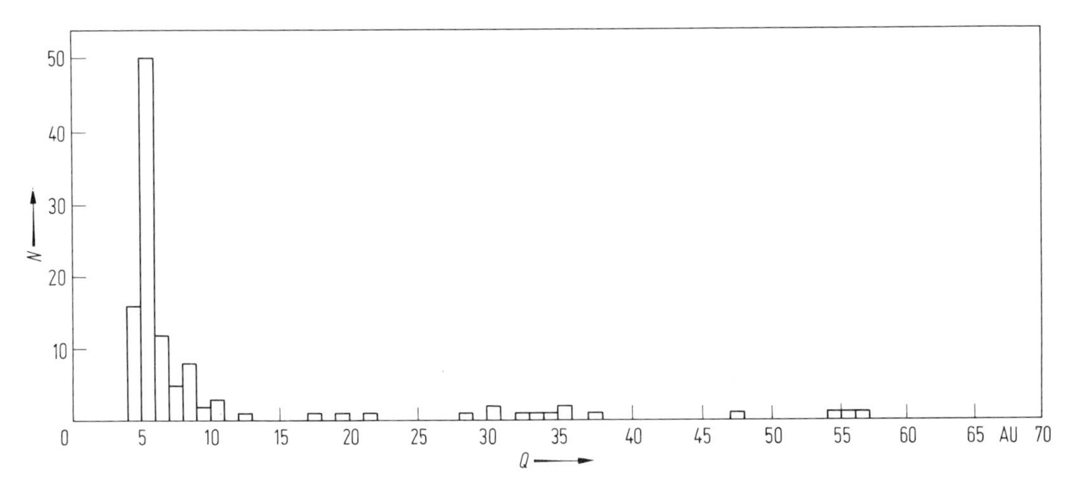

Fig. 6. Distribution of aphelion distance, Q in [AU], for short-period comets.

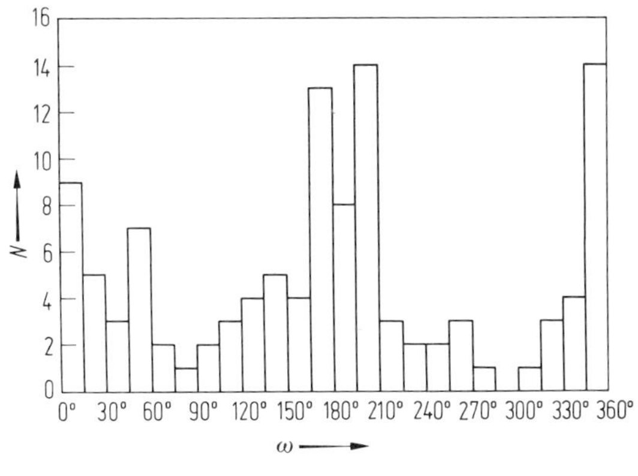

Fig. 7. Distribution of argument of perihelion, ω, for short-period comets.

Perihelion direction

Short-period comets (Fig. 8): ecliptical latitudes of perihelion practically all $<10°$, longitudes of perihelion show concentration toward that of Jupiter (13°). Long-period comets: perihelion directions have (nonrandom) distribution over whole sphere [45].

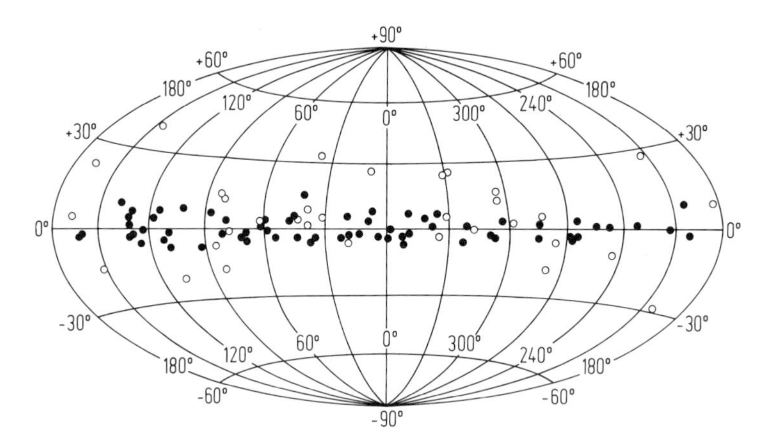

Fig. 8. Distribution of ecliptical longitude and latitude of perihelion for short-period comets [r]; full circles = comets of the Jupiter family; open circles = other short-period comets.

Variations of orbital elements

Short-period comets: Period distribution shows "Kirkwood gaps" [24]. Single encounters with Jupiter are inadequate for producing significant numbers of short-period orbits with low i from parabolic orbits with random i, and multiple encounters are important [12,f]. Long-period comets: changes $\Delta(1/a)$ in the reciprocal major axes for 392 long-period comets between 1800 and 1970 [13]. The representative value of $(1/a)$ for the original orbit is $+0.00004\,\mathrm{AU}^{-1}$ for a new comet, corresponding to $Q=5\cdot10^4$ AU and $P=4\cdot10^6$ a. Nearly 50% of new comets with $(1/a)<10^{-4}\,\mathrm{AU}^{-1}$ are lost to the solar system after perihelion passage.

Especially strong orbital changes due to close Jupiter approaches: [z].

Distribution of "original" orbits of nearly-parabolic comets: [32].

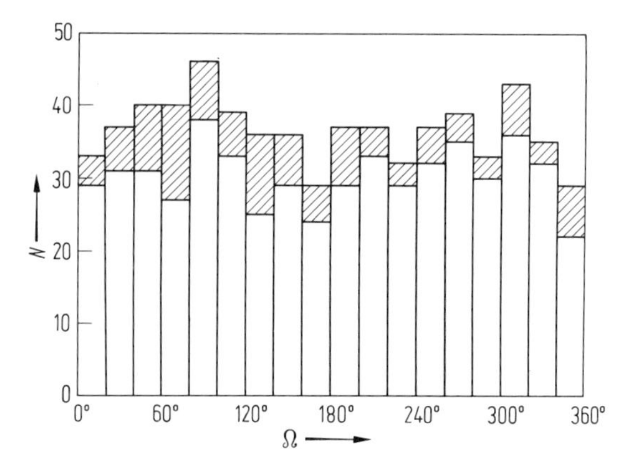

Fig. 9. Distribution of longitude of ascending node, ☊; short-period comets shaded.

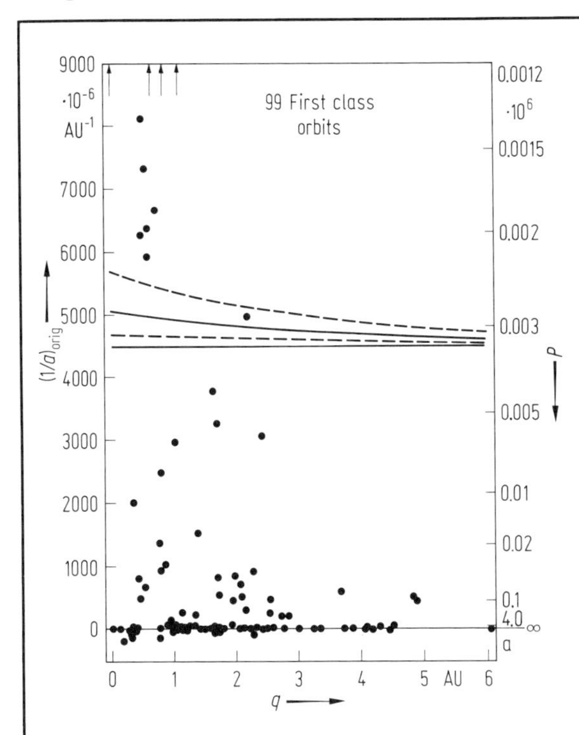

Fig. 10. Plot of $10^6\ (1/a)_{orig}$ vs q for 99 comet orbits of first class quality. The errors in the positions of the points are rarely larger than the points themselves. The horizontal line near the center of the diagram represents the fact that because of planetary perturbations, 50% of the values of $1/a$ can be expected to increase during one revolution of the comets about the sun, and 50% to decrease. The solid curve above it shows the average expected change of $1/a$ of the comets for which this quality increases, while the broken curves indicate the perturbations which should be exceeded by 20% and 80% of these comets respectively. The root-mean-square perturbation extends from the upper broken curve near $q=0$ to the solid curve near $q=5$ AU. The right-hand scale gives the equivalent revolution period P [r].

Comet groups

Only one group ("sun-grazers") has more than 2 convincing members. Comets 1882 II and 1965 VIII are almost certainly members of a comet that split at its previous perihelion passage (near beginning of 12th century) [29].

Split comets

More than a dozen comets have split at varying solar distances not related to perihelion [a, 41]; see Fig. 11.

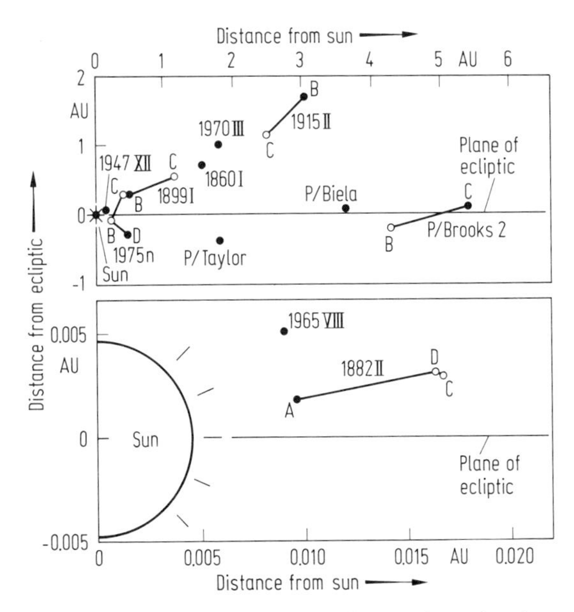

Fig. 11. Heliocentric distance vs. distance from the plane of the ecliptic for the points of splitting. The sungrazing comets are plotted separately on an enhanced scale. The primary breakups are represented by full circles; the subsequent fractures are marked by open circles interconnected by lines to indicate the chronological sequence of splitting [41].

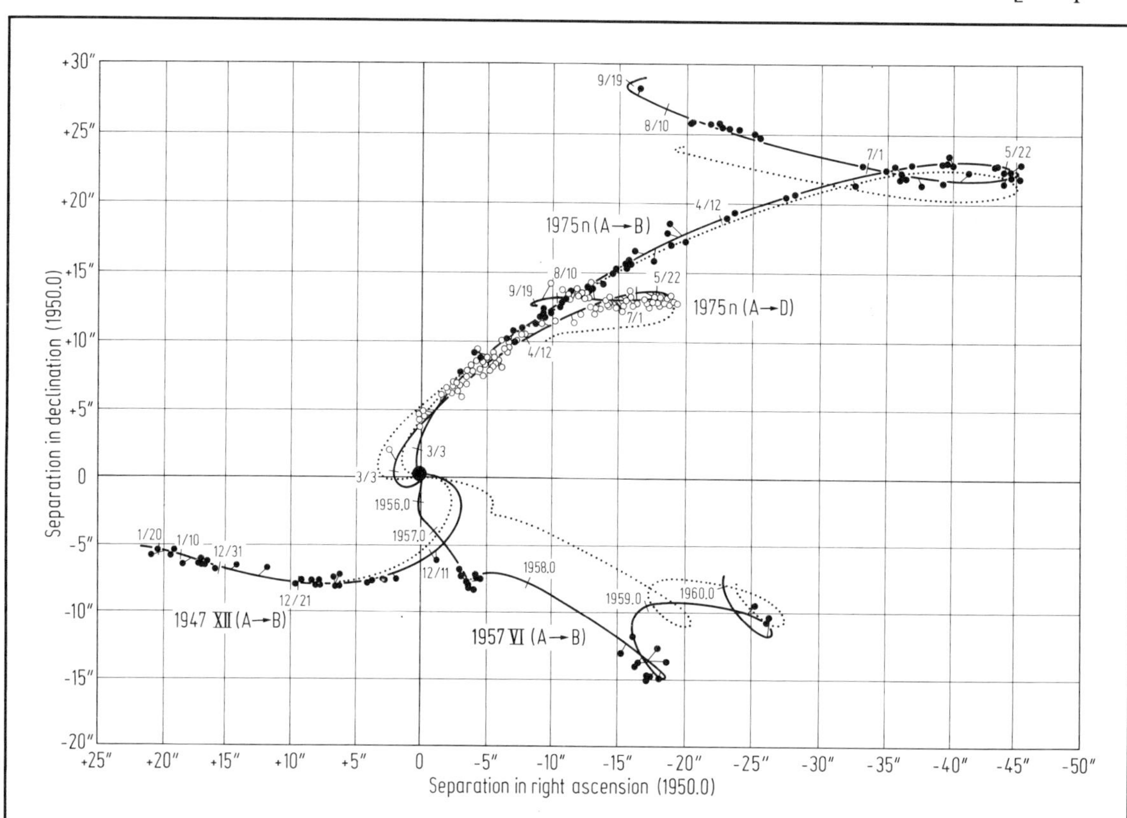

Fig. 11a. Apparent separations (in [arc seconds], equinox 1950.0) of three split comets in right ascension [multiplied by cos (declination)] and in declination. The comets are West 1975n, Wirtanen 1957 VI, Southern Comet 1947 XII [41].
Dots: observed separations A→B (i.e., B relative to A) for 1975n, A→B for 1957 VI, and A→B for 1947 XII.
Open circles: observed separations A→D for 1975n.
Solid circle: the position of the principal nucleus.
Heavy curves are least-squares fits based on the multiparameter model: Solution SDRTN for A→B of 1975n and A→B of 1957 VI, SDTN for A→D of 1975n, SDN for A→B of 1947 XII.
Short lines connect the observed separations with their calculated positions.
Marks give the calculated position for the standard dates in 1976 for 1975n, in 1947/48 for 1947 XII, and in the specified year for 1957 VI.
Dotted curves are least-squares fits based on the two-parameter model.

Lost comets

Of the 72 comets observed with more than one approach, only 4 (P/Biela, P/Brorsen, P/Tempel-Swift, P/Neujmin) seem to have been lost; one of them (P/Biela) had split, and two others showed sudden large changes in their non-gravitational parameters.

Non-gravitational forces

The observed non-gravitational effect on the motion of comets is taken as evidence for the existence of a solid icy nucleus. It arises when gas evolves asymmetrically from the rotating nucleus and is, e.g., noticeable in a change in the time of perihelion passage on successive returns.

A general discussion of cometary orbits is given in [33].
"New" comets show usually stronger deviations than "old" short-period comets. Comets with and without measurable deviation from Newtonian motion: [23]. Influence on P/Encke: [31]. Examples for practically only gravitational motions: P/Schwassmann-Wachmann 1 (q=5.4 AU), P/Oterma (q=3.4 AU), (see Table 8), P/Neujmin 1 (q=1.5 AU), P/Arend-Rigaux (q=1.4 AU). The latter two comets have mostly practically asteroidal appearance.

Table 8. Osculating elements for comet P/Oterma [30].
Symbols: see 3.3.3.1 (orbital elements).

T	1918 Dec. 30.4	1942 Aug. 21.1	1950 July 16.5	1958 June 10.5	1983 June 18.2
ω	241°.8	354°.7	354°.8	354°.9	55°.9
Ω	35°.1	155°.2	155°.1	155°.1	331°.2
i	3°.07	3°.99	3°.99	3°.99	1°.94
q [AU]	5.79	3.39	3.40	3.39	5.47
e	0.1604	0.1447	0.1428	0.1445	0.2430
P [a]	18.11	7.89	7.92	7.88	19.43

3.3.3.2 Photometric observations; polarimetry

Definitions: see 3.3.3.1.

Brightness variation with r
In [5], intrinsic brightness, J_0, at $r=\Delta=1$ AU is related to corrected observed brightness, J, by $J=J_0\cdot\Delta^{-2}r^{-n}$, where n is derived for each comet by curve fitting. $\bar{n}\approx 4$, but n is not constant for a given comet, sometimes changing abruptly, often near perihelion. Values of n are given in Table 9.
In [v] $m=H_y+5\log\Delta+y\log r$ is used with $y=-2.5\,n$, $\bar{y}\approx 10$ (ranging from -5 to $+28$); H_y is a constant depending only on y, $\bar{H}_{10}\approx +6$ (ranging from -3 to $+16$). From 100 comets before and after perihelion $\bar{n}$ is found to be statistically constant ($\bar{n}=3.4$) after perihelion for all comet classes with $P>25$a [50].

Table 9. Law of brightening for comets [w].
N = Number of comets
$\bar{n}$ = mean exponent of brightening law (see text)

Limits on a [AU] or e	N	$\bar{n}$
All comets		
$a<10$	12	7.42
$10<a<250$	23	4.96
$a\geqq 250$, $e<1$	26	3.80
$e\geqq 1$	56	4.33
all	117	4.65
Comets with continuum		
$10<a<250$	5	3.64
$a\geqq 250$, $e<1$	9	3.63
$e\geqq 1$	11	2.56
all	25	3.16

The effects of different observational techniques on published magnitudes and of secular changes in m of periodic comets are discussed in [25, f]. The photometric brightness of a few short-period comets appears to show some secular decrease, although the amount is a topic of some dispute.

Photographic photometry
Extrafocal method [1]; equidensitometry [A]; isophotometry [34, 38].

Photoelectric photometry, spectrophotometry
Extensively discussed e.g. in [f, u, 19].

"Nuclear" brightness
Derivation of nuclear radii from "nuclear" magnitude estimates [a].

Intrinsic brightness variations (bursts)
Exceeding 1···2 magnitudes, frequent among all classes of comets. P/Schwassmann-Wachmann 1 ($P=15$ a, $q=5.4$ AU) brightens by several magnitudes about 2 or 3 times a year [39]; P/Tuttle-Giacobini-Kresak ($P=5.6$ a, $q=1.2$ AU) increased by 9 magnitudes twice in 1973 [26].
For reviews on photometry, see [q, zz].

Polarimetric observations
Usually 15⋯20% (at 1 μm) and perpendicular to the tail direction. Degree of polarization changes with phase angle α and wavelength. Positive polarization maximum amounting to 20⋯30% at $\alpha=90°$, inversion of polarization sign at $\alpha\simeq 20°$, maximum depth of anomalous branch up to several % at $\alpha\simeq 10°$. For small α, the cometary polarization-phase angle dependence is similar to that for minor planets [11]. IR-polarization seems to be of the same order of magnitude as in the optical region [f, y].

3.3.3.3 Spectroscopic observations

Optical observations
Continuum produced by sunlight reflected by the nucleus and scattered by dust particles; energy distribution often (not always) redder than sunlight. Superimposed are emissions due to neutral radicals and atoms (coma) and ions (tail), most of them excited by resonance-fluorescence. Near the sun, metallic lines appear. Influence of Doppler-shifted Fraunhofer lines in the solar spectrum [43]; influence on the internal motions in the coma [18].

Infrared observations
[f]. Many comets contain silicates (10 and 18 μm emission present), but strength of silicate feature varies from comet to comet and also in the same comet. For 1973 XII, the particles in coma and tail were small (0.2 μm $<d<$ 2 μm) and temperature exceeded black body temperature; particles in the anti-tail were large ($d \geqq 10$ μm) and temperature was close to black body temperature; the dust albedo, γ, turns out to be high (≈ 0.2), assuming that the visible and IR radiation comes from the same type of particle [g 1]. See Fig. 15.

Radio observations
3 stable molecules identified: H_2O, HCN, CH_3CN. Unidentified lines and molecules not detected in microwave spectral line searches [x]. Connection to interstellar molecules [o].

UV observations
Several atoms, molecules and ions detected; O, H, and OH abundances and production rates indicate that H_2O is one of the main constituents of cometary nucleus [f, t].

Spectroscopic results: see Table 10⋯18, Figs. 12 and 13.

Reviews of spectroscopic observations in [c, f, i, zz, 44].

Table 10.

Spectral identification in comets	Method of observation
Head (coma + nucleus)	
CN, C_2, C_3, CH, $C^{12}C^{13}$, NH, NH_2, [O I], OH, Na, Ca, Cr, Mn, Fe, Ni, Cu, K, Co, H, (Al, Si, He)*)	optical
H, C, O, S, OH, CO, C_2, CS	UV
CH_3CN, HCN, H_2O, OH, CH	radio
CO^+, CH^+, CO_2^+, N_2^+, OH^+, Ca^+, H_2O^+	optical
CO^+, CO_2^+, CN^+, C^+	UV
Reflected sunlight	
Thermal emission, silicate features	IR
Plasma tail	
CO^+, CH^+, CO_2^+, N_2^+, OH^+, H_2O^+	optical
CO^+, CO_2^+, CN^+, C^+	UV
Dust tail	
Reflected sunlight	
Thermal emission, silicate features	IR

*) Tentative identification.

Table 11. Molecular and atomic transitions observed in comets.

Emitter	Transition	Wavelength range Å (if not otherwise noted)
I. Coma		
(1) Radicals		
CO	$A^1\Pi - X^1\Sigma^+$	1430···1700
C_2	$D^1\Sigma_u^+ - X^1\Sigma_g^+$	2314
	$A^3\Pi_g - X^3\Pi_u$	4350···6200
	$A^1\Pi_u - X^1\Sigma_g^+$	7000···8400
$C^{12}C^{13}$	$A^3\Pi_g - X^3\Pi_u$	4740···4840
CS	$A^1\Pi - X^1\Sigma^+$	2507···2663
OH	$A^2\Sigma^+ - X^2\Pi_i$	2819···2837
		3070···3160
		3450···3490**)
	$^2\Pi_{3/2}$, $J=3/2$, $F=2-2$, $1-1$	180 mm
	$X^2\Pi$ 5−2	1.08 m
NH	$A^3\Pi_i - X^3\Sigma^-$	3350···3680
CN	$B^2\Sigma^+ - X^2\Sigma^+$	3555···3595
		3845···3885
		4175···4215
	$A^2\Pi - X^2\Sigma^+$	7800···11000
CH	$B^2\Sigma - X^2\Pi$	3885···3925
	$A^2\Delta - X^2\Pi$	4260···4350
	$^2\Pi_{1/2}$, $J=1/2$, $F=1-1$	90 mm
C_3	$\tilde{A}^1\Pi_u - X^1\Sigma_g^+$ *)	3750···4100
NH_2	$\tilde{A}^2A_1 - X^2B_1$	4500···8150
(2) Atoms		
H	$1s^2S - 2p^2P^0$	1216
	$3p^2P^0 - 2s^2S$	6562
C	$2p3s^3P^0 - 2p^2\ ^3P$	1657
	$2s2p^3\ ^3D^0 - 2s^22p^2\ ^3P$	1561
	$2p^2\ ^1D - 2p3s^1P^0$	1931
O	$2p^33s^3S^0 - 2p^4\ ^3P$	1304
	$2p^33s^5S^0 - 2p^4\ ^3P$**)	1356
[O I]	$^3P - ^1D$	6300···6364
S	$3p^4\ ^3P - 3p^34s^3S^0$	1807···1826
Na	$3^2S - 3^2P^0$	5890···5896
Fe	$a^5F - z^5D^0$	5260···5430
He**)	$3^1P - 2^1S$	5015
Cr, Mn, Ni, Cu, Co	resonance and low excitation lines	3200···5500
K	resonance lines	7665···7699
Al**)	resonance lines	3961
Si**)	resonance lines	3904
Ca	resonance lines	4227
(3) Molecules		
CH_3CN	$v_8=1$, $J_K=6_3-5_3$, 6_0-5_0	2.7 mm
HCN	$J=1-0$	3.4 mm
H_2O	$J_{K_{-1}K_{+1}}=6_{16}-5_{23}$	13.5 mm

continued

Table 11 (continued).

Emitter	Transition	Wavelength range Å (if not otherwise noted)
II. Tail		
(1) Molecular ions		
CO_2^+	$\tilde{B}^2\Sigma_u^+ - \tilde{X}^2\Pi_g$	2890
	$A^2\Pi_u - X^2\Pi_g$	3370···3840
CO^+	$B^2\Sigma^+ - X^2\Sigma^+$	2040···2450
	$A^2\Pi_i - X^2\Sigma^+$	3400···6200
	$B^2\Sigma^+ - A^2\Pi_i$	3500···4240
H_2O^+	$\tilde{A}^2A_1 - \tilde{X}^2B_1$	5480···7540
OH^+	$A^3\Pi_i - X^3\Sigma^-$	3565···3620
N_2^+	$B^2\Sigma^+ - X^2\Sigma^+$	3540···4280
CH^+	$A^1\Pi - X^1\Sigma^+$	3950···4260
CN^+	$f^1\Sigma - a^1\Sigma$	3185···3263
	$c^1\Sigma - a^1\Sigma$	2181
(2) Atomic ions		
Ca^+	$4^2S - 4^2P^0$	3934···3968
C^+	$2p^2P^0 - 2p^2\ ^2D$	1335

*) Several vibronic transitions [16].
**) Tentative identification.

Table 12. Lifetimes τ of upper states of cometary bands [n].

Molecule	State	τ [ns]	Molecule	State	τ [ns]
OH	$A^2\Sigma(v=0)$	690	CN	$A^2\Pi(v=1\cdots9)$	680
NH	$A^3\Pi(v=0)$	404		$B^2\Sigma^+(v=0)$	65.6
CH	$A^2\Delta(v=0)$	530	NH_2	$\tilde{A}2A$	8300
	$B^2\Sigma^-(v=0)$	380	CO^+	$A^2\Pi(v=1)$	3490
C_2	$d^3\Pi(v=0)$	170		$B^2\Sigma^+(v=0)$	55
	$D^1\Sigma_u^+(v=0)$	14.6	N_2^+	$A^2\Pi_u(v=1)$	13900
				$B^2\Sigma_u^+(v=0)$	60.5

Table 13. Filter characteristics (wavelength λ/bandwidth $\Delta\lambda$) for cometary observations. (Suggested by IAU-Commission 15 in 1979)

Species	Band filter λ[Å]/$\Delta\lambda$[Å]
$CN(B^2\Sigma - X^2\Pi;\ \Delta v=0)$	3870/50
$C_3(^1\Pi_u - ^1\Sigma_g^+)$	4060/70
$C_2(d^3\Pi_g - a^3\Pi_u;\ \Delta v=0)$	5120/100
Continuum	3650/100
	4850/100

Table 14. The dust-to-gas mass ratio μ [1].

Object	μ	Remarks
Solar abundances	0.0075	Dust: silicates (with all metals) and S; Gas: H, He, Ne, C, N, and about 2/3 O.
Primitive mixture	0.54···0.76	Primitive mixture: starting from dust and gas (as described above), uncombined H and all He and Ne have been removed from the gaseous phase.
Comet Bennett 1970 II	0.5 (mean)	Ratio averaged over 45 days.
Comet Arend–Roland 1957 III	1.7 (mean)	Ratio averaged over 20 days.
Time variation in comet Arend–Roland		Ratios averaged over 3 days. Day −6 coincides with violent outburst of no more than a day, with production rates of gas and dust multiplied by five at least. The ratio is not much influenced during the outburst, but diminishes drastically afterwards.

Days after perihelion:	−9	−6	−3	0	+3	+6	+9
Dust-to-gas ratio:	6.2	5.6	1.4	1.2	1.2	1.0	0.8

Table 15. Production rates of major constituents, reduced to $r = 1$ AU [1].

Species	Comet			
	Tago–Sato–Kosaka 1969 IX	Bennett 1970 II	Kohoutek 1973 XII	West 1976 VI
	Production rate [10^{28} s^{-1}]			
H(Lα)	4.3	54···65	34	46
OH(X$^2\Pi_i$)	1.3	30	20	14
O(^{3}P)		6	2.7···6.9	16
O(^{1}D)		17	1.1	
C(^{3}P)			0.6···1.6	2.8
C(^{1}D)				1.9
CO(X$^1\Sigma^+$)				3.9

Table 16. Production rates of minor constituents, reduced to $r = 1$ AU [1].

	Production rate in [10^{28} radicals s^{-1}]	
Comet Kohoutek 1973 XII	C_2	CN
Before perihelion	0.10	0.04
After perihelion	0.04	0.01

Table 17. Observed isotope ratios of (^{12}C/^{13}C) [46].

Object	^{12}C/^{13}C
Terrestrial	89
Solar system	90
Interstellar matter	30···50
Comets:	
Ikeya 1963 I	70±15
Tago–Sato–Kosaka 1969 IX	100±20
Kohoutek 1973 XII	115+30/−20; 135+65/−45
Kobayashi–Berger–Milon 1975 IX	110+20/−30

Table 18. Depletions relative to solar abundances normalized to metal abundance [1]

Element	$N_{Co}/N_\odot$	$N_{Ch}/N_\odot$	N_{Co}/N_{Ch}
H	$5\cdot10^{-4}$	$5\cdot10^{-5}$	10:1
C	0.24	0.06	4:1
N	0.03	0.014	2:1
O	1.00	0.35	3:1
Metals	1.00	1.00	1:1

N_{Co} abundance in comets
N_{Ch} abundance in CI-chondrites
$N_\odot$ solar abundance

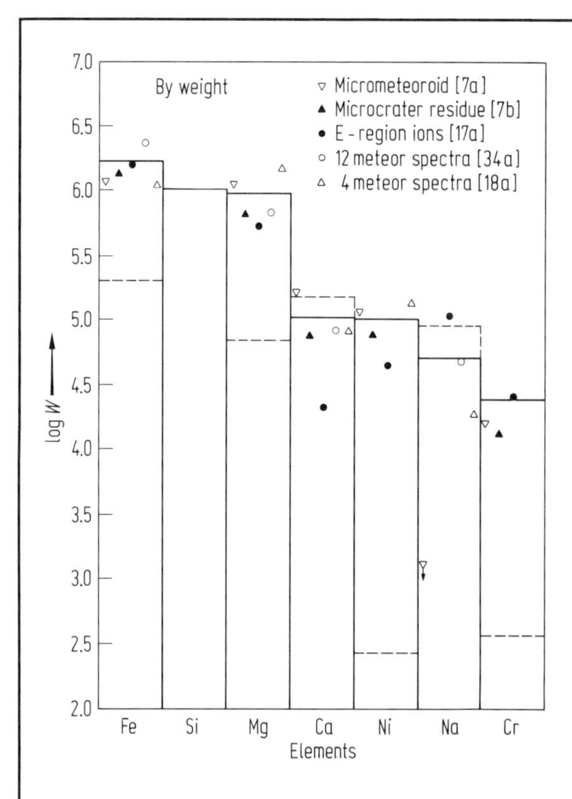

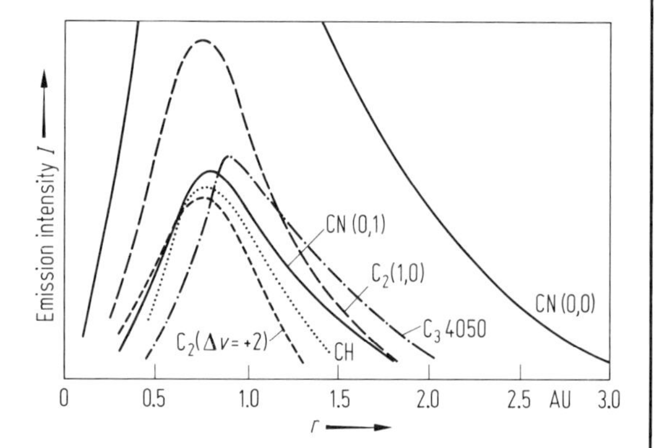

Fig. 13. Evolution of cometary emissions as a function of heliocentric distance (schematic) [c].

Fig. 12. The relative abundances, by weight, of 7 elements commonly found in cometary meteoroids. Values are plotted as log weight and normalized to Si at 6.00. Since Si values for meteor spectra are not available, normalization in this case was to the mean of Fe and Mg [l].

Full line: solar system, Cameron, 1973, also true for carbonaceous chondrites, type C1.
Dotted line: earth's crust [0].

3.3.3.4 Nucleus

Structure

"Icy conglomerate" model [47] generally favored over "sand bank" model [27]: the nucleus consists of intimate mixture of ices, clathrates and grains of rocky (meteoroidal) material ["clathrates" = compounds in which small atoms or molecules (e.g., noble gases, CH_4, SO_2, C_2H_2) are trapped in relatively large cavities of the honeycomb structure formed by the crystallization of certain substances such as e.g., water; clathrates formed by water with gases or volatile substances are also called hydrates]. Under activation of solar radiation and solar wind, the rate of sublimation determines the loss of gases and the loss of icy and rocky grains carried away by the gases, and the solar wind is essential for the occurrence and nature of the plasma phenomena. Gases and meteoroids are lost forever from the comet. True nucleus rarely, if ever, observed from the ground [f]. Tensile strength very weak (derived from splitting of sun-grazing comets, brightness outbursts, brittleness and low density of bright meteors that presumably are cometary debris) [a, l]. New comets might have thin outer frost layer over an upper mantle of ices which are more volatile than H_2O ice. The lower mantle has normal ice composition and could cover a kernel which might be asteroidal [l, w].

Dimension, mass

Brightness measurements when inactive at great solar distances give AS (A = Bond albedo, S = surface area) [a]; $(1-A)S$ from vaporization rate of H_2O production [10] (Table 19);
dimensions also by combining the comet's non-gravitational force with its brightness at maximum solar distance (Table 20) [w].

Range of radii

0.3···4 km, typically 1···2 km for short-period comets (Table 21) and up to an order of magnitude larger for long-period comets. Much larger comets with 50···60 km radius and masses of about $10^{21\cdot 2}$ g exist; e.g., the sun-grazers or the Great Comet of 1729 that could be seen with the naked eye at 4 AU from earth and sun [l, w]. Easily visible comets have masses of $10^{14}\cdots10^{16}$ g. The amount of dust left along the orbit (measured by meteor techniques) give masses for the meteor stream producing comets of $10^{13}\cdots10^{16}$ g.

Table 19. Nuclear radii by H_2O production [w, f].
A = Bond albedo
S = surface area in [km^2]
R = radius in [km].

Comet	$(1-A)S$ [km^2]	AS [km^2]	A	R [km]
Tago–Sato–Kosaka 1969 IX	5.88	9.55*)	0.63	2.20±0.05
Bennett 1970 II	15.71	29.31*)	0.66	3.76±0.08
Encke 1971 II	0.09***)	1.06*)	0.2**)	1.3
Kohoutek 1973 XII preperihelion	6.06	151	0.97	7.0
postperihelion	3.2	9.5	0.73	2.0
combined, all by O'Dell	6.06	9.5	0.60	2.2

*) Lambert's Law, Delsemme and Rudd.
**) Albedo assumed by author.
***) Only a small portion of the surface is active.

Table 20. Calculated radii, R, of short-period comets; $q \leqq 1.5$ AU [w].

The radii are calculated by combining the comet's radial non-gravitational acceleration, A_1, with its brightness when inactive at maximum solar distance, r_0; this brightness provides the quantity $R_1 = R \cdot A^{1/2}$, where A is the Bond albedo (fraction of total light reflected) and R the radius of the spherical nucleus.

Comet	q [AU]	r_0 [AU]	R_1 [km]	A_1	$R_1 A_1$	A (calc.)	R[km]
Honda–Mrkos–Pajdusakova	0.6	1.16	0.48	0.10	0.048	0.017	1.07*)
Giacobini–Zinner	1.0	2.52	0.70	0.2	0.14	0.333	1.21
Finlay	1.1	1.99	0.26	0.26	0.068	0.035	0.58*)
Tuttle–Giacobini–Kresák	1.1	1.68	0.14	0.66	0.092	0.070	0.31*)
Schaumasse	1.2	2.86	0.88	0.4	0.35		1.52**)
Borelly	1.4	2.99	0.66	0.09	0.059	0.026	1.48*)
Jackson–Neujmin	1.4	1.89	0.20	0.8	0.16	0.333	0.35**)

*) $A = 0.2$ assumed.
**) $A = 1/3$ assumed.

Table 21. Short-period comets observed to $r > 3.2$ AU [w].
N = number of appearences, other symbols, see Table 20.

Comet	N	q [AU]	r_0 [AU]	A	R[km]
Oterma	3	3.4	4.72	0.8	0.9
van Houten	1	3.9	4.10	0.8	3.9
Gunn	1	2.4	4.66	0.8	1.6
Kearns–Kwee	2	2.2	4.04	0.8	1.3
Ashbrook–Jackson	5	2.3	3.51	0.8	1.1
Whipple	6	2.5	3.32	0.4	1.7
Brooks 2	1	2.0	3.76	0.4	0.6
van Biesbroeck	2	2.4	4.56	0.15	4.3
Slaughter Burnham	1	2.5	3.66	0.15	1.7
Schwassmann–Wachmann 2	3	2.2	3.40	0.15	2.6
Comas Solá	3	1.8	3.69	0.15	3.5
Faye	2	1.6	3.41	0.15	2.6
Tempel 2	1	1.4	3.55	0.15	1.4
Encke	2	0.3	3.74	0.15	1.7

Mass distribution: unknown.

Mass loss: 0.1···1% per revolution.

Layer loss rate

At perihelion ≈1 cm d^{-1} for P/Encke. Each orbit will result in ≈30 cm layer peeling off P/Encke [50a].

Average surface temperature, T, in [K]

Varies as function of heliocentric distance, r in [AU]. With no gas emission, $T \propto r^{-1/2}$; with large gas emission near perihelion, T almost constant.

Spin rate, axis, precession

About as many prograde as retrograde spins (from non-gravitational forces); rotation axis orientations: tendency toward 90° obliquity of spin axis with orbital plane; periods: hours to days [z]; precession: discussed in [50a].

Cometary debris

Mass of solid particles about equal to that of the gases [49, w], and the total cometary mass crossing $r = 1.2$ AU from the sun is $2 \cdot 10^{10}$ g s^{-1}. If 1% is lost per passage, the loss rate is $2 \cdot 10^{8}$ g s^{-1}, a value about 10 times higher than the input of meteoritic material to the interplanetary complex required for quasi-stable equilibrium [48].

Reviews [f, w, x, zz, 27].

3.3.3.5 Coma

Phenomenology

(1) inner, molecular, chemical, or photochemical coma (dominated by chemically stable parent molecules): about 10^4 km diameter at 1 AU;

(2) radical or visible coma (gives rise to most of the observed molecular ions and radicals): several 10^5 km;

(3) atomic or UV coma: up to several 10^7 km.

Dust usually pervades the inner and visible coma, especially in "new" comets. Variations of size: see Fig. 14.

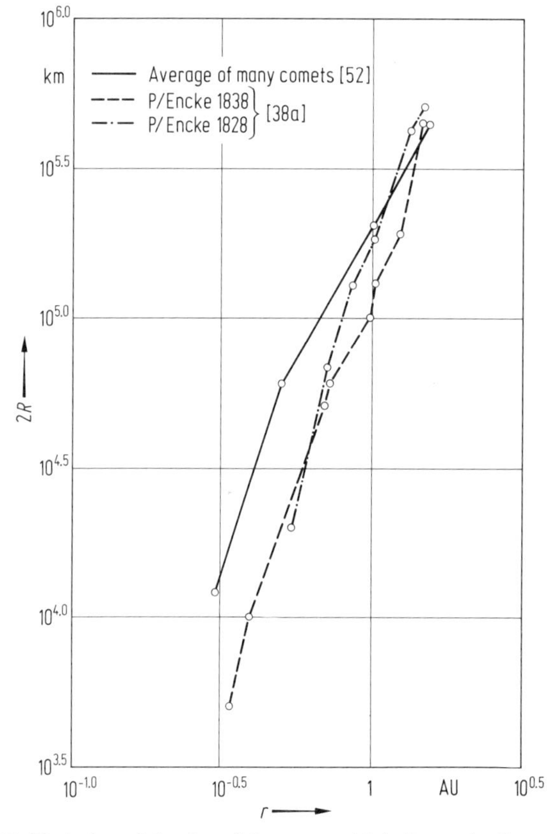

Fig. 14. Variation of the size of the coma with heliocentric distance [s].

Coma and nucleus make up the head of the comet.

Properties, chemical and physical processes [f, o, x, zz].

Production rates
Theory [20, 21, a, x];
derivation of production rates based
- on partial densities in [51],
- on solution of the energy balance and Clausius-Claperon equation in [20, a],
- on analysis of dust tails in [14],
- on analysis of the CN violet bands in [28],
- on analysis of various other emissions in [zz].

Mass loss associated with gas production can account for the non-gravitational forces.

3.3.3.6 Tails

Dust (Type II) tail and grains

Mechanical theory initiated by [3], extended by [7, 14]. Particle released from nucleus is then repulsed by solar radiation pressure, the radiation force at the particle's distance being $(1-\mu)$ in units of solar gravitation $\left(\text{i.e. } \mu=\frac{F_{\text{grav}}-F_{\text{rad}}}{F_{\text{grav}}}\right)$. $(1-\mu)<1$: orbit concave (hyperbolic) to sun; $(1-\mu)>1$: convex; $(1-\mu)=1$: straight line. Particles with the same $(1-\mu)$-value released continuously describe a curve called syndyname (or syndyne); particles with different $(1-\mu)$-values released simultaneously (e.g., in an outburst) describe a synchrone.

Mechanical concepts apply well to dust tails, not to ion tails which are controlled by solar wind and associated magnetic fields.

Theories successfully applied to several comets. E.g., Comet Bennett 1970 II; particle diameters a few µm; dust production rate at 0.56 AU: $2\cdot10^{7}$ g s^{-1}; nuclear radius 2.6 km [42].

20···30% of nearly parabolic comets show an anti-tail when geometrical conditions are favorable. A few short-period comets have also been found to display an anti-tail [f]. Particle diameters in anti-tails are of a few 10 µm to mm [y, 15], in agreement with the observed polarization [17].

Energy flux of comet Kohoutek: Fig. 15.
Reviews [f, 1].

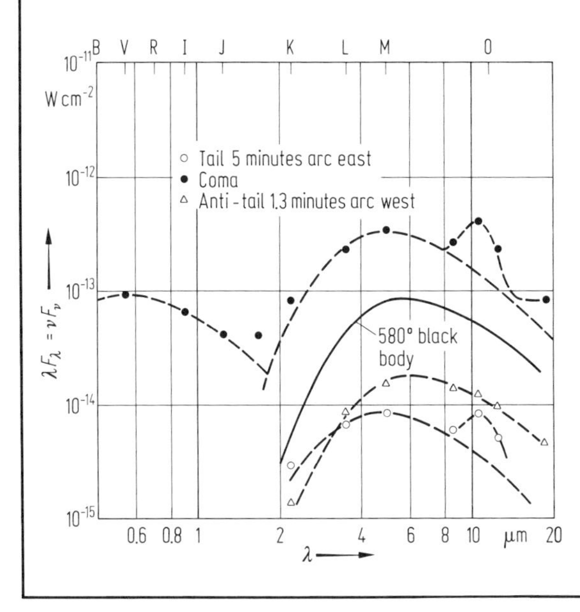

Fig. 15. Energy flux λF_λ in W cm^{-2} vs. wavelength for the coma, tail, and anti-tail of comet Kohoutek 1973 XII, measured on January 1.7, 1974. Coma and tail show the silicate feature at 10 µm; they have very similar energy distributions and are warmer than the black-body. The anti-tail does not show the 10 µm excess and is very near the black-body temperature. This result is interpreted to indicate that the particles in coma and tail are smaller (diameter ≦2 µm) than those in the anti-tail (≧10 µm). Adapted from [f].

Plasma (ion, Type I) tail

Nearly straight, nearly anti-solar directed (Fig. 16), typically 3000···4000 km wide, up to 10^8 km long. Knots and irregularities appear to propagate with $v=10\cdots100\ \mathrm{km\,s^{-1}}$, accelerations of $(1-\mu)\approx100$ last for a few hours, rarely days.

CO^+ density $N\approx10^3\ \mathrm{cm^{-3}}$ in rays [37], velocity of parent molecules are low ($v\approx0.5\ \mathrm{km\,s^{-1}}$) but high for ions. Larmor radius for CO^+ ions at 10^4 K in a field of 2γ is less than 10^3 km, sufficient to keep CO^+ in rays. Reviews and present status [j, k, ss].

Statistics of lag angles for plasma and dust tails [2] with relevant data for 60 comets from about 1600 observations. Interpretations [6].

The average long-term global solar-wind velocity field resulting from a sample of 809 observations is

$W_r=(400\pm11)\ \mathrm{km\,s^{-1}}$, $W_\theta=2.3\sin2B$, $W_\Phi=6.7\dfrac{(\cos B)^{2.315}}{r}$ with B = solar latitude, r = heliocentric distance [k].

Correlation with solar wind events observed from space probes [4, 8, 22, 35, 36].

For solar wind, see also 3.3.5.2.

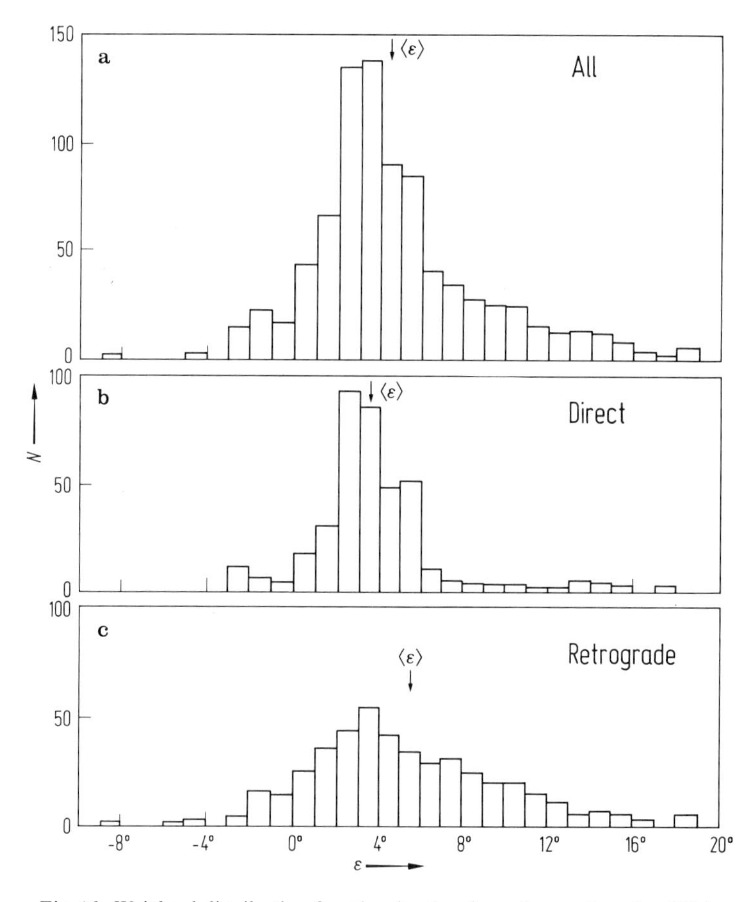

Fig. 16. Weighted distribution function for the aberration angle, ε, for 607 ion tail measures: a) total; b) comets in direct motion; c) retrograde comets [j].

3.3.3.7 The nature of cometary dust

Ref.: [e]

Aspect	Source of information	Basic information
Chemical composition	meteor spectra	H, O, C, N, Fe, Si, Mg, Ca, Ni, others (see Fig. 12)
	infrared data	silicates
	particle collections	chondritic, FSN, olivine/pyroxene, others
Physical structure	meteor studies	average density 0.3 and 0.8 g/cm^3, respectively, for photographic and radio meteoroids (millimeter to centimeter size range); strength $\approx 10^{5\pm}$ dyn/cm^2; irregular shape?
	particle collections	some irregular, very porous aggregates of 0.1 µm sized grains; others compact ablation products
Optical properties	spectrophotometry in visible and infrared	albedo low, perhaps $\gamma \approx 0.2$; effect of reddening (on an average, $+0\overset{m}{.}2$ in $B-V$); effect of phase angle
	polarimetric data	generally high polarization, occasionally negative
Size distribution	dust-tail studies	sizes from <1 µm to $\gg 1$ mm; number varies inversely as 4th to 5th power of size; probably a relatively sharp cutoff at lower limit but tailing off toward upper limit
	infrared data	micron sized particles; also larger than 10 µm
	polarimetric data	submicron sized particles; possibly steep slope
Dust inside nucleus	comet models	distribution very uncertain; various patterns possible
Amount of dust	dust-tail studies	in dust-rich comets near sun: dust $\doteqdot$ gas in mass
Fragmentation	studies of structures in dust tails	evidence of break-up of strongly non-spherical particles into submicron grains [g 1].

3.3.3.8 Laboratory studies and space experiments relevant to comets

Reviews in [f, l, p, 9, 23].

3.3.3.9 References for 3.3.3

Acknowledgements: The contributions of M. F. A'Hearn, B. Donn, H. Fechtig, P. Feldman, G. Herzberg, Z. Sekanina, and V. Vanysek are gratefully acknowledged.

(Listed are only references not included in Landolt-Börnstein, NS, Vol. VI/1, 1965.)

I. Proceedings of meetings

a Nature et Origine des Comètes, Coll. Internat. Univ. de Liège. Mem. Liège 5th. Ser., Vol. XII, 1966.

b The Motion, Evolution of Orbits, and Origin of Comets, IAU Sympos. **45**, (Chebotarev, G.A., Kazimirchak-Polonskaya, E.I., Marsden, B.G., eds.), Reidel Publ. Co., Dordrecht 1972.

c Comets: Scientific Data and Missions, Tucson Conference (Kuiper, G.P., Roemer, E., eds.), Univ. of Arizona, Tucoson 1972.

d Asteroids, Comets, Meteoric Matter, IAU Coll. **22** (Cristescu, C., Klepczynski, W.J., Milet, B., eds.), Acad. Rep. Romania, Bucharest 1974.

e Proc of the Shuttle Based Cometary Science Workshop, Marshall Space Flight Center, Huntsville, Alabama (Gary, G.A., ed.) NASA SP-355, Washington 1976, p. 142.

f The Study of Comets, IAU Coll. **25** (Donn, B., Mumma, M., Jackson, W.M., A'Hearn, M., Harrington, R., eds.), Washington 1976.
g Conference on "Cometary Missions" (Axford, W.I., Fechtig, H., Rahe, J., eds.), Remeis Observ. Publ. Vol. XXII, Bamberg 1979.
g1 Solid Particles in the Solar System (Halliday, I., McIntosh, D.A., eds.), Reidel, Dordrecht, 1981
g2 Workshop on Application of Modern Observational Techniques to Comets (Brandt, J.C., Donn, B., Greenberg, M., Rahe, J., eds.) NASA-SP, Washington, 1981.
g3 Comets: Gases, Ices, Grains and Plasma (Wilkening, L.L., ed.), Univ. of Arizona press, 1982.

II. Reviews and monographs

(Reviews published in Proceedings of meetings are not separately listed)

h Antrack, D., Biermann, L., Luest, Rh.: Annu. Rev. Astron. Astrophys. **2** (1964) 327.
i Arpigny, C.: Annu. Rev. Astron. Astrophys. **3** (1965) 351.
j Brandt, J.C.: Annu. Rev. Astron. Astrophys. **6** (1968) 267.
k Brandt, J.C., Mendis, D.A. in: Solar System Plasma Physics (Kennel, C.F. et al., eds.), Amsterdam, North-Holland (1978).
l Delsemme, A.H. (ed.): Comets, Asteroids, Meteorites, University of Toledo (1977) (=IAU Coll. **39**).
m Dobrovolsky, O.V.: Kometi (in Russian), Moscow (1966).
n Huber, K.P., Herzberg, G.: Molecular Spectra and Molecular Structure VI, Van Nostrand Reinhold Co., New York (1979).
o Jackson, W.M.; Molecular Photochemistry **4** (1972) 135.
p Keller, H.U.: Space Sci. Rev. **18** (1976) 641.
q Konopleva, V.P., Nazarchuk, V.P., Shul'man, L.M.: The Surface Brightness of Comets, Naukova Dumka, Kiev (1976).
r Marsden, B.G.: Annu. Rev. Astron. Astrophys. **12** (1974) 1.
s Mendis, D.A., Ip, W.-H.: Astrophys. Space Sci. **39** (1976) 335.
ss Mendis, D.A., Ip. W.-H.: Space Sci. Rev. **20** (1977) 145.
t Rahe, J.: Proc. 2nd European IUE Conf. Tübingen (1980) 15.
u Vanysek, V. in: Proc. 21st Nobel Sympos. (Elvius, A., ed.), Almqvist and Wiksell, Stockholm (1972).
v Vsekhsvyatskij, S.K.: Physical Characteristics of Comets, Nauka, Moscow (1958, 1966, 1967, 1974).
w Whipple, F.L. in: Cosmic Dust (McDonnell, J.A.M., ed.), Wiley and Sons (1978).
x Whipple, F.L., Huebner, W.F.: Annu. Rev. Astron. Astrophys. **14** (1976) 143.
y Icarus **23** (1974) (Whole issue devoted to Comet Kohoutek 1973 XII).
z Wachmann, A., Houziaux, L.: Landolt-Börnstein, NS Vol. VI/1 (1965) Chapt. 4.3.
zz Transactions IAU, Reports of Commission 15 (Physical Study of Comets, Minor Planets and Meteorites); appears every 3 years.

III. Atlases

A Cometas-Viento Solar, Observ. Astron. Univ. Cordoba, Argentina (1973).
B Hoegner, W., Richter, N.B.: Isophotometrischer Atlas der Kometen, Pts. I, II. Barth, Leipzig (1969).
C Knigge, R., Rahe, J.: The Helwan Photographs of Comet Halley 1910 II, Remeis Observ. Publ. Vol. XXIII, Bamberg (1979).
D Rahe, J., Brandt, J.C., Donn, B.: Atlas of Comet Halley 1910 II – Photographs and Spectra, NASA-SP, Washington (1981).
E Rahe, J., Donn, B., Wurm, K.: Atlas of Cometary Forms, NASA SP-198, Washington (1969).
F Swings, P., Haser, L.: Atlas of Representative Cometary Spectra, Astrophys. Inst. Univ. Liège (1965).

IV. Catalogues

G Marsden, B.G.: Catalogue of Cometary Orbits, Smithsonian Astrophys. Observ., Cambridge, Mass. (1979). This catalogue appears every few years.

Rahe

References mentioned in the text

0 Ahrens, L.H.: Distribution of the Elements in our Planet, McGraw-Hill Book Co., New York (1965) p. 96.
1 Baldet, F.: Mem. Soc. R. Sci. Liège **13** (1953) 64.
2 Belton, M.J.S., Brandt, J.C.: Astrophys. J. Suppl. **13** (1966) 125.
3 Bessel, F.W.: Astron. Nachr. **13** (1836) 185.
4 Biermann, L.: Z. Astrophys. **29** (1951) 274.
5 Bobrovnikoff, N.T.: Contrib. Perkins Observ. **15** (1941).
6 Brandt, J.C., Harrington, R.S., Roosen, R.G.: Astrophys. J. **184** (1973) 27.
7 Bredikhin, T., see Jaegermann: Kometenformen, St. Petersburg (1903).
7a Brownlee, D.E., Hörz, F., Tomandl, D.A., Hodge, P.W.: The Study of Comets (Donn, B., Mumma, M., Jackson, W.M., A'Hearn, M.F., Harrington, R., eds.), NASA SP-393, Washington, D.C. (1976) p. 962.
7b Brownlee, D.E., Tomandl, D.A., Hodge, P.W., Hörz, F.: Nature **252** (1974) 667.
8 Burlaga, L.F., Rahe, J., Donn, B., Neugebauer, M.: Sol. Phys. **30** (1973) 211.
9 Danielson, R.L., Kasai, G.H.: J. Geophys. Res. **73** (1968) 259.
10 Delsemme, A.H., Rudd, D.A.: Astron. Astrophys. **28** (1973) 1.
11 Dobrovolsky, O.V.: Unpublished (1979).
12 Everhart, E.: Astrophys. Lett. **10** (1972) 131.
13 Everhart, E., Raghavan, N.: Astron. J. **75** (1970) 258.
14 Finson, M.L., Probstein, R.F.: Astrophys. J. **154** (1968) 327.
15 Gary, G.A., O'Dell, C.R.: Icarus **23** (1974) 519.
16 Gausset, L., Herzberg, G., Lagerqvist, A., Rosen, B.: Astrophys. J. **142** (1965) 45.
17 Giese, R.H. in: Interplanetary Dust and Zodiacal Light (Elsaesser, H., Fechtig, H., eds.), Springer-Verlag, Berling, Heidelberg (1976).
17a Goldberg, R.A., Aikin, A.C.: Science **180** (1973) 294.
18 Greenstein, J.L.: Astrophys. J. **128** (1958) 106.
18a Harvey, G.A.: J. Geophys. Res. **78** (1973) 3913.
19 A'Hearn, M.: Astron. J. **80** (1976) 861.
20 Huebner, W.F.: Z. Astrophys. **63** (1965) 22.
21 Huebner, W.F.: Z. Astrophys. **65** (1967) 185.
22 Jockers, K., Luest, Rh.: Astron. Astrophys. **26** (1973) 1.
23 Kaimakov, E.A., Sharkov, V.I.: Proc. Winter School Space Phys. (Vernov, S.N., Kocharov, G.E., eds.), Vol. 2 (1969).
24 Kresak, L.: Bull. Astron. Inst. Czech. **16** (1965) 292.
25 Kresak, L.: Bull. Astron. Inst. Czech. **24** (1974) 264.
26 Kresak, L.: Bull. Astron. Inst. Czech. **25** (1974) 293.
26a Landgraf, W.: Submitted to Acta Astron.
27 Lyttleton, R.A.: Astrophys. Space Sci. **31** (1974) 385; **34** (1975) 491.
28 Malaise, D.J.: Mem. Soc. R. Sci. Liège, Ser. 5, **12** (1970) 199.
29 Marsden, B.G.: Astron. J. **72** (1967) 1170.
30 Marsden, B.G.: Astron. J. **75** (1970) 75.
31 Marsden, B.G., Sekanina, Z.: Astron. J. **79** (1974) 413.
32 Marsden, B.G., Sekanina, Z., Everhart, E.: Astron. J. **83** (1978) 64.
33 Marsden, B.G., Sekanina, Z., Yeomans, D.K.: Astron. J. **78** (1973) 211.
34 Miller, F.D.: Astron. J. **60** (1955) 173.
34a Millman, P.M.: From Plasma to Planet (Elvius, A., ed.), Almqvist and Wiksell, Stockholm (1972) p. 157.
34b Millman, P.M.: J. R. Astron. Soc. Canada **66** (1972) 201.
35 Niedner, M.B., Brandt, J.C.: Astrophys. J. **223** (1978) 655.
36 Niedner, M.B., Rothe, E.D., Brandt, J.C.: Astrophys. J. **221** (1978) 1014.
37 Rahe, J.: Astron. Nachr. **292** (1970) 31.
38 Rahe, J., McCracken, C.W., Hallam, K.L., Donn, B.: Astron. Astrophys. Suppl. **23** (1976) 1, 13.
38a Richter, N.B.: The Nature of Comets, Methuen & Co., London (1963).
39 Roemer, E.: Publ. Astron. Soc. Pac. **74** (1962) 351.
40 Sekanina, Z.: Astron. J. **74** (1969) 1223.
41 Sekania, Z.: Icarus **30** (1977) 574; **33** (1978) 173; **38** (1979) 300.
42 Sekanina, Z., Miller, F.D.: Science **179** (1973) 565.
43 Swings, P.: Lick Obs. Bull. **19** (1941) 131.
44 Swings, P.: Quart. J. R. Astron. Soc. **6** (1965) 26.

45 Tyror, J.G.: Mon. Not. R. Astron. Soc. **117** (1957) 370.
46 Vanysek, V., Rahe, J.: Moon and Planets **18** (1978) 441.
47 Whipple, F.L.: Astrophys. J. **111** (1950) 375; **113** (1951) 464.
48 Whipple, F.L. in: Zodiacal Light and the Interplanetary Medium (Weinberg, J.L., ed.), NASA-SP 150, Washington (1967).
49 Whipple, F.L.: Liège Intern. Coll., Ser. 6, **9** (1976).
50 Whipple, F.L.: Moon and Planets **18** (1978) 343.
50a Whipple, F.L., Sekanina, Z.: Astron. J. **84** (1979) 1894.
51 Wurm, K.: Z. Astrophys. **52** (1961) 285.
52 Wurm, K.: Astrophys. J. **89** (1939) 312.

3.3.4 Interplanetary dust and zodiacal light

3.3.4.1 Introduction

Interplanetary dust is the complex of dust grains in the solar system. The grain sizes are smaller than those of meteors and radio-meteors. The most definite limit might be given by $m < 10^{-5}$ g, equivalent to grain radii of approx. < 100 μm. Zodiacal light, as sun light scattered by the cloud of interplanetary dust particles, provides information on the average properties of very many particles in a large volume. It is a sensitive measure of the spatial distribution. The information on the physical characteristics and the orbital motion of the particles is less direct. The inner zodiacal light, apparently superimposed on the solar corona, is known as F-corona (= reflected Fraunhofer spectrum of the sun), the enhancement around the antisolar point is called Gegenschein.

Direct methods to study the interplanetary dust are:
- direct measurements of individual dust particles on board satellites and space probes;
- investigations of micrometeoroid impact craters on lunar surface samples;
- collection of dust particles using suitable collectors on board aircrafts, balloons and rockets with subsequent laboratory investigations.

It is evident that dynamical data are preferentially obtained from these direct methods. Composition analyses are yielded by space experiments and from laboratory investigations of collected dust. At present (1980) there is no serious discrepancy between direct measurements and zodiacal light observations.

Review articles: zodiacal light [L5, W4], direct methods [M5, A2, F3, B10], dynamics [D8].
Conference reports: [2, 3, 4].

3.3.4.2 Methods of measurements

3.3.4.2.1 Direct methods

The following methods are considered reliable:
a) Penetration detectors: Dust particles at cosmic velocities (i.e. between 1 and 70 km s^{-1}) hitting a foil produce holes if the energy of the individual projectiles exceeds a minimum energy. From penetration experiments and crater depth measurements Pailer and Grün [P1] have published the formula

$$T = \frac{1.0}{\varepsilon^{0.06}\varrho_T^{0.50}} m_{pr}^{0.40} (v_{pr} \cos\alpha)^{0.88} \varrho_{pr}^{0.33}$$

with T = crater depth in [cm], maximal penetration thickness, respectively
m_{pr} = projectile mass [g]; ϱ_{pr} = projectile density [g cm^{-3}];
v_{pr} = projectile velocity [km s^{-1}]; ϱ_T = target density [g cm^{-3}];
ε = ductility (% elongation); α = impact direction relative to target normal.

Particles penetrating the foil are triggering a detector. In use are pressurized cells and capacitors: from the cells the pressure drops down and is registered; thin capacitors are discharged through the flight path within the insulator [N3, N4, H11].

The penetration detectors flown on satellites Pegasus I, II, III [A1] and Explorer XVI [H2], XXIII [O1] and on space probes Pioneer 10, 11 [H11, H12] have used different shieldings. The shieldings, the resulting sensitivities and the detector surfaces are given in Table 1. Fig. 1 shows the sensitivities of penetration detectors and impact plasma detectors in comparison.

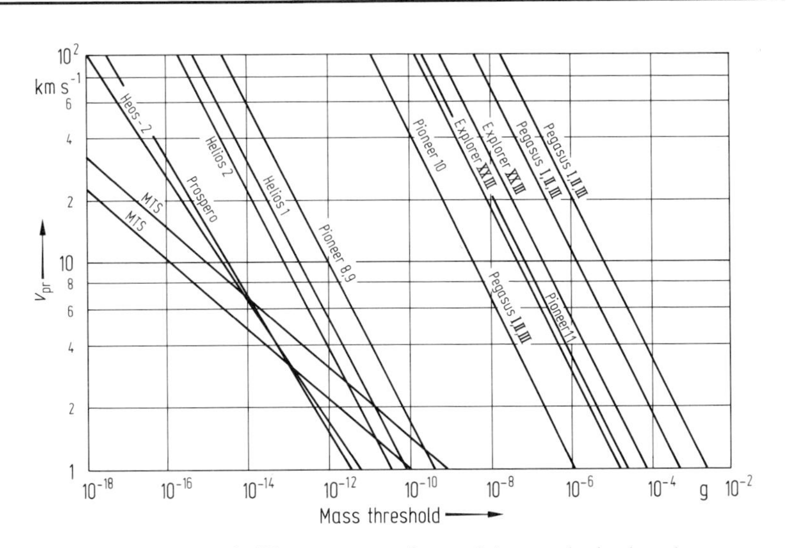

Fig. 1. Sensitivity of different penetration and impact ionization detectors [F3]. For details of shielding, see Table 1.

Table 1. Sensitivities of detectors for projectiles of density ϱ_{pr} and velocity $v_{pr} = 20$ km s^{-1}: Mass thresholds, m_{min}, for the Pegasus, Explorer, Pioneer, Prospero, HEOS-2, Helios, and Meteoroid Technology Satellite (MTS) detectors [F3].

Satellite launch	Type	m_{min} g	ϱ_{pr} g cm^{-3}	Detector surface m^2
Explorer XVI	25 μm Cu–Be	$8 \cdot 10^{-9}$	0.5	1
(Dec. 62)	50 μm Cu–Be	$3 \cdot 10^{-8}$	0.5	0.4
Explorer XXIII	25 μm steel	$8 \cdot 10^{-9}$	0.5	0.7
(Nov. 64)	50 μm steel	$3 \cdot 10^{-8}$	0.5	1.4
Pegasus, I, II, III	38 μm Al	$1 \cdot 10^{-9}$	0.5	7.5
(Feb., May, July 65)	230 μm Al	$2 \cdot 10^{-7}$	0.5	16
	410 μm Al	$1 \cdot 10^{-6}$	0.5	171
Pioneer 10	25 μm steel	$8 \cdot 10^{-9}$	0.5	0.26
(March 72)		$2 \cdot 10^{-9}$	3	
Pioneer 11	50 μm steel	$3 \cdot 10^{-8}$	0.5	0.57
(April 73)		$1 \cdot 10^{-8}$	3	
Pioneer 8 (Dec. 67)	impact plasma	$2 \cdot 10^{-13}$	8	0.01
Pioneer 9 (Nov. 68)	impact plasma	$2 \cdot 10^{-13}$	8	0.01
Prospero (Dec. 71)	impact plasma	$7 \cdot 10^{-16}$	8	0.002
HEOS-2 (Feb. 72)	impact plasma	$2 \cdot 10^{-16}$	8	0.01
MTS	capacitor discharge (1 μm)	$2 \cdot 10^{-17}$	8	0.07
(Aug. 72)	capacitor discharge (0.4 μm)	$2 \cdot 10^{-18}$	8	0.07
Helios 1 (Dec. 74)	impact plasma	$3 \cdot 10^{-14}$	8	0.01
Helios 2 (Jan. 76)	impact plasma	$2 \cdot 10^{-14}$	8	0.01

b) An impact plasma detector works on the basis of the registration of positive ions and electrons produced during high velocity impacts of dust grains onto a target surface. The collected charge, C, is $C \propto m_{pr}\, v_{pr}^{\alpha}$, with $\alpha = 2.5 \cdots 3.5$ depending on detector geometry. The velocity of the projectile is being determined from the empirical function for the rise time $t_{rise} \propto (1/v_{pr})^{\beta}$, with $\beta = 0.45 \cdots 0.75$. The accuracies for v_{pr} are within a factor of 2, for m within a factor of 10 [B3, D4].

c) Lunar microcraters: Larger bodies without atmospheres in space (moon, asteroids, etc.) are hit by interplanetary dust particles at cosmic velocities. Hence, lunar surface samples are exhibiting numerous craters in the micron- and submicron-size range. By comparison with simulated craters and related morphological studies [F3] fluxes, densities and dynamical data are obtained (see 3.3.4.3c).

d) Particle collections are only possible at relative low velocities ($v_{pr} < 1\ \mathrm{km\,s^{-1}}$). This condition is met within the atmosphere where dust grains are decelerated during entry into the atmosphere. Principally all masses are collectable. The major question is which particles are extraterrestrial in origin and which are terrestrial aerosols [H3, B10].

3.3.4.2.2 Zodiacal light photometry

The characteristic difficulty of this low-light-level photometry is the separation of unwanted sources of light. In sunlit space experiments stray light suppression is essential [L6]. Far from the sun the separation of the star background (see 8.3.1.2) limits the accuracy. For ground-based observations, additional atmospheric effects (extinction, scattering [S15] and airglow [D9, S17]) have to be corrected for.

Observations of the F-corona ($\varepsilon \leq 5°$) require special techniques [B7, M1].

3.3.4.3 Direct measurements of interplanetary dust

a) Flux of dust particles between 0.3 AU and 5 AU

The radial dependence of the interplanetary dust flux Φ has been measured by the Helios 1 (HE1) [G11, G12] and the Pioneer 10 (PI10) and 11 (PI11) spacecrafts [H11, H12] from 0.3 AU to 5 AU (see Table 2). The corresponding lower mass limits, m_{min}, at an impact speed of $20\ \mathrm{km\,s^{-1}}$ are for HE1: $m_{min} = 3 \cdot 10^{-13}$ g ($\varrho_{pr} = 8\ \mathrm{g\,cm^{-3}}$; factor 10 above the detector threshold), for PI10: $m_{min} = 2 \cdot 10^{-9}$ g and for PI11: $m_{min} = 1 \cdot 10^{-8}$ g ($\varrho_{pr} = 3\ \mathrm{g\,cm^{-3}}$); see Table 1.

Table 2. Radial dependence of the interplanetary dust flux Φ [G11, H11, H12]. The error limits Φ_{min} and Φ_{max} correspond to the 1σ confidence level.

Radius interval[2] AU	Dust particle flux $[\mathrm{m^{-2}\,s^{-1}}]$			Spacecraft[1]
	Φ_{min}	Φ	Φ_{max}	
0.3···0.4	$2.2 \cdot 10^{-4}$	$3.0 \cdot 10^{-4}$	$3.8 \cdot 10^{-4}$	HE1
0.4···0.5	$6.5 \cdot 10^{-5}$	$1.3 \cdot 10^{-4}$	$1.9 \cdot 10^{-4}$	HE1
0.5···0.6	$6.1 \cdot 10^{-5}$	$1.2 \cdot 10^{-4}$	$1.8 \cdot 10^{-4}$	HE1
0.6···0.7	$1.6 \cdot 10^{-5}$	$5.5 \cdot 10^{-5}$	$9.4 \cdot 10^{-5}$	HE1
0.7···0.8	$1.6 \cdot 10^{-5}$	$5.6 \cdot 10^{-5}$	$9.5 \cdot 10^{-5}$	HE1
0.8···0.9	0	$1.7 \cdot 10^{-5}$	$3.4 \cdot 10^{-5}$	HE1
0.9···1.0	$9.8 \cdot 10^{-6}$	$2.3 \cdot 10^{-5}$	$3.7 \cdot 10^{-5}$	HE1
1.0···1.2	$1.2 \cdot 10^{-5}$	$1.6 \cdot 10^{-5}$	$2.0 \cdot 10^{-5}$	PI10
1.4···1.6	$1.0 \cdot 10^{-5}$	$1.7 \cdot 10^{-5}$	$2.4 \cdot 10^{-5}$	PI10
1.6···2.6	$4.3 \cdot 10^{-6}$	$5.9 \cdot 10^{-6}$	$7.5 \cdot 10^{-6}$	PI10
2.6···3.6	$3.3 \cdot 10^{-6}$	$4.9 \cdot 10^{-6}$	$6.5 \cdot 10^{-6}$	PI10
3.6···5.0	$2.1 \cdot 10^{-6}$	$2.9 \cdot 10^{-6}$	$3.7 \cdot 10^{-6}$	PI10
1.0···1.2	$4.9 \cdot 10^{-6}$	$7.9 \cdot 10^{-6}$	$1.1 \cdot 10^{-5}$	PI11
2.3···3.6	$9.5 \cdot 10^{-7}$	$1.6 \cdot 10^{-6}$	$2.3 \cdot 10^{-6}$	PI11
3.6···5.0	$8.7 \cdot 10^{-7}$	$1.4 \cdot 10^{-6}$	$1.9 \cdot 10^{-6}$	PI11

[1]) HE = Helios.
PI = Pioneer.

[2]) No impact recorded between 1.2···1.4 AU (PI10) and 1.2···2.3 AU (PI11); see [H11, H12].

Power law approximation: $\Phi \propto r^{-\gamma}$, with $\gamma = 2.5 \pm 0.8$, for $r = 0.3 \cdots 1.0$ AU, and $\gamma = 1.4 \pm 0.5$ for $r = 1.0 \cdots 5.0$ AU. The difference to the continued strong decrease of zodiacal light intensity outside 1 AU (see 3.3.4.4.5) can be explained by either a change in the albedo of dust particles [C4] or by a change of the slope of the size distribution [S14].

b) Dust fluxes at 1 AU

Early measurements using acoustical microphones are reported in 4.6.1 of the previous volume, VI/1, of the New Series of Landolt-Börnstein. Doubts have been raised about the high fluxes of dust particles [N5, S6, C2, C3, L4]. Fig. 2 shows the currently accepted cumulative fluxes at 1 AU as a function of particle mass for the various types of detectors (see Table 1) in comparison with the results from the lunar microcrater investigations [F1]. From the Pioneer 8/9 and HEOS-2 dust experiments, information is available concerning the directional distribution of dust [B4, H6]. Most particles are in the submicron range, coming to the sensor in hyperbolic orbits from the direction of the sun (Fig. 3). These particles are the so-called β-meteoroids (see 3.3.4.6.3). The next most abundant components are the so-called apex-meteoroids in closed orbits around the sun.

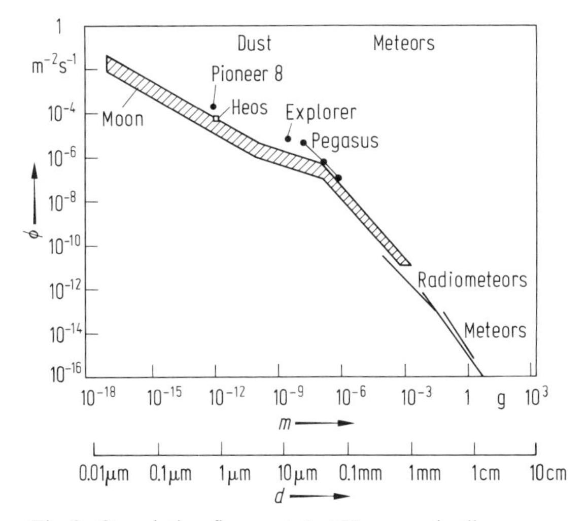

Fig. 2. Cumulative fluxes at 1 AU vs. projectile mass m and diameter d.

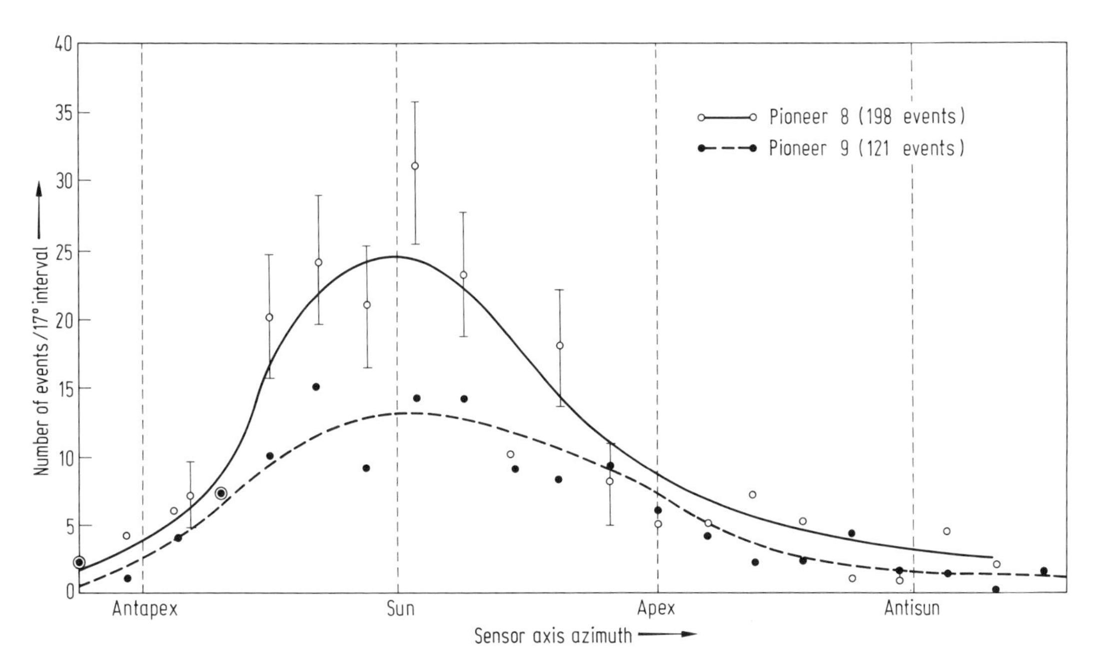

Fig. 3. The azimuthal distributions of 319 events recorded during 7 years of data acquisition from Pioneers 8 and 9 are shown as bold solid and dashed curves, respectively. Error bars are shown for the Pioneer 8 data only.

c) Lunar microcraters

Lunar surface samples exhibit impact craters from submicron up to mm sizes in diameter (Fig. 4) [H5, F2, A2]. The phenomenology of impact craters on (lunar) silicates shows a fractured zone surrounding the central crater (pit): it is called the spallation zone [H5, F2]. The slope of the plot in Fig. 4 shows 3 different parts: a steep one above 100 μm representing the undisturbed spatial distribution, a medium one between 100 μm and 2 μm representing the distribution influenced by the Poynting–Robertson effect [W11] and another steep section below 2 μm being governed by the β-meteoroids [S3, M12].

From particular samples it is possible to determine the exposure age of the pitted surface by applying the heavy ion track method [F1]. Knowing the average impact velocities of meteoroids it is possible to calculate fluxes of dust from crater size distributions [F1]. The flux Φ is given by $\Phi = N/(Ft)$, with N = effective numbers of craters, F = area, t = exposure time (Fig. 2). The fluxes in Fig. 2 are converted by [G6] from lunar microcraters into number densities $n(s)\mathrm{d}s$, [cm^{-3}]: see Table 3.

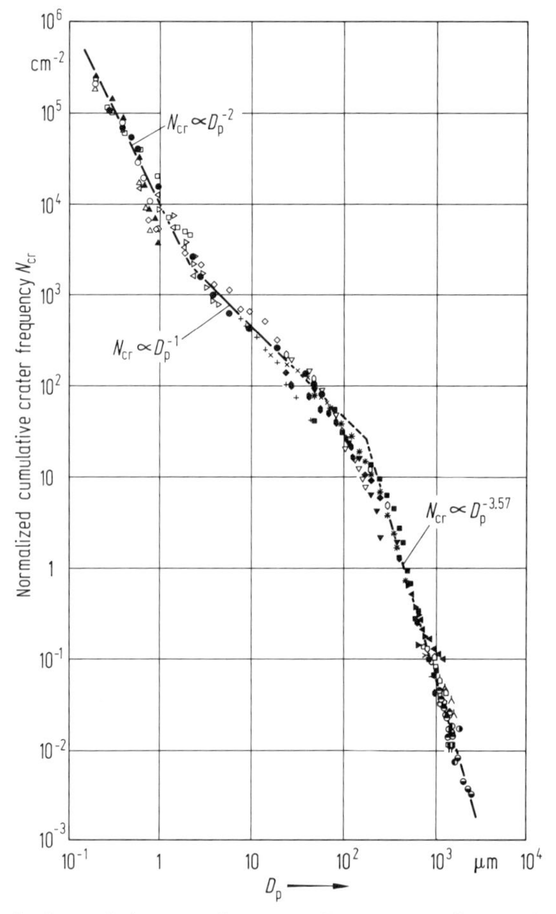

Fig. 4. Normalized cumulative crater frequency, N_{cr}, vs. crater diameter (central pit), D_p.

Table 3. Differential size distribution $n(s)\mathrm{d}s = n_0\, s^{-\varkappa}\,\mathrm{d}s$ for radii $s_1 < s < s_2$ [G6].

Maximum-model				Minimum-model			
s_1 cm	s_2 cm	$\varkappa$	n_0 cm^{-4}	s_1 cm	s_2 cm	$\varkappa$	n_0 cm^{-4}
$8.0 \cdot 10^{-7}$	$1.35 \cdot 10^{-5}$	2.70	$6.61 \cdot 10^{-23}$	$8.0 \cdot 10^{-7}$	$4.0 \cdot 10^{-5}$	2.70	$1.29 \cdot 10^{-23}$
$1.35 \cdot 10^{-5}$	$2.88 \cdot 10^{-3}$	2.00	$1.78 \cdot 10^{-19}$	$4.0 \cdot 10^{-5}$	$2.14 \cdot 10^{-3}$	2.50	$7.57 \cdot 10^{-22}$
$2.88 \cdot 10^{-3}$	$3.39 \cdot 10^{-2}$	4.33	$2.19 \cdot 10^{-25}$	$2.14 \cdot 10^{-3}$	$3.39 \cdot 10^{-2}$	4.33	$4.37 \cdot 10^{-26}$

Crater morphology: Simulation experiments [S8, B11, N1, N2] show that the ratio of crater diameter D to crater depth T is independent of impact velocity in the high velocity range ($>5\,\mathrm{km\,s^{-1}}$). D/T depends basically, however, on the density of projectiles and target. Using the same target, Fig. 5 shows the simulation results compared with the observations: iron meteoroids, stony meteoroids and a low density component. Iron meteoroids show predominantly submicron-sized craters [L13].

d) Dust near the earth

Dust experiments on the British satellite Prospero [B1] and on the ESRO-satellite HEOS 2 have observed bursts of particles [H7]. From the properties of the observed bursts, "groups" (bursts within $40\,\mathrm{min} < t < 13\,\mathrm{h}$) and "swarms" (bursts within $t < 40\,\mathrm{min}$) could be differentiated: Fig. 6. Groups are interpreted as lunar ejecta particles produced upon meteoroid impacts on the lunar surface [D7]. Swarms occur only within 10 earth radii leading to the interpretation that they are being produced as fragmentation products of larger bodies [F4]. It is believed that fragmentation occurs by electrostatic disruption when loosely bound material of possibly cometary origin ("fluffy" type) penetrates the auroral zones at 10 earth radii distance [F4]. 15 swarms of a total duration of 5.5 h have been observed within 2.5 years. The flux of small particles ($10^{-15} \cdots 10^{-12}$ g) is locally and for short time periods enhanced by up to 4 orders of magnitude (for example for 10^{-14} g particles from normally $10^{-4}\,\mathrm{m^{-2}\,s^{-1}}$ to $1\,\mathrm{m^{-2}\,s^{-1}}$).

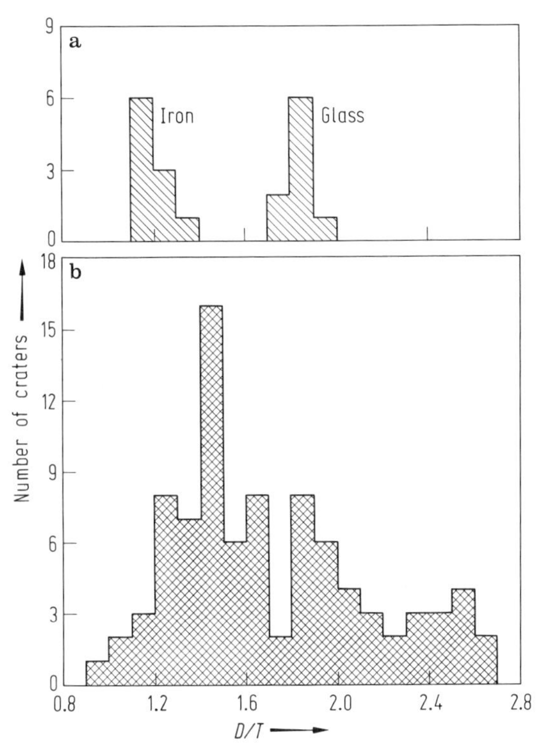

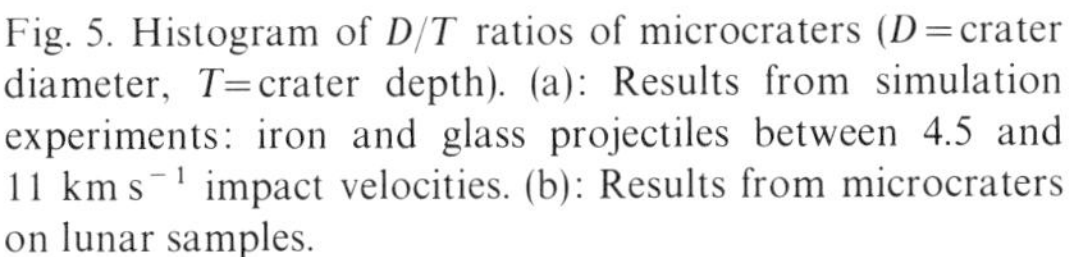

Fig. 5. Histogram of D/T ratios of microcraters (D = crater diameter, T = crater depth). (a): Results from simulation experiments: iron and glass projectiles between 4.5 and $11\,\mathrm{km\,s^{-1}}$ impact velocities. (b): Results from microcraters on lunar samples.

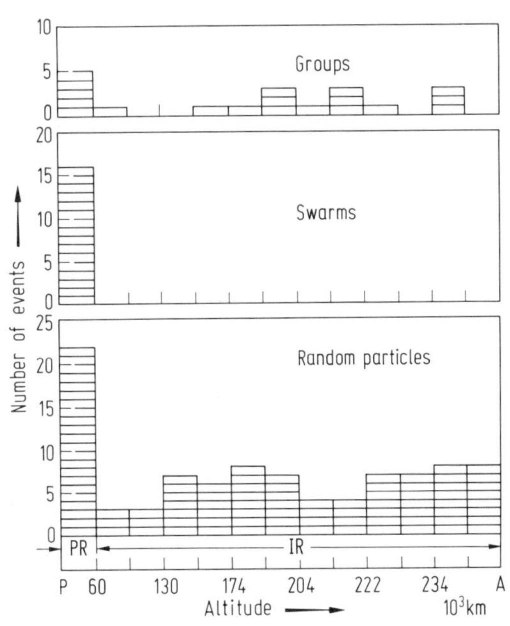

Fig. 6. Rate of groups, swarms, and random particles as a function of the altitude (HEOS-2 Exp. S 215 Feb. 7, 1972⋯ Aug. 2, 1974) [F4]. (P = perigee, A = apogee, PR = perigee region, IR = interplanetary region).

e) Particle collections in the atmosphere and the dust influx on the earth

Particle collection efforts have been undertaken for a long time. Ground collections and early collections in the atmosphere suffered from contamination. Hemenway [H3] used inflight shadowing techniques and reported on submicron-sized particle collections in the upper atmosphere. Brownlee [B10] collected dust particles $>10\,\mu\mathrm{m}$ diameter using balloons and U2 aircraft. He reported on fluffy type, presumably cometary dust. He was the first to prove the extraterrestrial origin of the collected particles by their trapped helium from the solar wind [B 12]. The reported fluxes [B12] are in agreement with the fluxes from lunar microcraters and modern in situ dust experiments. The chemical composition shows agreement with carbonaceous chondrites [B12] (see 3.3.2.2.4).

The particle influx on the earth has been calculated by Hughes [H9]. A total of approx. 40 tons per day has been found from integration of the flux curves at 1 AU. The results from the HEOS 2 dust experiment [F4] show that the moon is approximately in mass balance. The average fluxes from group events (see d) are the same order of magnitude as the flux of sporadic particles.

3.3.4.4 Observations of the zodiacal light

3.3.4.4.1 Definitions

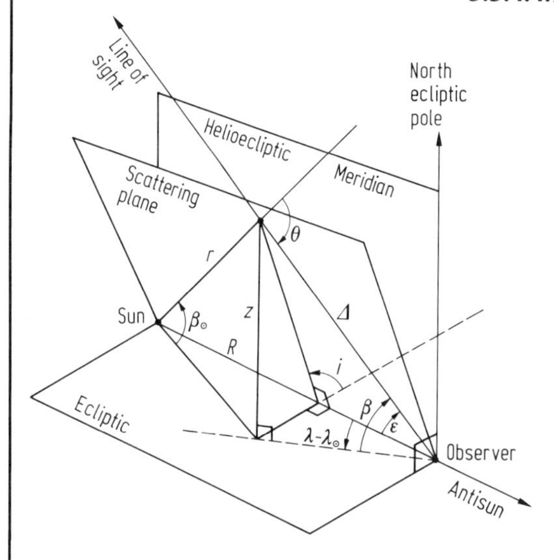

Fig. 7. Geometry for an observer in the ecliptic [D14].
R = heliocentric distance of observer
r = heliocentric distance of particle
Δ = observer-particle distance
z = height above ecliptic plane
$\beta_\odot$ = heliocentric ecliptic latitude
θ = scattering angle
β = ecliptic latitude of line of sight
λ = ecliptic longitude of line of sight
$\lambda_\odot$ = ecliptic longitude of the sun
$\lambda-\lambda_\odot$ = helioecliptic longitude
ε = elongation
i = inclination of scattering plane with respect to the ecliptic

Units of surface brightness related to the solar irradiance:
$B/\overline{B_\odot}$ = brightness in units of average brightness of the solar disk
S_{10} = solar-type star of visual magnitude $V=10.0$ per square degree ($=1/3282.8$ sr)
Following the recommendations of [S12] the following conversion is valid at 505 nm:
$1\,S_{10}=1.27\cdot10^{-9}\,\mathrm{erg\,cm^{-2}\,s^{-1}\,sr^{-1}\,\AA^{-1}}=4.54\cdot10^{-16}\,B/\overline{B_\odot}$ (sun at 1 AU) $=4.05\cdot10^{-3}$ Rayleigh/Å.
Especially in surface brightness photometry of the Milky Way the unit $S_{10}(\lambda)$ is used, which refers to stars of magnitude 10.0 in the indicated spectral range. For example, $1\,S_{10}(B)=1.79\,S_{10}$.

Polarization: The zodiacal light may be assumed to be linearly polarized either perpendicular or parallel to the scattering plane. This allows a simplified presentation of linear polarization.

I, I_1, I_2 = intensity and its components polarized perpendicular or parallel to the scattering plane, respectively.
$p=(I_1-I_2)/(I_1+I_2)$ = degree of polarization. Negative polarization ($p<0$) means that the electric vector is in the scattering plane.
q = degree of circular polarization; positive, if snapshot of tip of electric vector shows a right-handed helix [C1].

3.3.4.4.2 Intensity and polarization in the visible (400···700 nm)

Table 4 gives a "best" estimate compiled from relevant sources and therefore may deviate from the information given in Fig. 8. The isophotes at 710 nm [F6] are similar. A more detailed photometric model is found in [L15]. The accuracy of single observations [D12] varies from ±5···10% for elongations 15···60° in the ecliptic to ±25% at high ecliptic latitudes or in the F-corona. For the degree of polarization a typical accuracy is ±0.01···0.02, except at $\varepsilon<5°$, where the measurement is extremely difficult and has been done only once [B6] (Fig. 8: see p. 236).

The zodiacal light is symmetric both to the ecliptic (north-south) and the corresponding helioecliptic meridian (east-west) with the reservation that the actual plane of symmetry deviates slightly from the ecliptic (see 3.3.4.4.6). The motion of the observer with respect to this plane of symmetry may cause apparent asymmetries in the zodiacal light, including a deviation of the brightness maxima from the ecliptic.

Table 4. Zodiacal light in the ecliptic and the helioecliptic meridian, $\lambda \approx 550$ nm.
Symbols: see 3.3.4.4.1

1 ε	2 $I(\varepsilon, i=0°)$ S_{10}	3 $p(\varepsilon, i=0°)$	4 $I(\varepsilon, i=90°)$ S_{10}	5 $p(\varepsilon, i=90°)$	6 $I(\varepsilon, i=0°)/I(\varepsilon, i=90°)$
1°	$3.9 \cdot 10^6$	0.000	$2.6 \cdot 10^6$	0.00	1.5
2	$8.6 \cdot 10^5$	0.001	$4.3 \cdot 10^5$	0.001	2.0
5	$1.2 \cdot 10^5$	0.012	$4.8 \cdot 10^4$	0.01	2.5
10	$2.4 \cdot 10^4$	0.070	8300	0.05	2.9
15	9100	0.137	2960	0.09	3.1
20	4800	0.152	1550	0.10	3.1
30	1930	0.165	620	0.13	3.1
40	920	0.180	290	0.17	3.2
50	572	0.190	173	0.19	3.3
60	394	0.197	118	0.21	3.3
70	296	0.197	93	0.21	3.2
80	239	0.186	74	0.21	3.2
90	202	0.166	63	0.20	3.2
100	175	0.144	59	0.18	3.0
110	158	0.120	60	0.14	2.6
120	147	0.095	66	0.11	2.2
130	141	0.077	76	0.08	1.9
140	139	0.058	89	0.06	1.6
150	140	0.027	105	0.03	1.3
160	147	0.006	127	0.01	1.2
170	161	−0.02	152	−0.02	1.1
175	169	−0.01	166	−0.01	1.0
180	180	0.000	180	0.00	1.0
Ref.	1···2° [B7] 15° [L8] >30° [D13]	1···5° [B7] 15···20° [L8] 30° [V1] 90° [S11] 110···160° [W1]	1···5°, 15···30° from columns 2, 6 50···180° [D14]	1···5° from column 3 15° [L7] 90° [S11] 150···180° from column 3	1···5° [G8] 10° from columns 2, 4 15···30° [L7] 40···180° from columns 2, 4

Unreferenced data points have been interpolated on basis of the observations summarized in [L5].

Detailed intensity distribution of the F-corona: [S1, B7]
In the region of minimum brightness around the ecliptic pole the brightness level was found lower by $10\,S_{10}$ than given in Table 4 by the satellite experiments [S11, L17].

Polarization at large elongations
Negative polarization near $\varepsilon=170°$: $p=-0.04\pm0.02$ [F6], also detected by [I1, W1, W8].
Neutral point ($p=0$) at $\varepsilon \approx 160°$ [D14, F6, W1] for $\lambda=500$ nm, displacement towards smaller elongations with increasing wavelength reported by [W2].

Circular polarization:
$q=0.0\pm0.001$ at $\varepsilon=30\cdots110°$ [S16],
$q=0.005\pm0.0015$ at $\lambda-\lambda_\odot=-160°$,
$q=-0.003\pm0.001$ at $\lambda-\lambda_\odot=180°$ [W9].

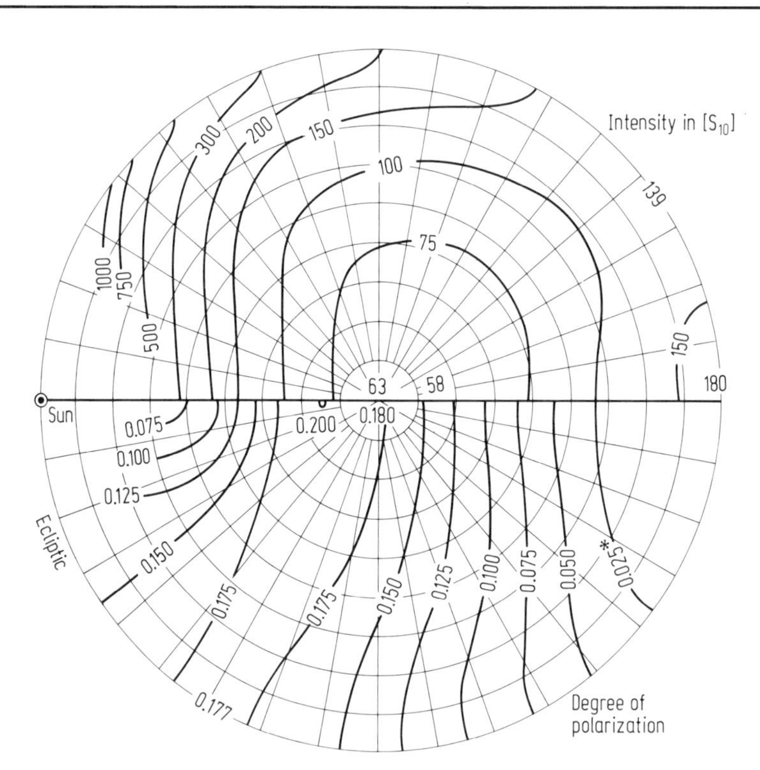

Fig. 8. Average distribution of zodiacal light (1964···1975, Teide) over the sky in helioecliptic coordinates [D14]. The ecliptic pole is at the center. $\lambda = 460$ and 502 nm, * means reduced reliability.

3.3.4.4.3 Spectrum and colour

The line spectrum is identical to the solar spectrum [B2, B5].

Doppler shifts

The measurements are difficult and partly give unexpected high shifts at $\varepsilon \leqq 50°$ [F8, H4, J1]. Probably less than 5% of the dust particles are in retrograde orbits [J1].

Colour

No systematic difference to solar colour from 220 nm to 2.4 µm [F7]. A reddening with respect to the sun of the order of 10% per 100 nm was found from space experiments at $\varepsilon \leqq 90°$ in the range $220\,\text{nm} \leqq \lambda < 900\,\text{nm}$ [F5, L7, L17, M11, P5].

Degree of polarization

Constant from 260 to 900 nm [D14, P5, V1], with possible increase at 260 nm [P5]. No wavelength dependence of polarized intensity from 400 to 860 nm [S13].

Ultraviolet excess for $\lambda < 200$ nm

A moderate excess [M2] or decrease [P4] relative to the solar spectrum are under discussion. Upper limits were given by [F5, M10].

3.3.4.4.4 Thermal emission

The wavelength at which scattering of solar radiation and thermal emission of interplanetary dust grains contribute equal amounts to the zodiacal light appears to be longer than 2.4 µm except for the F-corona, where it is less. Measurements of the thermal emission in the zodiacal light at 11 µm and 20 µm from $\varepsilon=35°\cdots65°$: [P6].

F-corona: Peaks between 3.4 $R_\odot$ and 9.2 $R_\odot$ detected at 2.2 µm [M1, P3], 3.5 µm [P3] and 10 µm [L12].

Table 5. Thermal emission in the zodiacal light at $\varepsilon=160°$.

λ µm	Emission W cm^{-2} sr^{-1} µm^{-1}	
	Observations [S9]	Model predictions [R3]
5···6	$(3\pm2)\cdot10^{-11}$	$1.8\cdot10^{-12}$
8···14	$6.5\cdot10^{-11}$*)	$7.8\cdot10^{-12}$
12···14	$(6\pm4)\cdot10^{-11}$	$8.0\cdot10^{-12}$
16···23	$(2.5\pm1.5)\cdot10^{-11}$	$5.2\cdot10^{-12}$
60		$1.6\cdot10^{-13}$
70···130	$9\cdot10^{-13}$	

*) At $\varepsilon=100°$, measured by [B9].

3.3.4.4.5 Radial gradient of intensity

Inner solar system: $I(R)\propto R^{-2.3\pm0.2}$ for 0.3 AU $\leqq R\leqq$ 1.0 AU [L9].

Outer solar system: $I(R,\varepsilon=140°)\propto R^{-2.5}$ to $R^{-3.0}$ for 1.0 AU $\leqq R\leqq$ 3.3 AU, no zodiacal light detectable beyond 3.3 AU [H1].

3.3.4.4.6 Symmetry plane of zodiacal light

Inside 1 AU: $\Omega=(85\pm5)°$, $i=(3.0\pm0.3)°$ [L11, M8].

Outside 1 AU: $\Omega=(96\pm15)°$, $i=(1.5\pm0.4)°$, close to the invariable plane (see 3.2.1.1) of the solar system [D15]. For a discussion of existing observations, see [L8, M7].

3.3.4.4.7 Temporal variations

Due to orbital motion with respect to the plane of symmetry: $\Delta I=10\cdots20\%$ (peak to peak) [D15, L11, L14].

Due to crossing of particle streams: ΔI up to 60 S_{10} [L14], not found by [B13, L9].

Solar cycle variation: less than $\pm10\%$ over 7 years [B13], less than 10% over solar cycle [D15].

For reports on short-time variations in polarization at large elongations, due to solar flares or dust in lunar libration regions, see the reviews.

3.3.4.5 Interpretation of zodiacal light observations

3.3.4.5.1 Spatial distribution

In general agreement with the observations the following assumption is made:
$n(r,z)=r^{-\nu}\cdot f(\beta_\odot)$ (n in [particles/cm^3], $\sin\beta_\odot=z/r$, see Fig. 7).

If the average scattering function $\sigma(\theta)$ of the particles is independent of r and z, ν follows from the observed radial gradient of intensity, $I(\varepsilon,R)\propto R^{-\nu-1}$.

Radial distribution:

0.09 AU $\leqq r\leqq$ 1 AU	$n(r)\propto r^{-1.3\pm0.2}$ [L9]	directly from $I(\varepsilon,R)$
1 AU $\leqq r\leqq$ 2.3 AU	$n(r)\propto r^{-\nu}$	$\nu=1.5$, $C=0.13$ or
2.3 AU $\leqq r\leqq$ 3.3 AU	$n(r)=n(1\text{ AU})(r^{-\nu}+C)$	$\nu=2.0$, $C=0.26$ [H1]
$r>3.3$ AU	$n(r)=0$	from model fits to $I(\varepsilon,R)$.

Out-of-ecliptic distribution:

$n(r,z)\propto r^{-1.2}\cdot(1+48(z/r)^2)^{-0.6}$ (ellipsoidal with axial ratio 1/7) [D11]
$n(r,z)\propto r^{-1.3}\cdot\exp(-2.1\cdot|z/r|)$ [L10]

Predictions for observations outside the ecliptic [G5].
The average orbital inclination of the particles producing the zodiacal light is $i \approx 30°$ [L8, S7].

Dust-free zone near the sun.
The observed peaks in infrared brightness (see 3.3.4.4.4) require an excess of material just outside the evaporation zones. Trajectory calculations for graphite and obsidian show that a long residence time of the dust particles in these regions is to be expected [L3, M13].

3.3.4.5.2 Size distribution

Effective radius $s \geqq 2.4\,\mu m$ [N6] from absence of blue colour. 70% of zodiacal light scattered by particles with $10\,\mu m \leqq s \leqq 80\,\mu m$ [G4], based on Fig. 2.

3.3.4.5.3 Scattering and absorption

a) Calculations for spherical particles (Mie's Theory)
See [H10, K 2] for theory and [D1, K1] for computational procedures.
Scattering functions for individual particles:
tables in [W7], figures in [G1, W7].
Scattering functions for size distributions:
tables in [G2, G3], figures in [D2, G1, G2, G3].
Absorption cross section: see [W7] for tables, [R3] for wavelength dependence 0.15···100 μm and [S4] for average over solar spectrum.

b) Measurements on irregular particles
Scattering functions deviate from Mie-theory predictions for equivalent spheres.
From light scattering: 5 μm SiO [H8], 0.5 μm NaCl and $(NH_4)_2SO_4$ [P2].
From microwave analog measurements: dielectric and absorbing particles [Z1], spheroid and "rough sphere" [G10].

c) Empirical volume scattering functions $S(\theta)$ and polarization p_s of the scattering function
Determined from inversion of brightness integral [D10, D11, L5] or from fit to the observed zodiacal light intensity in the ecliptic with an assumed spatial distribution [D13, L8]
$S(90°) = (7.8 \pm 1.6) \cdot 10^{-23}$ (cm² sr⁻¹) cm⁻³,
$p_s(90°) = 0.34 \pm 0.07$ [D10].

d) Gegenschein
Observations far from the earth [W3] have shown that it is due to back-scattering by interplanetary material. Earlier theories are covered in the bibliography [R4].

e) Temperature of interplanetary dust grains: see Fig. 9 and [L2, M13, R3, S4].

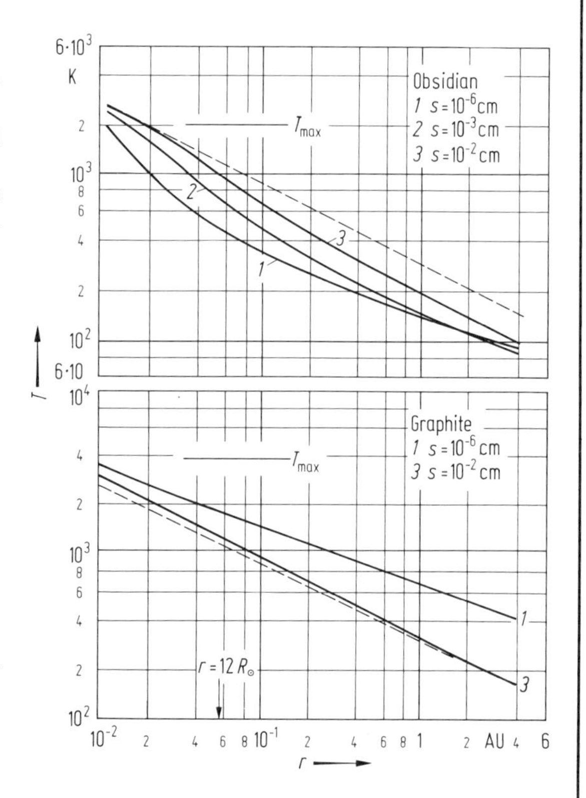

Fig. 9. Grain temperature, T, vs. distance from the sun, r, for different particle sizes s [R3]. The melting temperatures: T_{max}; the black-body dependence (dashed line) $T = 280 \cdot (1\,AU/r)^{1/2}$ K [L5].

3.3.4.5.4 Models

Because spatial distribution and size distribution are determined separately, the models try to determine structure and material of the particles. Spherical, mostly absorbing, particles have been proposed based on Mie's theory [R3], or loosely-structured ("fluffy") absorbing particles based on scatter experiments [G7]. At present there is no need to assume a special shape (e.g. elongated particles).

3.3.4.6 Dynamics of interplanetary dust

3.3.4.6.1 Sources

Dust production by short-period comets: $\lesssim 250$ kg s^{-1} in bound orbits [D3, R2, S5]
by long-period comets: ≈ 20000 kg s^{-1} [D3], ≈ 200000 kg s^{-1} [W6], mostly in hyperbolic orbits.
Dust production by collisions between asteroids appears to be a minor effect [D6].

With the interstellar gas flowing into the solar system [M6] interstellar dust would be expected at a mass ratio $\approx 1{:}100$, corresponding to $\approx 10^{-27}$ g cm^{-3} or $\approx 10^{-13}$ particles/cm^3 ($s = 0.1$ µm), but probably could not penetrate into the inner solar system [G9, L16, M9a, G13].

3.3.4.6.2 Forces

The Poynting-Robertson drag, resulting from the aberration of the impinging solar radiation [B14], makes a perfectly absorbing particle in initial circular orbit with radius r spiral into the sun in

$$t = 7 \cdot 10^2\, s \cdot \varrho \cdot r^2 \text{ years [W11], with } s \text{ in [µm]}, \varrho \text{ in [g cm}^{-3}], r \text{ in [AU]}.$$

The effect on the spatial distribution of the particles was treated by [B8, S10] leading to $n(r) \propto r^{-1}$ at least inside the region of injection. Ion impact drag increases the Poynting-Robertson drag by about 25%.

The Lorentz force, for the particles charged to typically 6 V [R1] or 12 V [W10] is effective in removing submicron particles from the solar system [S2], while negligible for particles with $s \gtrsim 10$ µm [M9]. Discussion of orbit perturbations [M9, C3a].

The Larmor radius for a 1 µm-particle at 1 AU is ≈ 70 AU.

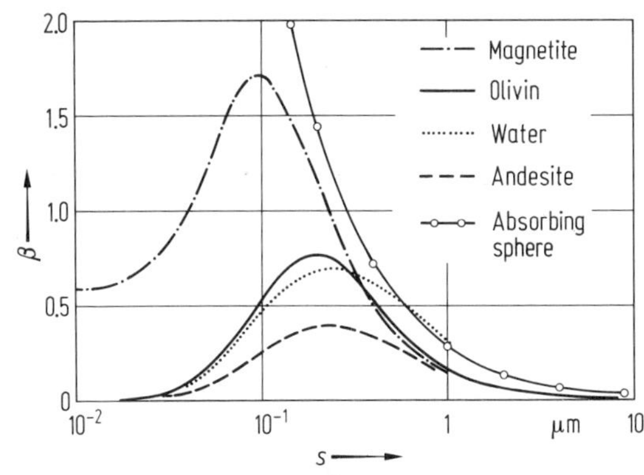

Fig. 10. Ratio β of radiation pressure to gravitation as function of grain radius, s, for magnetite ($\varrho = 5.2$ g cm^{-3}), olivin, andesite ($\varrho = 3.3$ g cm^{-3}), water ice ($\varrho = 1.0$ g cm^{-3}) [S4], and a perfectly absorbing sphere ($\varrho = 2.0$ g cm^{-3}).

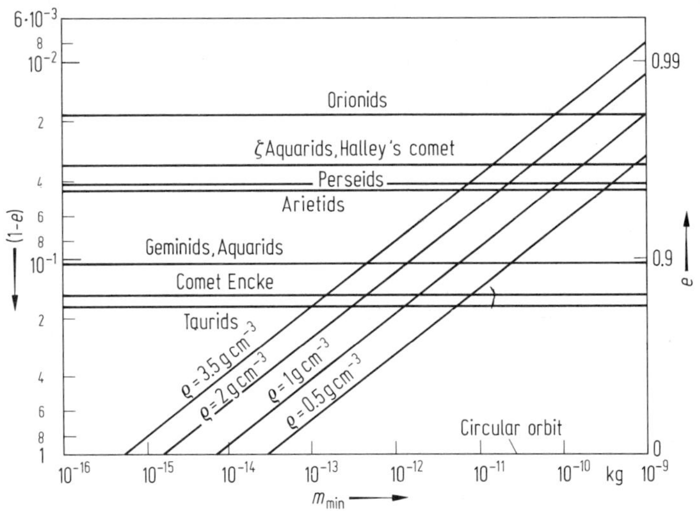

Fig. 11. Minimum mass, m_{min} necessary for perfectly absorbing meteoroids to remain in bound orbits under the action of radiation pressure, when released at perihelion of comet orbit [D8].

Table 6. Comparison of various forces for typical interplanetary conditions.
Symbols:

r = distance from the sun [AU]
n_e = electron density [cm^{-3}]
v_{sw} = solar wind speed [$km\,s^{-1}$]
V_0 = electrostatic potential of dust particle [V]
B = magnetic field strength [nT].
ϱ = density of the particles [$g\,cm^{-3}$]
s = particle radius [cm]
Q_{rp} = efficiency factor for radiation pressure; = 1 for perfectly absorbing sphere
v_t = tangential velocity of dust particle [$km\,s^{-1}$]
w = particle velocity in solar wind rest frame [$km\,s^{-1}$]

Forces in [dyn]:

gravity $F_g = 2.49 \cdot s^3 r^{-2}$
radiation pressure $F_{rp} = 4.54 \cdot 10^{-5} \cdot Q_{rp} s^2 r^{-2}$
Poynting-Robertson drag $F_{PR} = (v_t/c) F_{rp} = 4.51 \cdot 10^{-9} \cdot Q_{rp} s^2 r^{-2.5}$
ion impact pressure $F_{ip} = 2.01 \cdot 10^{-14} \cdot n_e v_{sw}^2 s^2 r^{-2}$
ion impact drag $F_{id} = (v_t/v_{sw}) F_{ip} = 1.20 \cdot 10^{-9} \cdot s^2 r^{-2.5}$
Coulomb pressure $F_{Cb} \leqq 3 \cdot 10^{-9} \cdot n_e s^2 V_0^2 v_{sw}^{-2} r^{-2}$
Lorentz force $F_L = 1.11 \cdot 10^{-13} \cdot s V_0 \, [w \times B] r^{-1}$

Pressure forces are acting in the radial, drag forces in the tangential direction.

Forces are given for typical interplanetary conditions: $r = 1$ AU, $n_e = 6\,cm^{-3}$, $v_{sw} = 400\,km\,s^{-1}$, $V_0 = 10$ V, $B = 5$ nT, $\varrho = 3.5\,g\,cm^{-3}$.

s	F_g	F_{rp}[D8]	F_{PR}	F_{ip}	F_{id}	F_{Cb}[M9]	F_L
cm	dyn						
10^{-5}	$-8.71 \cdot 10^{-15}$	$1.43 \cdot 10^{-14}$	$1.42 \cdot 10^{-18}$	$5.04 \cdot 10^{-18}$	$3.77 \cdot 10^{-19}$	10^{-21}	$1.57 \cdot 10^{-14}$
10^{-4}	$-8.71 \cdot 10^{-12}$	$1.43 \cdot 10^{-12}$	$1.42 \cdot 10^{-16}$	$5.04 \cdot 10^{-16}$	$3.77 \cdot 10^{-17}$	10^{-19}	$1.57 \cdot 10^{-13}$
10^{-3}	$-8.71 \cdot 10^{-9}$	$1.43 \cdot 10^{-10}$	$1.42 \cdot 10^{-14}$	$5.04 \cdot 10^{-14}$	$3.77 \cdot 10^{-15}$	10^{-17}	$1.57 \cdot 10^{-12}$
10^{-2}	$-8.71 \cdot 10^{-6}$	$1.43 \cdot 10^{-8}$	$1.42 \cdot 10^{-12}$	$5.04 \cdot 10^{-12}$	$3.77 \cdot 10^{-13}$	10^{-15}	$1.57 \cdot 10^{-11}$

3.3.4.6.3 Sinks

The main effects are summarized in Fig. 12, additional information is given below.

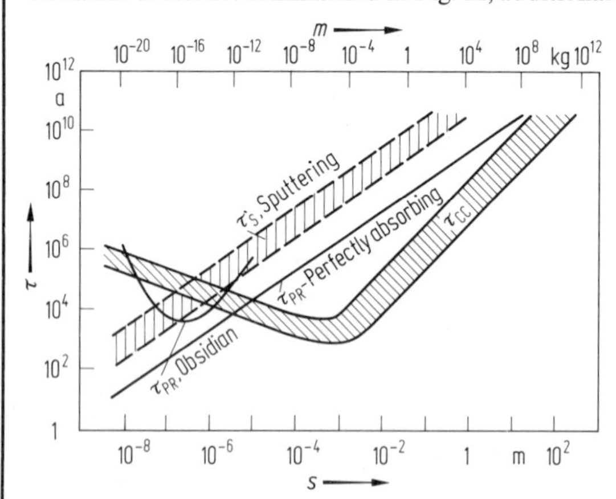

Fig. 12. Particle survival times, τ [D8]. Assumed sputtering rate $\Delta s = (0.05 \cdots 0.5)$Å/a [M3, M4]. The time τ_{cc} for destruction by catastrophic collisions was calculated for basalt (upper curve) and glass. Index PR refers to Poynting-Robertson effect.

Evaporation limit: Iron $\approx 24\,R_\odot$ [L1]; andesite, olivine, magnetite $\approx 10\,R_\odot$ [L1, R3]; obsidian, graphite $2.2 \cdots 4\,R_\odot$ [L1, M13, R3].
Inward mass flow due to Poynting-Robertson effect: 300 $kg\,s^{-1}$ [L8], 30 $kg\,s^{-1}$ [L13].

Beta-meteoroids: Particles reduced to submicron size by collisions, possibly also by evaporation or sputtering, blown out of the solar system by radiation pressure [Z2].
Outward flux at 1 AU: 1200 $kg\,s^{-1}$ [L8], $\approx 30000\,kg\,s^{-1}$ [L13].

Capture by planets: important for the planetary surfaces [A2] rather than for the interplanetary dust cloud.

Total loss: $10000 \cdots 20000\,kg\,s^{-1}$ [W5], $\approx 25000\,kg\,s^{-1}$ [D5], mainly by collisions.

3.3.4.7 References for 3.3.4

1 McDonnell, J.A.M. (ed.): Cosmic Dust, Wiley, Chichester (1978).

2 Weinberg, J.L. (ed.): The Zodiacal Light and the Interplanetary Medium. NASA SP-150, Washington (1967).

3 Elsässer, H., Fechtig, H. (eds.): Interplanetary Dust and Zodiacal Light, Lect. Notes Phys. **48** Springer, Berlin (1976).

4 Halliday, I., McIntosh, B.A. (eds.): Solid Particles in the Solar System, Reidel, Dordrecht (1980).

A1 Anonymous: Interim Report NASA X-53629 (1967).

A2 Ashworth, D.G. in: [1] p. 427.

B1 Bedford, D.K., Adams, N.G., Smith, D.: Planet. Space Sci. **23** (1975) 1451.

B2 Beggs, D.W., Blackwell, D.E., Dewhirst, D.W., Wolstencroft, R.D.: Mon. Not. R. Astron. Soc. **127** (1964) 329.

B3 Berg, O.E., Richardson, F.F.: Rev. Sci. Instrum. **40** (1969) 133.

B4 Berg, O.E., Grün, E.: Space Research XIII (1973) 1047.

B5 Blackwell, D.E., Ingham, M.F.: Mon. Not. R. Astron. Soc. **122** (1961) 129.

B6 Blackwell, D.E., Petford, A.D.: Mon. Not. R. Astron. Soc. **131** (1966) 399.

B7 Blackwell, D.E., Dewhirst, D.W., Ingham, M.F.: Adv. Astron. Astrophys. **5** (1967) 1.

B8 Briggs, R.E.: Astron. J. **67** (1962) 710.

B9 Briotta, D.A., jr.: Thesis, Cornett University (1976).

B10 Brownlee, D.E. in: [1] p. 295.

B11 Brownlee, D.E., Hörz, F., Vedder, J.F., Gault, D.E., Hartung, J.B.: Proc. Lunar Sci. Conf. 4th, Geochim. Cosmochim. Acta, Suppl. 4, **3** (1973) 3197.

B12 Brownlee, D.E., Tomandl, O.A., Olszewski, E.: Proc. Lunar Sci. Conf. 8th, Pergamon Press (1977) 149.

B13 Burnett, G. in: [3] p. 53.

B14 Burns, J.A., Lamy, P.L., Soter, S.: Icarus **40** (1979) 1.

C1 Clarke, D. in: Planets, Stars, and Nebulae studied with Photopolarimetry (Gehrels, T., ed.), University of Arizona, Tucson (1974) p. 45.

C2 Colombo, G., Lautman, D.A., Shapiro, I.I.: J. Geophys. Res. **71** (1966) 5705.

C3 Colombo, G., Shapiro, I.I., Lautman, D.A.: J. Geophys. Res. **71** (1966) 5719.

C3a Consolmagno, G.: Icarus **38** (1979) 398.

C4 Cook, A.F.: Icarus **33** (1978) 349.

D1 Dave, J.V.: Appl. Opt. **8** (1969) 155.

D2 Deirmendjian, D.: Electromagnetic Scattering on Spherical Polydispersions, Elsevier, New York (1969).

D3 Delsemme, A.H. in: [3] p. 314.

D4 Dietzel, H., Eichhorn, G., Fechtig, H., Grün, E., Hoffmann, H.-J., Kissel, J.: J. Phys. E (Scientific Instrum.) **6** (1973) 209.

D5 Dohnanyi, J.S.: Icarus **17** (1972) 1.

D6 Dohnanyi, J.S. in: [3] p. 187.

D7 Dohnanyi, J.S.: Space Research XVII (1977) 623.

D8 Dohnanyi, J.S. in: [1] p. 527.

D9 Dumont, R.: Ann. Astrophys. **28** (1965) 265.

D10 Dumont, R.: Planetary Space Sci. **21** (1973) 2149.

D11 Dumont, R. in: [3] p. 85.

D12 Dumont, R., Sanchez, F.: Astron. Astrophys. **38** (1975) 397.

D13 Dumont, R., Sanchez, F.: Astron. Astrophys. **38** (1975) 405.

D14 Dumont, R., Sanchez, F.: Astron. Astrophys. **51** (1976) 393.

D15 Dumont, R., Levasseur-Regourd, A.C.: Astron. Astrophys. **64** (1978) 9.

F1 Fechtig, H., Hartung, J.B., Nagel, K., Neukum, G., Storzer, D.: Proc. Lunar Sci. Conf. 5th, Pergamon Press (Gose, W.A., ed.) **3** (1974) 2463.

F2 Fechtig, H., Gentner, W., Hartung, J.B., Nagel, K., Neukum, G., Schneider, E., Storzer, D.: Proc. Soviet American Conf. on Cosmochem. of Moon and Planets, Translation: NASA-SP-370 (1976) 585.

F3 Fechtig, H., Grün, E., Kissel, J.: [1] p. 607.

F4 Fechtig, H., Grün, E., Morfill, G.: Planet. Space Sci. **27** (1979) 511.

F5 Feldman, P.D.: Astron. Astrophys. **61** (1977) 635.

F6 Frey, A., Hofmann, W., Lemke, D., Thum, C.: Astron. Astrophys. **36** (1974) 447.

F7 Frey, A., Hofmann, W., Lemke, D.: Astron. Astrophys. **54** (1977) 853.

F8 Fried, J.W.: Astron. Astrophys. **68** (1978) 259.

G1 Giese, R.H.: Space Sci. Rev. **1** (1963) 589.

G2 Giese, R.H.: Forschungsbericht BMBW-FB W71-23, ZLDI, München (1971).
G3 Giese, R.H.: Forschungsbericht BMBW-FB W72-19, ZLDI, München (1972).
G4 Giese, R.H.: J. Geophys. **42** (1977) 705.
G5 Giese, R.H.: Astron. Astrophys. **77** (1979) 223.
G6 Giese, R.H., Grün, E. in: [3] p. 135.
G7 Giese, R.H., Weiss, K., Zerull, R.H., Ono, T.: Astron. Astrophys. **65** (1978) 265.
G8 Gillett, F.C.: Thesis, University of Minnesota (1966).
G9 Greenberg, J.M. in: Evolutionary and Physical Properties of Meteoroids (Hemenway, C.L., et al., eds.) NASA SP 319, Washington (1973) p. 375.
G10 Greenberg, J.M. in: Planets, Stars, and Nebulae studied with Photopolarimetry (Gehrels, T., ed.), University of Arizona, Tucson (1974) p. 107.
G11 Grün, E., Fechtig, H., Kissel, J., Gammelin, P.: J. Geophys. **42** (1977) 717.
G12 Grün, E., Pailer, N., Fechtig, H., Kissel, J.: Planetary Space Sci. **28** (1980) 333.
G13 Gustafson, B.A.S., Misconi, N.Y.: Nature **282** (1979) 276.
H1 Hanner, M.S., Sparrow, J.G., Weinberg, J.L., Beeson, D.E. in: [3] p. 29.
H2 Hastings, E.C.: NASA TMX-810, NASA TMX-824, NASA TMX-899 (1963).
H3 Hemenway, C.L. in: The Dusty Universe (Field, G.B., Cameron, A.G.W., eds.), Neal Watson Academic (1975) p. 211.
H4 Hicks, T.R., May, B.H., Reay, N.K.: Mon. Not. R. Astron. Soc. **166** (1974) 439.
H5 Hörz, F., Brownlee, D.E., Fechtig, H. Hartung, J.B., Morrison, D.A., Neukum, G., Schneider, E., Vedder, J.F., Gault, D.E.: Planet. Space Sci. **23** (1975) 151.
H6 Hoffmann, H.J., Fechtig, H., Grün, E., Kissel, J.: Planet. Space Sci. **23** (1975) 215.
H7 Hoffmann, H.J., Fechtig, H., Grün, E., Kissel, J.: Planet. Space Sci. **23** (1975) 985.
H8 Holland, A.C., Gagne, G.: Appl. Opt. **9** (1970) 1113.
H9 Hughes, D.W.: Space Research XV (1975) 531.
H10 Hulst van de, H.C.: Light Scattering by Small Particles, Wiley, New York (1957).
H11 Humes, D.H., Alvarez, J.M., O'Neal, R.L., Kinard, W.H.: J. Geophys. Res. **79** (1974) 3677.
H12 Humes, D.H., Alvarez, J.M., Kinard, W.H., O'Neal, R.L.: Science **188** (1975) 473.
I1 Ingham, M.F., Jameson, R.F.: Mon. Not. R. Astron. Soc. **140** (1968) 473.
J1 James, J.F., Smeethe, M.J.: Nature **227** (1970) 588.
K1 Kattawar, G.W., Plass, G.N.: Appl. Opt. **6** (1967) 1377.
K2 Kerker, M.: The Scattering of Light and other Electromagnetic Radiation, Academic Press, New York and London (1969).
L1 Lamy, Ph.L.: Astron. Astrophys. **33** (1974) 191.
L2 Lamy, Ph.L.: Astron. Astrophys. **35** (1974) 197.
L3 Lamy, Ph.L. in: [3] p. 437.
L4 Lautman, D.A., Shapiro, I.I., Colombo, G.: J. Geophys. Res. **71** (1966) 5733.
L5 Leinert, C.: Space Sci. Rev. **18** (1975) 281.
L6 Leinert, C., Klüppelberg, D.: Appl. Opt. **13** (1974) 556.
L7 Leinert, C., Link, H., Pitz, E.: Astron. Astrophys. **30** (1974) 411.
L8 Leinert, C., Link, H., Pitz, E., Giese, R.H.: Astron. Astrophys. **47** (1976) 221.
L9 Leinert, C., Pitz, E., Hanner, M., Link, H.: J. Geophys. **42** (1977) 699.
L10 Leinert, C., Hanner, M., Pitz, E.: Astron. Astrophys. **63** (1978) 183.
L11 Leinert, C., Hanner, M., Richter, I., Pitz, E.: Astron. Astrophys. **82** (1980) 328.
L12 Lena, P., Viala, Y., Mondellini, J., Hall, D., McCurnin, T.W., Soufflot, A., Darpentigny, C., Belbéoch, J.: Astron. Astrophys. **37** (1974) 75.
L13 Le Sergeant, L.B., Lamy, Ph.L.: Nature **276** (1978) 800.
L14 Levasseur, A.C., Blamont, J. in: [3] p. 58.
L15 Levasseur-Regourd, A.C., Dumont, R.: Astron. Astrophys. **84** (1980) 277.
L16 Levy, E.H., Jokipii, J.R.: Nature **264** (1976) 423.
L17 Lillie, C.F. in: NASA SP-310 (Code, A.D., ed.), Washington (1972) p. 95.
M1 MacQueen, R.M.: Astrophys. J. **154** (1968) 1059.
M2 Maucherat-Joubert, M., Cruvellier, P., Deharveng, J.M.: Astron. Astrophys. **74** (1979) 218.
M3 Maurette, M., Price, P.: Science **187** (1975) 121.
M4 McDonnell, J.A.M., Ashworth, D.G.: Space Research XII (1972) 333.
M5 McDonnell, J.A.M. in: [1] p. 337.
M6 Meier, R.R.: Astron. Astrophys. **55** (1977) 211.
M7 Misconi, N.Y.: Astron. Astrophys. **61** (1977) 497.

M8 Misconi, N.Y., Weinberg, J.L.: Science **200** (1978) 1484.
M9 Morfill, G.E., Grün, E.: Planetary Space Sci. **27** (1979) 1269.
M9a Morfill, G.E., Grün, E.: Planetary Space Sci. **27** (1979) 1283.
M10 Morgan, D.H.: Astron. Astrophys. **70** (1978) 543.
M11 Morgan, D.H., Nandy, K., Thompson, G.I.: Mon. Not. R. Astron. Soc. **177** (1976) 531.
M12 Morrison, D.A., Zinner, E. in: [3] p. 227.
M13 Mukai, T., Yamamoto, T.: Publ. Astron. Soc. Japan **31** (1979) 585.
N1 Nagel, K., Neukum, G., Eichhorn, G., Fechtig, H., Müller, O., Schneider, E.: Proc. Lunar Sci. Conf. 6th (1975) 3417.
N2 Nagel, K., Neukum, G., Dohnanyi, J.S., Fechtig, H., Gentner, W.: Proc. Lunar Sci. Conf. 7th (1976) 596.
N3 Naumann, R.J.: NASA-TN D 3717 (1966).
N4 Naumann, R.J., Jex, D.W., Johnson, C.L.: NASA-TR R 321 (1969).
N5 Nilsson, C.S.: Science **153** (1966) 1242.
N6 Nishimura, T.: Publ. Astron. Soc. Japan **25** (1973) 375.
O1 O'Neale, R.L.: NASA Technical Note TD-d 4284 (1968).
P1 Pailer, N., Grün, E.: Planetary Space Sci. **28** (1980) 321.
P2 Perry, R.: MA Thesis, University of Arizona (1977).
P3 Peterson, A.W.: Astrophys. J. **148** (1967) L 37.
P4 Pitz, E., Leinert, C., Schulz, A., Link, H.: Astron. Astrophys. **69** (1978) 297.
P5 Pitz, E., Leinert, C., Schulz, A., Link, H.: Astron. Astrophys. **74** (1979) 15.
P6 Price, S.D., Murdock, T.L., Marcotte, L.P.: Astron. J. **85** (1980) 765.
R1 Rhee, J.W. in: [2] p. 291.
R2 Röser, S. in: [3] p. 319.
R3 Röser, S., Staude, H.J.: Astron. Astrophys. **67** (1978) 381.
R4 Roosen, R.G.: Icarus **13** (1970) 523.
S1 Saito, K.: Ann. Tokyo Astron. Obs., 2nd Series, XII, 2, (1970) p. 53.
S2 Schmidt, Th., Elsässer, H. in: [2] p. 287.
S3 Schneider, E., Storzer, D., Hartung, J.B., Fechtig, H., Gentner, W.: Proc. Lunar Sci. Conf. 4th, Geochim. Cosmochim. Acta, Pergamon Press, Suppl. 4, **3** (1973) 3277.
S4 Schwehm, G., Rohde, M.: J. Geophys. **42** (1977) 727.
S5 Sekanina, Z., Schuster, H.E.: Astron. Astrophys. **68** (1978) 429.
S6 Shapiro, I.I., Lautman, D.A., Colombo, G.: J. Geophys. Res. **71** (1966) 3695.
S7 Singer, S.F., Bandermann, L.W. in: [2] p. 379.
S8 Smith, D., Adams, N.G., Khan, H.A.: Nature **252** (1974) 101.
S9 Soifer, B.T., Houck, J.R., Harwit, M.: Astrophys. J. **168** (1971) L 73.
S10 Southworth, R.B.: Ann. N.Y. Acad. Sci. **119** (1964) 54.
S11 Sparrow, J.G., Ney, E.P.: Astrophys. J. **174** (1972) 705.
S12 Sparrow, J.G., Weinberg, J.L. in: [3] p. 41.
S13 Sparrow, J.G., Weinberg, J.L., Hahn, R.C. in: [3] p. 45.
S14 Stanley, J.E., Singer, S.F., Alvarez, J.M.: Icarus **37** (1979) 457.
S15 Staude, H.J.: Astron. Astrophys. **39** (1975) 325.
S16 Staude, H.J., Schmidt, T.: Astron. Astrophys. **20** (1972) 163.
S17 Sternberg, J.R., Ingham, M.F.: Mon . Not. R. Astron. Soc. **159** (1972) 1.
V1 Vande Noord, E.L.: Astrophys. J. **161** (1970) 309.
W1 Weinberg, J.L.: Ann. Astrophys. **27** (1964) 718.
W2 Weinberg, J.L., Mann, H.M.: Astrophys. J. **152** (1968) 665.
W3 Weinberg, J.L., Hanner, M.S., Mann, H.M., Hutchinson, P.B., Fimmel, R.: Space Research XIII (1973) 1187.
W4 Weinberg, J.L., Sparrow, J.G. in: [1] p. 75.
W5 Whipple, F.L. in: [2] p. 409.
W6 Whipple, F.L. in: [1] p.1.
W7 Wickramasinghe, N.C.: Light Scattering Functions for Small Particles, Hilger, London (1973).
W8 Wolstencroft, R.D., Rose, L.J.: Astrophys. J. **147** (1967) 271.
W9 Wolstencroft, R.D., Kemp, J.C.: Astrophys. J. **177** (1972) L 137.
W10 Wyatt, S.P.: Planet. Space Sci. **17** (1969) 155.
W11 Wyatt, S.P., jr., Whipple, F.L.: Astrophys. J. **111** (1950) 134.
Z1 Zerull, R.H.: Beitr. Phys. Atm. **49** (1976) 168.
Z2 Zook, H.A., Berg, O.E.: Planet. Space Sci. **23** (1975) 183.

3.3.5 Interplanetary particles and magnetic field

3.3.5.1 Interplanetary gas of non-solar origin (neutral hydrogen and neutral helium)

Neutral particles (H- and He-atoms) of the local interstellar gas penetrate the solar system where they execute collisionless Keplerian trajectories under the influence of the solar gravitational field and the solar radiation pressure and where they are subject to ionization loss processes. The distribution of these particles in the interplanetary space is determined by observations of the backscattering of the two solar emission lines H I 1216 Å and He I 584 Å and by model calculations. Typical numbers for the local interstellar gas (outside the solar system) are:

number densities for H-atoms [2, 3, 8, 9, 14, 15]	$N(\mathrm{H}) \approx 0.04 \cdots 0.15\ \mathrm{cm}^{-3}$
number densities for He-atoms [2, 2a, 3, 8a, 12, 15]	$N(\mathrm{He}) \approx 0.004 \cdots 0.02\ \mathrm{cm}^{-3}$
gas temperature [1, 2, 2a, 3, 4a, 9, 12, 15]	$T \approx 5 \cdot 10^3 \cdots 2 \cdot 10^4\ \mathrm{K}$
bulk velocity relative to the sun [1, 3, 4a, 8, 9, 12]	$V \approx 19 \cdots 25\ \mathrm{km\ s}^{-1}$
direction of solar motion relative to the interstellar gas [2, 4a, 7, 9, 18]	right ascension $\alpha \approx 252° \pm 5°$ declination $\delta \approx -17° \pm 5°$

Sky maps for backscatter radiation (H I: 1216 Å; He I: 584 Å):

from OGO 5	H I	[4, 9, 16, 17]
Mariner 9	H I	[6]
Mariner 10	H I, He I	[7, 15]
STP satellite	He I	[9, 18]
Nike Tomahawk	He I	[13]

Table 1. Ionization rates for H- and He-atoms in interplanetary space near the orbit of earth [9] Table 1.

Reaction	Ionization rate $10^{-7}\ \mathrm{s}^{-1}$	Remarks
Photoionization		Photoionization rates correspond to photon fluxes of about $2 \cdot 10^{10}$ photons $\mathrm{cm}^{-2}\ \mathrm{s}^{-1}$ in the spectral range from 910···375 Å, and $3.6 \cdot 10^{10}$ photons $\mathrm{cm}^{-2}\ \mathrm{s}^{-1}$ in the range from 665···150 Å, which are appropriate for low to moderate solar activity.
$\mathrm{H} + h\nu \rightarrow \mathrm{H}^+ + \mathrm{e}$	0.88	
$\mathrm{He} + h\nu \rightarrow \mathrm{He}^+ + \mathrm{e}$	0.80	
Charge exchange		Charge exchange rates correspond to solar wind flux densities of $3.3 \cdot 10^8$ particles $\mathrm{cm}^{-2}\ \mathrm{s}^{-1}$ for H^+ and $1.5 \cdot 10^7$ particles $\mathrm{cm}^{-2}\ \mathrm{s}^{-1}$ for He^{++}.
$\mathrm{H} + \mathrm{H}^+ \rightarrow \mathrm{H}^+ + \mathrm{H}$	6.6	
$\mathrm{H} + \mathrm{He}^{++} \rightarrow \mathrm{H}^+ + \mathrm{He}^+$	0.033	
$\mathrm{He} + \mathrm{H}^+ \rightarrow \mathrm{He}^+ + \mathrm{H}$	0.050	
$\mathrm{He} + \mathrm{He}^{++} \rightarrow \mathrm{He}^+ + \mathrm{He}^+$	0.0017	
Electron collisional ionization		Collisional ionization rates correspond to solar wind electron number density of about 8 electrons cm^{-3} and an electron temperature of about 10^5 K.
$\mathrm{H} + \mathrm{e} \rightarrow \mathrm{H}^+ + 2\mathrm{e}$	0.19	
$\mathrm{He} + \mathrm{e} \rightarrow \mathrm{He}^+ + 2\mathrm{e}$	0.037	

Summary of density and temperature determinations for the local interstellar gas [15]. Ionization rates for H- and He-atoms at 1 AU, see Table 1 [3, 9].

Model calculations for the distribution of neutral gas in the interplanetary space and/or the intensity of backscattered radiation [3, 5, 9···12, 15, 19]. Table for hydrogen densities in interplanetary space [15].

Spectral distribution of the scattered intensity taking into account the shape of the solar emission line [12].

References for 3.3.5.1

1 Adams, T.F., Frisch, P.C.: Astrophys. J. **212** (1977) 300.
2 Ajello, J.M.: Astrophys. J. **222** (1978) 1068.
2a Ajello, J.M., Witt, N., Blum, P.W.: Astron. Astrophys. **73** (1979) 260.
3 Axford, W.I.: in Solar Wind (Sonett, C.P., Coleman, P.J., jr., Wilcox, J.M., eds.), NASA-SP-308 (1972) p. 609.
4 Bertaux, J.L., Blamont, J.E.: Astron. Astrophys. **11** (1971) 200.
4a Bertaux, J.L., Blamont, J.E., Tabarié, N., Kurt, W.G., Bourgin, M.C., Smirnow, A.S., Dementeva, N.N.: Astron. Astrophys. **46** (1976) 19.
5 Blum, P.W., Pfleiderer, J., Wulf-Mathies, C.: Planet. Space Sci. **23** (1975) 93.
6 Bohlin, R.C.: Astron. Astrophys. **28** (1973) 323.
7 Broadfoot, A.L., Kumar, S.: Astrophys. J. **222** (1978) 1054.
8 Fahr, H.J.: Space Sci. Rev. **15** (1974) 483.
8a Freeman, J., Paresce, F., Bowyer, S.: Astrophys. J. **231** (1979) L 37.
9 Holzer, T.E.: Rev. Geophys. Space Phys. **15** (1977) 467.
10 Johnson, H.E.: Planet. Space Sci. **20** (1972) 829.
11 Joselyn, J.A., Holzer, T.E.: J. Geophys. Res. **80** (1975) 903.
12 Meier, R.R.: Astron. Astrophys. **55** (1977) 211.
13 Paresce, F., Bowyer, C.S., Kumar, S.: Astrophys. J. **187** (1974) 633.
14 Thomas, G.E.: Rev. Geophys. Space Phys. **13** (1975) 1063, 1081.
15 Thomas, G.E.: Annu. Rev. Earth Planet. Sci. **6** (1978) 173.
16 Thomas, G.E., Krassa, R.F.: Astron. Astrophys. **11** (1971) 218.
17 Thomas, G.E., Krassa, R.F.: Astron. Astrophys. **30** (1974) 223.
18 Weller, C.S., Meier, R.R.: Astrophys. J. **193** (1974) 471.
19 Witt, N., Ajello, J.M., Blum, P.W.: Astron. Astrophys. **73** (1979) 272.

3.3.5.2 Interplanetary plasma and magnetic field (solar wind)

3.3.5.2.1 Introduction

The hot plasma of the solar corona expands into interplanetary space, where the plasma flow, referred to as the solar wind, becomes supersonic a few solar radii above the solar surface. Coronal magnetic flux is carried together with the coronal plasma into interplanetary space giving rise to the interplanetary magnetic field (IMF). The field lines are stretched radially outward and twisted into a spiral by the combined action of the solar wind motion and solar rotation. The IMF is structured into sectors with respect to the magnetic field direction such that within a sector the sense of the magnetic field direction (the magnetic polarity) points either towards the sun along the spiral direction or away from it. The expansion of the solar wind occurs in streams of either high or low velocities (high and low speed streams), where each high speed stream is usually embedded within a magnetic sector with a single dominant magnetic polarity. The solar wind is turbulent over a broad spectrum of wavelengths. The fluctuations are predominantly due to the interaction of high and low speed streams, shock waves generated by solar activity, large amplitude Alfvén waves, rotational discontinuities, tangential discontinuities and higher frequency phenomena. The plasma of the solar wind is basically of a collisionless nature and the velocity distribution functions of the particles deviate drastically from thermodynamic equilibrium [1···11].

References for 3.3.5.2.1 (Monographs and conference proceedings)

1 Brandt, J.C.: Introduction to the Solar Wind, W.H. Freeman and Company, San Francisco, USA (1970).
2 Dyer, E.R., Roederer, J.G., Hundhausen, A.J. (eds.): The Interplanetary Medium. Part II of Solar-Terrestial Physics/1970, D. Reidel Publishing Company, Dordrecht, Holland (1972).
3 Hundhausen, A.J.: Coronal Expansion and Solar Wind, Springer-Verlag, Berlin-Heidelberg-New York (1972).
4 Mackin, R.J., jr., Neugebauer, M. (eds.): The Solar Wind, Pergamon Press, Oxford (1966).
5 Parker, E.N.: Interplanetary Dynamical Processes, Interscience Publ., New York-London (1963).
5a Parker, E.N., Kennel, C.F., Lanzerotti, L.J. (eds.): Solar System Plasma Physics. North-Holland Publishing Company, Amsterdam-New York-Oxford, Vol. 1 (1979).
6 Russell, C.T. (ed.): Solar Wind Three, published by Institute of Geophysics and Planetary Physics, University of California, Los Angeles (1974).
7 Schindler, K. (ed.): Cosmic Plasma Physics, Plenum Press, New York-London (1972) pp. 61···122.
8 Shea, M.A., Smart, D.F., Wu, S.T. (eds.): Study of Travelling Interplanetary Phenomena 1977, D. Reidel Publishing Company, Dordrecht, Holland (1977).
9 Sonett, C.P., Coleman, P.J., jr., Wilcox, J.M. (eds.): Solar Wind, NASA-SP-308 (1972).

10 Williams, D.J. (ed.): Physics of Solar Planetary Environments, published by AGU, Washington, D.C., USA, (1976) pp. 270···463.
11 Zirker, J.B. (ed.): Coronal Holes and High Speed Wind Streams, Colorado Associated University Press, Boulder, USA (1977).

3.3.5.2.2 In-situ observations

Summary of observations performed by space probes near the orbit of earth including tables summarizing solar wind parameters [21].

a) Average bulk parameters
Proton parameters (density, bulk velocity, temperature) [2, 21, 27, 30, 31, 36, 38, 42, 43, 46, 59]
Electron parameters (temperature, heat flux) [20···22, 30, 31, 40, 54]
Alpha-particle parameters (density, bulk velocity, temperature) [1, 3, 21, 28, 30, 31, 44]
Magnetic field data [4, 21, 27, 36, 38, 39, 41···43, 58].

Table 2. Plasma and magnetic field data for various types of solar wind flow as observed near the orbit of earth [21]. The data are represented in the form $A=\bar{A}\pm\sigma$ where $\bar{A}$ is the mean value and σ is the rms variation.

N_p = Proton number density
V = bulk velocity
θ_V = angle between the bulk velocity $\boldsymbol{V}$ and the ecliptic plane, where $\theta_V>0$ if the component of $\boldsymbol{V}$ perpendicular to the ecliptic is pointing to the north
ϕ_V = angle between the projection of the bulk velocity $\boldsymbol{V}$ into the ecliptic plane and the radial direction pointing away from the sun, where $\phi_V>0$ denotes a flow corotating with the sun
$\langle\delta V^2\rangle^{1/2} = 1.5\langle\delta V_{ecl}^2\rangle^{1/2}$, where $\langle\delta V_{ecl}^2\rangle^{1/2}$ is the rms value (for each 3-h interval) of deviations of the ecliptic component V_{ecl} of the bulk velocity V from 3-h average values
N_pV = proton flux
N_α/N_p = ratio of alpha-particle number density to proton number density
T_p = proton temperature
T_α = alpha-particle temperature
T_e = electron temperature
$\boldsymbol{Q}_p\cdot\hat{r}$ = radial component of the proton heat flux
$\boldsymbol{Q}_e\cdot\hat{r}$ = radial component of the electron heat flux
B = magnetic field strength
θ_B = angle between the magnetic field direction and the ecliptic plane
ϕ_B = angle between the projection of the magnetic field into the ecliptic and the radial direction pointing away from the sun. If the magnetic field points away from (towards) the sun then the average spiral direction implies $\phi_B<0$ ($\phi_B>0$).

Parameter	Average solar wind	Low speed solar wind $V\leqq 350\ \mathrm{km\,s^{-1}}$ but V not increasing with time	High speed solar wind $V\geqq 650\ \mathrm{km\,s^{-1}}$ but beyond the speed maximum of individual high speed streams	Estimated errors in determining the parameters
$N_p[\mathrm{cm^{-3}}]$	8.7 ± 6.6	11.9 ± 4.5	3.9 ± 0.6	$\pm30\%$
$V[\mathrm{km\,s^{-1}}]$	468 ± 116	327 ± 15	702 ± 32	$\pm2\%$
$\theta_V[°]$	-0.45 ¹)			$\pm1.5°$
$\phi_V[°]$	-0.6 ± 2.6	$+1.6\pm1.5$	-1.3 ± 0.4	$\pm1.5°$
$\langle\delta V^2\rangle^{1/2}[\mathrm{km\,s^{-1}}]$	20.5 ± 12.1	9.6 ± 2.9	34.9 ± 6.2	
$N_pV[\mathrm{cm^{-2}\,s^{-1}}]$	$(3.8\pm2.4)\cdot10^8$	$(3.9\pm1.5)\cdot10^8$	$(2.7\pm0.4)\cdot10^8$	
N_α/N_p	0.047 ± 0.019	0.038 ± 0.018	0.048 ± 0.005	$\pm15\%$
$T_p[\mathrm{K}]$	$(1.2\pm0.9)\cdot10^5$	$(0.34\pm0.15)\cdot10^5$	$(2.3\pm0.3)\cdot10^5$	$\pm15\%$
$T_\alpha[\mathrm{K}]$	$(5.8\pm5.0)\cdot10^5$	$(1.1\pm0.8)\cdot10^5$	$(14.2\pm3.0)\cdot10^5$	$\pm20\%$
$T_e[\mathrm{K}]$	$(1.4\pm0.4)\cdot10^5$	$(1.3\pm0.3)\cdot10^5$	$(1.0\pm0.1)\cdot10^5$	$\pm15\%$
$\boldsymbol{Q}_p\cdot\hat{r}[\mathrm{erg\,cm^{-2}\,s^{-1}}]$	$(1.3\pm2)\cdot10^{-4}$	$(2.9\pm2.0)\cdot10^{-5}$	$(2.3\pm0.9)\cdot10^{-4}$	$\pm30\%$
$\boldsymbol{Q}_e\cdot\hat{r}[\mathrm{erg\,cm^{-2}\,s^{-1}}]$	$(43\pm30)\cdot10^{-4}$	$(27\pm10)\cdot10^{-4}$	$(32\pm6)\cdot10^{-4}$	$\pm30\%$
$B[10^{-9}\,\mathrm{T}]$	6.2 ± 2.9			$\pm0.2\cdot10^{-9}\,\mathrm{T}$
$\theta_B[°]$	0.3 ± 25 ²)			2°
$\phi_B[°]$	-43 ± 40 ²)			2°

¹) This value differs from zero by less than instrumental systematic uncertainties.
²) θ_B and ϕ_B were averaged only over those intervals when $\boldsymbol{B}$ was pointing in the general direction away from the sun.

As observed near the orbit of earth the solar wind proton temperature T_p increases with the bulk velocity V (on the average of several solar rotations) according to

$$T_p^{1/2} = aV + b, \text{ with } a = 0.032\cdots0.036 \text{ and } b = -3.3\cdots-5.6 \text{ if } T_p \text{ in } [10^3\,\text{K}] \text{ and } V \text{ in } [\text{km s}^{-1}]$$ [11].

Variation of solar wind parameters with distance r from the sun:
bulk velocity [13, 14, 37, 42, 56b],
proton parameters (flux density, bulk velocity, temperature) [14, 42, 56b],
magnetic field data [4, 39, 42, 55, 56b].

No significant variation of the average bulk velocity with distance from the sun could be detected (for $0.7\,\text{AU} \leqq r \leqq 5\,\text{AU}$) [13, 42]. However, a slight change of flow direction from $\phi_V = 1.^\circ5$ at $r = 0.7$ AU to $\phi_V = 0.^\circ2$ at $r = 1$ AU has been observed [37, 42]. Furthermore the daily velocity variations decrease with distance from the sun (from $r = 1$ AU to $r = 5$ AU) [13].
Variation of plasma parameters (density, bulk velocity) with heliographic latitude [16, 21, 42, 45, 48].

The latitude dependence of bulk velocity and density is somewhat controversial. At least during the initial portion of a solar cycle an increase of bulk velocity ($7\cdots15\,\text{km s}^{-1}\,\text{degree}^{-1}$) and decrease of density ($-0.5\,\text{cm}^{-3}\,\text{degree}^{-1}$) have been observed (see also 3.3.5.2.3).

Table 3. Variation of several solar wind parameters with distance r (in [AU]) from the sun.
For definition of $N_p V$, T_p, B: see Table 2.
B_r, B_θ, B_ϕ = the magnetic field components in the direction of spherical coordinate unit vectors $\hat{r}$, $\hat{\theta}$, $\hat{\phi}$ with respect to a spherical coordinate system centered at the sun and the polar axis pointing to the north ecliptical pole.
B_n = the magnetic field component normal to the solar equatorial plane

Parameter	Mean (solar rotation average)	Radial distance range AU	Ref.
$N_p V$ [$\text{cm}^{-2}\,\text{s}^{-1}$]	$(2.4\cdot10^8)r^{-2}$	0.7···5	42
T_p [K]	$(8\cdot10^4)r^{-1}$	1···3	42
B [10^{-9} T]	$5(1+r^2)^{1/2}r^{-2}$	0.3···3	4
$\lvert B_r\rvert$ [10^{-9} T]	$2.89\,r^{-2.13}$	0.46···5	4
$\lvert B_\theta\rvert$ [10^{-9} T]	$2.93r^{-1.4}$	1···5	4 [1])
$\lvert B_n\rvert$ [10^{-9} T]	$0.82r^{-1.4}$	0.46···1	4 [2])
$\lvert B_\phi\rvert$ [10^{-9} T]	$3.17r^{-1.12}$	0.46···5	4

[1]) Pioneer 10. [2]) Mariner 10.

b) Large-scale structures
The large-scale features of the solar wind are usually dominated by the presence of high and low speed streams, interacting with one another because of the solar rotation, as well as by the basic spiral pattern of the IMF and the sector structure with respect to magnetic field direction. There is a close association between the photospheric magnetic field, the IMF sector structure, the solar wind streams and the coronal holes, where coronal holes are regions of unusually low density and low temperature in coronal regions of unipolar magnetic field.

Morphology of high speed streams [46, 53].
Variation of solar wind parameters through high speed streams and stream-stream interaction regions [4, 9, 9a, 10, 21, 22, 25].
Radial dependence of high speed streams [4, 9a, 13, 26, 53, 56b].
Time variation of high speed streams [15, 21].

IMF sector structure [4, 15, 21, 41, 51, 52, 56a, 56b, 58].
Radial dependence of the IMF sector structure [4, 55].
Latitude dependence of the IMF sector structure [47, 49].
Correlation between high speed streams and the IMF sector structure [4, 15, 21, 25, 56a, 59].
Correlation between high speed streams and coronal holes [25, 33, 53, 56a, 60].

c) Fluctuations

The power spectra of plasma and magnetic field fluctuations increase with period in the range from seconds to about one day and level off for larger periods (except for a power spectrum of the radial velocity component which continues to increase up to a period of 10 days). The shapes of the power spectra do not show a striking change with heliocentric distance from 0.7 AU to 5 AU (for periods larger than about one minute) but there is a general decrease of power with distance. The power spectra are roughly consistent with a power law $\propto|\boldsymbol{k}|^{-\alpha}$ ($\boldsymbol{k}$=three-dimensional wave vector) or $\propto f^{2-\alpha}$ (f =frequency of the fluctuations as seen by a spacecraft) with $3<\alpha\lesssim 4$ in the range 10^{-5} Hz$\lesssim f \lesssim 10^{-2}$ Hz for magnetic field fluctuations and 10^{-4} Hz$\lesssim f \lesssim 10^{-2}$ Hz for density and velocity fluctuations. The power in the magnetic field components is roughly an order of magnitude larger than the power in the magnetic field magnitude fluctuations (at 1 AU).

Power spectra (density and velocity) [24, 34, 35, 57].
Power spectra or rms deviations (magnetic field) [4, 8].
Radial dependence of power spectra (density and velocity) [34].
Radial dependence of power spectra or rms deviations (magnetic field) [4, 39, 42, 55].
Tables for radial dependence of rms deviations (magnetic field) [4].
Physical nature of fluctuations in various solar wind conditions [4···8, 9a, 21].
Morphology and kinematic motion of shocks [4, 12, 17, 18, 21, 34a, 56, 56b].
Plasma and magnetic field properties of shocks [4, 7, 9, 17, 18, 21, 29, 31, 32].
Tables for dynamical properties of shocks [21, 32].
Association of shocks with solar activity [12, 17, 25, 32].
Occurrence rate for discontinuities [4, 8].
Higher frequency phenomena [50, 57].

d) Kinetic properties

Tables for kinetic properties which may be derived from parameters given in Table 2 (plasma frequency, gyroradii etc.) [21, 27, 38].
Properties derived from the velocity distribution functions (e.g. temperature anisotropy):
 protons [19, 21, 30, 31, 59].
 electrons [20···22, 40, 46].
Contour lines of the velocity distribution functions:
 protons [19, 30, 31, 59].
 electrons [40, 46].
Difference of bulk velocities between alpha-particles and protons [1, 21].

e) Ion composition (see Table 4)

He-abundances [3, 21, 31, 44].
He-abundances associated with shocks or solar flares [25, 29].
Heavy ions [3, 21, 23, 28].
Tables for ion abundances [3, 21].

Table 4. Solar wind average abundances ([21] Table 9). The abundances are represented in the form N(A)/N(H) where A is the chemical symbol of the element (isotope) and N is the number density of A.

A	N(A)/N(H)	A	N(A)/N(H)
H	1	Ne^{21}	$1.7\cdot10^{-7}$
He^3	$1.7\cdot10^{-5}$	Ne^{22}	$5.1\cdot10^{-6}$
He^4	$4.0\cdot10^{-2}$	Si	$7.5\cdot10^{-5}$
O	$5.2\cdot10^{-4}$	Ar	$3.0\cdot10^{-6}$
Ne	$7.5\cdot10^{-5}$	Fe	$5.3\cdot10^{-5}$
Ne^{20}	$7.0\cdot10^{-5}$		

References for 3.3.5.2.2

0 Sonett, C.P., Coleman, P.J., jr., Wilcox, J.M. (eds.): Solar Wind, NASA-SP-308 (1972).
1 Asbridge, J.R., Bame, S.J., Feldman, W.C., Montgomery, M.D.: J. Geophys. Res. **81** (1976) 2719.
2 Axford, W.I.: Space Sci. Rev. **8** (1968) 331.
3 Bame, S.J.: in [0] p. 535.
4 Behannon, K.W.: Rev. Geophys. Space Phys. **16** (1978) 125.
5 Belcher, J.W.: J. Geophys. Res. **80** (1975) 4713.
6 Belcher, J.W., Davis, L., jr.: J. Geophys. Res. **76** (1971) 3534.
7 Burlaga, L.F.: Space Sci. Rev. **12** (1971) 600.

8 Burlaga, L.F.: in [0] p. 309.
9 Burlaga, L.F.: Space Sci. Rev. **17** (1975) 327.
9a Burlaga, L.F.: Space Sci. Rev. **23** (1979) 201.
10 Burlaga, L.F., Ogilvie, K.W., Fairfield, D.H., Montgomery, M.D., Bame, S.J.: Astrophys. J. **164** (1971) 137.
11 Burlaga, L.F., Ogilvie, K.W.: J. Geophys. Res. **78** (1973) 2028.
12 Chao, J.K., Lepping, R.P.: J. Geophys. Res. **79** (1974) 1799.
13 Collard, H.R., Wolfe, J.H.: in Solar Wind Three (Russell, C.T., ed.) (1974) p. 281 (cf. [6] in refs. for 3.3.5.2.1).
14 Cuperman, S., Levush, B., Dryer, M., Rosenbauer, H., Schwenn, R.: Astrophys. J. **226** (1978) 1120.
15 Davis, L., jr. in: [0] p. 93.
16 Dobrowolny, M., Moreno, G.: Space Sci. Rev. **18** (1976) 685.
17 Dryer, M.: Space Sci. Rev. **15** (1974) 403.
18 Dryer, M.: Space Sci. Rev. **17** (1975) 277.
19 Feldman, W.C., Asbridge, J.R., Bame, S.J., Montgomery, M.D.: Rev. Geophys. Space Phys. **12** (1974) 715.
20 Feldman, W.C., Asbridge, J.R., Bame, S.J., Montgomery, M.D., Gary, S.P.: J. Geophys. Res. **80** (1975) 4181.
21 Feldman, W.C., Asbridge, J.R., Bame, S.J., Gosling, J.T.: in The Solar Output and its Variation (White, O.R., ed.), Colorado Associated University Press, Boulder, USA (1977) p. 351.
22 Feldman, W.C., Asbridge, J.R., Bame, S.J., Gosling, J.T., Lemons, D.S.: J. Geophys. Res. **83** (1978) 5285.
23 Geis, J.: in [0] p. 559.
24 Goldstein, B., Siscoe, G.L.: in [0] p. 506.
25 Gosling, J.T.: Rev. Geophys. Space Phys. **13** (1975) 1053, 1072.
26 Gosling, J.T., Hundhausen, A.J., Bame, S.J.: J. Geophys. Res. **81** (1976) 2111.
27 Haerendel, G.: Landolt-Börnstein, NS, Vol. VI/1 (1965) p. 233.
28 Hirshberg, J.: Rev. Geophys. Space Phys. **13** (1975) 1059, 1077.
29 Hirshberg, J., Asbridge, J.R., Robbins, D.E.: Sol. Phys. **18** (1971) 313.
30 Hundhausen, A.J.: Space Sci. Rev. **8** (1968) 690.
31 Hundhausen, A.J.: Rev. Geophys. Space Phys. **8** (1970) 729.
32 Hundhausen, A.J.: in [0] p. 393.
33 Hundhausen, A.J.: in Coronal Holes and High Speed Wind Streams (Zirker, J.B., ed), Colorado Associated University Press, Boulder, USA (1977) p. 225.
34 Intriligator, D.S.: in Solar Wind Three (Russell, C.T., ed.) (1974) p. 294, 368 (cf. [6] in refs. for 3.3.5.2.1).
34a Intriligator, D.S.: Space Sci. Rev. **19** (1976) 629.
35 Jokipii, J.R.: Annu. Rev. Astron. Astrophys. **11** (1973) 1.
36 Kovalevski, J.V.: Space Sci. Rev. **12** (1971) 187.
37 Lazarus, A.J., Goldstein, B.E.: Astrophys. J. **168** (1971) 571.
38 Lüst, R.: in Solar-Terrestrial Physics (King, J.W., Newman, W.S., eds.), Academic Press, London-New York (1967) p. 1.
39 Mariani, F., Ness, N.F., Burlaga, L.F., Bavassano, B., Villante, U.: J. Geophys. Res. **83** (1978) 5161.
40 Montgomery, M.D.: in Cosmic Plasma Physics (Schindler, K., ed.), Plenum Press, New York-London (1972) p. 61.
41 Ness, F.: Annu. Rev. Astron. Astrophys. **6** (1968) 79.
42 Neugebauer, M.: Space Sci. Rev. **17** (1975) 221.
43 Neugebauer, M.: J. Geophys. Res. **81** (1976) 4664.
44 Ogilvie, K.W., Wilkerson, T.D.: Sol. Phys. **8** (1969) 435.
45 Rhodes, E.J., jr., Smith, E.J.: J. Geophys. Res. **80** (1975) 917.
46 Rosenbauer, H., Schwenn, R., Marsch, E., Meyer, B., Miggenrieder, H., Montgomery, M.D., Mühlhäuser, K.H., Pilipp, W., Voges, W., Zink, S.M.: J. Geophys. **42** (1977) 561.
47 Rosenberg, R.L., Coleman, P.J., jr.: J. Geophys. Res. **74** (1969) 5611.
48 Rosenberg, R.L., Winge, C.R.: in Solar Wind Three (Russell, C.T., ed.) (1974) p. 300 (cf. [6] in refs. for 3.3.5.2.1).
49 Rosenberg, R.L., Kivelson, M.G., Hedgecock, P.C.: J. Geophys. Res. **82** (1977) 1273.
50 Scarf, F.L.: Space Sci. Rev. **11** (1970) 234.
51 Schatten, K.H.: Rev. Geophys. Space Phys. **9** (1971) 773.
52 Schatten, K.H.: in [0] p. 65.
53 Schwenn, R., Montgomery, M.D., Rosenbauer, H., Miggenrieder, H., Mühlhäuser, K.H., Bame, S.J., Feldman, W.C., Hansen, R.T.: J. Geophys. Res. **83** (1978) 1011.
54 Scudder, J.D., Lind, D.L., Ogilvie, K.W.: J. Geophys. Res. **78** (1973) 6535.
55 Smith, E.J.: in Solar Wind Three (Russell, C.T., ed.) (1974) p. 257 (cf. [6] in refs. for 3.3.5.2.1).
56 Smith, E.J.: Space Sci. Rev. **19** (1976) 661.
56a Smith, E.J.: Rev. Geophys. Space Phys. **17** (1979) 610.

56b Smith, E.J., Wolfe, J.H.: Space Sci. Rev. **23** (1979) 217.
57 Unti, T.W.J., Neugebauer, M., Goldstein, B.E.: Astrophys. J. **180** (1973) 591.
58 Wilcox, J.M.: Space Sci. Rev. **8** (1968) 258.
59 Wolfe, J.H.: in [0] p. 170.
60 Zirker, J.B.: Rev. Geophys. Space Phys. **15** (1977) 257.

3.3.5.2.3 Ground-based observations

Information on the solar wind properties at locations not yet accessible to space probes comes mainly from studies of ionic comet tails and of radio radiation from various discrete astronomical sources [1, 2, 10, 19].

a) Ionic comet tails (see also 3.3.3.6)
The solar wind bulk velocity has been inferred from the observed direction of ionic comet tails (type I tails) and the observed motion of the comets, where the tails are assumed to lie in the direction of the solar wind velocity as seen by a hypothetical observer riding on the comet [1···3, 10, 12, 19].

No dependence of the solar wind velocity on the heliographic latitude has been found [3].

b) Radio radiation from discrete sources
The apparent scintillation (i.e. fluctuations of signal intensity) and angular broadening of discrete radio sources due to small-scale electron density fluctuations in the solar wind has provided information on the spectrum of interplanetary density fluctuations (for wave numbers $k \gtrsim 10^{-8}\,\mathrm{cm}^{-1}$ and for heliocentric distances r between $0.3\,\mathrm{AU} \lesssim r \lesssim 1\,\mathrm{AU}$) and on the solar wind bulk velocity (for heliocentric distances r between $0.025\,\mathrm{AU} \lesssim r \lesssim 1\,\mathrm{AU}$ and heliographic latitudes λ between $-60° \lesssim \lambda \lesssim 80°$) [4, 6, 13, 16, 21].
The shape of the power spectra for density fluctuations δN is controversial (Gaussian law versus power law) [6, 13···15].
Most observations are roughly consistent with a power law $\sim |\boldsymbol{k}|^{-\alpha}$ ($\boldsymbol{k}$ = three-dimensional wave vector) with $2.5 \lesssim \alpha \lesssim 3.5$ in the range $10^{-8}\,\mathrm{cm}^{-1} \lesssim |\boldsymbol{k}| \lesssim 2 \cdot 10^{-7}\,\mathrm{cm}^{-1}$ [4, 6, 15, 16].
Power spectra including both radio and in-situ observations [8, 15, 16].
$\langle(\delta N)^2\rangle^{1/2}$ decreases with distance r as r^{-2} or $r^{-2.5}$ [13, 14].
No significant variation of the bulk velocity V with heliocentric distance r for $r \gtrsim 0.3\,\mathrm{AU}$ has been found [4, 6, 13, 17].
For $r \gtrsim 0.025\,\mathrm{AU}$ increase of V with r and a random velocity of the order of $100\,\mathrm{km\,s^{-1}}$ up to $r \lesssim 0.15\,\mathrm{AU}$ has been detected [4, 11].
An average increase of V with the heliographic latitude λ (for $0.3\,\mathrm{AU} \lesssim r \lesssim 1\,\mathrm{AU}$ and $-60° \lesssim \lambda \lesssim 80°$) has been observed in the years 1966 and 1971 to 1975 with $\frac{\mathrm{d}V}{\mathrm{d}\lambda} \approx 2.1\,\mathrm{km\,s^{-1}\,degree^{-1}}$ [4···7, 9, 10, 17].
No variation of V with λ has been observed during the years 1967···1969 [10, 13, 17].
Corotating stream structures have been observed for heliographic latitudes $|\lambda| \lesssim 40°$ [4, 6, 10, 15, 17].
There is a correlation between high speed streams and coronal structures (coronal holes) [17, 20, 22].
Large-scale properties of flare-associated disturbances [15, 17, 18].

References for 3.3.5.2.3

1 Axford, W.I.: Space Sci. Rev. **8** (1968) 331.
2 Brandt, J.C.: Introduction to the Solar Wind, W.H. Freeman and Company, San Francisco, USA (1970) pp. 103···115.
3 Brandt, J.C., Harrington, R.S., Roosen, R.G.: Astrophys. J. **196** (1975) 877.
4 Coles, W.A.: Space Sci. Rev. **21** (1978) 411.
5 Coles, W.A., Maagoe, S.: J. Geophys. Res. **77** (1972) 5622.
6 Coles, W.A., Rickett, B.J., Rumsey, V.H.: in Solar Wind Three (Russell, C.T., ed.) (1974) p. 351 (cf. [6] in refs. for 3.3.5.2.1).
7 Coles, W.A., Rickett, B.J.: J. Geophys. Res. **81** (1976) 4797.
8 Cronyn, W.M.: Astrophys. J. **171** (1972) L 101.
9 Dennison, P.A., Hewish, A.: Nature, London **213** (1967) 343.
10 Dobrowolny, M., Moreno, G.: Space Sci. Rev. **18** (1976) 685.
11 Ekers, R.D., Little, L.T.: Astron. Astrophys. **10** (1971) 310.
12 Haerendel, G.: Landolt-Börnstein, NS, Vol. VI/1 (1965) pp. 233···253.

13 Hewish, A.: in Solar Wind (Sonett, C.P., Coleman, P.C., jr., Wilcox, J.M., eds.) NASA-SP-308 (1972) p. 477.
14 Hollweg, J.V.: J. Geophys. Res. **75** (1970) 3715.
15 Houminer, Z.: in Study of Travelling Interplanetary Phenomena 1977 (Shea, M.A., Smart, D.F., Wu, S.T., eds.) (1977) p. 119.
16 Jokipii, J.R.: Annu. Rev. Astron. Astrophys. **11** (1973) 1.
17 Kakinuma, T.: in Study of Travelling Interplanetary Phenomena 1977 (Shea, M.A., Smart, D.F., Wu, S.T., eds.) (1977) p. 101.
18 Kakinuma, T., Watanabe, T.: Space Sci. Rev. **19** (1976) 611.
19 Lüst, R.: in Solar-Terrestrial Physics (King, J.W., Newman, W.S., eds.), Academic Press, London-New York (1967) p. 1.
20 Rickett, B.J., Sime, D.G., Sheeley, N.R., jr., Crockett, W.R., Tousey, R.: J. Geophys. Res. **81** (1976) 3845.
21 Vitkevich, V.V.: in Interplanetary Medium. Part II of Solar Terrestrial Physics 1970 (Dyer, E.R., Roederer, J.G., Hundhausen, A.J., eds.), D. Reidel, Dordrecht, Holland (1972) p. 49.
22 Watanabe, T., Shibasaki, K., Kakinuma, T.: J. Geophys. Res. **79** (1974) 3841.

3.3.5.3 Energetic particles in interplanetary space

3.3.5.3.1 Modulation of galactic cosmic rays

Galactic cosmic rays are observed at 1 AU with energies in the range $1 \cdots 10^{15}$ MeV. Below $\approx 10^4$ MeV the cosmic rays are considerably modulated: While they diffuse inwards into the solar system they are continuously convected out and change their energy in the radially expanding solar wind. At energies <200 MeV/nucleon the modulated H, He spectra become insensitive to the shape of the respective interstellar spectra [20, 6, 14, 22].

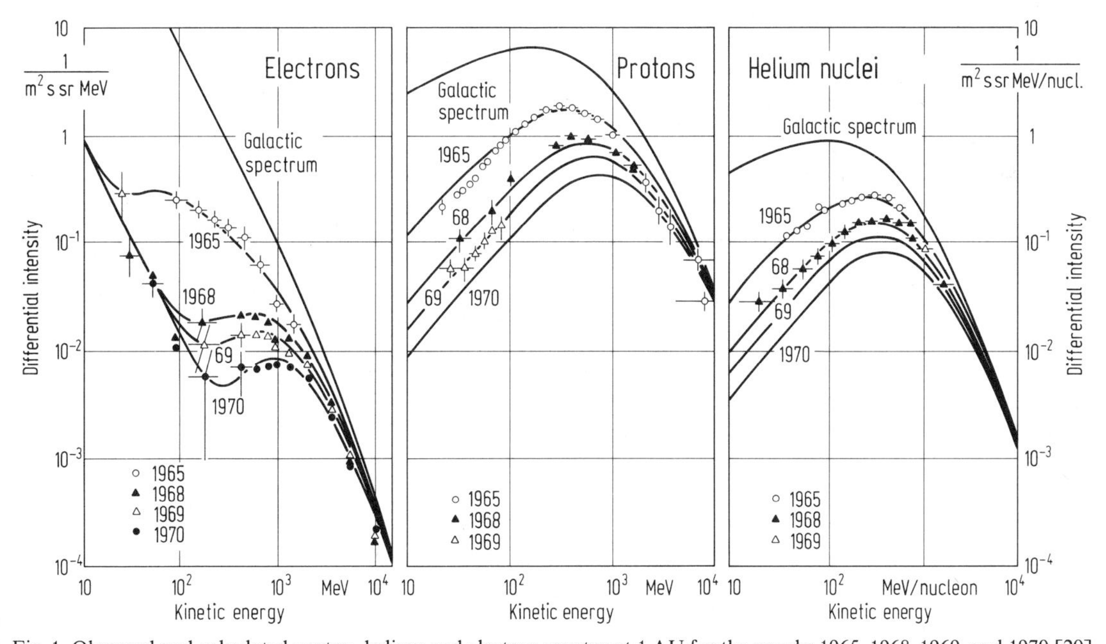

Fig. 1. Observed and calculated proton, helium and electron spectra at 1 AU for the epochs 1965, 1968, 1969, and 1970 [20].

Solar cycle variation of galactic cosmic rays [14, 17]. Galactic cosmic ray electron spectrum and change with time [13]. Annual mean nucleonic intensity at neutron monitor energies (≈ 10 GeV) is inversely correlated with sunspot number R_Z. At 10 GeV, 20% variation in solar cycle 19, 15% variation in cycle 20. Time lag of ≈ 1 year between cosmic ray intensity peak and sunspot minimum. Hysteresis effects in the modulation [21, 19, 3].

Anisotropies: North-south anisotropy due to radial gradient. At neutron monitor energies, between 1966 and 1968, a north-south anisotropy of ≈ 0.2% [2, 7]. Azimuthal anisotropy in ecliptic plane leads to diurnal cosmic ray variation. At neutron monitor energies longterm average amplitude of the anisotropy is ≈ 0.4% [15, 18]. Explanation in terms of corotation anisotropy [16].

Semidiurnal component of the cosmic ray intensity [8]. 20 year wave in the diurnal anisotropy [16]. Measurements of the azimuthal anisotropy at >480 MeV/nucleon H, He up to 3 AU [1].

Radial gradient: Derived from north-south anisotropy at neutron monitor energies [15].
Direct measurements with Pioneer 10, 11: At 29⋯67 MeV (protons) (6.8 ± 0.7)% per AU between 1 and 11.3 AU, at >70 MeV (protons) (3.4 ± 0.4)% per AU between 1 and 5 AU [11, 12]. Energy spectra and radial gradient of galactic cosmic ray helium [10a].

Forbush decreases: Large, sudden asymmetrical depressions in the cosmic ray flux lasting several days. Decrease phase 12⋯24 h long, recovery several days up to several weeks. Review of experimental observations [10, 4].

References for 3.3.5.3.1

1 Axford, W.I., Fillius, W., Gleeson, L.J., Ip, W.-H., Mogro-Campero, A.: Proc. 14th Int. Cosmic Ray Conf. **4** (1975) 1519.
2 Bercovich, M.: Proc. 12th Int. Cosmic Ray Conf. **2** (1971) 579.
3 Burger, J.J., Swanenburg, B.N.: J. Geophys. Res. **78** (1973) 292.
4 Cini-Castagnoli, G.: Proc. 13th Int. Cosmic Ray Conf. **5** (1973) 3706.
5 Forbush, S.E.: J. Geophys. Res. **78** (1973) 7933.
6 Goldstein, M.L., Fisk, L.A., Ramaty, R.: Phys. Rev. Lett. **25** (832) 1970.
7 Iucci, N., Storini, M.: Nuovo Cimento **10** (1972) 325.
8 Kane, R.P.: J. Geophys. Res. **80** (1975) 470.
9 Levy, E.H.: Proc. 14th Int. Cosmic Ray Conf. **4** (1975) 1215.
10 Lockwood, J.A.: Space Sci. Rev. **12** (1971) 658.
10a McDonald, F.B., Lal, N., Trainor, J.H., Van Hollebeke, M.A.I., Webber, W.R.: Astrophys. J. **216** (1977) 930.
11 McKibben, R.B., Pyle, K.R., Simpson, J.A., Tuzzolino, A.J., O'Gallagher, J.J.: Proc. 14th Int. Cosmic Ray Conf. **4** (1975) 1512.
12 McKibben, R.B., O'Gallagher, J.J., Pyle, K.R., Simpson, J.A.: Proc. 15th Int. Cosmic Ray Conf. **3** (1977) 240.
13 Meyer, P., Schmidt, P.J., L'Heureux, J.: Proc. 12th Int. Cosmic Ray Conf. **2** (1971) 548.
14 Mooral, H.: Space Sci. Rev. **19** (1976) 845.
15 O'Gallagher, J.J.: Rev. Geophys. Space Phys. **10** (1972) 821.
16 Pomerantz, M.A., Duggal, S.P.: Space Sci. Rev. **12** (1971) 75.
17 Pomerantz, M.A., Duggal, S.P.: Rev. Geophys. Space Phys. **12** (1974) 343.
18 Rao, U.R.: Space Sci. Rev. **12** (1971) 719.
19 Schmidt, P.J.: J. Geophys. Res. **77** (1972) 3295.
20 Urch, I.H., Gleeson, L.J.: Astrophys. Space Sci. **17** (1972) 426.
21 Van Hollebeke, M.A.I., Wang, J.R., McDonald, F.B.: J. Geophys. Res. **77** (1972) 6881.
22 Völk, H.J.: Rev. Geophys. Space Phys. **13** (1975) 547.

3.3.5.3.2 Anomalous component of low energy cosmic rays

Hump of the cosmic ray energy spectrum of N, O, Ne in the energy range 3⋯20 MeV/nucleon; flat energy spectrum of He between 10⋯80 MeV/nucleon at quiet times.

Measurements of energy spectra and relative abundances of elements with nuclear charge $1 \leqq Z \leqq 26$ [4, 7, 10, 5].

Table 1. Relative abundances, $N(A)/N(O)$, in the anomalous component from [5] compared with the galactic cosmic ray and solar system abundance, normalized to oxygen density, $N(O)$. A = chemical symbol of the element.

	$N(A)/N(O)$			
	Quiet time[1])		Galactic cosmic rays	Solar system
A	"hump" 3.4···11.4 MeV/nuc	11.4···24 MeV/nuc	250···850 MeV/nuc	
He	9.8 ±1.7[2])	—	≈30	103
C	0.06 ±0.03	0.18 ±0.09	1.08	0.55
N	0.26 ±0.15	0.27 ±0.11	0.28	0.17
O	1.0	1.0	1.0	1.0
Ne	0.074±0.03	0.024±0.024	0.19	0.16
Mg	0.026±0.02	0.018±0.018	0.20	0.049
Si	0.027±0.02	<0.017	0.13	0.047
S···Ca	<0.02[3])	<0.034[3])	0.08	0.03
Fe	0.03 ±0.02	<0.034[3])	0.1	0.039

[1]) Nov 73···Mar 75.
[2]) 3.4···6 MeV/nuc.
[3]) 95% confidence level.

Isotopic composition: Predominantly ^{4}He, ^{14}N, ^{16}O [11].

Heliocentric gradient of anomalous He between 1 and 9 AU:
(20±4)% per AU at 10···20 MeV/nucleon,
(10±2)% per AU at 30···56 MeV/nucleon [8].

Time variation: Anticorrelation with longterm solar activity [12, 13].

Ionization state (indirectly inferred) [9].

As a possible source are discussed:
a) Very low energy galactic cosmic rays having low charge states further accelerated within the heliosphere [1].
b) Interstellar neutral particles ionized in and picked up by the solar wind [2]; a small fraction of these newly ionized particles is further accelerated in the outer solar system by magnetosonic turbulence [3].
Numerical solution of the convection-diffusion equation including acceleration [6].

References for 3.3.5.3.2

1 Durgaprasad, N.: Astrophys. Space Sci. **217** (1977) L113.
2 Fisk, L.A., Kozlovsky, B., Ramaty, R.: Astrophys. J. **190** (1974) L35.
3 Fisk, L.A.: Astrophys. J. **206** (1976) 333.
4 Garcia-Munoz, M., Mason, G.M., Simpson, J.A.: Astrophys. J. **182** (1973) L81.
5 Klecker, B., Hovestadt, D., Gloeckler, G., Fan, C.Y.: Astrophys. J. **212** (1977) 290.
6 Klecker, B.: J. Geophys. Res. **82** (1977) 5287.
7 McDonald, F.B., Teegarden, B.J., Trainor, J.H., Webber, W.R.: Astrophys. J. **185** (1974) L105.
8 McDonald, F.B., Lal, N., Trainor, J.H., Van Hollebeke, M.A.I., Webber, W.R.: Astrophys. J. **216** (1977) 930.
9 McKibben, R.B.: Astrophys. J. **217** (1977) L113.
10 Mewaldt, R.A., Stone, E.C., Vidor, S.B., Vogt, R.E.: Proc. 14th Int. Cosmic Ray Conf. **2** (1975) 798.
11 Mewaldt, R.A., Stone, E.C., Vidor, S.B., Vogt, R.E.: Astrophys. J. **205** (1976) 931.
12 Mewaldt, R.A., Stone, E.C., Vogt, R.E.: Proc. 14th Int. Cosmic Ray Conf. **2** (1975) 804.
13 von Rosenvinge, T.T., McDonald, F.B.: Proc. 14th Int. Cosmic Ray Conf. **2** (1975) 792.

3.3.5.3.3 Interplanetary propagation of solar cosmic rays

Flare produced energetic particles are scattered by the magnetic irregularities in the solar wind and are convected with the solar wind. Depending on the scattering strength the propagation can be described as a diffusive process [17], or as an almost scatter-free process [3, 4, 13].

Mean free paths and diffusion coefficients of protons [6, 22, 10].
Compilation of mean free paths at different rigidities [24, 23].
Radial dependence of mean free paths [6, 22].
Mean free path parallel to the magnetic field of a 10 MeV proton at 1 AU solar distance: $\approx 0.1 \cdots 0.2$ AU.

Scatter-free propagation of protons [22, 16].
Scatter-free propagation of electrons [9].

Anisotropies [7, 12].

ESP-events (=energetic storm particles-events): long lasting (several hours) intensity enhancements of ≈ 1 MeV protons before the arrival of the flare-produced shock [2, 18, 11].
Change of proton to He ratio in ESP-events [8, 19].
Explanation in terms of acceleration of preexisting flare particles at interplanetary shock [5, 20].

Shock-spike events: Short-lived intensity enhancements (≈ 10 min) at sub-MeV energies associated with the passage of the shock disturbance [21, 14, 15].
Explanation in terms of acceleration at quasi-perpendicular shocks [1].

References for 3.3.5.3.3

1 Armstrong, T.P., Chen, G., Sarris, E.T., Krimigis, S.M., in: Study of Travelling Interplanetary Phenomena (Shea, M.A., Smart, D.F., Wu, S.T., eds.) (1977) 367 (=Astrophysics and Space Science Library Vol. 71, Reidel Publ. Comp., Dordrecht).
2 Bryant, D.A., Cline, T.L., Desai, U.D., McDonald, F.B.: Astrophys. J. **141** (1965) 478.
3 Earl, J.A.: Astrophys. J. **205** (1976) 900.
4 Earl, J.A.: Astrophys. J. **206** (1976) 301.
5 Fisk, L.A.: J. Geophys. Res. **76** (1971) 1662.
6 Hamilton, D.C.: J. Geophys. Res. **82** (1977) 2157.
7 Innanen, W.G., Van Allen, J.A.: J. Geophys. Res. **78** (1973) 1019.
8 Lanzerotti, L.J., Robbins, M.F.: Sol. Phys. **10** (1969) 212.
9 Lin, R.P.: Space Sci. Rev. **16** (1974) 189.
10 Ma Sung, L.S., Earl, J.A.: Astrophys. J. **222** (1978) 1080.
11 McCracken, K.G., Rao, U.R., Bukata, R.P.: J. Geophys. Res. **72** (1967) 4293.
12 McCracken, K.G., Rao, U.R., Bukata, R.P., Keath, E.P.: Sol. Phys. **18** (1971) 100.
13 Nolte, J.T., Roelof, E.C.: Proc. 14th Int. Cosmic Ray Conf. **5** (1975) 1722.
14 Ogilvie, K.W., Arens, J.F.: J. Geophys. Res. **76** (1971) 13.
15 Palmeira, R.A.R., Allum, F.R., Rao, U.R.: Sol. Phys. **21** (1971) 204.
16 Palmer, I.D., Palmeira, R.A.R., Allum, F.R.: Sol. Phys. **40** (1975) 449.
17 Parker, E.N.: Interplanetary Dynamical Processes, Interscience Publishers, New York (1975).
18 Rao, U.R., McCracken, K.G., Bukata, R.P.: J. Geophys. Res. **72** (1967) 4325.
19 Scholer, M., Hovestadt, D., Häusler, B.: Sol. Phys. **24** (1972) 475.
20 Scholer, M., Morfill, G.: Sol. Phys. **45** (1975) 227.
21 Singer, S., Montgomery, M.D.: J. Geophys. Res. **76** (1971) 6628.
22 Wibberenz, G. in: Study of Travelling Interplanetary Phenomena (cf. [1]), (1977) 323.
23 Zwickl, R.D., Webber, W.R.: Sol. Phys. **54** (1977) 457.
24 Zwickl, R.D., Webber, W.R.: J. Geophys. Res. **83** (1978) 1157.

3.3.5.3.4 Coronal propagation and injection

Variation of the number of solar flare particle events with solar longitude:
Non-relativistic electrons in [3], relativistic electrons in [11], 20···80 MeV protons in [12], high energy protons in [7]. Variation is due to coronal diffusion of particles from the flare site and escape of particles into interplanetary space [8, 1, 5]. Probability distribution for an event on the visible disk of the sun is symmetrical about $(45 \pm 15)°$ W heliocentric longitude.

Variation of spectral index γ with the associated flare heliocentric longitude [12]:
In the 20···80 MeV proton range γ can be represented by $\gamma \approx 2.7\ (1 + \Delta\lambda/2)$ where $\Delta\lambda$ is the separation angle in radians between flare longitude and 50° W.

Size distribution of flare-associated proton events [12]: $\mathrm{d}N/\mathrm{d}I = AI^{-\alpha}$, where I is maximum differential intensity in particles/(cm^2 s sr MeV), $\mathrm{d}N$ the number of events in the range of intensity $\mathrm{d}I$, $\alpha = 1.15 \pm 0.05$. A is a function of longitude, for (20···60)° W $A \approx 2.7$.

Coronal diffusion coefficients [2, 5, 4].
Existence of a fast propagation or preferred connection region of $\pm 30°$ in longitude around the flare site [2, 4, 9].

Coronal propagation and escape are rigidity independent [4, 6].

Solar injection time constants: Flare particles are injected near the preferred connection region not instantaneously but over some finite time interval (up to 40 h at 1 MeV, 4 h at 100 MeV). Compilation of time constants [13].
Effect of finite injection time on intensity: time profiles and anisotropies [10].

References for 3.3.5.3.4

1 Axford, W.I.: Planet. Space Sci. **13** (1965) 1301.
2 Lanzerotti, L.J.: J. Geophys. Res. **78** (1973) 3942.
3 Lin, R.P.: Space Sci. Rev. **16** (1974) 189.
4 Ma Sung, L.S., Van Hollebeke, M.A.I., McDonald, F.B.: Proc. 14th Int. Cosmic Ray Conf. **5** (1975) 1767.
5 Ng, C.K., Gleeson, L.J.: Sol. Phys. **46** (1976) 347.
6 Perron, C., Domingo, V., Reinhard, R., Wenzel, K.P.: J. Geophys. Res. **83** (1978) 2017.
7 Pomerantz, M.A., Duggal, S.P.: Rev. Geophys. Space Phys. **12** (1974) 343.
8 Reid, G.C.: J. Geophys. Res. **69** (1964) 2659.
9 Reinhard, R., Wibberenz, G.: Sol. Phys. **36** (1974) 473.
10 Schulze, B.M., Richter, A.K., Wibberenz, G.: Sol. Phys. **54** (1977) 207.
11 Simnett, G.M.: Space Sci. Rev. **16** (1974) 257.
12 Van Hollebeke, M.A.I., Ma Sung, L.S., McDonald, F.B.: Sol. Phys. **41** (1975) 189.
13 Wibberenz, G. in: Study of Travelling Interplanetary Phenomena (Shea, M.A., Smart, D.F., Wu, S.T., eds.) (1977) 323 (=Astrophys. and Space Sci. Library, Vol. 71, Reidel Publ. Dordrecht).

3.3.5.3.5 Solar flare particle composition and charge state

Average abundance of elements from He to Ni in solar flare particles above ≈ 15 MeV/nucleon approximates (within factor 2) photospheric or coronal abundance [4].
At energies below ≈ 10 MeV/nucleon enrichment in heavy elements increasing with charge number and decreasing with energy [9, 6, 17].
Below 10 MeV/nucleon large variations from flare to flare [2] and within each individual flare [15, 3, 10].

The p/He ratio in solar flares is statistically correlated with the maximum p intensity of the event [16].

The variation of the composition within each event is possibly due to a rigidity-dependent interplanetary propagation [15, 13].

Abnormally large abundance of ^3He in weak flares ($^3He/^4He \gtrsim 0.5$) [7, 8, 14].

Explanation of the overabundance of ^3He is given in terms of nuclear spallation [1, 11] or resonant heating of ^3He due to an ion cyclotron instability [5].

Ionization state abundance of solar flare particles (C, O, Fe) below 600 keV/nucleon [12].

At ≈ 1 MeV/nucleon ionization state of Fe is indirectly inferred to be $\approx 10 \cdots 15$ (near solar wind) [10, 13].

Isotopic abundances of solar flare nuclei [4a, 8a]: $^{20}Ne/^{22}Ne \approx 7.7$.

References for 3.3.5.3.5

1 Anglin, J.D.: Astrophys. J. **198** (1975) 733.
2 Armstrong, T.P., Krimigis, S.M.: J. Geophys. Res. **76** (1971) 4230.
3 Armstrong, T.P., Krimigis, S.M.: Proc. 14th Int. Cosmic Ray Conf. **5** (1975) 1592.
4 Crawford, H.J., Price, P.B., Cartwright, B.G., Sullivan, J.D.: Astrophys. J. **195** (1975) 213.
4a Dietrich, W.F., Simpson, J.A.: Astrophys. J. **231** (1979) L 91.
5 Fisk, L.A.: Astrophys. J. **224** (1978) 1048.
6 Hovestadt, D., Vollmer, O., Gloeckler, G., Fan, C.Y.: Proc. 13th Int. Cosmic Ray Conf. **2** (1973) 1498.
7 Hovestadt, D., Klecker, B., Gloeckler, G., Fan, C.Y.: Proc. 14th Int. Cosmic Ray Conf. **5** (1975) 1613.
8 Hurford, G.J., Stone, E.C., Vogt, R.E.: Proc. 14th Int. Cosmic Ray Conf. **5** (1975) 1624.
8a Mewaldt, R.A., Spalding, J.D., Stone, E.C., Vogt, R.E.: Astrophys. J. **231** (1979) L 97.
9 Mogro-Campero, A., Simpson, J.A.: Astrophys. J. **171** (1972) L 5.
10 O'Gallagher, J.J., Hovestadt, D., Klecker, B., Gloeckler, G., Fan, C.Y.: Astrophys. J. **209** (1976) L97.
11 Ramaty, R., Kozlovsky, B.: Astrophys. J. **193** (1974) 729.
12 Sciambi, R.K., Gloeckler, G., Fan, C.Y., Hovestadt, D.: Astrophys. J. **214** (1977) 316.
13 Scholer, M., Hovestadt, D., Klecker, B., Gloeckler, G., Fan, C.Y.: J. Geophys. Res. **83** (1978) 3349.
14 Serlemitsos, A.T., Balasubrahmanyan, V.K.: Astrophys. J. **198** (1975) 195.
15 Van Allen, J.A., Venkatarangan, P., Venkatesan, D.: J. Geophys. Res. **79** (1974) 1.
16 Van Hollebeke, M.A.I.: Proc. 14th Int. Cosmic Ray Conf. **5** (1975) 1563.
17 Webber, W.R.: Proc. 14th Int. Cosmic Ray Conf. **5** (1975) 1997.

3.3.5.3.6 Corotating energetic particle events

Long-lived streams of low energy ($\approx 0.1 \cdots 5$ MeV/nucleon) nucleons near the corotating interaction regions (resulting from the interaction of a fast solar wind stream running into a slow solar wind stream) [2, 1, 4].

Duration $\approx 4 \cdots 10$ days
Flux of 1 MeV protons at 1 AU:
≈ 10 p (cm^{-2} s^{-1} sr^{-1} MeV^{-1}).

Heliocentric gradient [6]:
between 0.3 and 1 AU: $\approx 350\%$ per AU
between 1 AU and 4 AU: $\approx 100\%$ per AU
between 4 AU and 9 AU: variable from -40% per AU to -100% per AU.

Composition of energetic particles in corotating events [3, 5].

References for 3.3.5.3.6

1 Barnes, C.W., Simpson, J.A.: Astrophys. J. **210** (1976) L91.
2 McDonald, F.B., Teegarden, B.J., Trainor, J.H., von Rosenvinge, T.T.: Astrophys. J. **203** (1976) L 149.
3 McGuire, R.E., von Rosenvinge, T.T., McDonald, F.B.: Astrophys. J. **224** (1978) L87.
4 Pesses, M.E., Van Allen, J.A., Goertz, C.A.: J. Geophys. Res. **83** (1978) 553.
5 Scholer, M., Hovestadt, D., Klecker, B., Gloeckler, G.: Astrophys. J. **227** (1979) 323.
6 Van Hollebeke, M.A.I., McDonald, F.B., Trainor, J.H., von Rosenvinge, T.T.: J. Geophys. Res. **83** (1978) 4723.

3.4 Abundances of the elements in the solar system

3.4.1 Introduction

The primordial nebula from which the sun, the planets, and other planetary objects formed had a well defined chemical composition. This composition corresponds closely to that of the present sun and of many stars. This follows from astronomical observations and from the constancy of the isotopic composition of the elements in meteoritic and terrestrial material.

The chemical and isotopic composition of primordial matter presents a mixture of several genetic components of matter that formed under different conditions through thermonuclear reactions [1···3].

Sections 3.4.2 through 3.4.4 summarize our present knowledge of the composition of the solar system, and of what is assumed to be the same, the primordial mixture from which the members of the solar system formed. Its non-volatile components are found in almost unaltered elemental ratios in certain types of meteorites.

The surface rocks of the earth and the moon and, to a lesser degree, most stony meteorites show deviations from the primordial mixture due to chemical fractionation during condensation and as a magma. The least differentiated meteorites are the Type I carbonaceous chondrites. The data given in the tables are based on the following empirical quantities: (1) chemical analyses of terrestrial and lunar rocks, and of five typical meteorites, (2) data from spectral analyses of the sun, (3) semi-empirical estimates of the primordial abundance distribution of the elements, their isotopes and their major genetic components as derived from certain abundance rules.

Not included here are recently discovered variations in the isotopic composition of certain elements that can only be explained by assuming incomplete mixing of the primordial genetic components or secondary thermonuclear reactions. Such deviations from the isotopic composition of the bulk solar system matter have been discovered in meteoritic inclusions for the rare gases, for O, Mg, Ca, U, and several other elements [4].

The tables are arranged in sequence of decreasing fractionation of solar matter. Relative elemental abundances in the primordial matter and relative abundances of the individual nuclear species and their respective genetic components are listed in 3.4.3 and 3.4.4.

References for 3.4.1

1 Burbidge, E.M., Burbidge, G.R., Fowler, W.A., Hoyle, F.: Rev. Mod. Phys. **29** (1957) 547.
2 Cameron, A.G.W.: Astrophys. J. **121** (1955) 144; **131** (1960) 521.
3 Amiet, J.P., Zeh, H.D.: Z. Phys. **217** (1968) 485.
4 Podosek, F.A.: Annu. Rev. Astron. Astrophys. **16** (1978) 293.

3.4.2 Terrestrial and lunar surface rocks and meteorites

Following a suggestion of H. Wänke and H. Palme data for typical, well-defined rocks and meteorites are listed, instead of average literature values for the different classes of objects as done previously. The data for Orgueil were compiled by H. Palme, B. Spettel and H. Wänke (to be published) on the basis of recent reliable, and in part unpublished, analytical results.

In Table 1 the elemental compositions, as obtained by chemical analyses, are listed
(a) for two types of terrestrial rocks: (1) a continental basalt, produced by partial melting of the mantle, and (2) ultramafic nodules, representing the composition of the upper mantle,
(b) for two types of lunar material: (1) a mare basalt and (2) lunar soil, and
(c) for two highly fractionated meteorites: (1) Juvinas and (2) Canyon Diablo.

The lunar soil samples listed in Table 1 are similar in composition to the average composition of the lunar crust. The mare basalt is in many respects chemically complementary to the highland samples and is thought to be a differentiation product of the lunar interior. Juvinas is a basalt from the eucrite parent body, the third differentiated planet from which samples are available. Canyon Diablo may be representative for a Fe-Ni-core of a differentiated meteorite parent body.

In Table 2 this composition is listed for three stony meteorites, belonging to three different types of chondrites, in sequence of decreasing degree of oxidation and increasing degree of fractionation from primordial solar matter. The first meteorite, the type 1 carbonaceous chondrite Orgueil, can be assumed to contain all the elements which are not volatile at ordinary temperatures in practically the same proportions as primordial matter.

Table 1. Elemental composition of terrestrial and lunar surface and meteorites (for details, see text). Elemental contents are given in gram per ton [ppm] unless percent [%] is indicated.

		Earth				Moon				Highly differentiated meteorites			
Z	Element	Basalt BCR-1		Upper mantle		Basalt 70215		Highland soil Apollo 16 Stat. 1, 10		Juvinas (eucrite)		Canyon Diablo (iron)	
			Ref.		Ref.		Ref.		Ref.		Ref.		Ref.
1	H												
2	He												
3	Li												
4	Be	1.7	F1										
5	B	5	F1										
6	C	330											
7	N	30											
8	O	45.48%											
9	F	470	F1	16.3	J1	72	W2	59	W1	19	W3		
10	Ne												
11	Na	24260	F1	2420	J1	2740	W2	3450	W1	2800	W3		
12	Mg	2.09%	F1	23.1%	J1	5.02%	W2	3.82%	W1	4.0%	W3		
13	Al	7.20%	F1	2.1%	J1	4.61%	W2	14.0%	W1	7.1%	W3		
14	Si	25.48%	F1	21.1%	J1	17.9%	W2	21.1%	W1	23.0%	W3	<0.003	W4
15	P	1570	F1			440	W2	530	W1	400	M4		
16	S	392	F1			1620	W2	650	D1			13	S1
17	Cl	50	F1	1.33	J1	4.5	W2	28	R1	18	W3		
18	Ar												
19	K	14100	F1	260	J1	370	W2	889	W1	222	W3		
20	Ca	4.95%	F1	2.50%	J1	7.60%	W2	11.4%	W1	7.7%	W3		
21	Sc	33	F1	17.0	J1	85.9	W2	8.7	W1	28.5	W3		
22	Ti	1.32%	F1	0.13%	J1	7.53%	W2	0.33%	W1	0.38%	W3		
23	V	399	F1	77	J1	50	D2	26	L1				
24	Cr	17.6	F1	3140	J1	2710	W2	685	W1	2090	W3	9.4	S1
25	Mn	1390	F1	1010	J1	1950	W2	506	W1	3990	W3		
26	Fe	9.42%	F1	6.08%	J1	15.44%	W2	4.30%	W1	14.5%	W3	91.7%	W4
27	Co	38	F1	105	J1	21.3	W2	28.5	W1	5.8	W3	0.49%	L2
28	Ni	15.8	F1	2110	J1	1	M1	420	W1	1.1	M2	7.06%	W4
29	Cu	18.4	F1	28	J1	6.4	W2	7.0	W1	1.65	W3	125	S1
30	Zn	120	F1	50	J1	3.0	W2	24	W1	1.1	W3	29	S1
31	Ga	20	F1	3.0	J1	3.6	W2	4.6	W1	2.6	W3	79.7	S1
32	Ge	1.54	F1	1.2	J1	1.7	M1	0.46	K1	0.004	M2	322	S1
33	As	0.70	F1	0.10	J1	0.046	W2	0.16	W1	0.18	W3	12.4	S1
34	Se	0.10	F1	0.041	J1	0.17	W2	0.23	K1	0.077	M2		
35	Br	0.15	F1	0.011	J1	0.025	W2	0.162	K1	0.16	M2		
36	Kr												
37	Rb	46.6	F1	0.81	J1	0.36	W2	2.0	K1	0.25	M2		
38	Sr	330	F1	28	J1	143	W2	153	W1	78	M4		
39	Y	37.1	F1	4.6	J1	58	W2	38.5	W1	16	M4		
40	Zr	190	F1	11	J1	176	W2	176	W1	46	M4		
41	Nb	13.5	F1	0.9	J1	18	W2	11	W1	2.7	E1		
42	Mo	1.1	F1									7.4	S1
43	Tc												
44	Ru	0.001	F1									6.4	N1
45	Rh	0.0002	F1									1.5	N1
46	Pd	0.012	F1	0.0045	M3			0.024	W1	0.0004	M2	3.6	S1

Table 1, continued

		Earth				Moon				Highly differentiated meteorites			
Z	Element	Basalt BCR-1	Ref.	Upper mantle	Ref.	Basalt 70215	Ref.	Highland soil Apollo 16 Stat. 1, 10	Ref.	Juvinas (eucrite)	Ref.	Canyon Diablo (iron)	Ref.
47	Ag	0.036	F1	0.005	M3	0.0011	M1	0.011	K1	0.1	M2	0.04	S1
48	Cd	0.12	F1	0.051	M3	0.0018	M1	0.13	K1	0.03	M2	0.02	S2
49	In	0.095	F1	0.012	M3			0.015	B1			0.012	S1
50	Sn	2.6	F1										
51	Sb	0.69	F1			0.00018	M1	0.0016	K1	0.042	M2	0.33	S1
52	Te	0.003	H1			0.0021	M1	0.015	K1	0.00001	M2	0.00009	G1
53	I	<1	F1									0.00003	G1
54	Xe												
55	Cs	0.95	F1			0.013	W2	0.98	K1	0.005	M2		
56	Ba	675	F1	6.9	J1	59	W2	120	W1	31	M4		
57	La	26	F1	0.63	J1	5.5	W2	13	W1	2.6	W3		
58	Ce	54	F1			19.6	W2	33	W1				
59	Pr	7	F1			3.0	W2	4.6	W1	0.94	W3		
60	Nd	29	F1			19	W2	20	W1	5.1	W3		
61	Pm												
62	Sm	6.6	F1	0.38	J1	6.7	W2	5.7	W1	1.5	W3		
63	Eu	1.94	F1	0.15	J1	1.40	W2	1.2	W1	0.61	W3		
64	Gd	6.6	F1			9.1	W2	7.4	W1	2.3	W3		
65	Tb	1.0	F1			1.8	W1	1.0	W1	0.60	W3		
66	Dy	6.3	F1			11.5	W2	7.1	W1	3.2	W3		
67	Ho	1.2	F1			2.5	W2	1.8	W1	0.42	W3		
68	Er	3.59	F1			7.2	W2			2.3	W3		
69	Tm	0.6	F1										
70	Yb	3.36	F1	0.42	J1	7.0	W2	4.1	W1	1.72	W3		
71	Lu	0.55	F1	0.064	J1	1.1	W2	0.55	W1	0.28	W3		
72	Hf	4.7	F1	0.35	J1	6.3	W2	4.0	W1	1.3	W3		
73	Ta	0.91	F1	0.04	J1	1.6	W2	0.49	W1	0.12	W3		
74	W	0.40	F1	0.0164	J1	0.09	W2	0.30	W1	0.041	W3	1.45	A1
75	Re	0.0008	F1	0.0001	J1	$2\cdot10^{-6}$	M1	0.00067	K1	$10\cdot10^{-6}$	M2	0.180	F2
76	Os	0.0001	F1	0.004	M3					$18\cdot10^{-6}$	M2	2.1	C1
77	Ir	$3.4\cdot10^{-6}$	K1	0.0035	J1	$3\cdot10^{-6}$	M1	0.015	W1	$28\cdot10^{-6}$	M2	2.0	C1
78	Pt	0.002	F1									8.0	C1
79	Au	0.00095	F1	0.0005	J1	$26\cdot10^{-6}$	M1	0.011	W1	0.007	M2	1.47	C1
80	Hg	0.011	F1										
81	Tl	0.3	F1	0.0015	M3	0.00016	M1	0.007	K1	0.0011	M2		
82	Pb	17.6	F1										
83	Bi	0.05	F1	0.002	M3	0.0001	M1	0.001	K1	0.0035	M2		
90	Th	6.0	F1	0.094	J1	0.34	W2	1.6	W1				
92	U	1.74	F1	0.026	J1	0.09	W2	0.48	K1	0.09	W3	<0.0006	G1

References for Table 1

A1 Amiruddin, A., Ehmann, W.D.: Geochim. Cosmochim. Acta **26** (1962) 1011.
B1 Boynton, W.V., Chou, C.L., Robinson, K.L., Warren, P.M., Wasson, J.T.: Proc. Lunar Sci. Conf. 7th (1976) 727.
C1 Crocket, T.H.: Geochim. Cosmochim. Acta **36** (1972) 517.
D1 Duncan, A.R., Erlank, A.J., Willis, J.P., Ahrens, L.H.: Proc. Lunar Sci. Conf. 4th (1973) 1097.
D2 Duncan, A.R., Erlank, A.J., Willis, J.P., Sher, M.K., Ahrens, L.H.: Proc. Lunar Sci. Conf. 5th (1974) 1147.
E1 Erlank, A.J., Willis, J.P., Ahrens, L.M., Gurney, J.J., McCarthy, T.S.: Proc. Lunar Sci. Conf. (Abstract) 3rd (1979) 239.
F1 Flanagan, F.J.: Geochim. Cosmochim. Acta **37** (1973) 1189.
F2 Fouche, K.F., Smales, A.A.: Chem. Geology **1** (1966) 329.
G1 Goles, G.G., Anders, E.: Geochim. Cosmochim. Acta **26** (1962) 723.
H1 Higuchi, H., Morgan, J.W.: Proc. Lunar Sci. Conf. 6th (1975) 1625.
J1 Jagoutz, E., Palme, H., Baddenhausen, H., Blum, M., Cendales, M., Dreibus, G., Spettel, B., Lorenz, V., Wänke, H.: Proc. Lunar Sci. Conf. 10th (1979) 2031.
K1 Krähenbühl, U., Ganapathy, R., Morgan, J.W., Anders, E.: Proc. Lunar Sci. Conf. 4th (1973) 1325.
L1 Laul, J.C., Schmitt, R.A.: Proc. Lunar Sci. Conf. 4th (1973) 1349.
L2 Lovering, J.F., Nichiporuk, W., Chodos, A., Brown, H.: Geochim. Cosmochim. Acta **11** (1957) 263.
M1 Morgan, J.W., Ganapathy, R., Higuchi, H., Krähenbühl, U., Anders, E.: Proc. Lunar Sci. Conf. 5th (1974) 1703.
M2 Morgan, J.W., Higuchi, H., Takahashi, H., Hertogen, J.: Geochim. Cosmochim. Acta **42** (1978) 27.
M3 Morgan, J.W., Wandless, R.K.: Proc. Lunar Sci. Conf. (Abstract) 10th (1979) 855.
M4 McCarthy, T.S., Erlank, A.J., Willis, J.P.: Earth Planet. Sci. Lett. **18** (1973) 433.
N1 Nichiporuk, W., Brown, H.: J. Geophys. Res. **70** (1965) 459.
R1 Jovanovic, S., Reed, G.W.,jr.: Proc. Lunar Sci. Conf. 4th (1973) 1313.
S1 Smales, A.A., Mapper, D., Fouche, K.F.: Geochim. Cosmochim. Acta **31** (1967) 673.
S2 Schmitt, R.A., Smith, R.H., Olehy, D.A.: Geochim. Cosmochim. Acta **27** (1963) 1077.
W1 Wänke, H., Baddenhausen, H., Dreibus, G., Jagoutz, E., Kruse, H., Palme, H., Spettel, B., Teschke, F.: Proc. Lunar Sci. Conf. 4th (1973) 1461.
W2 Wänke, H., Palme, H., Baddenhausen, H., Dreibus, G., Jagoutz, E., Kruse, H., Palme, C., Spettel, B., Teschke, F., Tracker, R.: Proc. Lunar Sci. Conf. 6th (1975) 1313.
W3 Wänke, H., Baddenhausen, M., Balacescu, A., Teschke, F., Spettel, B., Dreibus, G., Palme, H., Quijano-Rico, M., Kruse, H., Wlotzka, F., Begemann, F.: Proc. Lunar Sci. Conf. 3rd (1972) 1251.
W4 Wai, C.M., Wasson, J.T.: Geochim. Cosmochim. Acta **33** (1969) 1465.

Table 2. Elemental composition of three stony meteorites of classes C1, H5, and E4 in gram per ton, unless percent is indicated.

Z	Element	Orgueil C1	Richardton H5	Ref.	Abee E4	Ref.
1	H					
2	He					
3	Li	1.45			2.3	T1
4	Be	0.025				
5	B	0.27	0.48	S3		
6	C	3.5%	800	B3	3700	B3
7	N					
8	O	47.0%				
9	F	54			64	D1
10	Ne					
11	Na	5020	7100	E2	6700	B1
12	Mg	9.36%	13.85%	M5	11.1%	M5
13	Al	0.82%	1.05%	M5	0.78%	M5
14	Si	10.68%	16.31%	M5	16.30%	M5
15	P	1010	1040	M5	1960	M5
16	S	5.8%	1.42%	K1	6.67%	K1
17	Cl	678			560	D1
18	Ar					
19	K	517	720	M5	868	T1
20	Ca	0.90%	1.15%	M5	0.87%	M5
21	Sc	5.9	9.8	B4	4.4	B1
22	Ti	0.044%	0.06%	M5	0.05%	M5
23	V	55.6			58	B1
24	Cr	2670	3200	N1	3500	B1
25	Mn	1820	2300	M5	1920	M5
26	Fe	18.3%	29.0%	M5	32.5%	M5
27	Co	501	811	N1	892	I1
28	Ni	1.08%	1.72%	N1	1.91%	B1
29	Cu	108			202	B1
30	Zn	347			454	I1
31	Ga	9.1			17.4	I1
32	Ge	31.3	6.95	S4	48	B1
33	As	1.85			2.4	B1
34	Se	18.9	7.7	L1	34.0	I1
35	Br	2.53			2.56	D1
36	Kr					
37	Rb	2.06	2.54	K2	3.45	T1
38	Sr	8.6	8.23	K2	6.9	T1
39	Y	1.44			1.02	S1
40	Zr	3.82	7.24	S2	7	M5
41	Nb	0.3				
42	Mo	0.92				
43	Tc					
44	Ru	0.69	0.82	B5	0.97	C1
45	Rh	0.13				
46	Pd	0.53			0.79	C1
47	Ag	0.21	0.051	L1	0.39	I1
48	Cd	0.77	0.020	S5	0.77	L1
49	In	0.08	0.00025	L1	0.11	I1
50	Sn	1.75	0.46	S4	2.15	H2
51	Sb	0.13	0.093	T2	0.23	B2
52	Te	2.34	0.74	G1	2.54	I1
53	I	0.56	0.038	G1	0.45	D1
54	Xe					
55	Cs	0.19	0.079	L1	0.237	T1
56	Ba	2.2	4.0	M3	2.25	T1
57	La	0.245	0.32	S1	0.15	S1
58	Ce	0.638	0.48	S1	0.48	S1
59	Pr	0.096	0.12	S1	0.054	S1
60	Nd	0.474	0.61	S1	0.24	S1
61	Pm					
62	Sm	0.154	0.20	S1	0.095	S1
63	Eu	0.058	0.081	S1	0.050	S1
64	Gd	0.204	0.34	S1	0.16	S1
65	Tb	0.037	0.053	S1	0.025	S1
66	Dy	0.254	0.34	S1	0.16	S1
67	Ho	0.057	0.068	S1	0.040	S1
68	Er	0.166	0.205	S1	0.131	S1
69	Tm	0.026	0.033	S1	0.014	S1
70	Yb	0.165	0.19	S1	0.094	S1
71	Lu	0.025	0.033	S1	0.019	S1
72	Hf	0.12	0.204	S2	0.10	E1
73	Ta	0.014				
74	W	0.089	0.13	A1	0.37	H1
75	Re	0.037	0.081	M1	0.051	M1
76	Os	0.49	0.814	M1	0.586	M1
77	Ir	0.48	0.77	M2	0.5	B1
78	Pt	1.05	1.7	B6	1.3	C1
79	Au	0.14	0.26	B6	0.44	B1
80	Hg	(5.3)				
81	Tl	0.14	0.0025	L1	0.1	I1
82	Pb	2.43			1.98	H1
83	Bi	0.11	0.0014	L1	0.134	I1
90	Th	0.029	0.038	B7	0.023	H1
92	U	0.0082	0.011	M4	0.0068	H1

References for Table 2

A1 Amiruddin, A., Ehman, W.D.: Geochim. Cosmochim. Acta **26** (1926) 1011.
B1 Baedecker, P.H., Wasson, J.T.: Geochim. Cosmochim. Acta **39** (1975) 735.
B2 Binz, C.M., Kurimoto, R.K., Lipschutz, M.E.: Geochim. Cosmochim. Acta **38** (1974) 1579.
B3 Belsky, T., Kaplan, I.R.: Geochim. Cosmochim. Acta **34** (1970) 257.
B4 Bate, G.L., Potratz, H.A., Huizenga, J.R.: Geochim. Cosmochim. Acta **18** (1960) 101.
B5 Bate, G.L., Huizenga, J.R.: Geochim. Cosmochim. Acta **27** (1963) 345.
B6 Baedecker, P.H., Ehman, W.D.: Geochim. Cosmochim. Acta **29** (1965) 329.
B7 Bate, G.L., Huizenga, J.R., Potratz, H.A.: Geochim. Cosmochim. Acta **16** (1959) 88.
C1 Crocket, J.H., Keays, R.R., Hsieh, S.: Geochim. Cosmochim. Acta **31** (1967) 1615.
D1 Dreibus, G., Spettel, B., Wänke, H. in: Origin and Distribution of the Elements 2 (Ahrens, L.H., ed.) Pergaman Press, Oxford, N.Y. (1979) 33.
E1 Ehman, W.D., Rebagay, T.V.: Geochim. Cosmochim. Acta **34** (1970) 649.
E2 Edwards, G., Urey, H.C.: Geochim. Cosmochim. Acta **7** (1955) 154.
G1 Goles, G.G., Anders, E.: Geochim. Cosmochim. Acta **26** (1962) 723.
H1 Hintenberger, H., Jochum, K.P., Seufert, M.: Earth Planet. Sci. Lett. **20** (1973) 391.
H2 Hamaguchi, H., Onuma, N., Hirao, Y., Yokoyama, H., Bando, S., Furukawa, M.: Geochim. Cosmochim. Acta **33** (1969) 507.
I1 Ikramuddin, M., Binz, C.M., Lipschutz, M.E.: Geochim. Cosmochim. Acta **40** (1976) 133.
K1 Kaplan, I.R., Hulston, J.R.: Geochim. Cosmochim. Acta **30** (1966) 479.
K2 Kaushal and Wetherill: J. Geophys. Res. **74** (1969) 2717.
L1 Laul, J.C., Ganapathy, R., Anders, E., Morgan, J.W.: Geochim. Cosmochim. Acta **37** (1973) 329.
M1 Morgan, J.W., Lovering, J.F.: Geochim. Cosmochim. Acta **31** (1967) 1893.
M2 Müller, O., Baedecker, P.A., Wasson, J.T.: Geochim. Cosmochim. Acta **35** (1971) 1121.
M3 Moore, C.B., Brown, H.: J. Geophys. Res. **68** (1963) 4293.
M4 Morgan, J.W., Lovering, J.F.: Talanta **15** (1968) 1079.
M5 Michaelis, V.H., Ahrens, L.H., Willis, J.T.: Earth Planet. Sci. Lett. **5** (1969) 387.
N1 Nichiporuk, W., Chodos, A., Helin, E., Brown, H.: Geochim. Cosmochim. Acta **31** (1967) 1911.
S1 Schmitt, R.A., Smith, R.H., Lasch, J.E., Mosen, A.W., Olehy, D.A., Vasilevskis, J.: Geochim. Cosmochim. Acta **27** (1963) 577.
S2 Shima, M.: Geochim. Cosmochim. Acta **43** (1979) 353.
S3 Shima, M.: Geochim. Cosmochim. Acta **27** (1963) 911.
S4 Shima, M.: Geochim. Cosmochim. Acta **28** (1964) 517.
S5 Schmitt, R.A., Smith, R.H., Olely, D.A.: Geochim. Cosmochim. Acta **27** (1963) 1077.
T1 Tera, F., Eugster, O., Burnett, D.S., Wasserburg, G.J.: Proc. Apollo 11 Lunar Science Conf. (1970) 1637.
T2 Tanner, J.T., Ehman, W.D.: Geochim. Cosmochim. Acta **31** (1967) 2007.

3.4.3 Relative atomic abundances *N* of the elements in a type 1 carbonaceous chondrite and in the solar photosphere

Revised data for the relative abundances of the elements in the solar photosphere obtained by spectral analyses were recently compiled by Ross and Aller [1] and discussed at the IAU Conference 1976 in Grenoble [2]. Recent revisions by H. Holweger [3] were also used in compiling the tables. The values for the type 1 carbonaceous chondrite Orgueil are derived from recent results of chemical analyses listed in the preceding table. The values for Orgueil and for the number of atoms in the solar photosphere are given relative to 10^6 atoms of Si as is customary in the cosmochemical literature. For comparison with astronomical data N for the sun is given relative to 10^{12} atoms of H, viz. on the astronomical scale which is a factor of $10^{1.6}$ different from the cosmochemical scale. Practically all values agree within reasonable limits of error.

Table 3. Relative atomic abundances of the elements normalized to N (Si)$=10^6$ for Orgueil and the solar photosphere. The logarithms of solar values are normalized to N(H)$=10^{12}$, (N(Si)$=10^{7.6}$).

		Orgueil C1 [4]		Sun		
		Cosmochemical normalization N(Si)$=10^6$			Astronomical normalization $\log N$(H)$=12$	
Z	Element	N	$\log N$	N	$\log N$	Ref.
1	H			$2.5\cdot10^{10}$	**12**	
2	He			$2.0\cdot10^{9}$	10.9	1
3	Li	55	1.74	0.25	1.0	1
4	Be	0.73	−0.14	0.35	1.15	1
5	B	6.6	0.82	5.0	2.3	3
6	C	$7.7\cdot10^{5}$	5.89	$7.9\cdot10^{6}$	8.5	3
7	N			$2.1\cdot10^{6}$	7.9	1
8	O	$7.3\cdot10^{6}$	6.86	$1.7\cdot10^{7}$	8.8	2
9	F	$7.1\cdot10^{2}$	2.85	10^{3}	4.6	1
10	Ne			$2.5\cdot10^{6}$	8.0	3
11	Na	$5.7\cdot10^{4}$	4.76	$5.2\cdot10^{4}$	6.32	2
12	Mg	$1.01\cdot10^{6}$	6.00	$9.3\cdot10^{5}$	7.57	2
13	Al	$8.00\cdot10^{4}$	4.90	$6.3\cdot10^{4}$	6.40	2
14	Si	**$1.00\cdot10^{6}$**	**6**	**$1.0\cdot10^{6}$**	7.60	
15	P	$8.58\cdot10^{3}$	3.93	$6.8\cdot10^{3}$	5.43	2
16	S	$4.8\cdot10^{5}$	5.68	$3.9\cdot10^{5}$	7.19	2
17	Cl	$5.0\cdot10^{3}$	3.70	$8\cdot10^{3}$	5.5	1
18	Ar			10^{5}	6.6	2
19	K	$3.48\cdot10^{3}$	3.54	$3.2\cdot10^{3}$	5.1	2
20	Ca	$5.91\cdot10^{4}$	4.77	$5.7\cdot10^{4}$	6.32	2
21	Sc	35	1.54	28	3.04	2
22	Ti	$2.42\cdot10^{3}$	3.38	$1.6\cdot10^{3}$	4.80	3
23	V	$2.90\cdot10^{2}$	2.46	$2.6\cdot10^{2}$	4.02	1
24	Cr	$1.35\cdot10^{4}$	4.13	$1.4\cdot10^{4}$	5.75	2
25	Mn	$8.72\cdot10^{3}$	3.94	$6.3\cdot10^{3}$	5.40	2
26	Fe	$8.60\cdot10^{5}$	5.93	$8.3\cdot10^{5}$	7.52	2
27	Co	$2.24\cdot10^{3}$	3.34	$2.3\cdot10^{3}$	4.97	2
28	Ni	$4.83\cdot10^{4}$	4.68	$4.7\cdot10^{4}$	6.27	1
29	Cu	$4.50\cdot10^{2}$	2.65	$3.6\cdot10^{2}$	4.16	2
30	Zn	$1.40\cdot10^{3}$	3.15	$6.6\cdot10^{2}$	4.42	2
31	Ga	34	1.54	16	2.80	1
32	Ge	$1.14\cdot10^{2}$	2.06	65	3.50	3
33	As	6.5	0.81			
34	Se	63	1.80			
35	Br	8	0.90			
36	Kr					
37	Rb	6.4	0.81	11	2.63	2
38	Sr	26	1.41	17	2.82	2
39	Y	4.3	0.63	2.5	2.00	2
40	Zr	11	1.04	1.4	2.75	1
41	Nb	0.85	−0.07	2.5	2.0	2
42	Mo	2.5	0.40	4	2.2	2
43	Tc					
44	Ru	1.8	0.26	2	1.9	2
45	Rh	0.33	−0.48	0.8	1.5	2

continued

Table 3, continued

		Orgueil C1 [4]		Sun		
		Cosmochemical normalization $N(Si)=10^6$			Astronomical normalization $\log N(H)=12$	
Z	Element	N	log N	N	log N	Ref.
46	Pd	1.32	0.12	0.8	1.5	2
47	Ag	0.50	−0.30	0.18	0.85	1
48	Cd	1.80	0.26	1.78	1.85	1
49	In	0.17	−0.77	1.12	1.7	2
50	Sn	3.88	0.59	2.51	2.0	2
51	Sb	0.27	−0.57	0.4	1.0	1
52	Te	4.83	0.68			
53	I	1.16	0.06			
54	Xe					
55	Cs	0.37	−0.43		< 2.1	
56	Ba	4.22	0.63	3.2	2.1	2
57	La	0.46	−0.34	0.34	1.13	2
58	Ce	1.20	0.079	0.90	1.55	2
59	Pr	0.18	−0.74	0.24	0.98	3
60	Nd	0.87	−0.06	0.45	1.25	3
61	Pm					
62	Sm	0.27	−0.57	0.14	0.76	3
63	Eu	0.10	−1.00	0.13	0.7	2
64	Gd	0.34	−0.47	0.32	1.1	1
65	Tb	0.061	−1.21			
66	Dy	0.41	−0.39	0.39	1.1	1
67	Ho	0.091	−1.04			
68	Er	0.26	−0.59	0.16	0.8	1
69	Tm	0.041	−1.39	0.035	0.15	3
70	Yb	0.25	−0.60	0.16	0.8	2
71	Lu	0.038	−1.42	0.16	0.8	2
72	Hf	0.18	−0.74	0.20	0.9	2
73	Ta	$2.1\cdot10^{-2}$	−1.68			
74	W	0.13	−0.89	0.16	0.8	2
75	Re	0.052	−1.28			
76	Os	0.68	−0.17	0.16	0.8	2
77	Ir	0.65	−0.19	4.0	2.2	2
78	Pt	1.42	0.15	1.6	1.8	2
79	Au	0.19	−0.72	0.14	0.75	1
80	Hg					
81	Tl	0.17	−0.77	0.2	0.9	2
82	Pb	3.09	0.49	2.0	1.90	3
83	Bi	0.14	−0.85			
90	Th	$3.2\cdot10^{-2}$	−1.49	0.040	0.20	3
92	U	$0.91\cdot10^{-2}$	−2.04		< 0.6	1

References for Table 3

1 Ross, J.E., Aller, L.M.: Sci. **191** (1976) 1223.

2 Intern. Astron. Union Conf., Grenoble 1976, compiled by Müller, E.A., Transactions **16b** (1977) 118, some data slightly revised by Holweger, H., (priv. commun.).

3 As [2], but revised by Holweger, H., to be published.

4 Palme, H., Spettel, B., Wänke, H., to be published.

3.4.4 Primordial abundances in the solar system

Constancy of the isotopic composition of the elements on earth, on the moon, and in meteorites, and also, as far as known, in the solar photosphere, demonstrates that the matter from which the solar system had formed had been sufficiently mixed prior to condensation of solid bodies to allow definition of the concept of "primordial elemental abundances". Variations due to natural radioactive decay or cosmic-ray irradiation are in general well understood. Recently discovered abnormal isotope ratios in peculiar parts of some meteorites (such as the "Allende inclusions") pertain to only small fractions of the total mass and do not affect this concept. It is, however, still an open question to what extent this concept – and thus, the data listed in the following tables – is valid outside of our solar system and, perhaps also, in the solar interior.

Empirical values for the relative elemental abundances have been derived for the first time by Goldschmidt [G1] by averaging data for chemical analyses of meteorites. Suess and Urey [S1] have shown that these data, together with the accurately known isotopic compositions of the elements, allow us to recognize regularities in the relative abundances of the individual nuclear species, which in turn allow estimations of semi-empirical elemental abundances ("cosmic abundances"). These semi-empirical abundance values have been found to agree surprisingly well with the analytically determined trace element contents of carbonaceous chondrites of type 1. This has then led to improved understanding of abundance regularities [B1, A1] and to further slight corrections of abundance values as deduced from data on carbonaceous type 1 chondrites [A1, S2, C1]. In this way it can be firmly concluded that the chemical composition of this type of meteorite represents the composition of primordial solar matter very accurately, if one disregards the elements which are volatile or form volatile compounds at ordinary temperatures, such as carbon, oxygen, nitrogen, and the noble gases.

The three most important abundance rules that obviously hold for nuclides heavier than $A \gtrsim 60$ are the following [A1, S2]: Essential smoothness with mass number exists for (1) the products of s-component abundance times neutron capture cross-section (in the $kT = 30$ keV range), (2) β^- ("r") – component abundances, and (3) β^+ ("p") – component abundances (see Fig. 1). A fourth rule valid for light nuclei postulates smoothness of the abundances of the multiple-α nuclei heavier than ^{16}O.

The "most probable abundances" listed in the following tables are primarily based on data determined by spectral or chemical analyses for the type 1 carbonaceous chondrite Orgueil and for the sun. Relative amounts and isotopic compositions of the rare gases were estimated from the composition of solar rare gas components in gas-rich meteorites [M1] and from solar wind data [G1].

The abundance values listed in the tables do not differ appreciably from those given in [S2]. For most elements differences from the widely used tables of Cameron [C1] are also negligible; for several elements, however, they appear to be significant and important.

As had been anticipated from abundance rules [A1, S2], the experimental values for the Zr and Hf abundances in Orgueil now have been found to be a factor of three lower than had been assumed previously. The lower new value of Mo had also been predicted. These facts eliminate the need for considering the possibility of type 2 carbonaceous chondrites being better representatives of primordial matter than type 1 [U1, S2]. The analytical abundance values of type 1 carbonaceous chondrites deviate only for very few elements from those expected from the viewpoint of abundance rules and further new and improved analytical data may well eliminate any discrepancy between the trace element content of these meteorites and the primordial abundance distribution.

As in [S2], the abundances of the three individual genetic components are listed in Table 5b for nuclei with $A \geqq 56$.

In the Sm mass region Allende anomalies have been used as a lead to determine the components [M2].

Fig. 1 and Tables 4 and 5 on the following pages.

References for 3.4.4

A1 Amiet, J.P., Zeh, H.D.: Z. Physik **217** (1968) 485.
B1 Burbidge, E.M., Burbidge, G.R., Fowler, W.A., Hoyle, F.: Rev. Mod. Phys. **29** (1957) 567.
C1 Cameron, A.G.W.: Spac. Sci. Rev. **15** (1973) 121.
C2 Conrad, J.: Thesis, Heidelberg (1976).
G1 Geiss, J., Bochsler, P.: Proc. 4th Solar Wind Conf. Burghausen (1979).
M1 Marti, K., Wilkening, L., Suess, H.E.: Astrophys. J. **173** (1972) 445.
M2 Marti, K., Lugmair, G., Scheinin, N.B.: Proc. Lunar Sci. Conf. 9th. (1978) 672.
S1 Suess, H.E., Urey, H.C.: Rev. Mod. Phys. **28** (1956) 53.
S2 Suess, H.E., Zeh, H.D.: Astrophys. and Space Sci. **23** (1973) 173.
U1 Urey, H.C.: Ann. N.Y. Acad. Sci. **194** (1969) 35.

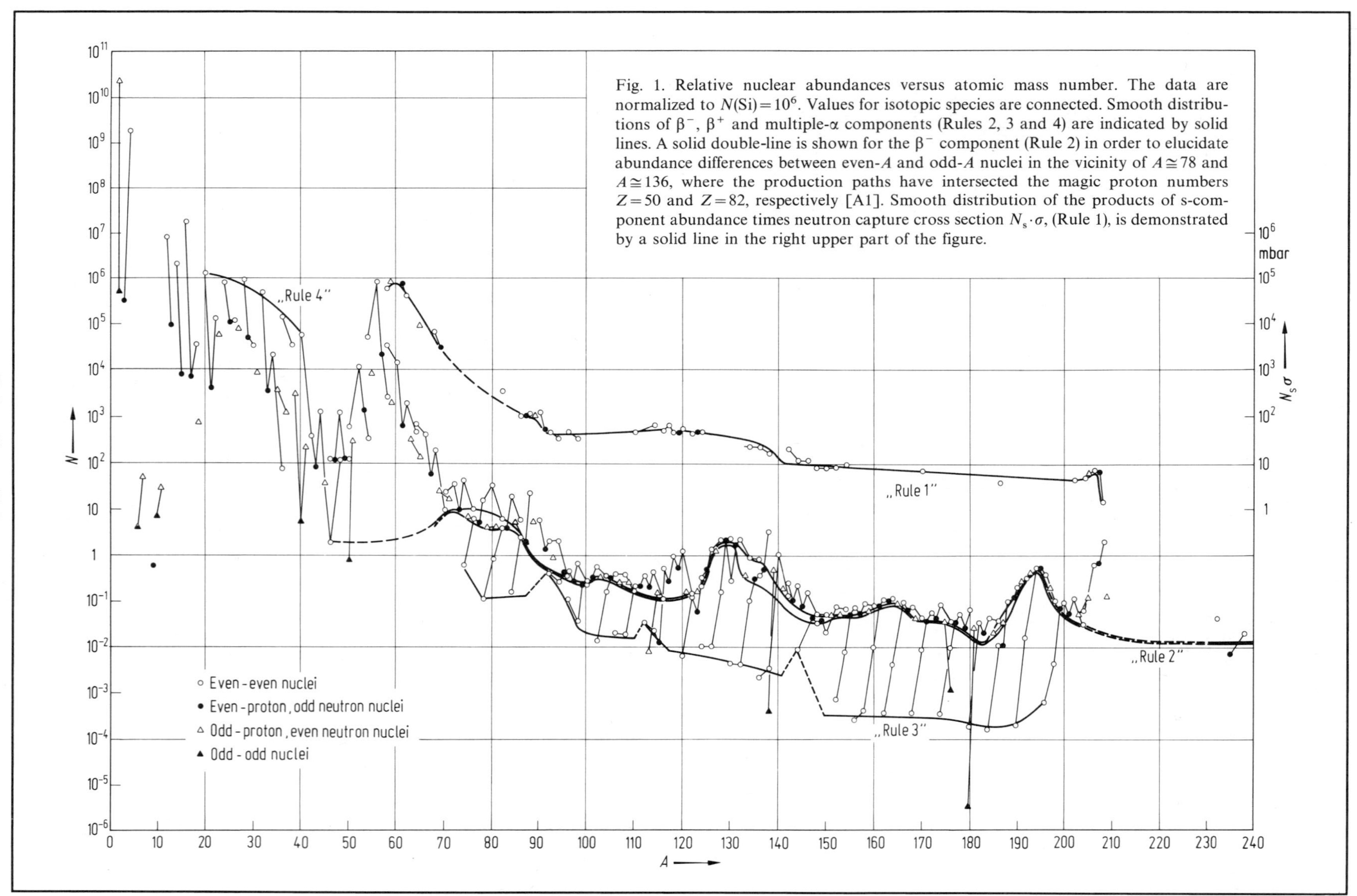

Fig. 1. Relative nuclear abundances versus atomic mass number. The data are normalized to $N(\mathrm{Si})=10^6$. Values for isotopic species are connected. Smooth distributions of β^-, β^+ and multiple-α components (Rules 2, 3 and 4) are indicated by solid lines. A solid double-line is shown for the β^- component (Rule 2) in order to elucidate abundance differences between even-A and odd-A nuclei in the vicinity of $A \cong 78$ and $A \cong 136$, where the production paths have intersected the magic proton numbers $Z=50$ and $Z=82$, respectively [A1]. Smooth distribution of the products of s-component abundance times neutron capture cross section $N_s \cdot \sigma$, (Rule 1), is demonstrated by a solid line in the right upper part of the figure.

Table 4. Most probable relative primordial solar system abundances of the elements (column 4) based on relevant empirical values (column 3). The empirical values are for type 1 carbonaceous chondrites (CC1), mainly for Orgueil as listed in Table 3, except for the elements H, He, C, N, O, for which solar photospheric values are listed. The empirical Ne/He ratio is derived from the solar wind composition [G1].
Data are numbers of atoms normalized to $N\,(\mathrm{Si})=10^6$.

Z	Element	Normalized abundance empirical	most probable	Z	Element	Normalized abundance empirical	most probable
1	H	$2.5 \cdot 10^{10}$	$2.5 \cdot 10^{10}$	44	Ru	1.8	1.8
2	He	$2.0 \cdot 10^{9}$	$2.0 \cdot 10^{9}$	45	Rh	$3.3 \cdot 10^{-1}$	$3.3 \cdot 10^{-1}$
3	Li	$5.5 \cdot 10$	$5.5 \cdot 10$	46	Pd	1.32	1.32
4	Be	$7.3 \cdot 10^{-1}$	$7.3 \cdot 10^{-1}$	47	Ag	$5.0 \cdot 10^{-1}$	$5.0 \cdot 10^{-1}$
5	B	6.6	6.6	48	Cd	1.80	1.30
6	C	$7.9 \cdot 10^{6}$	$7.9 \cdot 10^{6}$	49	In	$1.74 \cdot 10^{-1}$	$1.74 \cdot 10^{-1}$
7	N	$2.1 \cdot 10^{6}$	$2.1 \cdot 10^{6}$	50	Sn	3.9	2.4
8	O	$1.7 \cdot 10^{7}$	$1.7 \cdot 10^{7}$	51	Sb	$2.7 \cdot 10^{-1}$	$2.7 \cdot 10^{-1}$
9	F	$7.1 \cdot 10^{2}$	$7.1 \cdot 10^{2}$	52	Te	4.8	4.8
10	Ne	$1.4 \cdot 10^{6}$	$1.4 \cdot 10^{6}$	53	I	1.16	1.16
11	Na	$5.7 \cdot 10^{4}$	$5.7 \cdot 10^{4}$	54	Xe	–	6.1
12	Mg	$1.01 \cdot 10^{6}$	$1.01 \cdot 10^{6}$	55	Cs	$3.7 \cdot 10^{-1}$	$3.7 \cdot 10^{-1}$
13	Al	$8.0 \cdot 10^{4}$	$8.0 \cdot 10^{4}$	56	Ba	4.2	4.2
14	**Si**	$\mathbf{1.00 \cdot 10^{6}}$	$\mathbf{1.00 \cdot 10^{6}}$	57	La	$4.6 \cdot 10^{-1}$	$4.6 \cdot 10^{-1}$
15	P	$8.6 \cdot 10^{3}$	$8.6 \cdot 10^{3}$	58	Ce	1.20	1.20
16	S	$4.8 \cdot 10^{5}$	$4.8 \cdot 10^{5}$	59	Pr	$1.8 \cdot 10^{-1}$	$1.8 \cdot 10^{-1}$
17	Cl	$5.0 \cdot 10^{3}$	$5.0 \cdot 10^{3}$	60	Nd	$8.7 \cdot 10^{-1}$	$8.7 \cdot 10^{-1}$
18	Ar	–	$2.2 \cdot 10^{5}$	62	Sm	$2.7 \cdot 10^{-1}$	$2.7 \cdot 10^{-1}$
19	K	$3.5 \cdot 10^{3}$	$3.5 \cdot 10^{3}$	63	Eu	$1.00 \cdot 10^{-1}$	$1.00 \cdot 10^{-1}$
20	Ca	$5.9 \cdot 10^{4}$	$5.9 \cdot 10^{4}$	64	Gd	$3.4 \cdot 10^{-1}$	$3.4 \cdot 10^{-1}$
21	Sc	$3.5 \cdot 10$	$3.5 \cdot 10$	65	Tb	$6.1 \cdot 10^{-2}$	$6.1 \cdot 10^{-2}$
22	Ti	$2.4 \cdot 10^{3}$	$2.4 \cdot 10^{3}$	66	Dy	$4.1 \cdot 10^{-1}$	$4.1 \cdot 10^{-1}$
23	V	$2.9 \cdot 10^{2}$	$2.9 \cdot 10^{2}$	67	Ho	$9.1 \cdot 10^{-2}$	$9.1 \cdot 10^{-2}$
24	Cr	$1.35 \cdot 10^{4}$	$1.35 \cdot 10^{4}$	68	Er	$2.6 \cdot 10^{-1}$	$2.6 \cdot 10^{-1}$
25	Mn	$8.7 \cdot 10^{3}$	$8.7 \cdot 10^{3}$	69	Tm	$4.1 \cdot 10^{-2}$	$4.1 \cdot 10^{-2}$
26	Fe	$8.6 \cdot 10^{5}$	$8.6 \cdot 10^{5}$	70	Yb	$2.5 \cdot 10^{-1}$	$2.5 \cdot 10^{-1}$
27	Co	$2.2 \cdot 10^{3}$	$2.2 \cdot 10^{3}$	71	Lu	$3.8 \cdot 10^{-2}$	$3.8 \cdot 10^{-2}$
28	Ni	$4.8 \cdot 10^{4}$	$4.8 \cdot 10^{4}$	72	Hf	$1.8 \cdot 10^{-1}$	$1.8 \cdot 10^{-1}$
29	Cu	$4.5 \cdot 10^{2}$	$4.5 \cdot 10^{2}$	73	Ta	$2.1 \cdot 10^{-2}$	$2.1 \cdot 10^{-2}$
30	Zn	$1.40 \cdot 10^{3}$	$1.40 \cdot 10^{3}$	74	W	$1.27 \cdot 10^{-1}$	$1.27 \cdot 10^{-1}$
31	Ga	$3.4 \cdot 10$	$4.4 \cdot 10$	75	Re	$5.2 \cdot 10^{-2}$	$5.2 \cdot 10^{-2}$
32	Ge	$1.13 \cdot 10^{2}$	$1.13 \cdot 10^{2}$	76	Os	$6.8 \cdot 10^{-1}$	$7.2 \cdot 10^{-1}$
33	As	6.5	6.5	77	Ir	$6.5 \cdot 10^{-1}$	$6.5 \cdot 10^{-1}$
34	Se	$6.3 \cdot 10$	$6.3 \cdot 10$	78	Pt	1.42	1.42
35	Br	8.0	8.0	79	Au	$1.9 \cdot 10^{-1}$	$1.9 \cdot 10^{-1}$
36	Kr	–	$2.5 \cdot 10$	80	Hg		$4.0 \cdot 10^{-1}$
37	Rb	6.4	6.4	81	Tl	$1.7 \cdot 10^{-1}$	$1.7 \cdot 10^{-1}$
38	Sr	$2.6 \cdot 10$	$2.6 \cdot 10$	82	Pb	3.1	3.1
39	Y	4.3	5.4	83	Bi	$1.36 \cdot 10^{-1}$	$1.36 \cdot 10^{-1}$
40	Zr	$1.10 \cdot 10$	$1.10 \cdot 10$				
41	Nb	$8.5 \cdot 10^{-1}$	$8.5 \cdot 10^{-1}$	90	Th	$3.2 \cdot 10^{-2}$	$3.2 \cdot 10^{-2}$
42	Mo	2.5	2.5	92	U	$9.1 \cdot 10^{-3}$	$9.1 \cdot 10^{-3}$

Note added in proof: Due to some last minute corrections in the tables, a few data points in Fig. 1 (B, Cd, Sn, Te, and Xe) slightly differ from their corresponding values in the tables.

Table 5. Relative primordial nuclear abundances. Normalization as in Table 4. Interpolated values (I) cannot be used for comparisons with theoretical expectations. Values marked by an upper index[a] are corrected for natural radioactive decay during the last $4.8 \cdot 10^9$ yr. Values in brackets are particularly uncertain.

Table 5a. Light nuclei $A \leqq 56$.

A	Isotope of	Normalized abundance	A	Isotope of	Normalized abundance	A	Isotope of	Normalized abundance
1	H	$2.5 \cdot 10^{10}$	23	Na	$5.7 \cdot 10^4$	41	K	$2.3 \cdot 10^2$
2	H	$(5.0 \cdot 10^5)$	24	Mg	$8.0 \cdot 10^5$	42	Ca	$3.8 \cdot 10^2$
3	He	$(3.0 \cdot 10^3)$	25	Mg	$1.0 \cdot 10^5$	43	Ca	$8.3 \cdot 10$
4	He	$2.0 \cdot 10^9$	26	Mg	$1.1 \cdot 10^5$	44	Ca	$1.23 \cdot 10^3$
6	Li	4.1	27	Al	$8.0 \cdot 10^4$	45	Sc	$3.5 \cdot 10$
7	Li	$5.1 \cdot 10$	28	Si	$9.2 \cdot 10^5$	46	Ca	1.8
9	Be	$7.3 \cdot 10^{-1}$	29	Si	$4.7 \cdot 10^4$		Ti	$1.9 \cdot 10^2$
10	B	1.3	30	Si	$3.1 \cdot 10^4$	47	Ti	$1.8 \cdot 10^2$
11	B	5.3	31	P	$8.6 \cdot 10^3$	48	Ca	$1.12 \cdot 10^2$
12	C	$7.8 \cdot 10^6$	32	S	$4.6 \cdot 10^5$		Ti	$1.8 \cdot 10^3$
13	C	$8.8 \cdot 10^4$	33	S	$3.6 \cdot 10^3$	49	Ti	$1.32 \cdot 10^2$
14	N	$2.1 \cdot 10^6$	34	S	$2.0 \cdot 10^4$	50	Ti	$1.27 \cdot 10^2$
15	N	$7.6 \cdot 10^3$	35	Cl	$3.8 \cdot 10^3$		V	$7.3 \cdot 10^{-1}$
16	O	$1.69 \cdot 10^7$	36	S	$7.2 \cdot 10$		Cr	$5.9 \cdot 10^2$
17	O	$6.6 \cdot 10^3$		Ar	I $1.8 \cdot 10^5$	51	V	$2.9 \cdot 10^2$
18	O	$3.5 \cdot 10^4$	37	Cl	$1.2 \cdot 10^3$	52	Cr	$1.13 \cdot 10^4$
19	F	$7.1 \cdot 10^2$	38	Ar	$3.7 \cdot 10^4$	53	Cr	$1.28 \cdot 10^3$
20	Ne	$1.27 \cdot 10^6$	39	K	$3.3 \cdot 10^3$	54	Cr	$3.2 \cdot 10^2$
21	Ne	$3.8 \cdot 10^3$	40	Ar	?		Fe	$5.0 \cdot 10^4$
22	Ne	$1.29 \cdot 10^5$		K	5.6[a]	55	Mn	$8.7 \cdot 10^3$
				Ca	$5.7 \cdot 10^4$	56	Fe	$7.9 \cdot 10^5$

Table 5b. Heavy nuclei ($A \geqq 56$) and abundance components. An S next to the value for the abundance component indicates that more than 10% is subtracted to account for other components with abundances derived by interpolation.

A	Isotope of	Normalized abundance					Remarks
		total	component				
			β^-	s	β^+	other	
56	Fe	$7.9 \cdot 10^5$				$7.9 \cdot 10^5$	exothermic reactions
57	Fe	18000.0					
58	Fe	2700.0		(2700.)	0		
	Ni	33000.0	0	0		(33000.0)	
59	Co	2200.0		(2200.)			
60	Ni	12700.0		I (2600.)		S (11000.0)	
61	Ni	557.0		(557.)			
62	Ni	1780.0		1780.			
63	Cu	311.0		311.0			s-branch; ^{63}Ni-decay
64	Ni	456.0		456.0			s-branch
	Zn	685.0		685.0			s-branch
65	Cu	139.0	I (5.0)	134.0			major s-branch
66	Zn	389.0	I (5.0)	384.0			
67	Zn	57.0	I (5.0)	52.0			
68	Zn	260.0	I (5.0)	255.0			
69	Ga	26.3	I (5.0)	21.3			
70	Zn	8.7	8.7	(0)			
	Ge	23.0	0.0	23.0			
71	Ga	17.7	S (7.7)	I (10.0)			
72	Ge	31.0	I 9.0	S 22.0			

continued

Table 5b, continued

A	Isotope of	Normalized abundance					Remarks
		total	component				
			β⁻	s	β⁺	other	
73	Ge	8.7	S 5.4	I (3.3)			
74	Ge	41.0	I 9.0	S 32.0			
	Se	0.57	0	0	0.57		
75	As	6.5	S 4.9	I (1.6)			
76	Ge	8.7	8.7	(0)	0		
	Se	5.7	0	5.4	I (0.3)		
77	Se	4.7	S 3.7	I (1.0)			
78	Se	14.8	I (7.5)	S (7.3)			
	Kr	0.09	0	0	0.09		
79	Br	4.1					minor s-branch; ^{79}Se-decay
80	Se	31.5	I (6.5)	S 25.0			major s-branch
	Kr	0.56	0	S 0.46	I (0.1)		minor s-branch
81	Br	3.9	S (3.0)	I (0.9)			
82	Se	5.7	5.7	(0.)	0		
	Kr	2.9	0	2.8	I (0.1)		
83	Kr	2.9	S 2.2	I 0.7			
84	Kr	14.2	I (5.0)	S 9.0	0		
	Sr	0.15	0	0	0.15		
85	Rb	4.6	S 3.9	I (0.7)			major s-branch; ^{85}Sr-decay
86	Kr	4.3					minor s-branch
	Sr	2.6	0	2.5	I (0.1)		major s-branch
87	Rb	1.8 [a]	1.8		0		minor s-branch
	Sr	1.8 [a]	0	1.8			major s-branch
88	Sr	21.5	I (1.0)	20.5			
89	Y	5.4	I (0.8)	S 4.6			
90	Zr	5.7	I (0.6)	S 5.1	?		
91	Zr	1.23	I (0.5)	S 0.73	?		
92	Zr	1.88	I (0.5)	S 1.3			
	Mo	0.37	0	0	0.37		
93	Nb	0.85					
94	Zr	1.93	I (0.4)	S 1.5	0		
	Mo	0.23	0	(0)	0.23		
95	Mo	0.40					
96	Zr	0.31	0.31	0	0		
	Mo	0.42	0	0.42	0		
	Ru	0.099	0	0	0.099		
97	Mo	0.24	S 0.13	I 0.11			
98	Mo	0.61	I 0.3	S 0.3	0		
	Ru	0.034	0	0	0.034		
99	Ru	0.23	S (0.16)	I (0.06)	I (0.01)		major s-branch; ^{99}Tc-decay
100	Mo	0.24	0.24	0	0		
	Ru	0.23	0	0.21	I 0.02		
101	Ru	0.31	S 0.26	I 0.05			
102	Ru	0.57	S 0.45	I 0.12	0		
	Pd	0.013	0	0	0.013		
103	Rh	0.33	S 0.28	I 0.05			
104	Ru	0.33	0.33	(0)	0		
	Pd	0.15	0	0.13	I 0.02		
105	Pd	0.29	S 0.25	I 0.04			
106	Pd	0.36			0		
	Cd	0.016	0	0	0.016		

continued

Table 5b, continued

A	Isotope of	Normalized abundance					Remarks
		total	component				
			β^-	s	β^+	other	
107	Ag	0.26	S 0.20	I 0.04	I 0.02		minor s-branch; ^{107}Pd-decay
108	Pd	0.35	I 0.18	S (0.17)	0		
	Cd	0.012	0	(0)	0.012		
109	Ag	0.24	S 0.17	I 0.06	I (0.01)		
110	Pd	0.156	0.156	(0)	0		
	Cd	0.16	0	0.14	I 0.02		
111	Cd	0.17	S 0.10	I (0.06)	I (0.01)		
112	Cd	0.31	I 0.15	S 0.16	0		
	Sn	0.024	0	0	0.024		
113	Cd	0.16	S (0.09)	I (0.06)	I (0.01)		
	In	0.0074	0	(0)	(0.0074)		
114	Cd	0.37	I 0.12	S 0.25	0		
	Sn	0.016	0	0	0.016		
115	In	0.167	S 0.12	I 0.04	I (0.01)		
	Sn	0.0084	0	(0)	(0.0084)		
116	Cd	0.10	0.10	(0)	0		
	Sn	0.35	0	0.33	0.02		
117	Sn	0.182	I 0.10	S (0.07)	I (0.01)		
118	Sn	0.58	I 0.11	S 0.45	I 0.02		
119	Sn	0.21	I 0.10	S (0.10)	I (0.01)		
120	Sn	0.79	I 0.11	S 0.68	0		
	Te	0.0043	0	0	0.0043		
121	Sb	0.155	S 0.115	I 0.04			
122	Sn	0.113	0.113	(0)	0		
	Te	0.115	0	0.11	I 0.004		
123	Sb	0.115	0.115	(0)	0		
	Te	0.042	0	0.04	I (0.002)		
124	Sn	0.139	0.139	0	0		
	Te	0.22	0	0.22	0		
	Xe	0.006	0	0	0.006		
125	Te	0.34	S 0.27	I 0.07			
126	Te	0.90	S 0.48	I 0.42	0		
	Xe	0.0055	0	0	0.0055		
127	I	1.16	1.1	I 0.04			
128	Te	1.53	1.53	(0)	0		
	Xe	0.116	0	0.116	0		
129	Xe	1.61	1.56	I 0.05			
130	Te	1.66	1.66	0	0		
	Xe	0.20	0	0.20	0		
	Ba	0.0042	0	0	0.0042		
131	Xe	1.29	1.25	I (0.04)			
132	Xe	1.65	S 1.35	I (0.3)	0		
	Ba	0.0040	0	0	0.0040		
133	Cs	0.37	S 0.33	I (0.04)			
134	Xe	0.64	0.64	(0)	0		
	Ba	0.10	0	0.10			major s-branch; ^{134}Cs-decay
135	Ba	0.27	S 0.23	I (0.04)			
136	Xe	0.54	0.54	0	0		
	Ba	0.33	0	0.33	0		
	Ce	0.0023	0	0	0.0023		
137	Ba	0.47					

continued

Table 5b, continued

A	Isotope of	Normalized abundance					Remarks
		total	component				
			β^-	s	β^+	other	
138	Ba	3.0	I (0.4)	2.6			
	La	0.0004	0	0	0	0.0004	
	Ce	0.0031	0	0	0.0031		
139	La	0.46					
140	Ce	1.06	I (0.25)	S 0.81			
141	Pr	0.18					
142	Ce	0.133	0.133	0	0		
	Nd	0.24	0	0.24			
143	Nd	0.104 [a]					
144	Nd	0.21	I (0.08)	S (0.13)			
	Sm	0.0084	0	0	0.0084		
145	Nd	0.072					
146	Nd	0.150	I 0.06	S 0.09			
147	Sm	0.042 [a]	S 0.034	I 0.008			major s-branch; ^{147}Pm-decay
148	Nd	0.050	0.050	0	0		
	Sm	0.030	0	0.030			
149	Sm	0.036	S 0.031	I 0.005			
150	Nd	0.049	0.049	0	0		
	Sm	0.020	0	0.020			
151	Eu	0.048	0.046	I 0.002			minor s-branch; ^{151}Sm-decay
152	Sm	0.072	0.054	0.018	0		major s-branch; ^{152}Eu-decay
	Gd	0.00068	0	(0)	$\leqq$0.0007		
153	Eu	0.052	0.049	I 0.003			
154	Sm	0.062	0.062	(0)	0		
	Gd	0.0075	0	0.0072	I (0.0003)		
155	Gd	0.051	0.048	I 0.003			
156	Gd	0.070	S 0.056	I 0.014	0		
	Dy	0.00025	0	0	0.00025		
157	Gd	0.053	0.048	I 0.005			
158	Gd	0.084	S 0.059	I 0.025	0		
	Dy	0.00041	0	(0)	$\leqq$0.00041		
159	Tb	0.061	0.057	I 0.004			
160	Gd	0.074	0.074	0	0		
	Dy	0.0094	0	0.0090	I 0.0004		
161	Dy	0.077	0.074	I 0.003			
162	Dy	0.105	S 0.09	I 0.016	0		
	Er	0.00036	0	0	0.00036		
163	Dy	0.102	0.097	I 0.005			s-branch; ^{163}Ho-decay
164	Dy	0.116	S 0.08	I 0.04	0		s-branch
	Er	0.0042	0	S 0.0039	I (0.0003)		s-branch
165	Ho	0.091	0.088	I 0.003			
166	Er	0.087	S 0.074	I 0.013			s-branch; ^{166}Ho-decay
167	Er	0.060	0.055	I 0.005			
168	Er	0.070	S 0.04	I 0.03	0		
	Yb	0.00035	0	0	0.00035		
169	Tm	0.041	S 0.036	I 0.005			
170	Er	0.039	0.039	(0)	0		
	Yb	0.0075	0	0.0075			major s-branch; ^{170}Tm-decay
171	Yb	0.036	S 0.031	I 0.005			
172	Yb	0.055	S 0.038	I 0.017			
173	Yb	0.041	S 0.033	I 0.008			

continued

Table 5b, continued

A	Isotope of	Normalized abundance					Remarks
		total	component				
			β^-	s	β^+	other	
174	Yb	0.080	S (0.04)	I 0.04	0		
	Hf	0.00032	0	0	0.00032		
175	Lu	0.037	S 0.032	I 0.005			
176	Yb	0.032	0.032	(0)	0		
	Lu	0.00113 [a]	0	0.0011	0		
	Hf	0.0092 [a]	0	0.009			major s-branch; ^{176}Lu*-decay
177	Hf	0.033	S 0.029	I (0.004)			
178	Hf	0.049	S 0.035	I (0.014)			
179	Hf	0.025	S 0.019	I (0.006)			
180	Hf	0.063	S 0.04	I 0.024	0		
	Ta	$3.\cdot 10^{-6}$	0	0	0	$3.\cdot 10^{-6}$	
	W	0.00017	0	0	0.00017		
181	Ta	0.021	S 0.013	I 0.008			
182	W	0.033	S (0.01)	I 0.023			
183	W	0.018	S (0.01)	I 0.01			
184	W	0.039			0		
	Os	0.00014	0	0	0.00014		
185	Re	0.019	S 0.015	I 0.004			
186	W	0.036	0.036	(0)	0		
	Os	0.012	0	0.012			
187	Re	0.035 [a]	0.035	(0)	0		minor s-branch
	Os	0.010 [a]	0	0.010			minor s-branch
188	Os	0.096	0.09	I (0.01)			
189	Os	0.116	0.11	I (0.006)			
190	Os	0.19	0.18	I (0.01)	0		
	Pt	0.00018	0	0	0.00018		
191	Ir	0.24	0.24	I (0.003)			
192	Os	0.30	S 0.26	I (0.04)	0		
	Pt	0.0111	0	0.0111			major s-branch; ^{192}Ir-decay
193	Ir	0.41	0.40	I (0.01)			
194	Pt	0.47	0.46	I (0.01)			
195	Pt	0.48	0.47	I (0.005)			
196	Pt	0.36	0.33	I (0.03)	0		
	Hg	0.0006	0	0	0.0006		
197	Au	0.190	0.18	I (0.01)			
198	Pt	0.102	0.10	(0)	0		
	Hg	0.040	0	0.040			
199	Hg	0.068	S 0.056	I (0.012)			
200	Hg	0.092	I (0.05)	S (0.04)			
201	Hg	0.053					
202	Hg	0.119	I (0.04)	S 0.08			
203	Tl	0.051					
204	Hg	0.027	0.027	(0)	0		
	Pb	0.06	0	0.06			
205	Tl	0.123	I (0.025)	S 0.10			minor s-branch; ^{205}Pb-decay
206	Pb	0.57	I (0.02)	S 0.46		I (0.09)	end of α-chain
207	Pb	0.63	I (0.02)	S 0.53		I (0.08)	end of α-chain
208	Pb	1.79	I (0.02)	1.7		I (0.06)	end of α-chain
209	Bi	0.136	I (0.02)	I (0.08)		S (0.04)	end of α-chain
232	Th	0.041 [a]	(0.015)	0			several β^- contributions
235	U	0.0074 [a]		0			several β^- contributions
238	U	0.019 [a]	(0.010)	0			several β^- contributions

3.5 Chronology of the solar system

3.5.1 Introduction

3.5.1.1 Dating techniques

The absolute time scale for solar system evolution is exclusively based on radiometric age dating [1, 2]. The amount of a decay product accumulated in a closed system (no loss or gain from the system of radioactive parent ^{k}P or "radiogenic" daughter isotope $^{i}D_r$) is translated into the accumulation time t by $t=\tau \ln [(^{i}D_r/^{k}P)+1]$; ($\tau=1/\lambda=T_{\frac{1}{2}}/\ln 2$ = mean life of radioisotope; λ = decay constant; D, P in number of atoms). Ideally, the event to be dated must involve the complete removal of the preexisting element D. Frequently, the fractionation between the chemical elements P and D as caused by the process to be dated is incomplete. Then mass spectrometric measurements of all isotopes of element D help to distinguish the radiogenic part $^{i}D_r$ from the non-radiogenic part $^{i}D_0$. An event can be dated only if it is accompanied by at least some parent-daughter fractionation. This is most frequently achieved by geochemical differentiation during condensation, rock formation, or metamorphism. In case of partially open systems, involved methods exist to account for loss of radiogenic isotopes (Pb-Pb method; ^{39}Ar-^{40}Ar method). The chronologically important radionuclides and the dating methods based on these isotopes are listed in Table 1.

Table 1. Radiometric dating methods.
spont.: spontaneous
fiss.: fission
K-capt.: K-capture

Nuclide	Isotopic abundance %	Decay type	Decay product	Half-life [3] a	Method	Standard references
^{87}Rb	27.83	β^-	^{87}Sr	$4.88\cdot10^{10}$	Rb-Sr	4
					initial Sr	5
^{232}Th	100	α	^{208}Pb	$1.40\cdot10^{10}$	Th-Pb	6, 7
^{238}U	99.28	α	^{206}Pb	$4.47\cdot10^{9}$	U-Pb	6, 7
					U-He	8
		spont. fiss.	$^{131\cdots136}Xe$	$8.2\cdot10^{15}$	U-Xe, Xe-Xe	9
			fiss. tracks		fiss. track	10
^{235}U	0.72	α	^{207}Pb	$7.04\cdot10^{8}$	U-Pb	6, 7
$^{238}U+^{235}U$		α	$^{206,\,207}Pb$	above	Pb-Pb	6, 7
^{40}K	0.0117	β^-	^{40}Ca	$\left\{ 1.25\cdot10^{9} \right.$	K-Ca	11
		K-capt.	^{40}Ar	$\left. \lambda_k/\lambda_\beta=0.117 \right.$	K-Ar, ^{39}Ar-^{40}Ar	12, 13, 14, 15
^{147}Sm	15.0	α	^{143}Nd	$1.06\cdot10^{11}$	Sm-Nd	16
^{187}Re	62.6	β^-	^{187}Os	$\approx5\cdot10^{10}$	Re-Os	17
Extinct nuclides						
^{129}I	–	β^-	^{129}Xe	$1.7\cdot10^{7}$	I-Xe	18, 19, 20
^{244}Pu	–	(α) spont. fiss.	$^{131\cdots136}Xe$	$\left\{ 8.2\cdot10^{7} \right.$	Pu-Xe	19, 21
			fiss. tracks	$\left. \lambda_f/\lambda_\alpha=1.25\cdot10^{-3} \right.$	Pu-track	22
^{26}Al	–	β^+	^{26}Mg	$7.4\cdot10^{5}$	Al-Mg	23
^{107}Pd	–	β^-	^{107}Ag	$6.5\cdot10^{6}$	Pd-Ag	24
^{146}Sm	–	α	^{142}Nd	$1.03\cdot10^{8}$		25
^{205}Pb	–	K-capt.	^{205}Tl	$2\cdot10^{7}$		26
^{248}Cm		(α) spont. fiss.	$^{131\cdots136}Xe$	$3.7\cdot10^{5}$		27
^{250}Cm		spont. fiss.	$^{131\cdots136}Xe$	$1.1\cdot10^{4}$		27

3.5.1.2 Types of "ages" (definitions [1] and abbreviations)

(MA) Model age:	Age for which an assumption must be made about the initial isotopic composition of D before the clock starts to operate.
(MA-WR) Whole rock age:	Model age based on isotopic ratios within a single average ("whole rock") sample.
(IA) Isochron age:	Based on isotopic ratios in various cogenetic samples (e.g., minerals from one rock) or differently bound reservoirs within one sample. No assumption, self-consistent recognition of radiogenic quantities.
(IA-M) Mineral isochron age:	IA measured on minerals from one rock.
(IA-WR) Whole rock isochron age:	IA measured on a cogenetic suite of whole rocks, ultimately derived from a common source.
(GR) Gas retention age:	Age involving a volatile daughter isotope and determining the time since this volatile was retained in the sample.
Pb-Pb age:	Age based on the combined application of the $^{238}U \rightarrow {}^{206}Pb$ and $^{235}U \rightarrow {}^{207}Pb$ decays.
(UI) Upper intercept age:	In Pb-Pb dating of partially open systems, a deduced figure which is corrected for Pb-loss and hence refers to the true formation age.
(LI) Lower intercept age:	In Pb-Pb dating of partially open systems, a deduced figure which indicates the time of a major Pb-loss (time of metamorphism).
(PA) ^{39}Ar-^{40}Ar plateau age:	K-Ar isochron gas retention age in which K is measured by internal neutron activation and partial ^{40}Ar-losses are accounted for in a self-consistent manner.
(CREA) Cosmic ray exposure age:	Such ages are not based on decay of primordial radionuclides but on the amount of nuclides produced by the interaction of primary galactic cosmic rays with targets exposed in space (unshielded by the earth's atmosphere). Cosmic rays penetrate only ≈ 1 m into solid rocks; hence, CREA measures the time since surface exposure or since the break-up of a larger object into meter-sized pieces in free space.

3.5.1.3 Time periods

We distinguish data relevant to

(a) absolute age of the solar system (section 3.5.2);
(b) duration of solar system formation (high resolution data around the time of solar system formation, 3.5.3);
(c) subsequent evolution during the history of the solar system (e.g., planetary evolution, 3.5.4).

For topics (a) and (b), meteorites are the most important sources of information because of their relatively primitive nature. For the genetic interrelations implied in the various petrological types, see 3.3.2. For topic (c), all accessible solid planetary materials are potential data sources (meteorites, lunar rocks, terrestrial rocks). In this article, emphasis is given to the age and duration of solar system formation (topics (a), (b)). The planetological evolution of individual objects must be dealt with in a rather crude way since it includes the whole fields of geochronology and lunar chronology. Here, we concentrate on events which are of exemplary nature for the planetological evolution in general.

3.5.1.4 References for 3.5.1

0 Jäger, E., Hunziker, J. (eds.): Lectures in Isotope Geology, Springer (1979).
1 Kirsten, T.: Origin. Sol. Syst. (Dermott, S.F., ed.) Wiley, London (1978) 267.
2 Jäger, E. in: [0] p. 1.
3 Steiger, R.H., Jäger, E.: Earth Planet. Sci. Lett. **36** (1977) 359.
4 Jäger, E. in: [0] p. 13.
5 Faure, G., Powell, J.L.: Strontium Isotope Geology, Vol. 5 of Minerals, Rocks and Inorganic Materials, Springer (1972).
6 Gebauer, D., Grünenfelder, M. in: [0] p. 105.
7 Köppel, V., Grünenfelder, M. in: [0] p. 134.
8 Meier, H.: Fortschr. Chem. Forschg. **7** (1966) 234.
9 Shukoljukov, J., Kirsten, T., Jessberger, E.K.: Earth Planet Sci. Lett. **24** (1974) 271.
10 Naeser, C.W. in: [0] p. 154.
11 Hamilton, E.J.: Applied Geochronology, Academic Press, London (1965) 48.
12 Schaeffer, O.A., Zähringer, J. (eds.): Potassium Argon Dating, Springer (1966).
13 Dalrymple, G.B., Lanphere, M.A.: Potassium Argon Dating, Freeman (1969).
14 Hunziker, J.C. in: [0] p. 52.
15 Dallmeyer, R.D. in: [0] p. 77.
16 Lugmair, G.W., Scheinin, N.B., Marti, K.: Proc. Lun. Sci. Conf. 6th **2** (1975) 1419.
17 Hamilton, E.J.: Applied Geochronology. Academic Press, London (1965) 121.
18 Reynolds, J.H.: Phys. Rev. Lett. **4** (1960) 8.
19 Kirsten, T.: Chem. Ztg. **98** (1974) 288.
20 Podosek, F.A.: Geochim. Cosmochim. Acta **34** (1970) 341.
21 Podosek, F.A.: Earth Planet. Sci. Lett. **8** (1970) 183.
22 Fleischer, R., Price, P., Walker, R.M.: Geochim. Cosmochim. Acta **32** (1968) 21.
23 Lee, T., Papanastassiou, D., Wasserburg, G.: Geophys. Res. Lett. **3** (1976) 41.
24 Kelly, W.R., Wasserburg, G.J.: Geophys. Res. Lett. **5** (1978) 1079.
25 Lugmair, G., Kurtz, J., Marti, K., Scheinin, N.: Earth Planet. Sci. Lett. **27** (1975) 79.
26 Huěy, J., Kohman, T.: Earth Planet. Sci. Lett. **16** (1972) 401.
27 Leich, D., Niemeyer, S., Michel, M.: Earth Planet. Sci. Lett. **34** (1977) 197.

3.5.2 Age of the solar system

In principle, the age t_s of the solar system may be inferred from mineral isochrons of primitive meteorites (especially, undifferentiated chondrites) which formed very early and remained undisturbed ever after. This is true since the time interval which must, in principle, be added to account for the time between the earliest condensation and the meteorite formation is $\ll t_s$ (see 3.5.3). Table 2 is a summary of such mineral isochron ages for meteorites. It includes also data for the oldest lunar relicts.

Table 2. Age data for individual meteorites which remained essentially undisturbed after their formation (I.). Also data for the oldest preserved lunar relicts (II.).

Type [1])	Meteorite	Method	Age type [2])	Age [3]) 10^9 a	Ref.
I.					
C 3	Allende	Pb-Pb	IA-M	4.565±0.004	2
		Pb-Pb	IA-M	4.553±0.004	3
		Pb-Pb	UI	4.57 ±0.02	2
		Th-Pb	IA-M	4.565	2
		K-Ar	PA	4.52 ±0.03	27
C 4	Karoonda	K-Ar	PA	4.52 ±0.1	4
L 4	Bjurböle	K-Ar	PA	4.52 ±0.03	5
L 6	Peace River	Rb-Sr	IA-M	4.47 ±0.03	6
LL 5	Krähenberg	Rb-Sr	IA-M	4.60 ±0.02	7
LL 5	Olivenza	Rb-Sr	IA-M	4.54 ±0.16	8
		K-Ar	PA	4.50 ±0.05	9
LL 5	St. Mesmin	K-Ar	PA	4.49 ±0.05	10
LL 6	St. Severin	Rb-Sr	IA-M	4.46 ±0.09	11
		Pb-Pb	IA-M	4.53 ±0.02	11
		K-Ar	PA	4.42 ±0.03	12
H 4	Forest Vale	K-Ar	MA-WR	4.53 ±0.1	13
H 4	Ochansk	K-Ar	PA	4.49	14
H 5	Allegan	K-Ar	PA	4.52 ±0.1	4
H 6	Guarena	K-Ar	PA	4.42 ±0.03	12
		Rb-Sr	IA-M	4.47 ±0.08	15
H 6	Mount Brown	K-Ar	PA	4.50 ±0.05	9
H 6	Queen's Mercy	K-Ar	PA	4.50 ±0.05	9
H	Nadiabondi	K-Ar	MA-WR	4.48 ±0.1	13
		U-He	MA-WR, GR	4.50 ±0.2	13
E 4	Adhi Kot	K-Ar	MA-WR	4.6 ±0.1	16
		U-He	MA-WR, GR	4.6 ±0.2	16
E 4	Indarch	Rb-Sr	IA-M	4.47 ±0.15	17
An	Norton County	Rb-Sr	IA-M	4.61 ±0.1	18
An	Angra dos Reis	Pb-Pb	IA-M	4.548±0.002	19
		U-Pb	IA-M	4.61 ±0.07	19
		Th-Pb	IA-M	4.57 ±0.1	19
Eu	Ibitira	Rb-Sr	IA-M	4.44 ±0.1	20
Eu	Juvinas	Rb-Sr	IA-M	4.51 ±0.07	20
		Sm-Nd	IA-M	4.56 ±0.08	21
HO	Kapoeta	Rb-Sr	IA-M	4.45 ±0.12	22
Iron	Mundrabilla (olivine)	K-Ar	PA	4.54 ±0.02	23
Iron IA	Toluca, silicate incl.	K-Ar	MA-WR	4.55 ±0.1	24
Iron IB	Four Corners	K-Ar	MA-WR	4.50 ±0.1	24
Iron IIE	Colomera	Rb-Sr	IA-M	4.53 ±0.04	25
II.					
Lunar dunite 72417		Rb-Sr	IA-M	4.46 ±0.1	26
troctolite 76535		Rb-Sr	IA-M	4.51 ±0.09	22

[1]) For explanation, see 3.3.2.

[2]) For definitions, see 3.5.1.2.

[3]) Frequent deviations of the tabulated ages from the values given in the cited papers are due to an adjustment of decay constants according to the recommended values of the Subcommission on Geochronology [1].

One may also determine meaningful lower limits of t_s from whole rock isochron ages of reservoirs which formed very early in the history of the solar system. Examples are whole meteorite parent bodies (sampled in individual meteorites of a given type) and the lunar anorthositic crust. Such data are given in Table 3, together with an estimate for the earth's age from Pb-isotope evolution models.

Table 3. Whole rock isochron ages (IA-WR) for the most ancient planetological reservoirs.

Objects [1])	Method	Age [2]) [10^9 a]	Ref.
L-chondrites	Rb-Sr	4.45 ±0.12	28
LL-chondrites	Rb-Sr	4.47 ±0.15	17
H-chondrites	Rb-Sr	4.59 ±0.14	29
C-chondrites	Rb-Sr	≈4.60	30
E-chondrites	Rb-Sr	4.45 ±0.13	31
Chondrites	Pb-Pb	4.505±0.008	32
Chondrites	Pb-Pb	≈4.57	33
Eucrites	Pb-Pb	4.59 ±0.02	34
Iron meteorites (silicate inclusions)	Rb-Sr	4.5 ±0.2	35
Irons (metal)	Re-Os	4 ±0.8	36
All meteorite classes	Pb-Pb	4.50 ±0.03	37
Lunar anorthositic rocks	Rb-Sr	4.51 ±0.15	38
Lunar soils 10084	U-Pb (UI)	≈4.6	39
Lunar troctolite and mare basalts	Sm-Nd	4.53 ±0.1	40
Terrestrial Galenas	Pb-isotopes	≈4.57	41 [3])

[1]) For explanation, see 3.3.2.
[2]) Frequent deviations of the tabulated ages from the values given in the cited papers are due to an adjustment of decay constants according to the recommended values of the Subcommission on Geochronology [1].
[3]) 2-stage model.

Summarizing the data of Tables 2 and 3, weighted mean values are $(4.52 \pm 0.02) \cdot 10^9$ a for individual isochron ages and $(4.54 \pm 0.03) \cdot 10^9$ a for whole rock isochron ages. Altogether, the best estimate for the age of the solar system deduced with all methods from all appropriate objects (with the new decay constants) is $(4.53 \pm 0.02) \cdot 10^9$ a.

References for 3.5.2

1 Steiger, R.H., Jäger, E.: Earth Planet. Sci. Lett. **36** (1977) 359.
2 Chen, J.H., Tilton, G.R.: Geochim. Cosmochim. Acta **40** (1976) 635.
3 Tatsumoto, M., Unruh, D., Desborough, G.: Geochim. Cosmochim. Acta **40** (1976) 617.
4 Podosek, F.A.: Geochim. Cosmochim. Acta **35** (1971) 157.
5 Turner, G.: Meteorite Res. (Millman, P., ed.) Reidel (1969) 407.
6 Gray, C., Papanastassiou, D., Wasserburg, G.J.: Icarus **20** (1973) 213.
7 Kempe, W., Müller, O.: Meteorite Research, Vienna (1968) 418.
8 Sanz, H.G., Wasserburg, G.J.: Earth Planet. Sci. Lett. **6** (1969) 335.
9 Turner, G., Cadogan, P.H.: Meteoritics **8** (1973) 447.
10 Cadogan, P.H., Turner, G.: Meteoritics **10** (1975) 375.
11 Manhes, G., Minster, J.F., Allegre, C.: Meteoritics **10** (1975) 451.
12 Podosek, F.A., Huneke, J.C.: Meteoritics **8** (1973) 64.
13 Kirsten, T., Krankowsky, D., Zähringer, J.: Geochim. Cosmochim. Acta **27** (1963) 13.
14 Turner, G.: Unpublished.
15 Wasserburg, G.J., Papanastassiou, D., Sanz, H.: Earth Planet. Sci. Lett. **7** (1969) 33.
16 Zähringer, J.: Geochim. Cosmochim. Acta **32** (1968) 209.
17 Gopalan, K., Wetherill, G.W.: J. Geophys. Res. **75** (1970) 3457.
18 Bogard, D., Burnett, D.S., Eberhardt, P., Wasserburg, G.: Earth Planet. Sci. Lett. **3** (1967) 179.
19 Adorables *): Lunar Sci. VII (1976) 443.

*) Abbreviation, substitute for names of the authors of the Angra dos Reis Consortium.

20 Birck, J.L., Minster, J.F., Allegre, C.: Meteoritics **10** (1975) 364.
21 Lugmair, G.W., Scheinin, N.B., Marti, K.: Proc. Lun. Sci. Conf. 6th **2** (1975) 1419.
22 Papanastassiou, D., Wasserburg, G.J.: Lunar Sci. VII (1976) 665.
23 Kirsten, T.: Meteoritics **8** (1973) 400.
24 Bogard, D., Burnett, D., Eberhardt, P., Wasserburg, G.J.: Earth Planet. Sci. Lett. **3** (1967) 275.
25 Sanz, H.G., Burnett, D.S., Wasserburg, G.J.: Geochim. Cosmochim. Acta **34** (1970) 1227.
26 Papanastassiou, D., Wasserburg, G.J.: Proc. Lun. Sci. Conf. 6th **2** (1975) 1467.
27 Jessberger, E.K., Dominik, B., Staudacher, Th., Wasserburg, G.J.: Icarus **42** (1980) 380.
28 Gopalan, G., Wetherill, G.W.: J. Geophys. Res. **76** (1971) 8484.
29 Kaushal, S., Wetherill, G.W.: J. Geophys. Res. **74** (1969) 2717.
30 Kaushal, S., Wetherill, G.W.: J. Geophys. Res. **75** (1970) 463.
31 Gopalan, G., Wetherill, G.W.: J. Geophys. Res. **74** (1969) 4349.
32 Huey, J.M., Kohman, T.: J. Geophys. Res. **78** (1973) 3227.
33 Tilton, G.R.: Earth Planet. Sci. Lett. **19** (1973) 321.
34 Silver, T., Duke, M.: EOS Trans. American Geophys. Union **52** (1971) 269.
35 Burnett, D.S., Wasserburg, G.J.: Earth Planet. Sci. Lett. **2** (1967) 397.
36 Herr, W., Hoffmeister, W., Hirt, B., Geiss, J., Houtermans, F.G.: Z. Naturforsch. **16a** (1961) 1053.
37 Kanasevich, E.R. (literature survey): Radiometric Dating for Geologists (Hamilton, E., Farquhar, R., eds.) Intersci. London (1968) 147.
38 Schonfeld, E.: Lunar Sci. VII (1976) 773.
39 Tatsumoto, M.: Proc. Lun. Sci. Conf. 1th **2** (1970) 1595.
40 Lugmair, G., Kurtz, J., Marti, K., Scheinin, N.: Lunar Sci. VII (1976) 509.
41 Stacey, J., Kramers, J.: Earth Planet. Sci. Lett. **26** (1975) 207.

3.5.3 Duration of solar system formation

To increase time resolution for events occurring $\approx 4.53 \cdot 10^9$ a ago:

Use of isotopes with mean lives $\approx 10^6 \cdots 10^8$ a (Table 1), now extinct, but present $4.53 \cdot 10^9$ a ago. Their decay products are still detectable in primitive objects. Example: ^{129}I, $\tau = 24.5 \cdot 10^6$ a, decay product ^{129}Xe (stable). Measure, via ^{129}Xe, the ^{129}I/^{127}I ratio at time of Xe-retention. (Stable ^{127}I is measured by n-activation leading to ^{128}Xe [1].) If this ratio was uniform in the solar nebula, differences between two objects reflect Xe-retention time differences δt. Reference object is chondrite Bjurböle, $t \approx (4.53 \pm 0.02) \cdot 10^9$a, $\delta t = 0$. Resolvable time differences $\delta t \approx 0.5 \cdot 10^6$ a.

Results:

Data in Table 4: Total time span (including all meteorite types) $\approx 25 \cdot 10^6$ a. Most of this scatter is explained as result of thermal metamorphism within meteorite parent bodies immediately after condensation and accretion. (Heating and subsequent cooling below Xe-retention temperature (≈ 650 °C) depends on size of parent body.) If one selects carbonaceous chondrites which were never reheated, the resulting time span of $\approx 5 \cdot 10^6$ a should give duration of condensation. However, some more evolved chondrites are "older" than "condensates" (Table 4).

Possible explanations:

a) Evolution of matter occurred at different times in different regions of the solar system (parent bodies cooling in one region while condensation was still going on in other regions). b) Solar nebula was not isotopically homogenized. Differences in ^{129}I/^{127}I ratios are in part due to initial differences, not to time differences. Not unlikely in view of other evidence for isotopic inhomogeneities [2, 3].

In the solar nebula the decay of ^{87}Rb has led to a monotonic increase of ^{87}Sr/^{86}Sr ratio with time. At the time of condensation the ^{87}Sr/^{86}Sr ratio is frozen in. It can be extracted from Rb-free minerals (no later radioactive growth of ^{87}Sr) or deduced from regular Rb-Sr mineral isochrons [4, 5]. Differences in $(^{87}\text{Sr}/^{86}\text{Sr})_{\text{initial}}$ ratios can be translated into time differences [4]. Fig. 1 summarizes relevant results.

Rb is depleted during condensation, hence Sr-isotopes in Rb-poor objects may be related to condensation. The corresponding time spread is $\approx 10 \cdot 10^6$ a. Moon and meteorites are contemporaneous within this range. More evolved ^{87}Sr/^{86}Sr ratios probably refer to metamorphism within parent bodies.

Table 4. I-Xe ages of individual meteorites relative to the chondrite Bjurböle.

Type [1]	Meteorite	I-Xe age [2] 10^6 a	Ref.	Type [1]	Meteorite	I-Xe age [2] 10^6 a	Ref.
C1	Orgueil	-7.0 ± 1.0	8	LL3	Chainpur matrix	$+3.1 \pm 0.7$	9
C2	Murchison	-7.2 ± 1.0	8	LL3	Chainpur chondrules	$+10.5 \pm 1.0$	9
C3	Allende inclus.	-2.4 ± 1.3	15	LL6	St. Severin	$+7.5 \pm 1.8$	9
C4	Karoonda	-3.9 ± 1.0	9	E4	Indarch	-0.8 ± 1.2	13
H4	Menow	-10.0 ± 1.6	10	E4	Abee	-0.7 ± 1.5	1
H4	Beaver Creek	$+0.8 \pm 2.4$	10	Au	Pena Blanca Spring	$+3.6 \pm 1.0$	9
H5	Nadiabondi	-14.9 ± 3.2	10	Au	Shallowater	-0.6 ± 0.1	9
H5	Allegan chondrules	-2.3 ± 1.8	9	Hexahedrite	El Taco silicate	$+3.8 \pm 0.8$	9
H5	Pantar dark	$+2.6 \pm 1.7$	4, 11	Iron I-A	Landes	$+2.61 \pm 0.62$	14
H5/6	Ambapur Nagla	$+4.5 \pm 1.7$	3, 10	Iron I-A	Copiapo	$+1.36 \pm 0.66$	14
H6	Kernouve	-12.6 ± 2.8	3, 10	Iron I-B	Woodbine	-3.68 ± 0.30	14
L5	Arapahoe	-9.9 ± 0.8	12	Iron	Mundrabilla silicate	-0.68 ± 0.59	14
L5	Ausson	-4.8 ± 3.2	10	Iron	Mundrabilla troilite	-10.2 ± 0.8	14
L6	Peetz	-6.9 ± 5.6	10				
L6	Bruderheim chondrule	$+0.6 \pm 1.4$	11				

[1]) For explanation, see 3.3.2. [2]) Relative to the standard meteorite Bjurböle (L4).

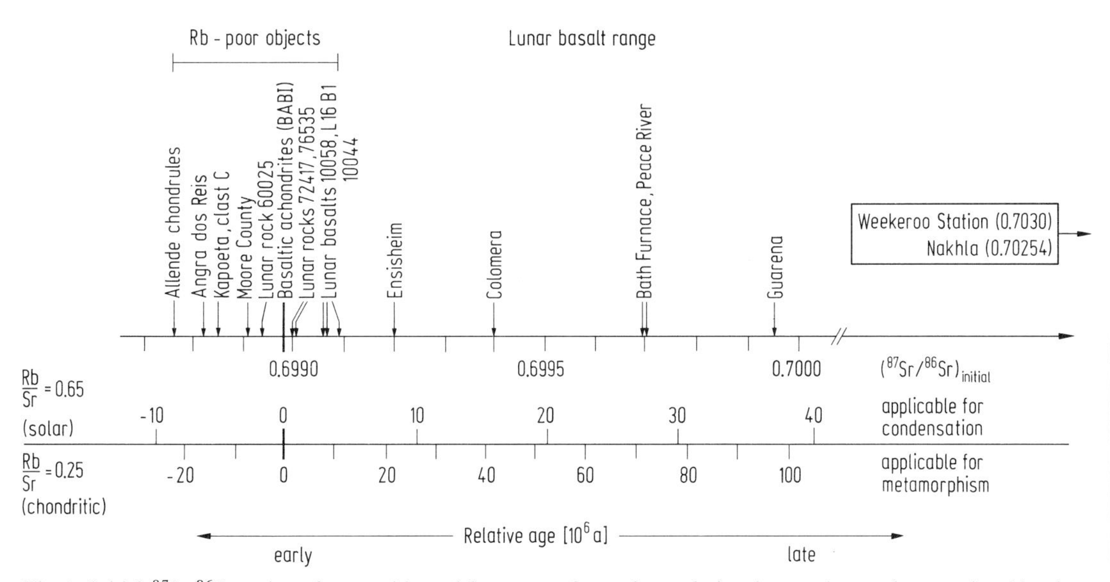

Fig. 1. Initial $^{87}Sr/^{86}Sr$ ratios of meteoritic and lunar samples and translation into a time scale assuming (a) solar Rb/Sr environment (Rb/Sr = 0.65, upper scale) and (b) chondritic Rb/Sr environment (Rb/Sr ≈ 0.25, lower scale). Times are measured relative to BABI (Basaltic Achondrite Best Initial). Data mainly from CALTEC, compiled in [6] and [4]. Bath Furnace and Ensisheim from [7].

References for 3.5.3

1 Jeffery, P.M., Reynolds, J.H.: J. Geophys. Res. **66** (1961) 3582.
2 Clayton, R.N., Onuma, N., Mayeda, T.: Earth Planet. Sci. Lett. **30** (1976) 10.
3 Podosek, F.A.: Annu. Rev. Astron. Astrophys. **16** (1978) 293.
4 Gray, C., Papanastassiou, D., Wasserburg, G.: Icarus **20** (1973) 213.
5 Faure, G., Powell, J.L.: Strontium Isotope Geology, Vol. 5 of Minerals, Rocks and Inorganic Materials, Springer (1972).
6 Papanastassiou, D., Wasserburg, G.J.: Lunar Sci. VII (1976) 665.
7 Wetherill, G.W., Mark, R., Lee-Hu, C.: Science **182** (1973) 281.

Kirsten

8 Lewis, R.S., Anders, E.: Proc. Nat. Acad. Sci. USA **72** (1975) 268.
9 Podosek, F.A.: Geochim. Cosmochim. Acta **34** (1970) 341.
10 Jordan, J., Kirsten, T., Richter, H.: Z. Naturforsch. **35**a (1980) 145.
11 Merrihue, C.: J. Geophys. Res. **71** (1966) 263.
12 Drozd, R.J., Podosek, F.A.: Earth Planet. Sci. Lett. **31** (1976) 15.
13 Hohenberg, C., Podosek, F.A., Reynolds, J.H.: Science **156** (1967) 202.
14 Niemeyer, S.: Geochim. Cosmochim. Acta **43** (1979) 843.
15 Podosek, F.A., Lewis, R.S.: Earth Planet. Sci. Lett. **15** (1972) 101.

3.5.4 Planetary evolution

3.5.4.1 Meteorites and their parent bodies

Radiometric meteorite ages significantly younger than $4.53 \cdot 10^9$ a indicate later events in the evolution of the individual parent bodies. They are best explained as the result of reheating in catastrophic collisions among meteorite bodies. Table 5 summarizes such data. It is evident that such events occurred during the whole age of the solar system and do still occur (see e.g., Clovis!). The period from $3.5 \cdots 4 \cdot 10^9$ a was rather "active", particularly for achondrites.

Table 5. Meteorite ages significantly below $4.53 \cdot 10^9$ a.

Type [1])	Meteorite	Method	Age type [2])	Age [3]) 10^9 a	Ref.
Ach.	Nakhla	Rb–Sr	IA–M	1.22 ± 0.01	3
		Rb–Sr	IA–M	1.34 ± 0.03	4
		K–Ar	PA	≈ 1.3	5
Iron	Kodaikanal (silicate incl.)	Rb–Sr	IA–M	3.72 ± 0.08	6
Ho	Kapoeta (clast)	Rb–Sr	IA–M	≈ 3.56	7
Ho	Malvern	K–Ar	PA	≈ 3.6	8
Ho	Bununu	K–Ar	PA	≈ 4.3	9
Ho	Bholghati	K–Ar	PA	≈ 3.35	10
Eu	Stannern	K–Ar	PA	≈ 3.65	11
Eu	Pasamonte	K–Ar	PA	4.0	11
Eu	Petersburg	K–Ar	[4])	< 2.2	11
LL6	Appley Bridge	K–Ar	PA	3.8	12
LL6	Mangwendi	K–Ar	PA	3.9	12
LL6	Ensisheim	K–Ar	[4])	< 1.5	13
L6	Colby, Wisc.	K–Ar	PA	3.8	14
L6	Peace River	K–Ar	PA	≈ 0.46	15
L6	Bruderheim	K–Ar	[4])	< 0.5	14
L6	Chateau Renard	K–Ar	[4])	< 0.31	14
L6	Rakovka	K–Ar	[4])	< 1.5	13
L6	Zavid	K–Ar	[4])	< 0.55	14
L6	Zemaitkiemis	K–Ar	[4])	< 0.53	14
L6	Zomba	K–Ar	[4])	< 0.44	14
L5	Wittekrantz	K–Ar	PA	≈ 0.35	13
L5	Ergheo	K–Ar	[4])	< 0.53	14
L4	Barratta	K–Ar	[4])	< 0.47	14
H6	Clovis 2	K–Ar	[4])	≈ 0.01	15
H	Darmstadt	K–Ar	[4])	< 0.32	15
C3	Allende	Pb–Pb	LI	0.28 ± 0.07	16
		Pb–Pb	LI	0.11 ± 0.07	17
L	Chondrites	K–Ar + U–He	WR–GR	0.52 ± 0.06	18
L	Chondrites	Rb–Sr	WR	$0.3 \cdots 0.5$	19

[1]) For explanation, see 3.3.2. [2]) For definitions, see 3.5.1.2.
[3]) Adjusted for new decay constants [20].
[4]) No plateau defined but deduced from behaviour of age spectrum (see [21] for explanation).

The collisional history of meteorites before earth capture determines the CREA-distribution (CREA = Cosmic ray exposure age, see 3.5.1.2). It is linked to the mean life of solid objects before destruction and/or planetary capture and hence to their primary orbits and the place of origin [1, 2]. Significant break-ups of parent bodies are deduced from CREA-clusters.

Table 6. Distribution of meteorite cosmic ray exposure ages (CREA).

Meteorite class or meteorite	Significance	Exposure ages 10^6 a	Ref.
Stony meteorites	typical range	1···30	22, 23
Iron meteorites	typical range	100···1000	22, 24, 25
Chondrites	majority	< 10	22, 23
Achondrites	majority	> 10	22
Farmington (chondr.)	minimum for stones	0.02	26
Mayo Bèlwa (achondr.)	maximum for stones	82	27
Pitts	minimum for irons	4	28
Deep Springs	maximum for irons	2300	24
Carbonaceous chondrites	cut-off	< 15	23
Stony irons	cut-off	< 200	29
Hexahedrites (irons)	cut-off	< 300	25
H-chondrites	cluster	4	23
L-chondrites	cluster (less pronounced)	5	23
LL-chondrites	cluster	8	23
Aubrites (achondr.)	cluster	40	30
Eucrites (achondr.)	cluster	5 + 11	30
Diogenites (achondr.)	cluster	15	30
Octahedrites type III (irons)	cluster	650	24, 25
Octahedrites type IV A (irons)	cluster	400	24, 25

References for 3.5.4.1

1 Wetherill, G.W.: Annu. Rev. Earth Planet. Sci. **2** (1974) 303.
2 Anders, E. in: Phys. Stud. Minor Planets, NASA SP-267 (1971) 429.
3 Gale, N., Arden, J., Hutchison, R.: Earth Planet. Sci. Lett. **26** (1975) 195.
4 Papanastassiou, D.A., Wasserburg, G.J.: Geophys. Res. Lett. **1** (1974) 23.
5 Podosek, F.A.: Earth Planet. Sci. Lett. **19** (1973) 135.
6 Burnett, D.S., Wasserburg, G.J.:Earth Planet. Sci. Lett. **2** (1967) 397.
7 Papanastassiou, D.A., Wasserburg, G.J.: Lunar Sci. VII (1976) 665.
8 Kirsten, T., Horn, P.: Sov. American Conf. Cosmochem. Moon and Planets, NASA SP-370 (1977) 525.
9 Rajan, R., Huneke, J., Smith, S., Wasserburg, G.J.: Earth Planet. Sci. Lett. **27** (1975) 181.
10 Leich, D.A., Moniot, R.: Lunar Sci. VII (1976) 479.
11 Podosek, F.A., Huneke, J.C.: Geochim. Cosmochim. Acta **37** (1973) 667.
12 Turner, G., Cadogan, P., Yonge, C.: Lunar Sci. V (1974) 807.
13 Turner, G., Cadogan, P.: Meteoritics **8** (1973) 447.
14 Turner, G.: Meteorite Research (Millman, P., ed.), Reidel, Dordrecht (1969) 407.
15 Turner, G.: Unpublished (1976).
16 Tatsumoto, M., Unruh, D., Desborough, G.: Geochim. Cosmochim. Acta **40** (1976) 617.
17 Chen, J.H., Tilton, G.R.: Geochim. Cosmochim. Acta **40** (1976) 635.
18 Taylor, G.J., Heymann, D.: Earth Planet. Sci. Lett. **7** (1969) 151.
19 Gopalan, G., Wetherill, G.W.: J. Geophys. Res. **76** (1971) 8484.
20 Steiger, R.H., Jäger, E.: Earth Planet. Sci. Lett. **36** (1977) 359.
21 Kirsten, T.: Origin Sol. Syst. (1978) 267.
22 Kirsten, T., Schaeffer, O.A. in: Elementary Particles in Science, Technology and Society (Yuan, L.C., ed.), Academic Press, N.Y. (1971) 76.
23 Schultz, L.: Meteoritics **11** (1976) 359.
24 Voshage, H.: Z. Naturforsch. **22a** (1967) 477.
25 Scott, E., Wasson, J.: Rev. Geophys. Space Phys. **13** (1975) 527.
26 Heymann, D., Anders, E.: Geochim. Cosmochim. Acta **31** (1967) 1793.
27 Heusser, G., Hampel, W., Kirsten, T., Schaeffer, O.A.: Meteoritics **13** (1978) 492.
28 Cobb, J.C.: Science **151** (1966) 1524.
29 Begemann, F., Weber, H., Vilcsek, E., Hintenberger, H.: Geochim. Cosmochim. Acta **40** (1976) 353.
30 Wasson, J.T.: Meteorites, Springer, Berlin (1974).

3.5.4.2 Lunar evolution

Fig. 2. summarizes significant epochs in lunar chronology.

Time before present	For definitions and methods, see 3.5.1.
$\approx 4.5 \cdot 10^9$ a:	Oldest relicts of lunar crust; see Table 2.
$\approx 4.3 \cdot 10^9$ a:	End of initial differentiation deduced from a) IA–WR–UI–U/Pb age of highland rocks and soils [1]; b) Upper cut-off of highland rock age distribution (IA–M K–Ar (PA), Rb–Sr), U–Pb; see [2] for references; c) IA–WR Rb–Sr for lunar soils [3, 4].
$4.3 \cdots 3.8 \cdot 10^9$ a:	Ages of lunar highland rocks (lunar crust). (IA–M K–Ar (PA), Rb–Sr); numerous analyses see [2] for references. Distribution dominated by impacts. Major basins formed between 3.8 and $3.9 \cdot 10^9$ a, Serenitatis basin: $(3.88 \pm 0.03) \cdot 10^9$ a [5].
$4.0 \cdots 3.1 \cdot 10^9$ a:	Mare volcanism (flooding of basins). (IA–M K–Ar (PA), Rb–Sr); numerous analyses, see [2] for references.
$3.1 \cdot 10^9$ a:	End of igneous activity (lower cut-off for mare basalt ages). Younger volcanism ($\approx 2.5 \cdot 10^9$ a) possible, but not sampled [6].
$< 1 \cdot 10^9$ a:	Local craters and regolith turnover and formation. Mainly from CREA-distribution for individual rocks. (Excavation time!) CREA range from $1 \cdots 700 \cdot 10^6$ a [7] ($\approx 2/3$ below $150 \cdot 10^6$ a). Example: Tycho $96 \cdot 10^6$ a[8]. Typical CREA for soils: $150 \cdots 450 \cdot 10^6$ a [7].

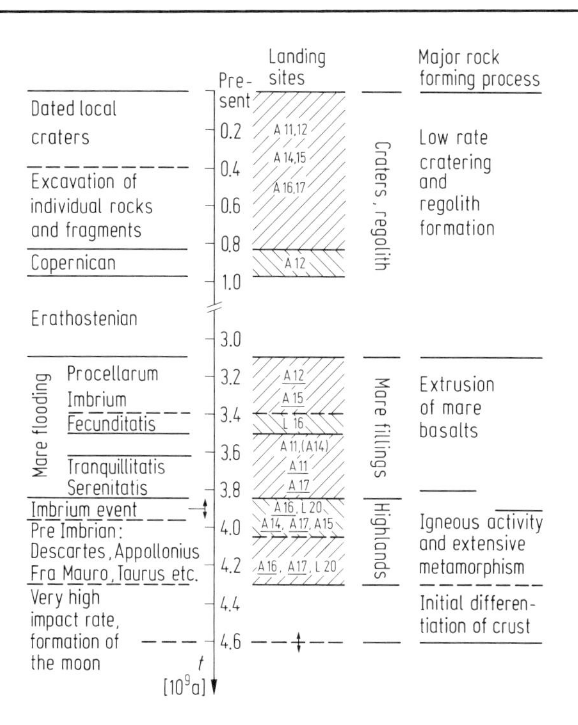

Fig. 2. Summary of lunar chronology.
A = Apollo-, L = LUNA-landing sites; t = time before present.

References for 3.5.4.2

1 Tera, F., Wasserburg, G.J.: Proc. Lunar Sci. Conf. 5th **2** (1974) 1571.
2 Kirsten, T.: Origin Sol. Syst. (1978) 267.
3 Papanastassiou, D., Wasserburg, G.J.: Earth Planet. Sci. Lett. **17** (1972) 52.
4 Papanastassiou, D., Wasserburg, G.J.: Earth Planet. Sci. Lett. **13** (1972) 368.
5 Jessberger, E.K., Staudacher, Th., Dominik, B., Kirsten, T.: Proc. Lunar Sci. Conf. 9th **1** (1978) 1655.
6 Head, J.W.: Rev. Geophys. Space Phys. **14** (1976) 265.
7 Arvidson, R., Crozaz, G., Drozd, R., Hohenberg, C., Morgan, C.: Moon **13** (1975) 259.
8 Arvidson, R., Drozd, R., Guiness, E., Hohenberg, C., Morgan, C.: Proc. Lunar Sci. Conf. 7th **3** (1976) 2817.

3.5.4.3 Geochronology

Large-scale differentiation $\approx 3.7 \cdots 3.9 \cdot 10^9$ a ago deduced from 2-stage Pb-isotope evolution model [1]; no rocks preserved predating this period. Oldest preserved rocks:

Amitsoq-type gneisses from Isua, Greenland and Godthaab area; Rb–Sr and Pb–Pb IA–WR; IA–M: $3.7 \cdots 3.8 \cdot 10^9$ a [2, 3, 4].

Gneisses Saglek Bay, Labrador, IA–WR, Rb–Sr: $(3.62 \pm 0.1) \cdot 10^9$ a [5].

Gneisses and granites, Rhodesian Basement, IA–WR, Rb–Sr: $3.5 \cdots 3.6 \cdot 10^9$ a [6, 7].

Subsequent orogenic activity has a cyclic time characteristic ("Megacycles" of $800 \cdot 10^6$ a; orogenic cycles of $\approx 200 \cdot 10^6$ a). Reflected in distribution of rock ages (Fig. 3).

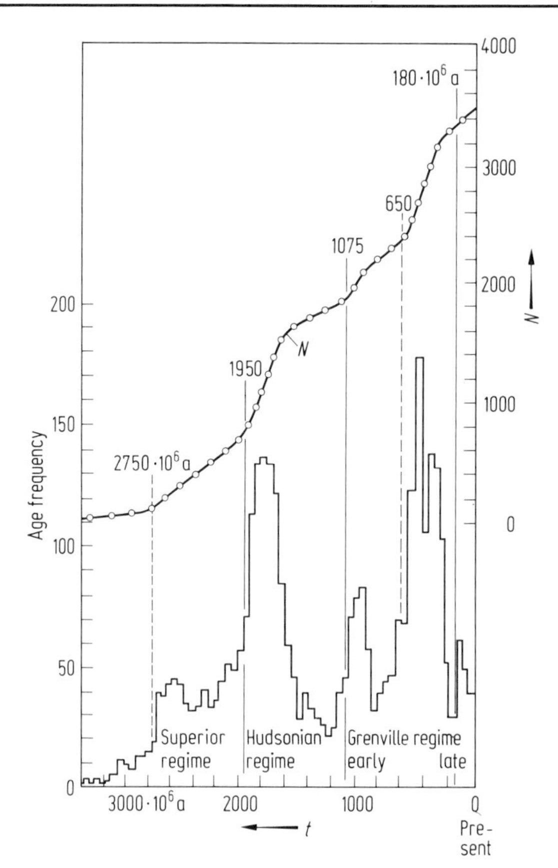

Fig. 3. Frequency histogram of radiometric ages t of more than 3000 metamorphic and igneous terrestrial rocks from all continents (lefthand scale) and cumulative age frequency $N(>t)$ (upper curve, righthand scale) [8]. Note the preponderance of certain age figures.

References for 3.5.4.3

1 Stacey, J., Kramers, J.: Earth Planet. Sci. Lett. **26** (1975) 207.
2 Moorbath, S., O'Nions, R., Pankhurst, R., Gale, N., McGregor, V.: Nature, Physic. Sci. **240** (1972) 78.
3 Moorbath, S., O'Nions, R., Pankhurst, R.: Nature, Physic. Sci. **245** (1973) 138.
4 Moorbath, S., O'Nions, R., Pankhurst, R.: Earth Planet. Sci. Lett. **27** (1975) 229.
5 Hurst, R., Bridgewater, D., Collerson, K., Wetherill, G.: Earth Planet. Sci. Lett. **27** (1975) 293.
6 Hickman, M.H.: Nature **251** (1974) 295.
7 Hawkesworth, C., Moorbath, S., O'Nions, R., Wilson, J.: Earth Planet. Sci. Lett. **25** (1975) 251.
8 Dearnley, R. in: The Application of Mod. Phys. to the Earth and Planet. Interiors. (Runcorn, S.K., ed.), Wiley-Interscience, London (1969) 103.

3.5.5 Summary and link to the ages of the elements

A summary of the cosmochronological time scale is given in Fig. 4. From the existence of extinct nuclides (^{129}I, ^{244}Pu, ^{26}Al, ^{107}Pd) it can be inferred that a significant proportion ($\approx 5\%$?) of elements constituting the solar system was synthesized within only a few 10^6 a before solar system formation [1···4; see also 5].

The mean age of the bulk of the heavy elements deduced from isotopic considerations including long-lived nuclides (^{238}U, ^{235}U, ^{232}Th, ^{187}Re) [6] depends on the time characteristic of the element production rate. Present best estimates are:

Mean element age: $11\cdot10^9$ a;
duration of nucleosynthesis: $10\cdot10^9$ a;
(before solar system formation)
age of Galaxy: $15\cdot10^9$ a [2].

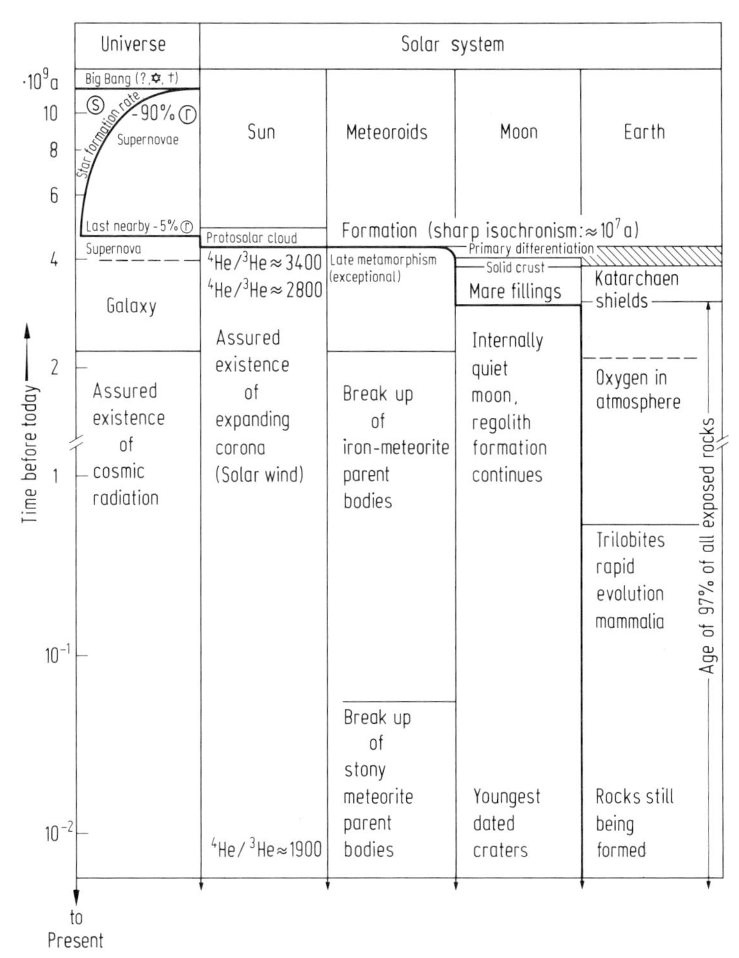

Fig. 4. Schematic summary of a cosmochronological time table (logarithmic scale). The heavy line separates periods of ongoing evolution of the respective objects from the quiet periods afterwards. ⓢ and ⓡ indicate nucleosynthesis by s, r process resp., e.g. 90% r means 90% of all nuclei produced in the r process are formed.

Finally, it shall be mentioned that there are now strong indications from isotopic anomalies [7] that small quantities of presolar condensates have escaped homogenization during solar system formation, and ages in excess of $4.53 \cdot 10^9$ a, the age of the solar system, are feasible. K–Ar plateau ages in excess of $5 \cdot 10^9$ a are reported for Allende inclusions [8] but their interpretation is pending.

References for 3.5.5

1 Schramm, D.N., Wasserburg, G.J.: Astrophys. J. **162** (1970) 57.
2 Schramm, D.N.: Annu. Rev. Astron. Astrophys. **12** (1974) 383.
3 Lee, T., Papanastassiou, D., Wasserburg, G.: Geophys. Res. Lett. **3** (1976) 41.
4 Kelly, W.R., Wasserburg, G.J.: Geophys. Res. Lett. **5** (1978) 1079).
5 Kirsten, T.: Origin. Sol. Syst. (1978) 267.
6 Burbridge, E.M., Burbridge, G.R., Fowler, W., Hoyle, F.: Rev. Mod. Phys. **29** (1957) 547.
7 Podosek, F.: Annu. Rev. Astron. Astrophys. **16** (1978) 293.
8 Jessberger, E.K., Dominik, B.: Nature **277** (1979) 554.

3.1.2.8 Radio emission of the disturbed sun

The radio emission of the disturbed sun originates mainly in distinct regions of the solar atmosphere known as "active regions", which are usually associated with sunspots and/or chromospheric plages. It is superimposed on the radiation of the undisturbed sun.

Nevertheless, if the radiation from these active regions is subtracted from the daily radio flux values of the whole sun, there remains a weak, long-period variation of the radiation flux, which is described by the term:

1. Basic component of solar radio emission (3.1.2.8.1)

The radio emission from the very active regions varies irregularly and, to some extent, has extraordinarily large amplitudes of short duration. On account of the properties of the variability and of the different spectra this radiation is subdivided into three further components:

2. The slowly varying component (3.1.2.8.2)
3. The "noise storms" of the m-wave region. (3.1.2.8.4)
4. The burst radiation (3.1.2.8.6)

The frequency ranges over which the components 2 to 4 mainly occur are represented in Fig. 1 [a].

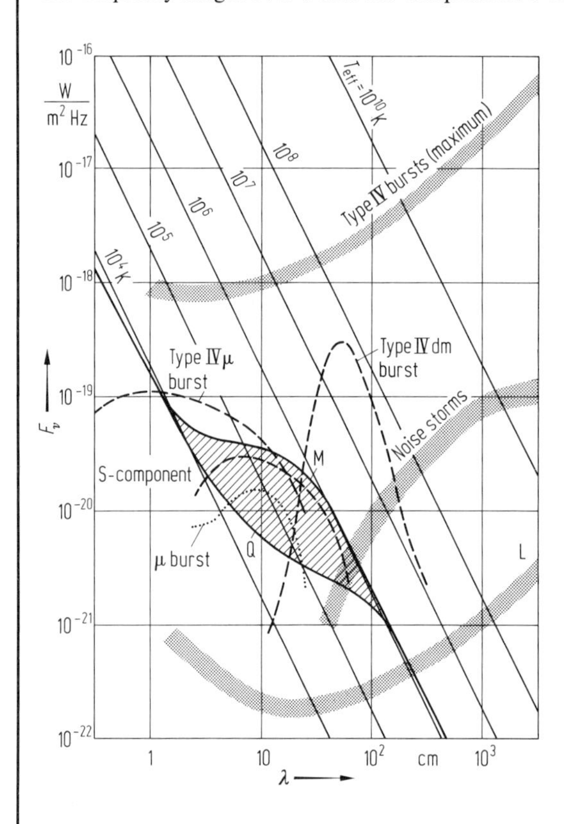

Fig. 1. Radio spectra of different radiation components. F_ν = solar radiation falling on the earth's surface
Q = quiet sun
M = daily flux at sunspot maximum
L = limit of detectability of single frequency patrol observations
S = slowly varying component
Parallel lines: thermal radiation for a given effective temperature, T_{eff}. For other symbols, see text.

3.1.2.8.1 The basic component of solar radio emission

If the radiation from the active regions is subtracted from the daily measured radiation flux, the remaining radiation of the undisturbed sun (Q in Fig. 1) shows a slow variation over a period of 11 years. At wavelength $\lambda = 10$ cm its amplitude is about 55% of the radiation of the quiet sun (see 3.1.1.6, Table 1) [a, 11]. The amplitude has a flat maximum in the lower dm-wavelength range (M in Fig. 1) [13, 24].

The radiation of this component is ascribed to an increased electron density which occurs in the lower corona at the time of the sunspot maximum. The electrons are trapped by arch structures of magnetic fields.

The basic component has also been described as a very slow decay of the radiation of the S-component from active regions (see Fig. 1 and 3.1.2.8.2); but finally this idea also leads to an increased electron density in the lower corona.

3.1.2.8.2 The slowly varying component

The slowly varying component or simply "S-component" originates in discrete regions of the solar atmosphere known as "active regions", which are usually associated with sunspots and/or chromospheric plages. The variable radio emission from these regions is closely correlated with the day-to-day variations in relative sunspot number or with the area of the plage regions. Because of this high correlation, the emission is also referred to as the "slowly varying sunspot component".

The position of the emission regions on the solar surface is determined daily at a number of observatories by either mapping the entire solar disk or by scanning the sun with a fan-beam parallel to the solar equator. These data are regularly published in [b, c] and also in special offprints from the various observatories.

The spectral range over which the S-component increases significantly the radiation flux of the quiet sun is

$1.5\ \text{cm} < \lambda < 70\ \text{cm}$.

The enhancement of the solar flux density in this spectral range is given in Table 1 (see also Fig. 1). Based on a constant relative sunspot number $R = 200$, it is apparent that the enhancement of the radio emission in the solar cycle 20 was of the order of 20···30% less than that in the preceding solar cycle 19.

The S-component can also be recognized as relative brightening of the active regions on solar radio maps taken in the mm-wavelength range [9]. Their radiation temperature at $\lambda = 1$ mm is on the average 300 K higher than their surroundings. This difference in temperature increases to ca. 600 K for $\lambda = 5$ mm and to 900 K for $\lambda \cong 8$ mm [15, 16, 19]. But the increased emission can hardly be detected when measuring the daily radio flux of the whole sun.

Table 1. Enhancement of the solar flux density $F_\odot$ by the radiation of the slowly varying sunspot component during the 19th and 20th sunspot cycles.
$F_{\odot\,max}$ = flux density during solar maximum (relative sunspot number $R = 200$)
$F_{\odot\,min}$ = flux density of the quiet sun (see 3.1.1.6.1 Table 1)
r = mean correlation coefficient of the flux and the sunspot number for each wavelength [d] p. 133.

λ cm	f MHz	Solar cycle 19 $F_{\odot\,max}/F_{\odot\,min}$	Solar cycle 20 $F_{\odot\,max}/F_{\odot\,min}$	r
100	300	(1)		
60	500	1.42	2.07	
30	1000	4.20	2.99	0.74
20	1500	4.72	3.36	0.78
15	2000	4.86	3.48	0.80
10	3000	3.92	3.26	0.86
6	5000	(2.7)	2.75	(0.80)
3	10000	1.57	1.37	0.71
1.5	20000		1.10	

Extent of the emission regions

Angular size 1′···8′
Average size 4′
Linear extent 200000 km
Mean height above photosphere (18000 ± 5000) km
(derived from the displacement of the source of the radio emission at $\lambda = 9.1$ cm with respect to its associated sunspot [6, 7, 25], or from rotation velocity [1]).

Upper bound above photosphere at λ = 3 cm: 70000 km
at λ = 20 cm: 140000 km
(derived from solar eclipse observations at the solar limb [8]).

Structure of the emission regions

The emitting regions consist of one or more bright emission centers that are surrounded by a more diffuse halo [5, 3, 17, 18]. This appearance is particularly evident for $\lambda=3$ cm. The emission centers tend to become covered by the halo as one moves to longer wavelengths.

The diameters of the emission centers are 10″ to 40″ (7000 to 30000 km). They are located above the regions of maximum magnetic field strength in the photosphere, i.e. primarily above sunspots. The radiation from the emission centers exhibits a high degree of circular polarization. If the emitting region is situated above a sunspot of north magnetic polarity (magnetic field pointing out of the sun), the emission is right circularly polarized. Bipolar sunspot pairs have two emission centers, one of which emits right circular polarized radiation, while the other is observed mainly in the left circular mode [4, 18]. The ratio of the radiation intensity of the emission centers is about 5···17% of the intensity of the halo at $\lambda=3$ cm and 5···40% at $\lambda=1.76$ cm [4].

The spectrum

The spectrum of the regions of activity is usually derived from synoptic radio flux measurements of the sun. The radio flux of the sun over a wide range of frequencies is first determined on a day when an isolated activity region is present on the solar disk. The background intensity of the solar disk devoid of sunspots is then subtracted from this measurement.

The spectrum obtained in this way usually shows a maximum in the interval 5 cm$<\lambda<$10 cm and sometimes a second maximum at longer wavelengths [12]. Usually the intensity falls monotonically from the maximum toward longer wavelengths. Many spectra decrease in brightness at the shorter wavelengths, although an occasional rise in intensity may also occur. The spectral profiles in this frequency range have led to a classification of the spectra into 5 or 6 different types as shown in Fig. 2 [23]. The frequencies of occurrence of these various spectral types (in [%]) are:

Spectral code	Occurrence frequency [%]
$A_{0,1}$	43
A_2	32
A_3	6
B	10
C	9

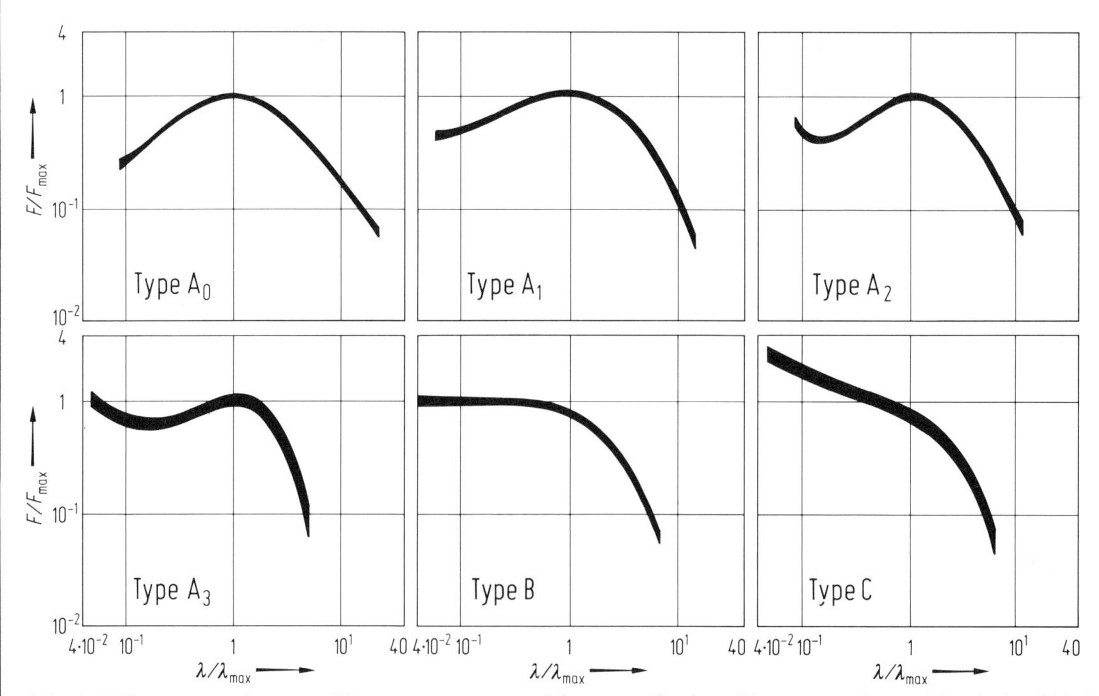

Fig. 2. Different types of spectra of S-component sources. (After normalization with respect to λ_{max} and F_{max}); the widths of the curves represent the errors, [23].

Theory of the emission

The emission of the S-component is generally interpreted to be thermal radiation. The two fundamental processes responsible for this radiation are Coulomb (collisional) bremsstrahlung and gyrosynchrotron emission at the lower harmonics of the gyrofrequency [a, 10, 20]. The fraction of the total emission of a given active region attributed to the gyrosynchrotron process is variable [22]. The intensity of this emission component depends on the strength of the magnetic field in the active region [22]. Present-day theories ascribe a strong gyrosynchrotron component to the bright emission centers, while the weaker centers and the halo emit primarily by the thermal bremsstrahlung process [4].

Long-term monitoring of the S-component

The high correlation of the radio emission of the S-component with soft X-rays [21, 14] and the resulting implications for geophysical phenomena underscore the significance of long-term monitoring of variations in the solar radio flux. The most important continuous observation programs have been conducted by the institutes in Table 2 (for other programs, see also [c]). The measurements of the S-component at 2.8 GHz during solar cycle 20 are shown in Fig. 3.

Table 2. Long-term solar radio observations in the microwave region.

Observing station	Latitude	Longitude	Frequency f [GHz]	Beginning
Algonquin Radio Observatory N.R.C., Ottawa, Canada	45°.95 N	78°.05 W	2.8	Febr. 1947
AFCRL Sagamore Hill Radio Obs., Hamilton, Mass., USA	42.63 N	70.82 W	15.4	Jan. 1968
			8.8	Jan. 1966
			4.99	Apr. 1966
			2.7	Jan. 1966
			1.415	Jan. 1966
			0.606	Jan. 1966
Gorky Univ. Research Inst. of Radiophys., Gorky, U.S.S.R.	56.1 N	44.3 E	9.1	Jan. 1964
			2.95	Jan. 1966
Heinrich-Hertz-Institute, Tremsdorf near Potsdam, GDR	52.28 N	13.13 E	9.5	Jan. 1956
			1.47	Jan. 1954
Manila Observatory, Manila, Philippines	14.63 N	121.10 E	8.8	Jan. 1968
			5.0	Jan. 1968
			2.7	Jan. 1968
			1.42	Jan. 1968
			0.606	Jan. 1968
Ondrejov Observatory, Ondrejov near Prague, Czech.	49.91 N	14.78 E	0.536	Jan. 1956
Toyokawa Obs. Nagoya Univ., Toyokawa, Japan	34.83 N	137.37 E	9.3	May 1956
			3.75	Nov. 1951
			2.0	June 1957
			1.0	March 1957

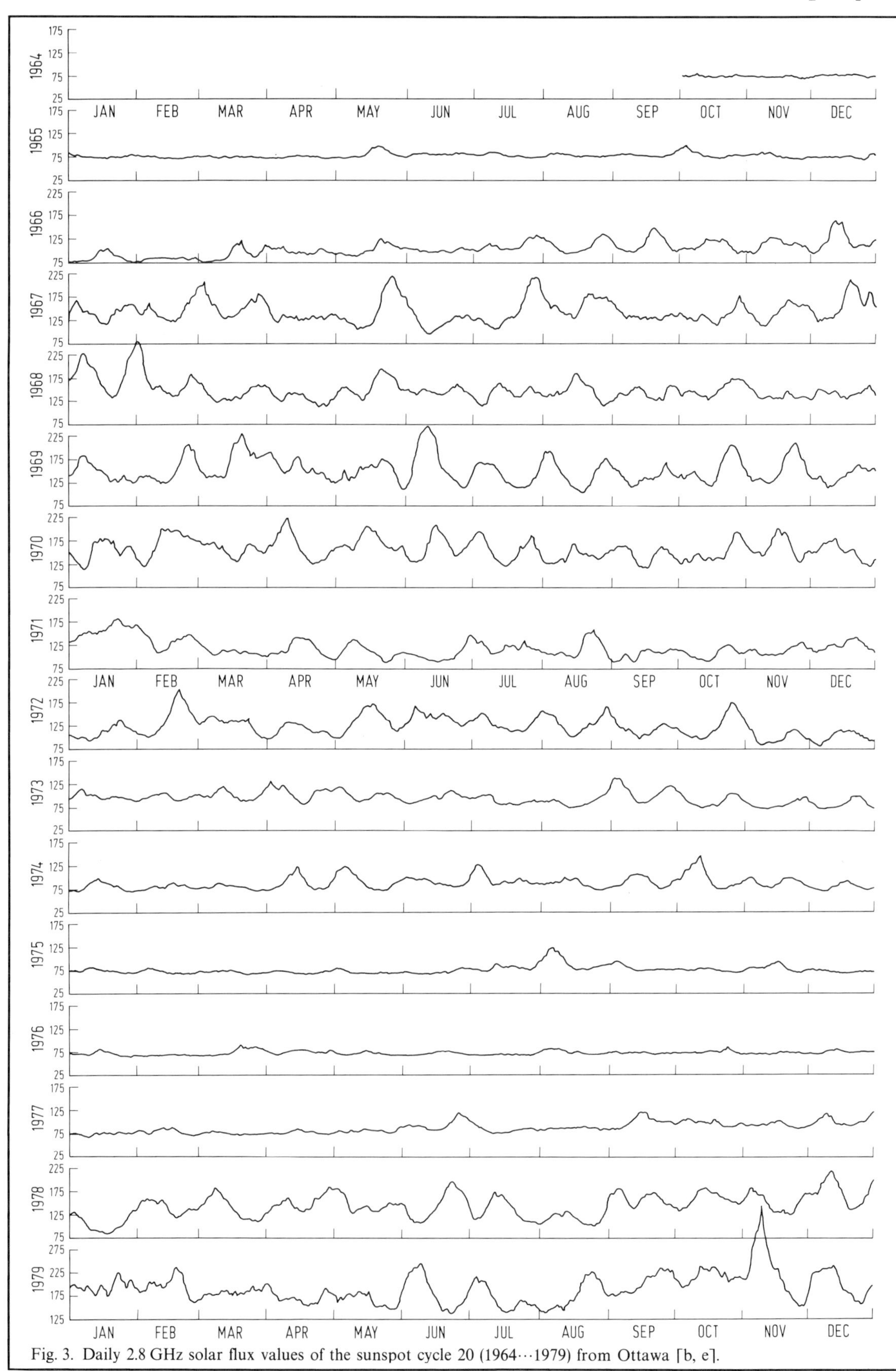

Fig. 3. Daily 2.8 GHz solar flux values of the sunspot cycle 20 (1964···1979) from Ottawa [b, e].

3.1.2.8.3 References for 3.1.2.8.1 and 3.1.2.8.2

General references

a Krüger, A.: Introduction to Solar Radioastronomy and Radio Physics, Reidel, Dordrecht (1979).
b Solar Geophysical Data (U.S. Department of Commerce).
c Quarterly Bulletin on Solar Activity.
d Landolt-Börnstein, N.S., Vol. VI/1 (1965).
e Covington, E.: A Working Collection of Daily 2800 MHz Solar Flux Values, Nat. Res. Coun. Canada, Report No. ARO – 5 (1977).

Special references

1 Cole, D.G., Mullaly, R.F.: Proc. Astron. Soc. Australia **1** (1969) 192.
2 Falchi, A.D., Felli, M., Tofani, G.: Sol. Phys. **48** (1976) 59.
3 Felli, M., Pampaloni, P., Tofani, G.: Sol. Phys. **37** (1974) 395.
4 Felli, M., Tofani, G., Fürst, E., Hirth, W.: Sol. Phys. **42** (1975) 377.
5 Fürst, E., Hachenberg, O., Zinz, W., Hirth, W.: Sol. Phys. **32** (1973) 445.
6 Graf, W., Bracewell, R.N.: Sol. Phys. **28** (1973) 425.
7 Graf, W., Bracewell, R.N.: Sol. Phys. **33** (1973) 75.
8 Hachenberg, O., Popowa, M., Prinzler, H.: Z. Astrophys. **58** (1963) 36.
9 Hachenberg, O., Steffen, P., Harth, W.: Sol. Phys. **60** (1978) 105.
10 Kakinuma, T., Swarup, G.: Astrophys. J. **136** (1962) 975.
11 Krüger, A., Krüger, W., Wallis, G.: Z. Astrophys. **59** (1964) 37.
12 Krüger, A., Michel, H.-St.: Nature **206** (1965) 601.
13 Krüger, A., Olmr, J.: Bull. Astron. Inst. Czech. **24** (1973) 202.
14 Krüger, A., Taubenheim, J., Entzian, G. in: Solar Flares and Space Research (de Jager, C., Svestka, Z. eds.), North-Holland Publ. Co. (1969) 181.
15 Kundu, M.R.: Sol. Phys. **13** (1970) 348.
16 Kundu, M.R.: Sol. Phys. **21** (1971) 130.
17 Kundu, M.R., Alissandrakis, C.E.: Nature **257** (1975) 465.
18 Kundu, M.R., Alissandrakis, C.E., Bregman, J.D., Hin, A.C.: Astrophys. J. **213** (1977) 278.
19 Kundu, M.R. in: Solar Magnetic Fields (Howard, R., ed.), Int. Astron. Union Symp. No. **43**, Reidel, Dordrecht (1971) 642.
20 Lantos, P.: Ann. Astrophys. **31** (1968) 105.
21 Michard, R., Ribes, E. in: Structure and Development of Solar Active Regions (Kiepenheuer, K.O., ed.), Int. Astron. Union Symp. **35**, Reidel, Dordrecht (1968) 420.
22 Shimabukuro, F.J., Chapman, G.A., Mayfield, E.B., Edelson, S.: Sol. Phys. **30** (1973) 163.
23 Steffen, P.: Sol. Phys. **67** (1980) 89.
24 Tanaka, H.: Proc. Res. Inst. Atmosph. Nagoya Univ. **11** (1964) 41.
25 Waldmeier, M.: Sol. Phys. **57** (1978) 369.

3.1.2.8.4 Noise storms

Noise storms, a very common manifestation of solar activity, are observed in the meter and decameter wavelength ranges. As indicators of solar activity, noise storms assume an intermediate position between those phenomena connected with the more gradual development of active regions, i.e. the S-component (3.1.2.8.2) and the short duration radio bursts (3.1.2.8.6). Detailed reviews are presented in [a···d].

Occurrence frequency of noise storms at sunspot maximum (f = 125 MHz)

Intensity [$W m^{-2} Hz^{-1}$]	Occurrence frequency ([%] of time observation)
$5 \cdots 40 \cdot 10^{-22}$	9.25
$40 \cdots 200 \cdot 10^{-22}$	1.97
$> 200 \cdot 10^{-22}$	2.0
Total	13.2%

Solar noise storms are thus recorded during 13.2% of the total observation time at f = 125 MHz.

Duration of noise storms

The average noise storm lasts for several days. Individual storms can last from only half an hour up to 14 days (half a solar rotation). It is difficult to determine the exact duration because of the strongly beamed radiation emanating from the storm region.

The location of emission regions on the solar disk

There is no doubt that noise storms are closely connected with active regions. Nevertheless, the emission region does not necessarily need to be situated directly over a sunspot group. Considerable displacement of the emission regions can occur because of the local magnetic field line configuration in the corona. As a result, the unambiguous association of a noise storm with one or the other of two close sunspot groups is not always possible.

The occurrence and location of noise storms is determined daily by a linear scan across the solar equator. These daily measurements are carried out, for example, with the multielement interferometer in Nancy, France, at $f = 169$ MHz. They are published regularly in the Quarterly Bulletin of Solar Activity.

Directivity of storm emission

The directivity of noise storm emission was derived from the center-to-limb variation of the storm intensity by various authors [16, 11]. Although this beaming effect is directly apparent from the observations, a quantitive description such as the half power width of the emission pattern cannot be made without additional assumptions. The mean half power width resulting from the available data is of the order of 80°.

Simultaneous measurements from two well-separated locations, for example with the help of satellites, have shown that some Type I noise peaks are beamed in much narrower patterns with widths $\cong 25°$. The strong variation in directivity of the many individual emission peaks, however, is responsible for the broader mean width mentioned above [3, 14].

Details of the noise storm phenomenon

The noise storm is basically composed of two emission components:

A. The noise storm continuum: a more or less slowly varying radiation with a broad continuous spectrum.

B. The storm bursts: short-lived, narrow-band emission superimposed on the noise continuum; mostly bursts of generic Type I or Type III.

Individual noise storms can be dominated by either the continuum or the burst component. It is difficult to distinguish and isolate the continuum component if the burst emission is very dominant. The existence of both components in all cases, however, seems to be well substantiated [a] (see also [d] p. 131).

A. Noise storm continua

The spectrum

The high frequency end of the continuum noise storm radiation is typically located at about $f = 300$ MHz. A maximum in intensity occurs between $f = 100$ and $f = 200$ MHz, followed by a relative minimum, usually between $f = 40 \cdots 60$ MHz. A further increase in radiation flux is detected toward lower frequencies, i.e. at decameter wavelengths. The intensity of the continuum component, the position of the maximum and the behaviour in the decameter range can all vary considerably from storm to storm. Examples of three different spectra with their maxima at 67, 100 and 170 MHz respectively are shown in Fig. 4 as reproduced from [13].

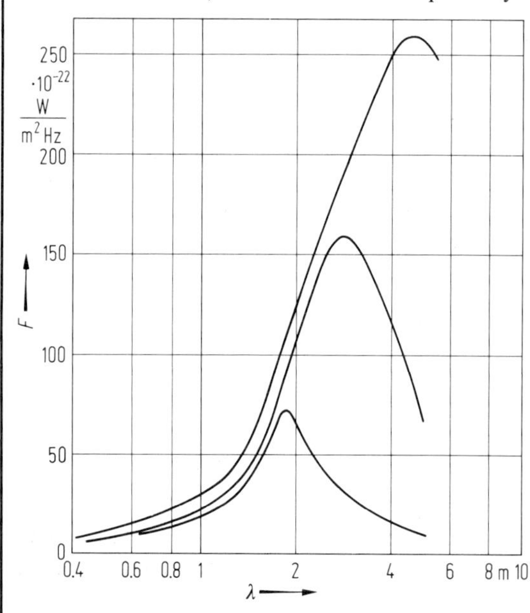

Fig. 4. Three different storm spectra with maximum intensity at $f = 175$, 100, and 67 MHz.
F = flux density above quiet-sun level
λ = wavelength

The size of the emission regions

The typical extent of the emission regions is observed to be between 1 and 10 arc minutes. The diameter appears to grow toward longer wavelengths [7, 8].

The height of the emission regions

Noise storm emission at meter wavelengths is found to originate in regions at altitudes from 0.1 to over 1.0 solar radii above the photosphere. Considerable differences in apparent height of the emission region for a given constant frequency are observed in various individual noise storms [4].

B. Storm bursts

Series of narrow-band, short-period emission peaks are continuously present together with the continuum radiation. In the meter wavelength range these are primarily Type I radio bursts. The behavior changes at longer wavelengths ($\lambda > 10$ m) and the Type I bursts are gradually replaced by burst events of Type III (see also 3.1.2.8.6).

Characteristics of Type I bursts

Duration: 0.1···2 s; increases toward longer wavelengths.
Bandwidth: 2···10 MHz; decreases toward longer wavelengths.
Emission region: usually cospatial with the source region of the continuum emission. Size of region for a single Type I burst $<1'$.
Emitted energy: $10^{11}\cdots10^{14}$ erg; strongly varying wavelength dependence for different noise storms.
Polarization of emission: strongly circularly polarized.

Investigations with high spectral and temporal resolution have shown that the Type I noise peaks can be categorized into a number of subclasses [c, d, 7]:

a) stationary bursts: The maximum intensity remains at a constant frequency (a in Fig. 5).
b) drifting bursts: The maximum intensity shifts with time to higher or to lower frequencies (b in Fig. 5).
c) split bursts: The maximum intensity splits into two or more emission peaks in the frequency and/or time domain (c in Fig. 5).
d) spike bursts: Very quickly drifting bursts, of very short duration when observed at a constant frequency [6] (d in Fig. 5).

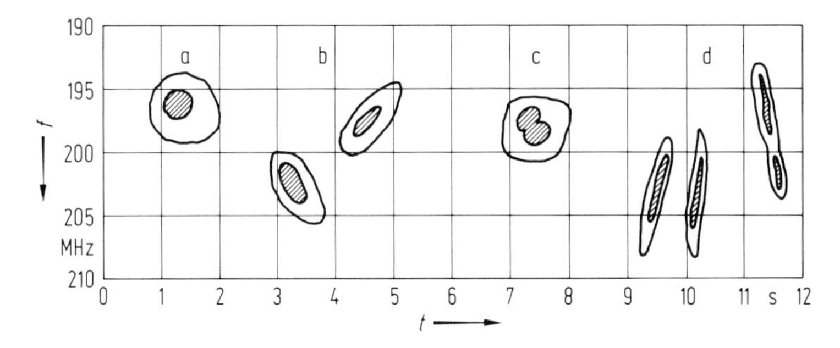

Fig. 5. Dynamic spectra of Type I bursts of different subclasses.

Occurrence frequency of the Type I subclasses

The subclasses of bursts observed can vary greatly for each individual noise storm. The mean occurrence frequencies reported in [c] are as follows:

a) stationary bursts 45···65 %
b) drifting bursts 22···28
c) split bursts 5··· 7
d) spike bursts 5···18

Chains of Type I bursts

Storm bursts do not always occur with a random distribution during a noise storm. They sometimes exhibit a tendency to cluster together or form "chains" which consist of 10 or more rapidly successive Type I bursts within a narrow frequency range. The chains sometimes also display a positive or negative drift in frequency. The radial velocities derived from these drifts lie in the range $v_r = \pm 750$ km s^{-1}. A histogram of the distribution of radial velocities, however, shows not only the strong maximum at $v_r = 0$ km s^{-1}, but also a definite preference for positive (away from the sun) motion with a velocity $v_r \cong 200$ km s^{-1}. Incidentally, the chains are very similar to the single Type I bursts with respect to their frequency dependence and regions of emission [c, d, 8, 19].

Type III storms at decameter wavelengths

The Type I bursts are replaced by bursts of Type III at frequencies less than about $f = 50$ MHz. The Type III bursts are narrow-band emission events that quickly drift across a wide range of frequencies (see also 3.1.2.8.6).

Type III decameter storms are only observed in association with a noise storm at meter wavelengths. As such they are commonly interpreted as the extension of the noise storm into the long wavelength range [1, 10]. In fact, Type III bursts are often observed to begin next to the low-frequency limit of the associated Type I activity [9, 15]. Type III emission can originate from the same source region as the meter noise storm, but it can also be significantly displaced from the main storm emission region [b]. A model of these burst events has been developed by [9, 15].

Theoretical interpretation

Although the observations provide a relatively comprehensive description of the noise storm event, the theoretical treatment still evolves from many diverse interpretations. One fundamental assumption, however, is that the emission originates in plasma clouds that are confined and held in place by closed magnetic arcs above or close to active regions (Fig. 6). The emission is not thermal, as evidenced by the high emission temperature and the directivity of the emitted radiation. The actual emission process has been assumed to be the result of stimulation of the ambient plasma by fast electron beams [17, 18] or, alternatively, gyrosynchrotron radiation [12]. Alfvén waves have been mentioned as drivers in the production of suprathermal or mildly relativistic electrons. The Type III bursts, in contrast to the above, usually originate in open magnetic configurations. A comprehensive review with an extensive bibliography may be found in [d].

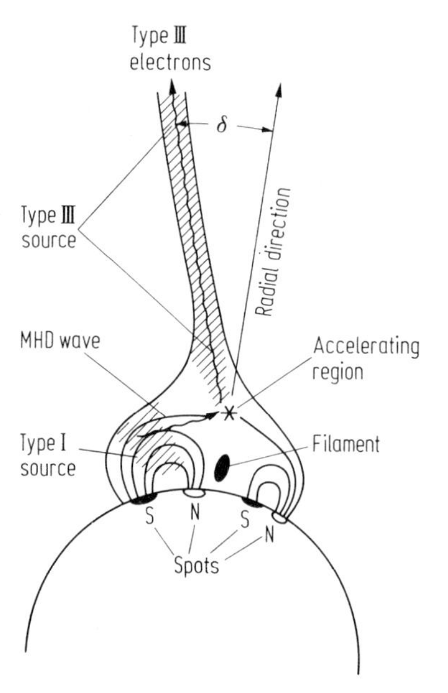

Fig. 6. Model for the positions of the Type I and Type III storm sources [15].
MHD wave = magneto hydrodynamic wave
δ = deviation from radial direction

3.1.2.8.5 References for 3.1.2.8.4

General references

a Fokker, A.D.: Ph. D. Thesis, Leiden (1960).
b Gergely, T.E.: Ph. D. Thesis, University of Maryland (1974).
c Elgarøy, O.: Solar Noise Storms, Pergamon Press (1977).
d Krüger, A.: Introduction to Solar Radio Astronomy and Radio Physics, Reidel, Dordrecht (1979).

Special references

1 Boischot, A., de la Noë, J., du Chaffaut, M., Rosolen, C.: C. R. Acad. Sci. Paris **272** (1971) 166.
2 Bougeret, J.L.: Astron. Astrophys. **24** (1973) 53.
3 Caroubalos, C., Steinberg, J.L.: Astron. Astrophys. **32** (1974) 245.
4 Clavelier, B.: Ann. Astrophys. **30** (1967) 895.
5 Daigne, G.: Nature **220** (1968) 5167.
6 De Groot, T.: Rech. Astron. Obs. Utrecht **18** (1966) 1.
7 Elgarøy, Ø.: Astrophys. Norvegica **7** (1961) 123.
8 Elgarøy, Ø., Ugland, O.: Astron. Astrophys. **5** (1970) 372.
9 Gordon, J.M.: Astrophys. Lett. **5** (1971) 251.
10 Hanasz, J.: Australian J. Phys. **19** (1966) 635.
11 Le Squeren, A.M.: Ann. Astrophys. **26** (1963) 97.
12 Mollwo, L.: Sol. Phys. **12** (1970) 125.
13 Smerd, S.F.: Ann. of the IGY (= Intern. Geophys. Year) **34** (1964) 331.
14 Steinberg, J.L., Coroubalos, C., Bougeret, J.L.: Astron. Astrophys. **37** (1974) 109.
15 Stewart, R.T., Labrum, N.R.: Sol. Phys. **27** (1972) 192.
16 Susuki, S.: Ann. Tokyo Astron. Obs. **7** (1961) 75.
17 Takakura, T.: Publ. Astron. Soc. Japan **15** (1963) 462.
18 Trakhtengerts, V.Yu.: Soviet Astron. **10** (1966) 281.
19 Wild, J.P. in: Radio Astronomy (van de Hulst, H.C., ed.), Int. Astron. Union Symp. **4** (1957) 321.

3.1.2.8.6 Solar radio bursts

Radio bursts from the sun are short-lived emission processes (typical duration less than a few hours), during which the radiation energy can attain values many times greater than that of the quiet sun. The radio bursts are more or less correlated with the solar flare phenomenon in the visible (see 3.1.2.7).

Radio bursts occur with substantially different temporal signatures over the entire range of radio frequencies from 10 MHz to 300 GHz. They are therefore generated at all levels of the solar atmosphere.

The different time profiles of the emission process in the various frequency ranges provide a basis for distinguishing fundamentally different types of bursts, the most prominent of which are:

	Burst type	Spectral range
3.1.2.8.6.1	microwave bursts	1000···100000 MHz
3.1.2.8.6.2	fast-drift bursts (Type III bursts)	10··· 600 MHz
3.1.2.8.6.3	slow-drift bursts (Type II bursts)	10··· 500 MHz
3.1.2.8.6.4	continuum bursts (Type IV bursts)	10··· 8000 MHz

The interrelation of the various types of bursts is schematically displayed in Fig. 7 in the form of a time-frequency diagram. Such diagrams are often used to describe the evolution of a burst event. The physics of the flare-burst phenomenon has become a fascinating subfield of modern astrophysics and has generated a correspondingly extensive literature over the past decade. Comprehensive reviews that are available include [a···d]. (Fig. 7: p. 296).

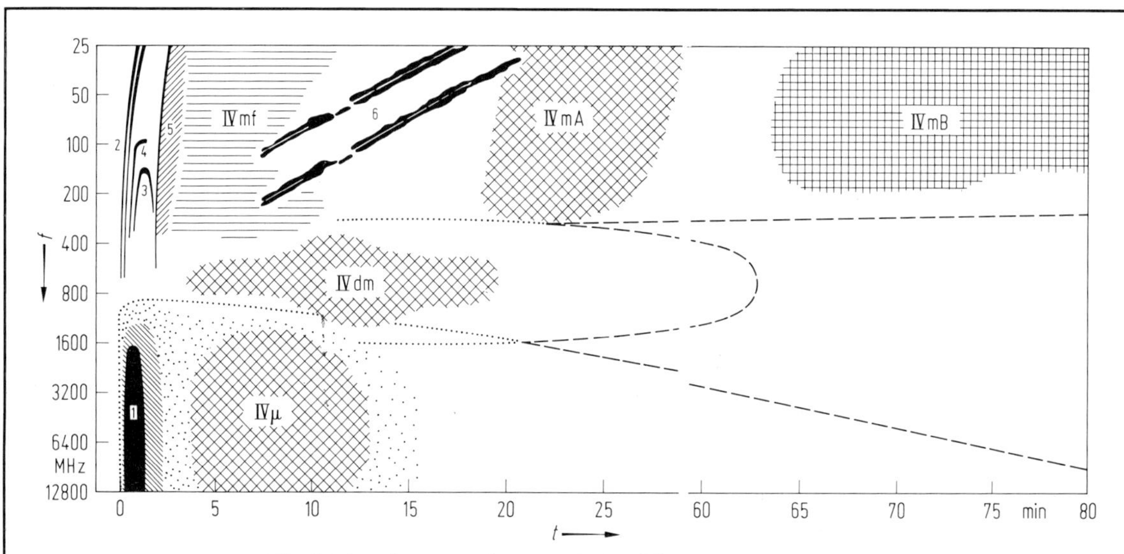

Fig. 7. Time-frequency-diagram of the different types of bursts.
1 Microwave bursts (3.1.2.8.6.1)
2 to *5* Fast-drift bursts (3.1.2.8.6.2)
2 Standard type III burst; *3* U-type burst; *4* J-type burst; *5* Type V burst; *6* Type II burst;
IV Type IV bursts (for details see 3.1.2.8.6.4)

3.1.2.8.6.1 Microwave bursts

All bursts included in this group occur at frequencies $f > 1000$ MHz. The temporal development of these bursts is similar over a wide range of the frequency spectrum [27]. It is generally smooth in contrast to the strong transient radiation peaks that characterize the bursts of Type I, Type II or Type III at meter wavelengths.

Morphology

In the past particular attention has been devoted to the time profiles of the bursts, i.e. to the variation of the received radiation intensity as a function of time. Classification schemes have been derived from many observations of such microwave bursts. One rough classification system differentiates between three fundamental types:

i) gradual bursts
ii) impulsive bursts
iii) complex bursts, Type IVμ bursts

These elementary burst types probably also differ in their respective emission mechanisms. Some brief comments for each type:

i): these include the bursts having a "gradual rise and fall of flux density" as well as those with a typical "pre- and post-burst increase".

ii): the impulsive bursts display a rapid increase in radiation intensity followed by a slower fall-off. The duration of the entire event is about 1···5 min. The flash phase in the visible, the peak in the "hard X-rays", and a series of Type III bursts at meter wavelengths (see below) all coincide with these impulsive bursts.

iii): the time signature of this burst-type appears to contain a combination of several impulsive bursts together with a broad-band continuum (Type IVμ). The unambiguous demarcation between a complex event and a burst of Type IVμ is not always possible. The Type IVμ radiation is associated primarily with very energetic events which correlate with the release of particles (proton flares) during the explosive phase.

Elaborate classification schemes have been devised and introduced in order to characterize more precisely the time variation of the solar radio bursts. One of the goals of this effort is to describe sufficiently the time profiles with a simple alphanumeric code. A convenient nomenclature would thus be created for the prompt dissemination of observational data (e.g. using URSIgrams), and would provide an adequate archival system for subsequent statistical investigations.

A first comprehensive classification was introduced for the International Geophysical Year (IGY) and is described in Landolt-Börnstein, NS, Vol. VI/1, p. 143. A newer ordering scheme was developed in [e,g]. Special number codes were introduced in [h] and [f]. Table 3 provides a comparison of the various codes.

Table 3. The key for identifying types of event by numerical SGD code and letter symbol, according to the Instruction Manual for Monthly Report (prepared by H. Tanaka).

SGD:Solar Geophysical Data [f]

URSI:International Union of Radio Science

SGD code	Letter symbol	URSIGRAM CODES		Morphological classification	Remarks
		URANE	URANO		
1	S	1	1	simple 1	small event
2	S/F	1	1	simple 1 F	F means fluctuations
3	S	2	1	simple 2	moderate event
4	S/F	2	1	simple 2 F	moderate event
5	S	2	1	simple	
6	S		0	minor	defined as simple rise and fall of minor bursts (1⋯2 min)
8	S		1	spike	self-evident by duration
20	GRF	3	1	simple 3	gradual rise and fall
21	GRF (GRF/A)	3	1	simple 3 A	A means underlying
22	GRF (GRF/F)	3	1	simple 3 F	fluctuations of short periods be listed separately
23	GRF	3	1	simple 3 AF	
24	R		8	rise	
25	R (R/A)		8	rise A	
26	FAL			fall	
27	RF	3		rise and fall	(steeper than GRF)
28	PRE	9		precursor	
29	PBI	4	2	post-burst increase	
30	PBI (PBI/A)	4	2	post-burst increase A	
31	ABS			post-burst decrease	
32	ABS	5		absorption	(temporal fall of flux 'negative burst')
40	F	7	4	fluctuations	
41	F	7	4	group of bursts	a group of minor bursts close to each other
42	SER	8	4	series of bursts	a series of bursts occur intermittently from base level with considerable time intervals between bursts
45	C	6	3	complex	
46	C	6	3	complex F	
47	GB	6	3	great burst	(type IVμ burst)

Spectral characteristics

Microwave bursts have a continuous distribution of intensity [27], which can be described by four different spectral types [27a]:

A. The spectrum starts in the cm-range, rises toward higher frequencies and reaches a maximum at mm-wavelengths.
B. The spectrum starts in the cm-range, rises toward lower frequencies and reaches a maximum at dm-wavelengths.
C. The spectrum has a maximum at cm-wavelengths.
D. The spectrum is extremely flat and broad-band.

The spectrum of a complex burst can assume various types over the duration of the burst. Refer to Landolt-Börnstein, NS, Vol. VI/1, p. 146, for diagrams of the various spectra and their relative occurrence frequencies.

Radiation intensity at burst maximum

The radiation intensity at the burst maximum runs from 10···1000 s.f.u. (s.f.u. = solar flux unit = 10^{-22} W m^{-2} Hz^{-1}). The strength of a burst is characterized by the peak flux I_{max}, which is grouped into three categories:

1. $I_{max} < 50$ s.f.u.
2. 50 s.f.u. $< I_{max} < 500$ s.f.u.
3. $I_{max} > 500$ s.f.u.

The letter describing the spectral type (A, B, C, D) and the number of the intensity group (1, 2 or 3) are applied together for the complete characterization of a microwave burst.

Size of the emission region

The emission region is roughly comparable with the size of the flaring region, i.e. mean diameters of the order of ≈2 arc min. Impulsive bursts have small emission cores (typical diameter <15 arc sec). After the maximum the burst cores expands [2, 39].

Polarization

The microwave burst emission is partially circularly polarized, the fraction of polarization ranging from 0 to 50%. The polarization may occasionally change its sense between left and right circular during a burst [19, 26]. A statistical variation with heliographic longitude is evident [38].

Theoretical models of the emission

The emission of the gradual bursts can be attributed to a thermalization process. This is surely not the case for the impulsive bursts, for which one possible mechanism would be gyrosynchrotron radiation of suprathermal electrons. The impulsive bursts are coincident with the flash phase of the optical flare, and their time profiles are remarkably similar to those of the hard X-ray emission. Certain discrepancies remain, however, from the attempt to explain both the gyrosynchrotron radiation and the hard X-rays as coming from suprathermal electrons in the same emission source region [43]. The microwave flux density would have to be several orders of magnitude greater than the observed one. It was therefore necessary to develop other specialized models [61, 34, 60, 3, 39].

3.1.2.8.6.2 Fast-drift bursts (Type III bursts)

Fast-drift bursts consist of narrow-band emission, the limiting and center frequencies of which drift rapidly with time to lower frequencies from starting frequencies $f_{initial} < 600$ MHz. They were first recognized and classified as Type III bursts in [66].

Fast-drift bursts are a frequently occurring mode of solar burst emission at meter wavelengths. One may observe three per hour, on the average, during sunspot maximum.

It is generally accepted today that these bursts are originally produced by a beam of energetic electrons moving radially outward through the corona at velocities from $0.25c$ to $0.6c$ (c= velocity of light). The radio emission is generated along the way at successively higher coronal levels where the electrons excite plasma oscillation at the local plasma frequency $f_p = \sqrt{4\pi N_e e^2/m}$ or its second harmonic (N_e= electron density; e= electron charge; m= electron mass). For recent summaries the reader is referred to [a, c, d].

The various burst types

The following burst types all belong to the general class of fast-drift bursts.

1. Standard Type III bursts

The sudden onset of the emission at a given frequency is quite typical and defines the frequency drift which is approximately proportional to the observing frequency. The emission can usually be followed down to the lowest possible frequencies. On the earth's surface the ionospheric cut-off occurs at about 10 MHz, but on satellites they have even been observed down to about 10 kHz [37].

2. U-type bursts

U-type bursts are a special subclass of the fast-drift bursts [46]. In contrast to the standard Type III burst, the drift rate reverses at a certain frequency and the emission is observed to move back to higher frequencies. A typical time-frequency diagram of this event assumes the form of an inverted "U" (see Fig. 7). It must be assumed that the fast electrons originally responsible for the emission first move outward and then change their direction, perhaps following a magnetic loop, to return to regions of higher plasma density. The temporal development of the emission process, the fast electron beam and the magnetic field are schematically displayed in Fig. 8. This theory is supported by the observation that the emission source regions of the ascending and descending legs of the event are physically displaced from each other [70].

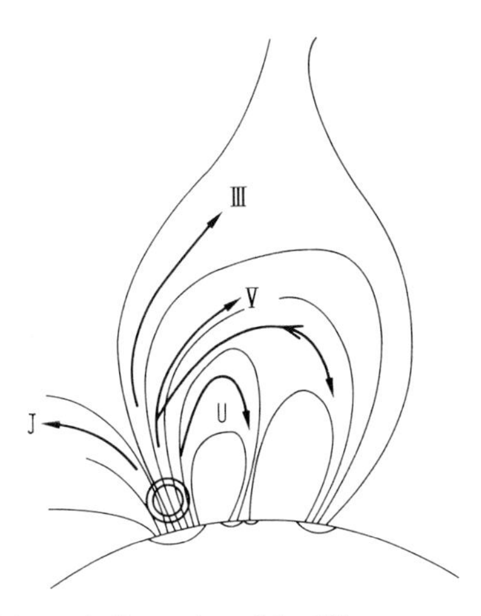

Fig. 8. Schematic illustration of the different types of bursts. Details are given in the text.

3. J-type bursts

The J-type bursts start off looking like a U-type burst, except that the second part of the emission event (the drifting back to higher frequencies) is missing [23]. The loss of this trailing leg leaves the time-frequency diagram in the form of an inverted "J". The probable evolution of the emission process in the corona is also pictured in Fig. 8. The disappearance of the downward leg of the emission can be caused either by a change in the structure of the magnetic loop [53, 13], or by scattering of the electron beam at the apex of the loop [10].

4. Type V bursts

A Type III burst is occasionally followed by wide-band continuum emission that typically lasts for 30 s to 3 min at frequencies $f<200$ MHz. This continuum has been classified as a Type V burst. Since the height of the emission source region in the corona (as measured by the emission frequency) and the intensity of the emission are approximately the same as the height and intensity of the preceding Type III burst, there appears to exist a close connection between the two burst types. It is presently assumed, in fact, that the Type V burst is generated by the same fast electrons responsible for the prior Type III burst, but that a part of the electron beam is trapped and stored in a large magnetic loop (see Fig. 8).

5. Type III storm bursts

Type III storms can be observed in the decameter range for $f<60$ MHz. A large number of the Type III bursts (e.g. Type III b and others) can be observed here as well. The Type III storms occur only when a noise storm is in progress at meter wavelengths. The existence of Type III storms provides evidence that electron acceleration processes initiated by noise storms can occur in localized regions of the outer corona [59].

6. Other fine structure

In addition to the burst types already mentioned, there exist various fast-drift events referred to as stria-bursts (Type III b), drift-pair bursts, shadow-type bursts, and others. A detailed description of these phenomena is contained in [c] or also [14].

Spectral characteristics of standard Type III bursts

Starting frequencies of Type III bursts range from about 600 MHz down to 10 MHz. The frequency drift rate at 100 MHz is about 48 MHz s^{-1} [18]. Over the range of frequencies from 60 kHz to 600 MHz it has been found [3] that the drift rate $\dot{f}$ in [MHz s^{-1}] is well represented by

$$\dot{f} = -\alpha f^{\beta}$$

with

$$\alpha = 0.01, \quad \beta = 1.84; \quad f \text{ in [MHz]}.$$

The bandwidth lies in the range from 10···50 MHz. The second harmonic is occasionally observed in addition to the fundamental emission. The ratio of frequencies of the harmonic to its fundamental mode is generally somewhat less than two (1.85···2.00). Using an acceptable model of the electron density distribution with height $N_e(r)$ and assuming that the Type III emission occurs at the local plasma frequency, one can derive the velocity of the exciting agent from the measured drift rate. One typically obtains velocities between 0.25 c and 0.6 c (rarely, even higher), with a mean value of 0.33 c [57].

The velocity stays approximately constant in the lower corona up to a height $h \simeq 0.5\, R_{\odot}$. A measurable deceleration is observed only at larger solar distances with $h > 1.0\, R_{\odot}$.

The emission source region

The angular diameter of a Type III burst is in the order of 4 arc min and seems to increase toward lower frequencies.

The time profile

At a fixed frequency the time profile generally rises steeply to a maximum followed by a slower, approximately exponential decline.

The duration

A standard Type III burst tends to last longer at lower frequencies. The duration t_d in [s] varies according to the power law [18]:

$$t_d = \alpha f^{-\beta}$$

with

$$\alpha \simeq 60, \quad \beta \simeq 0.66; \quad f \text{ in [MHz]}.$$

The polarization

The elliptical polarization found in earlier observations could not be confirmed by more recent studies [44, 8].

Observations from satellites

Measurements from satellites have extended the range of observation of Type III bursts to below $f \simeq 10$ MHz, i.e. the lower limit of transparency of the terrestrial ionosphere. Type III bursts were observed in this way at first down to 200 kHz [20, 21], and later even down to 10 kHz [37]. The emission source regions of these low frequency bursts are located well out into the interplanetary space between sun and earth. Satellites have also enabled the first direct measurement of the "exciting electron streams" with energies between 10···100 keV at the earth [37]. These experiments thereby confirmed the above-mentioned theory that the Type III bursts are produced by a beam of fast electrons. A further result of these measurements was the discovery that the second harmonic is preferentially excited at frequencies $f < 3$ MHz. The satellite experiments are summarized in Table 4. Further details may be found in [22].

Table 4. Most important solar space radio astronomy experiments.

Satellite	Equipment	Results, remarks	Ref.
Alouette-I	swept frequency receiver 1.5···10 MHz	first spaceborne detection of solar Type III bursts	31
Alouette-II	swept frequency receiver 0.1···15 MHz		32, 30
Zond-3	fixed frequencies 20 kHz, 210 kHz, 2 MHz		48, 49
Luna-11	fixed frequencies	first utilization of spin modulation	50
OGO-3	swept frequencies 2···4 MHz, 25···100 kHz		28, 17
OGO-5	eight frequencies between 50 kHz and 3.5 MHz	comparison with electron events	4, 29
ATS-II	six frequencies between 450 kHz and 3.5 MHz	comparison with streamer models	1
RAE-1 (−2)	swept frequency receiver 200 kHz···5.4 MHz	observations of Type III storms	21
IMP-6	fixed frequencies between 50 kHz and 3.5 MHz		5, 25, 36
Mars-3	STEREO-1 experiment		11, 12
Helios-A -B		First measurement of Type III bursts directivity at low frequencies	64

Theoretical discussion

The suggestion that fast electron beams (10···100 keV) penetrate into the corona and excite plasma oscillation close to the local plasma frequency f_p, a part of which is then converted to electromagnetic radiation, has received strong support from the observations. Some of the theoretical aspects that need to be mentioned within the framework of this model are:

(i) the acceleration of the electrons
(ii) the excitation of the plasma oscillations by the electron beam
(iii) the conversion of the plasma oscillations into electromagnetic radiation
(iv) the propagation of the electromagnetic radiation in the corona

A comprehensive review with an extensive bibliography may be found in [c] as well as in [56]. Brief comments on each of the above aspects:

(i): Both the acceleration mechanism and the exact location of the acceleration are essentially unknown.
(ii): The excitation of the plasma oscillations was recently treated in [73, 54, 56].
(iii): The conversion into microwave radiation is discussed in [55] and [33].
(iv): A detailed description of the propagation, refraction, and scattering of the radio bursts in the corona may be found in [c].

3.1.2.8.6.3 Slow-drift bursts (Type II bursts)

Slow-drift bursts (Type II) consist of narrow-band burst emission similar to that of the Type III bursts described in the preceding part. In contrast to these, the Type II bursts drift much more slowly through the spectrum toward lower frequencies. This drift rate is of the order of 0.25 to 1.0 MHz s^{-1}. The second harmonic of this radiation is often seen along with the fundamental emission. The time-frequency diagram for a slow-drift burst is given in Fig. 7.

Type II bursts are singular events that occur only in connection with large flares (see 3.1.2.7). They are therefore relatively rare. This phenomenon was recognized and first classified as a Type II burst in [66]. Recent reviews may be found in [c] as well as in [68, 15, 42].

If one assumes, in analogy with the Type III bursts, that the frequency drift is a result of the outward displacement of the excitation agent through the corona, thereby inducing microwave emission at the local plasma frequency or its harmonics, then one can derive the velocity of the radial motion from a model of the coronal electron density. This velocity usually lies in the range 200···2000 $\text{km}\cdot\text{s}^{-1}$, which is much greater than the speed of sound in the corona and also greater than the propagation velocity of Alfvén waves. A likely excitation mechanism is therefore the shock front propagation through the corona away from the flare site.

Temporal evolution of a Type II burst

Type II bursts are detected about 2···15 min after the start of a flare and last about 10···15 min in the frequency range $f > 30$ MHz. The burst begins abruptly at a frequency around 100 MHz and is joined simultaneously by its second harmonic at about 250 MHz. At a fixed frequency $f < 100$ MHz the time profile is generally complex, displaying peaks with brightness temperature $T_b > 10^{12}$ K.

The emission source region

Many detailed results concerning the emission region have been obtained with the Culgoora heliograph. Type II bursts have been seen with this instrument to have extended emission regions with a diameter $\simeq 0.5\,R_\odot$ and often display remarkable spatial and temporal variations [69, 35, 16, 58]. The sources occasionally seem to consist of isolated emission centers distributed over a spherical shell centered on the flare site [51]. The shock front evidently propagates away from the flare in a spherical wave with a large aperture angle ($\simeq 200°$).

The spectral structure

The spectrum at a fixed time t_0 is generally narrow-band but complex. One conspicuous characteristic is the appearance of the second harmonic, which occurs in about 50···60% of the bursts. The intensity of the second harmonicsis comparable with that of the fundamental emission. Higher harmonics are not detected. It is still an open question whether the emission from the fundamental and its harmonic occur at the same place or at different heights in the corona. Frequency splitting can also occasionally be seen in the spectra. This effect is the appearance of two distinct maxima separated from each other by an amount equal to about 10% of the mean frequency, i.e. 10 MHz for a fundamental at 100 MHz and a corresponding split of 20 MHz for the second harmonic. This split persists during the frequency drift over a large part of the frequency range. Possible causes for this phenomenon such as the magnetic field [62], a Doppler shift [24] or a geometrical effect [42] have been investigated.

Theoretical model

The propagation velocity of the Type II bursts in the corona v_{II} (300···2000 km s^{-1}) exceeds the velocity of sound v_s but is smaller than the thermal velocity v_{th}:

$$v_s < v_{\text{II}} < v_{\text{th}}\,.$$

The velocity v_{II} thus lies in the permitted range for magneto-hydrodynamic (MHD-)shock generation, and a most important prerequisite for the assumption of emission resulting from the passage of an MHD-shock front is thereby fulfilled [63]. Recent model calculations have been performed by [72, 74], and also in [62]. Nonradial propagation of the shock waves has been considered by [51]. A review with further references is contained in [c].

3.1.2.8.6.4 Continuum bursts (Type IV bursts)

Type IV bursts are defined as a slowly evolving emission process with a broad continuum spectrum that occurs at some part of the radio frequency range after a large flare. Like the Type II bursts, these continuum bursts are relatively rare, since they only occur in connection with an unusually intense flare event. Continuum bursts were first detected at meter wavelengths and classified as Type IV in [6, 7]. They are discussed in the more recent reviews [9] and [b].

Classifications

Many different forms of continuum bursts have been discovered across the radio frequency spectrum and various ordering schemes have been devised and applied by as many various authors. The classification by Krüger [c] will be adopted here. It divides the continuum bursts into five subclasses:

1. Type IVμ bursts
2. Type IVdm bursts
3. Type IVmF bursts (≙ IV mf in Fig. 7)
4. Type IVmA bursts (moving Type IV)
5. Type IVmB bursts (stationary Type IV)

These burst types are schematically displayed in Fig. 1. Time-frequency-diagram is shown in Fig. 7.

Type IVμ bursts

The Type IVμ microwave bursts are usually clearly separated from the other continuum bursts by a minimum in intensity in the lower decimeter range. The Type IVμ spectrum spans frequencies from 1···100 GHz with a maximum at about 10 GHz. The flux density at the maximum can exceed 10^4 s.f.u. (see 3.1.2.8.6.1). Gyrosynchrotron radiation is the commonly assumed emission process. The Type IVμ bursts cannot be unambiguously separated from the complex μ-bursts.

Type IVdm bursts

An emission maximum ($I_{max} > 10^4$ s.f.u.) with a wide continuum spectrum sometimes appears in the frequency band from 200···1500 MHz. This emission is clearly separated from both the Type IVμ bursts and the continuum bursts at meter wavelengths (see below) in the time-frequency diagram. Differences in the polarization from that of the Type IVμ are also apparent. A distinct subclass for the Type IVdm burst is therefore justified.

Type IVmF bursts

These continuum bursts in the meter range appear almost simultaneously with the optical flare following a series of Type III bursts. They are therefore recognized as a unique form of emission and are defined as a separate subclass of the Type IV bursts. These bursts have been identified by the various authors with such descriptions as "Type IVmA$_1$" in [b], or as "flare continuum" in [70], or as "first stationary source S$_1$IV" in [9]. The terminology applied here will be "flare continuum" with the abbreviation IVmF (IV mf in Fig. 7).

The spectrum extends from less than 10 MHz up to 1 GHz. The duration of the burst is typically 30 min. The emission coincides in the time-frequency diagram with possible Type V bursts (see Fig. 7). A discrimination between these two types is only possible because of the longer duration of the Type IVmF bursts. The emission mechanism is presumed to be gyrosynchrotron radiation. Examples of overlapping with Type II bursts are discussed by [46].

Type IVmA bursts (moving Type IV)

A second group of continuum bursts in the meter range is characterized by its rapidly moving emission source in the corona. The motion is usually radially outward, but can occasionally be parallel to the solar limb. Bursts of this type were first unambiguously identified after the clear demonstration of the source's location with interferometers [65]. Many further details were revealed by the radio images generated by the Culgoora heliograph [68, 52, 47]. The spectrum, which runs from less than 10 MHz to about 200 MHz, does not change during the displacement of the source. The motion can therefore be observed at a single fixed frequency in contrast to the situation with Type II or Type III bursts. The emission mechanism can be assumed from this fact to be a gyrosynchrotron process.

The moving Type IV bursts can be explained on the basis of four different physical processes in the corona [42]:

physical form	expansion velocity
expanding magnetic arches	$\simeq$ 300 km s^{-1}
ejection of plasma clouds	$\simeq$ 300 km s^{-1}
shock fronts	>1000 km s^{-1}
jets	$\simeq$ 0.5 c

Type IVmB bursts (stationary Type IV)

Type IVmB is emission that does not significantly change its source position during the burst event. It appears during the later stages of a large flare and remains observable long after the flare is extinguished. The typical duration is of the order of several hours. The spectral range of this burst type extends from 60···600 MHz. The emission occurs at a height in the corona that corresponds approximately to that where $f = f_p$ (local plasma frequency). The radiation has a strong component of circular polarization and is beamed in a narrow cone. The Type IVmB bursts display a certain resemblance to the continuum component of noise storms and are occasionally even preempted by the commencement of a real noise storm.

3.1.2.8.6.5 References for 3.1.2.8.6

General references, current data series

a Wild, J.P., Smerd, S.F.: Annu. Rev. Astron. Astrophys. **10** (1972) 159.
b Krüger, A.: Physics of Solar Continuum Radio Bursts, Akademie-Verlag, Berlin (1972).
c Krüger, A.: Introduction to Solar Radio Astronomy and Radio Physics, Reidel, Dordrecht (1979).
d Radio Physics of the Sun; Int. Astron. Union Symp. **86** (Kundu, M.R., Gergely, T.E., eds.) Reidel, Dordrecht (1980).
e Solar Radio Emission Instruction Manual for Monthly Report, ed.: Toyokawa Observatory, Japan.
f Solar Geophysical Data, No. 426 (Supplement) (1980) National Geophysical and Solar Terrestrial Data Center, Boulder USA.
g Quarterly Bulletin on Solar Activity, Zürich.
h Synoptic Codes for Solar and Geophysical Data, ed.: The International Ursigramm and World Data Service, Boulder USA.

Special references

1 Alexander, J.K., Malitson, H.H., Stone, R.G.: Sol. Phys. **8** (1969) 388.
2 Alissandrakis, C.E., Kundu, M.R.: Sol. Phys. **41** (1975) 119.
3 Alvarez, H., Haddock, F.T.: Sol. Phys. **30** (1973) 175.
4 Alvarez, H., Haddock, F.T., Lin, R.P.: Sol. Phys. **26** (1972) 468.
5 Alvarez, H., Haddock, F.T., Potter, W.H.: Sol. Phys. **34** (1974) 413.
6 Boischot, A.: Compt. Rend. Acad. Sci. Paris **244** (1957) 1326.
7 Boischot, A. in: Paris Symposium on Radio Astronomy (Bracewell, R.N., ed.), Stanford Univ. Press (1959) p. 187.
8 Boischot, A., Lecacheux, A.: Astron. Astrophys. **40** (1975) 55.
9 Boischot, A. in: Corona Disturbances, Int. Astron. Union Symposium **57** (Newkirk, G., ed.), Reidel, Dordrecht (1974) p. 423.
10 Caroubalos, C., Steinberg, J.L.: Astron. Astrophys. **32** (1974) 245.
11 Caroubalos, C., Pick, M., Rosenberg, H., Slottje, C.: Sol. Phys. **30** (1973) 473.
12 Caroubalos, C., Poquérusse, M., Steinberg, J.L.: Astron. Astrophys. **32** (1974) 255.
13 Daene, H., Formichev, V.V.: Heinrich-Hertz-Institut Suppl. Ser. Solar Data II 6 (1971) 241.
14 de la Noë, J., Boischot, A.: Astron. Astrophys. **20** (1972) 55.
15 Dryer, M.: Space Sci. Rev. **15** (1974) 403.
16 Dulk, G.A.: Proc. Astron. Soc. Australia **1** (1970) 308.
17 Dunkel, N., Helliwell, R.A., Vesecky, J.: Sol. Phys. **25** (1972) 197.
18 Elgarøy, Ø., Lyngstad, E.: Astron. Astrophys. **16** (1972) 1.
19 Enomé, S., Kakinuma, T., Tanaka, H.: Sol. Phys. **6** (1969) 428.
20 Fainberg, J., Stone, R.G.: Sol. Phys. **15** (1970) 222 u. 433.

21 Fainberg, J., Stone, R.G.: Sol. Phys. **17** (1971) 392.
22 Fainberg, J., Stone, R.G.: Space Sci. Rev. **16** (1974) 145.
23 Fokker, A.D.: Sol. Phys. **8** (1969) 376.
24 Fomichev, V.V., Chertok, I.M.: Soviet Astron. **11** (1967) 396.
25 Frank, L.A., Gurnett, D.A.: Sol. Phys. **27** (1972) 446.
26 Fürst, E., Hachenberg, O., Hirth, W.: Sol. Phys. **28** (1973) 533.
27 Hachenberg, O.: Z. f. Astrophys. **46** (1958) 67.
27a Hachenberg, O., Krüger, A.: Z. f. Astrophys. **59** (1964) 261.
28 Haddock, F.T., Graedel, T.E.: Astrophys. J. **160** (1970) 293.
29 Haddock, F.T., Alvarez, H.: Sol. Phys. **29** (1973) 183.
30 Hakura, Y., Nishiraki, R., Tao, K.: J. Radio Res. Lab. Japan **16** (1969) 215.
31 Hartz, T.R.: Ann. Astrophys. **27** (1964) 831.
32 Hartz, T.R.: Planetary Space Sci. **17** (1969) 267.
33 Heyvaerts, J., Verdies de Genduillac, G.: Astron. Astrophys. **30** (1974) 211.
34 Holt, S.S., Ramaty, R.: Sol. Phys. **8** (1969) 119.
35 Kai, K.: Sol. Phys. **10** (1969) 460.
36 Kai, K.: Sol. Phys. **11** (1970) 310.
37 Kellogg, P.J., Lai, J.C., Cartwright, D.G.: US. Dept. Commerce, Spec. Report UAG-28 Boulder (1973) 288.
38 Krüger, A.: Phys. Solarterr. **1** (1976) 7.
39 Kundu, M.R. in: [d] p. 157.
40 Lin, R.P., Evans, L.G., Fainberg, J.: Astrophys. Letters **14** (1973) 191.
41 Maxwell, A., Swarup, G.: Nature **181** (1958) 36.
42 McLean, D.J. in: Coronal Disturbances, IAU Symposium **57** (Newkirk, G. ed.), Reidel, Dordrecht (1974) 301.
43 Peterson, L.E., Winkler, J.R.: J. Geophys. Res. **64** (1959) 697.
44 Pick, M., Raoult, A., Vilmer, N. in: [d] p. 235.
45 Riddle, A.C.: Sol. Phys. **13** (1970) 448.
46 Robinson, R.D., Smerd, S.F.: Proc. Astron. Soc. Australia **2** (1975) 1.
47 Schmal, E.J.: Australian J. Phys. Astrophys. Suppl. **29** (1973) 1.
48 Slysh, V.I.: Soviet Astron. **11** (1967) 72.
49 Slysh, V.I.: Soviet Astron. **11** (1967) 389.
50 Slysh, V.I.: Cosmic Res. **5** (1967) 759.
51 Smerd, S.F.: Proc. Astron. Soc. Australia **1** (1970) 305.
52 Smerd, S.F., Dulk, G.A. in: Solar Magnetic Fields, Int. Astron. Union Symp. **43** (Howard, R. ed.), Reidel, Dordrecht (1971) p. 616.
53 Smith, D.F.: Sol. Phys. **13** (1970) 444.
54 Smith, D.F.: Sol. Phys. **33** (1973) 213.
55 Smith, D.F.: Sol. Phys. **34** (1974) 393.
56 Smith, D.F.: Space Sci. Rev. **16** (1974) 91.
57 Stewart, R.T.: Australian J. Phys. **18** (1965) 67.
58 Stewart, R.T., Sheridan, K.V., Kai, K.: Proc. Astron. Soc. Australia **1** (1970) 313.
59 Stewart, R.T., Labrum, N.R.: Sol. Phys. **27** (1972) 192.
60 Takakura, T., Scalise, E.: Sol. Phys. **11** (1970) 434.
61 Takakura, T., Kai, K.: Publ. Astron. Soc. Japan **18** (1966) 57.
62 Tidman, D.A., Birmingham, T.T., Stainer, H.M.: Astrophys. J. **146** (1966) 207.
63 Uchida, Y.: Publ. Astron. Soc. Japan **12** (1960) 373.
64 Weber, R.R., Fitzenreiter, R.J., Novaco, J.C., Fainberg, J.: Sol. Phys. **54** (1977) 431.
65 Weiss, A.A.: Australian J. Phys. **16** (1963) 526.
66 Wild, J.P., McCready, L.L.: Australian J. Sci. Res. **A3** (1950) 387.
67 Wild, J.P., Sheridan, K.V., Trent, G.H. in: Paris Symposium on Radio Astronomy (Bracewell, R.N., ed.), Stanford Univ. Press (1959) p. 176.
68 Wild, J.P., Smerd, S.F.: Annu. Rev. Astron. Astrophys. **10** (1972) 159.
69 Wild, J.P.: Proc. Astron. Soc. Australia **1** (1969) 181.
70 Wild, J.P.: Proc. Astron. Soc. Australia **1** (1970) 365.
71 Wild, J.P.: Sol. Phys. **9** (1969) 260.
72 Zaitsev, V.V.: Soviet Astron. **12** (1969) 610.
73 Zaitsev, V.V., Mityakov, N.A., Rapaport, V.O.: Sol. Phys. **24** (1972) 444.
74 Zheleznyakov, V.V.: Soviet Astron. **9** (1965) 191.